Adaptive Filtering Under Minimum Mean p-Power Error Criterion

Wentao Ma and Badong Chen

T0336098

CRC Press
Taylor & Francis Group
Boca Raton London New York

CRC Press is an imprint of the
Taylor & Francis Group, an **informa** business
A CHAPMAN & HALL BOOK

Designed cover image: Wentao Ma and Badong Chen

First edition published 2024
by CRC Press
2385 NW Executive Center Drive, Suite 320, Boca Raton FL 33431

and by CRC Press
4 Park Square, Milton Park, Abingdon, Oxon, OX14 4RN

CRC Press is an imprint of Taylor & Francis Group, LLC

ISBN: 9781032001654 (hbk)
ISBN: 9781032008622 (pbk)
ISBN: 9781003176114 (ebk)

DOI: 10.1201/9781003176114

Typeset in Palatino
by codeMantra

Contents

Symbols and Abbreviations

The main symbols and abbreviations used throughout the text are listed as follows:

k	discrete and continuous time index
t	continuous time index
$E\{\cdot\}$	the statistical-expectation operator
$\|\| \cdot \|\|$	vector or matrix norm
$I \in \mathbb{R}^{N \times N}$	the identity matrix
$(\cdot)^{\mathrm{T}}$	transpose of a vector or a matrix
$\mathrm{Tr}[\cdot]$	the trace of a matrix
$\lambda_i\{\cdot\}$	the *i*th eigenvalue of a matrix
$\mathrm{Re}(\cdot)$	the real part of a complex number
$\mathbf{u}()$	input signal vector
$\mathbf{R}_{\mathbf{uu}}$	correlation matrix
$\mathbf{P}_{\mathbf{u}d}$	cross-correlation vector
$F_{XY}(x, y)$	joint distribution function
$\kappa(.,.)$	Mercer kernel
$\langle .,. \rangle_{\mathcal{H}}$	inner product
$\mathrm{sign}(\cdot)$	sign function
$\mathrm{diag}[\cdot]$	diagonal matrix
$\mathbf{A}^{-1}$	inverse of matrix **A**
$\mathbb{R}^n$	*n*-dimensional real Euclidean space
$\mathbf{w}_o$	optimal weight
$\tilde{\mathbf{w}}$	weight error power
$v(k)$	additive noise at time instant *k*
$\boldsymbol{\omega}$	weight vector in feature space **F**
$\tilde{\boldsymbol{\omega}}$	weight error power in feature space **F**
$\boldsymbol{\varphi}$	a nonlinear feature mapping
$\mathbf{F}$	a high-dimensional feature space
$\boldsymbol{\Psi}(\cdot)$	weighted-auto-correlation matrix
$\boldsymbol{\Phi}(\cdot)$	weighted-cross-correlation vector
γ	forgetting factor
MSE	mean square error
LAFs	linear adaptive filters
RMSE	root mean square error
EMSE	excess mean square error
MPE	mean *p*-power error
MFE	mean fourth error
MSD	mean square deviation
LAD	least absolute deviation
LMS	least mean square

LMF	least mean fourth
RLS	recursive least squares
FLOM	fractional lower order moment
LMP	least mean p-power
ALMP	adaptive least mean p-power
NLMP	normalized least mean p-power
SLMP	smoothed least mean p-power
VNLMP	variable normalization least mean p-power algorithm
ALMP	adaptive least mean p-power
PLMP	proportionate least mean p-power
KMPE	kernel mean p-power error
KMMPE	kernel mixture mean p-power error
LMMN	least mean mixed norm
RMN	robust mixed norm
GMN	generalized mixed norm
DGMN	diffusion generalized mixed norm
FIR	finite impulse response
KAF	kernel adaptive filtering
RKHS	reproducing kernel Hilbert spaces
KLMS	kernel least mean square
KRLS	kernel recursive least squares
KLMP	kernel least mean p-power
KRMN	kernel recursive mixed norm
RFF-EX-KRLP	random Fourier features extended KRLP
DAF	diffusion adaptive filtering
DLMS	diffusion least mean square
DRLS	diffusion recursive least squares
DLMP	diffusion least mean p-power
DNLMP	diffusion normalized least mean p-power
RDNLMP	robust diffusion normalized least mean p-power
ELM	extreme learning machine
RLS-ELM	recursive least square extreme learning machine
SRLS-ELM	sparse recursive least square extreme learning machine
LMPELM	least mean p-power extreme learning machine
RLMP-ELM	recursive least mean p-power extreme learning machine
SRLMP-ELM	sparse recursive least mean p-power extreme learning machine
BLS	broad learning system
LP-BLS	least p-power broad learning system
MN-BLS	mixed norm broad learning system
L2LP	mixed l_2-l_p adaptive algorithm
SNR	signal-to-noise ratio
GSNR	generalized SNR

Preface

Over the past few decades, adaptive filters have found widespread application in various scenarios, including system identification, echo cancellation, channel equalization, time series prediction, and so on. To ensure the efficient design of an adaptive filter, it is crucial to select an appropriate loss function (or criterion function) that can enhance the filter's convergence performance. The classical adaptive filtering algorithms, such as least mean square (LMS) and recursive least squares (RLS), are primarily developed based on the well-known minimum mean square error (MMSE) criterion, which performs exceptionally well when signals follow Gaussian distributions. However, the mean square error (MSE) loss function only captures the second-order statistics in the data and may result in suboptimal filtering performance in non-Gaussian situations, particularly when the underlying system is affected by noises of heavy-tailed or multimodal distributions.

To enhance the filtering performance in the presence of non-Gaussian noises, adaptive filters have been developed using various non-MSE (non-quadratic) loss functions. These include the mean p-power error (MPE) loss, Huber's loss, risk-sensitive loss, correntropy loss, error entropy loss, and others. Among these, the minimum MPE (MMPE) criterion is particularly noteworthy as it is a natural extension of MMSE and can capture higher order ($p > 2$) or lower order ($0 < p < 2$) statistics while being mathematically and computationally simple. The MMPE encompasses special cases such as the least absolute deviation (LAD) ($p = 1$), MMSE ($p = 2$), and least mean fourth (LMF) ($p = 4$). In practical applications, the MPE has demonstrated superior performance compared to the conventional MSE when used as a loss function in adaptive filtering. For a finite impulse response (FIR) filter, the MPE loss can yield a more accurate solution than the Wiener solution of MSE. In addition, by selecting an appropriate p, MPE-based adaptive filters can achieve faster and more robust convergence performance under heavy-tailed or light-tailed non-Gaussian noises.

To date, numerous adaptive filtering algorithms have been developed under the MMPE criterion. This book aims to consolidate all of these algorithms, along with their corresponding analysis and numerical results, into a single comprehensive resource. Many of the contents of this book were originally published in previous papers by the authors. This book is divided into eight chapters, with Chapter 1 providing an introduction to the background and outline of the book. Chapter 2 reviews classical adaptive filtering algorithms under the MMSE criterion, while Chapter 3 covers the basic definition and properties of MMPE as well as several extended versions of MMPE. Chapter 4 focuses primarily on gradient-based (LMS type) adaptive filtering algorithms under MMPE criterion, while Chapter 5 presents recursive

(RLS type) adaptive filtering algorithms under MMPE. Chapter 6 introduces some nonlinear adaptive filtering algorithms under MMPE criterion, and Chapter 7 focuses on adaptive filtering algorithms under mixture MMPE criterion. Finally, Chapter 8 discusses adaptive filtering under kernel MMPE criterion.

This book is a valuable resource for graduates, professionals, and researchers seeking to enhance the performance of adaptive filtering algorithms and design new adaptive algorithms under MMPE. It is also an excellent reference for those interested in adaptive system training and machine learning. In addition, this book can be used as a reference textbook for graduate or undergraduate students majoring in electronics communication, electrical or computer engineering.

The authors are grateful to the National Key R&D Program of China and National Natural Science Foundation of China, which have funded this book. We also acknowledge the support and encouragement from our colleagues and friends.

1

Introduction

1.1 Basic Knowledge of Adaptive Filtering Algorithms

Classical filters, such as the Wiener filter [1,2] and Kalman filter [3,4], require accurate estimation of the correlation coefficient and noise power of input signals for effective application. However, this is often difficult to achieve in practice, and inaccurate estimation can significantly impact filtering performance. In addition, the parameters of these filters are typically fixed and cannot be adjusted in response to changing input signals, limiting their real-time processing capabilities. To address these limitations and meet the demands of signal processing, adaptive filters (AFs) have been developed as a class of optimal filtering methods from the Wiener and Kalman filters. Unlike classical filters, AFs incorporate a feedback channel between the output and filter system, allowing for dynamic adjustment of filter coefficients based on the output and expected signal at a given time [5,6].

AFs are capable of automatically adjusting the filtering structure in digital signal processing, whereas nonadaptive filters have static filter coefficients that result in fixed transfer functions. In many applications, adaptive coefficients are required for processing due to the lack of prior knowledge of the parameters to be operated, such as the characteristics of some noise signals. In such cases, AFs are typically utilized to adjust the filter coefficients and frequency responses with feedback. With the development and maturity of adaptive filtering technologies, AFs have become widely used as effective tools in various fields, including signal processing [7–12], control [13,14], and image processing [15,16]. This is due to the stronger adaptability and better filtering performance of AFs.

AFs can be categorized into two types based on their structure: linear adaptive filters (LAFs) and nonlinear adaptive filters. Nonlinear adaptive filters (NAFs), such as Voetlrra filters [17,18], kernel filters [19,20], and neural network-based AFs [21,22], have stronger signal-processing capabilities. Owing to their low computational complexity, LAFs are still widely used in most practical applications. The LAFs built with a linear combiner are designed for sequential learning [20]. They are equipped with a mechanism that enables the filter to adjust its free parameters automatically in response

DOI: 10.1201/9781003176114-1

to statistical variations in the environment in which it operates. This capability has led to a wide range of applications of AFs in diverse fields, such as adaptive equalization in communication receivers, adaptive noise cancelation in active noise control, adaptive beamforming in radar and sonar, system identification, and adaptive control.

1.1.1 AF Framework

AFs mainly involve three elements: filter structure, cost function, and optimization algorithm. In general, AFs rely on error-correction learning for their adaptive capability. A common filtering configuration is depicted in Figure 1.1, where a tapped-delay-line (transversal) is used as the filter for adaptation. The filter has a set of adjustable parameters (weights) denoted by the vector $\mathbf{w}(k-1)$, where k denotes discrete time. An input signal vector $\mathbf{u}(k)$, applied to the filter at time k, produces the actual response $y(k)$, which is compared with an externally supplied desired response $d(k)$ to produce the error signal $e(k)$. This error signal is, in turn, used to produce an adjustment to the parameter vector $\mathbf{w}(k-1)$ of the filter by an incremental amount denoted by $\Delta\mathbf{w}(k)$. The adjustment is made to minimize the cost function $J(\mathbf{w})$, which measures the difference between the actual and desired responses. The optimization algorithm determines the incremental adjustment $\Delta\mathbf{w}(k)$ that minimizes the cost function $J(\mathbf{w})$. Accordingly, the updated parameter vector of the filter can be expressed by

$$\mathbf{w}(k) = \mathbf{w}(k-1) + \Delta\mathbf{w}(k) \tag{1.1}$$

On the next iteration at time $k+1$, $\mathbf{w}(k)$ becomes the latest value of the parameter vector to be updated. The adaptive filtering process is continually repeated in this manner until the filter reaches a condition, whereafter the parameter adjustments become small enough to stop the adaptation. As is clear, the weights here embody the hypothesis in the definition of sequential learning. Overall, the filter structure, cost function, and optimization algorithm work together to enable AFs to adapt to changing input signals and achieve optimal filtering performance.

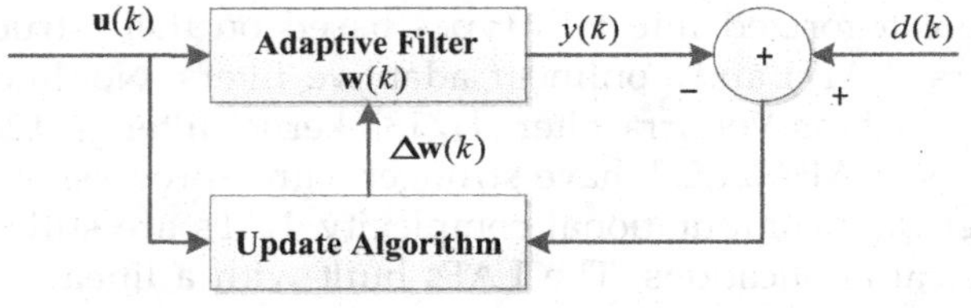

FIGURE 1.1
Block diagram of adaptive filters.

As previously mentioned, AFs utilize feedback to adjust filter coefficients and frequency responses, and the adaptive process involves update algorithms that use a cost function to determine how to change the filter coefficients to reduce the cost of the next iteration process. The adaptive algorithm generates a correction factor based on input and error signals, which is used to update the coefficients according to the defined cost function to obtain an estimation of the desired signal. The least mean square (LMS) [5] and recursive least square (RLS) [23] are two outstanding coefficient updating algorithms. The cost function is another key element and serves as the evaluation criterion for the best performance of the filter, such as the ability to reduce the noise component in the input signal.

Compared with other types of filters, the key to the better effect of AFs is the feedback structure of AFs. The adaptive process of adaptive filtering is to adjust the coefficient of FIR or IIR filter by adaptive algorithm (update algorithm) according to a suitable cost function (or adaptive learning criterion) so that the error signal is close to zero. In the following subchapters, the classical adaptive learning criteria and corresponding update algorithms are reviewed.

1.1.2 Adaptive Criteria

An essential aspect of designing AFs is the availability of an adaptive criterion. Traditional criteria for AFs include the least square (LS) criterion [24], minimum mean square error (MMSE) criterion [25,26], least absolute deviation (LAD) [27–29], and higher order error criteria [30–32]. The LS criterion is mathematically tractable and has a closed-form solution, defined by minimizing the sum of squared errors between observed and fitted values. Usually, a regularized version of the LS solution may be preferred [33]. Many AFs have been developed using the LS criterion, such as the RLS and its variants [34–37]. The MMSE criterion is commonly used as a measure of estimation quality in statistics and signal processing, which minimizes the mean square error (MSE) between the filter output and a desired variable. In AFs, the MMSE is often used as a cost function for stochastic approximation methods, which are a family of iterative algorithms that attempt to find the extrema of functions that cannot be computed directly and only estimated through noisy observations. The LMS algorithm [38–42], proposed in 1960 by Bernard Widow and Ted Hoff, is a typical stochastic gradient descent algorithm with MMSE criterion.

The LS and MSE criteria, which rely on the assumption that the error follows a Gaussian distribution, may be inadequate for practical data due to the likelihood of non-Gaussian interferences or outliers. To address this issue, robust regression methods are necessary to mitigate the bad effects of outliers on the estimation. In addition, research has shown that modeling expression data using heavy-tailed distributions [27], like the Laplacian distribution, can lead to more accurate results. Some authors have proposed a microarray normalization method that assumes errors follow a Laplacian model [43–45], using LAD regression as an optimization technique to compute normalization coefficients and avoid the effects of outliers in the original data. The LAD algorithm, also known as the sign-error LMS or pilot LMS [46–48], calculates the sum of absolute residuals $\sum_{i=1}^{k} |e(i)|$ and searches for the minimum value, making it more robust to anomalous points with large deviations in the data compared to the LMS algorithm, which may cause relatively large fluctuations after squaring.

Moreover, several studies have indicated that AFs based on higher order moments of the error signal can outperform those using MSE in certain critical applications [49,50]. Notably, the mean fourth error (MFE) criterion has been adopted as a cost function in adaptive filtering fields due to its convexity with respect to the weight vector. By stochastic gradient method, the least mean fourth (LMF) algorithm was developed to minimize the MFE and obtain the optimal weight [51]. Research has shown that the LMF algorithm can outperform the LMS algorithm in cases involving additive non-Gaussian noise, resulting in smaller excess MSE at the same convergence speed [31].

1.1.3 Typical Algorithms

1.1.3.1 Linear Adaptive Filtering Algorithms

In recent times, a plethora of adaptive filtering algorithms (AFAs) have emerged from diverse origins, each possessing unique characteristics. Researchers are particularly interested in AFAs that exhibit fast convergence, low computational complexity, and good numerical stability. Linear adaptive filters and their corresponding algorithms are more commonly employed in practical applications compared to NAFs due to their uncomplicated structure and low computational complexity. Notable algorithms under the MMSE criterion include the LMS, RLS, affine projection [52,53], and sub-band decomposition algorithms [54–56], among others, as documented in [6].

Over the past few decades, numerous AFAs have been proposed for various applications. Despite this, the LMS and RLS algorithms remain classical algorithms that have been thoroughly examined for optimality and stability. Since their inception, they have garnered significant attention, with research focusing on convergence analysis [57–62], performance enhancement [63–67], and the development of several LMS or RLS algorithms with unique methodologies, such as sparse AFAs [68–89], diffusion AFAs [90–107], constrained AFAs [108–115], and kernel AFAs [116–131]. These algorithms have been utilized for sparse system identification, distributed estimation, nonlinear time-series prediction, and other applications. In the following section, we will review these special algorithms.

1.1.3.1.1 Sparsity-Aware AFs

Sparsity-aware AFs have been developed for sparse system identification and can be categorized into two types: sparsity constraint AFAs [68–78] and proportionate AFAs [68–78]. Sparsity constraint AFAs are designed by integrating a sparsity constraint, such as an *lp*-norm constraint, into the cost function of the classical LMS or RLS algorithm [68–78]. The zero attraction term added to the update equation of the filter tap-weight vector aims to accelerate the identification speed by attracting small coefficients toward zero. However, the steady-state performance and instantaneous behavior of these algorithms depend on the selection of the zero attractor, which should be set according to the power of the measurement noise signal to ensure good steady-state mean square performance [70,71]. An adaptive strategy is proposed to select the zero attractor in the l_0-norm constraint LMS algorithm [74]. In addition to the zero attractor LMS-aware algorithms, l_1-norm regularized RLS adaptive algorithms have also been suggested in [75–78]. The SPARLS algorithm [75] presents an expectation-maximization approach for sparse system identification. The authors of [76] propose the application of an online coordinate descent algorithm together with the least-squares cost function penalized by an l_1-norm term. Another RLS algorithm for sparse system identification is proposed in [77], where the RLS cost function is regularized by adding a weighted norm of the current system estimate. Eksioglu further considers the regularization of the RLS cost function in a manner alike to the approach as outlined in [77]; the regularization term is defined as a general convex function of the system estimate, and an update algorithm is developed for the convex regularized RLS using results from subgradient calculus [77].

By expediting the elimination of inactive taps that correspond to system sparsity, sparsity-constrained AFs can deliver substantial performance improvements compared to their conventional counterparts, particularly during the steady-state phase. Nevertheless, there are also systems that are not strictly sparse but exhibit a relatively sparse (non-uniform) structure, where a small number of taps contribute to a significant portion of the energy [79]. In situations where sparsity is a crucial factor, proportionate-type algorithms have emerged as an important class of sparsity-aware AFAs. These algorithms employ the proportionate updating mechanism to update the filter coefficients. The pioneer in this field was Duttweiler, who introduced the proportionate normalized least mean square (PNLMS) algorithm [79]. This algorithm updates the filter coefficients by assigning a gain proportional to the magnitude of the current coefficient. The PNLMS algorithm outperforms the LMS and NLMS algorithms when applied to a sparse impulse response. However, its effectiveness diminishes when the impulse response is dispersive. To address this issue, several enhanced PNLMS algorithms have been proposed in the literature [80–82] to enhance the algorithm's resilience against time-varying sparsity. A different set of algorithms was developed by searching for a condition that would lead to the quickest overall convergence when all the coefficients approach their true values simultaneously. This approach gave rise to the μ-law PNLMS (MPNLMS) [83] and its variant [84]. The MPNLMS algorithm tackles the problem of assigning excessive update gain to large coefficients, which is a common issue with PNLMS algorithms. However, the convergence rate becomes unacceptably slow when dealing with correlated input conditions, such as speech. The Wavelet domain MPNLMS (WMPNLMS) [85] algorithm effectively tackles the problem of input decorrelation while preserving the sparsity of the impulse response. It achieves this by generating the conditional probability density function of the current weight deviations based on the preceding weight deviations, using a range of proportionate-type LMS algorithms [86]. Despite extensive research on the proportionate update mechanism in the context of NLMS-based AFs, the efficient design of sparse RLS algorithms using this mechanism remains an open issue. In a previous study [87], a natural recursive least squares (NRLS) algorithm was proposed, which utilized a proportionate matrix on the input vector to exploit system sparsity. However, this approach may render NRLS more sensitive to the condition number of the input covariance matrix than the standard RLS in certain scenarios [88]. In another study [89], a proportionate recursive least squares (PRLS) algorithm was introduced, which applies a proportionate matrix on the (Kalman) gain vector of the standard RLS.

1.1.3.1.2 Diffusion AFs

In the field of signal processing, distributed estimation has become a fundamental problem in recent years [90]. Typically, a group of nodes distributed across a geographical area work together to estimate an unknown model parameter based on linear measurements received by all nodes. There are three main methods for distributed estimation: incremental, consensus, and diffusion strategies. Among these, the diffusion strategy has been found to offer more advantages [90]. Diffusion-based algorithms are widely employed for distributed estimation, wherein neighboring nodes diffuse their estimates and measurements to adapt and combine their estimates. Among these algorithms, diffusion LMS (DLMS) is a fundamental method that utilizes the MSE criterion and diffusion structure [90–99]. Thanks to the exponentially weighted least squares (EWLS) criterion, the diffusion RLS (DRLS) algorithm has been enhanced to achieve rapid convergence, even for colored signals [100,101]. This algorithm aims to solve the network-wide LS estimation problem in a distributed adaptive manner, approaching the optimal LS without the need to transmit or invert any matrix. Mateos et al. introduced a novel distributed RLS algorithm for solving the EWLS problem using the alternating direction method of multipliers [102]. To minimize computation and communication expenses, this algorithm was further refined by censoring observations with small innovations, resulting in several variants [103–105]. To address the issue of biased estimation due to noisy input signals, various non-cooperative bias-eliminating algorithms have been proposed that utilize a bias-compensated mechanism [106,107].

1.1.3.1.3 Constrained AFs

Linearly constrained adaptive filtering (CAF) algorithms have gained significant attention and have been effectively utilized in various applications such as system identification, interference suppression, and array signal processing [108]. The primary advantage of CAFs is their ability to prevent error accumulation resulting from error correction, making them a preferred choice in these applications. Among the linearly constrained AFAs, the constrained LMS (CLMS) stands out as a simple stochastic gradient-based adaptive algorithm [109,110]. Initially designed as an adaptive solution to a linearly constrained minimum variance filtering problem in antenna array processing, CLMS has become a popular choice in various applications. Although the CLMS algorithm is simple and computationally efficient, it suffers from low convergence speed, particularly when the input signal is correlated. To address this issue, the constrained RLS (CRLS) algorithm was introduced in [111], albeit at the cost of higher computational complexity. Further improvements to the CRLS algorithm can be found in [112,113], while several constrained affine projection (CAP) algorithms were also developed in [114,115]. However, these constrained AFs with MSE criterion tend to perform poorly, especially when the signals involve non-Gaussian noises or outliers. This is mainly because the MSE criterion only captures the second-order moment.

In addition to classical LMS and RLS family AFs, non-MSE (or non-second-order moments)-based AFs have also demonstrated exceptional performance under certain conditions. For instance, the LMF and the LAD algorithms have shown remarkable results. The MFE criterion is a convex function (and thus unimodal) of the weight vector [51,132], which can outperform AFs with MSE for non-Gaussian additive noise, such as uniform and sinusoidal noise distributions. In such cases, the LMF algorithm has been found to yield smaller excess MSE for the same convergence speed. Various stability issues, tracking behaviors, and convergence analyses of the LMF algorithm have been explored in [49,50,133–138]. The Normalized LMF (NLMF) algorithm has been found to outperform the NLMS algorithm, particularly in low SNR scenarios, resulting in better steady-state performance [138]. In recent times, there has been a surge of interest in sparse NLMF algorithms. These algorithms incorporate different sparse penalty functions, such as zero-attracting (ZA), reweighted zero-attracting (RZA), reweighted l_1-norm, l_p-norm, and l_0-norm, leading to the development of various sparse NLMF-type algorithms [139–144]. Moreover, there are proportionate-aware LMF algorithms available to estimate the parameters of a sparse system with precision [145–147]. |In addition, ref. [148] has introduced a diffusion LMF (DLMF) algorithm that utilizes the diffusion strategy to improve the performance of distributed estimation in strong, non-Gaussian noise environments. To strike a balance between fast convergence rate and low steady-state misalignment, a variable step-size method has been incorporated into the DLMF. Furthermore, a sparse diffusion LMF algorithm has been proposed for estimating sparse parameters in Gaussian mixture noise environments [149]. The behavior of the DLMF algorithm has been analyzed in terms of mean and mean square in [150].

Overall, AFAs utilizing the MSE and MFE criteria have demonstrated exceptional performance in the realm of adaptive signal processing. Nevertheless, their convergence capabilities may be compromised when confronted with measurement noise that contains impulsive interferences. Accordingly, to combat impulsive interferences [151–155], the LAD criterion, which is based on l_1-norm minimization, has been proposed. The algorithms that incorporate LAD are referred to as least absolute deviation [151,152] or sign AFAs. In recent times, there has been extensive research on the steady-state and tracking analysis of signed-aware AFAs under various assumption conditions [156–158]. Considering the robustness of the signed aware algorithms, the sparsity sign subband AF (SSAF) [159,160] and diffusion sign algorithms [161–164] have been found to be highly robust signed aware algorithms for sparse system identification and distributed estimation. These algorithms minimize the l_1-norm of the sub-band a posteriori error vector, making them effective in handling sparse data.

1.1.3.2 Nonlinear AFAs

Linear adaptive filters have gained popularity in practical applications due to their straightforward structure and low computational complexity. However, their limited signal-processing capacity has restricted their use in certain applications. Nonlinear adaptive filters, such as Volterra filters, neural network-based adaptive filters, and kernel adaptive filters (KAFs), have emerged as a promising research area in adaptive signal processing due to their robust signal-processing capabilities. This book primarily concentrates on the KAFs and neural networks with random weights (NNRW)-based nonlinear adaptive filters.

The KAF has garnered significant attention in the fields of machine learning and signal processing as a powerful tool for solving nonlinear problems [20]. By transforming input data into higher or even infinite-dimensional reproducing kernel Hilbert spaces (RKHS), KAFs based on the conventional linear framework in RKHS have been extensively researched to address a wide range of nonlinear applications, including pattern classification, system identification, time-series prediction, and channel equalization. The Kernel Recursive Least-Squares (KRLS) algorithm, which can be considered as the RLS algorithm in RKHS, was initially developed in [116]. Several variants of KRLS have been proposed in a sequential manner, including sliding-window KRLS, extended KRLS algorithms, sparse KRLS, and quantized KRLS, as documented in [117–121]. Liu et al. further developed LMS algorithm in RKHS, called kernel least-mean-square (KLMS) algorithm [122]. Moreover, its theoretical convergence behavior was analyzed and derived because of its inherent simplicity and robustness in [123] and [124]. To reduce the computational complexity of the KLMS, the quantized KLMS (QKLMS) [125] and KLMS with promoting sparsity strategy [126] were proposed by quantized and constrained growth method. Some improved versions of KLMS were presented in [127–131]. An overview of kernel adaptive filtering is referred to [20]. While traditional KAF algorithms are effective in minimizing the widely used MSE, they are primarily designed to handle Gaussian noises. Unfortunately, real-world environments often contain non-Gaussian noises, which can cause KAF algorithms to become less robust. This is because MSE only captures the second-order statistics of the error signal, leaving KAF algorithms vulnerable to the limitations of this approach.

Neural networks (NNs) have been extensively researched as effective NAFs for system identification and noise cancelation, as evidenced by numerous studies [21,165–168]. The NNRW with a non-iterative learning mechanism is a feedforward neural network that employs a random learning algorithm to select input layer parameters and obtain output layer parameters through non-iterative calculation, resulting in an exceptionally fast learning speed. According to the different network structures and the degree of randomness, the current mainstream research methods for NNRWs include Random Vector Functional Link (RVFL) networks, Extreme Learning Machine (ELM), and Broad Learning System (BLS). In this book, we focus on reviewing ELM and BLS. Both ELM and BLS share the common feature of random weight and bias from the input layer (or feature layer) to the middle layer, while the weight and bias from the middle layer to the input layer are obtained by seeking the pseudo-inverse. The key difference between BLS and ELM lies in whether the feature layer (or input layer) is connected to the output layer (BLS: yes, ELM: no) and whether the input layer directly inputs data or feature (ELM: data, BLS: feature).

The ELM [169] is a novel fast learning algorithm designed to train a single-layer feedforward network (SLFN) with hidden neuron weights that are randomly initialized and fixed. This approach differs significantly from traditional training algorithms, such as the back-propagation (BP) algorithm and its improved versions [170,171], which require the tuning of hidden neuron weights. ELM, on the other hand, offers fast learning speed [172], universal approximation capability [172,173], and a unified learning paradigm for regression and classification [173]. For online identification problems, data samples often arrive in a time-ordered sequence. To address this, Liang et al. [174] proposed the online sequential ELM (OS-ELM), which can learn data one-by-one or chunk-by-chunk with fixed or varying chunk sizes. In addition, several improvements have been proposed and successfully applied in various applications [175–179]. In ELM, OS-ELM and many variants of them, the MSE criterion is usually adopted to construct their cost functions. Since the MSE criterion only takes into account the second-order statistics, it makes sense in the signal processing with Gaussian assumption. Consequently,

ELM suffers from two drawbacks: (1) MSE minimization learning can easily suffer from overfitting. The problem will be serious if the characteristics of the learned dataset can't be represented by the training data [180,181]. (2) ELM may perform poorly in the data under nonlinear and non-Gaussian situations, as it only captures the second-order statistics in the samples [182].

BLS [183,184] is a shallow neural network model that has emerged as a promising discriminative learning method. It has demonstrated the potential to outperform some deep neural network-based learning methods, including the multi-layer perceptron (MLP)-based method [185], deep belief networks (DBNs) [186], and stacked autoencoders (SAEs) [187]. To create a BLS, there are several essential steps that must be taken. First, the input data must be transformed into general mapped features using feature mappings. These generated mapping features are then connected by nonlinear activation functions to form the "enhancement nodes". The mapped features and the "enhancement nodes" are then sent together into the output layer, and the corresponding output weights are obtained through the use of pseudoinverse. One of the benefits of BLS is that all weights and biases of the hidden layer units can be randomly generated and remain unchanged. This means that only the weights between the hidden layer and the output layer need to be trained, which greatly simplifies the training process. Furthermore, in the event that new samples are introduced or the network requires expansion, a number of practical incremental learning algorithms have been developed to ensure that the system can be quickly remodeled without the need for a complete retraining process from the beginning [183]. As a result of these appealing features, BLS has garnered increasing attention [188–195] and has been successfully implemented in various applications, including image recognition, face recognition, and time-series prediction. In addition, several variants of BLS, such as fuzzy BLS [196], graph regularized BLS [197], recurrent BLS [198], and structured manifold BLS [199], have been developed from different perspectives.

The standard BLS algorithm employs the minimum mean square error (MMSE) criterion as its default optimization criterion for training the network output weights. However, like other MMSE-based methods mentioned earlier, it may experience a decline in performance in complex noise environments, particularly when the data are tainted by outliers.

1.2 AFAs under MMPE Criterion

While commonly used cost functions like LS and MSE are reliable in most practical situations and remain the go-to for adaptive filters, they do have limitations. For instance, they only capture second-order statistics in the data, which can be a poor approximation criterion in nonlinear and non-Gaussian scenarios, such as heavy-tail or finite-range distributions. To address this issue, researchers have explored non-MSE (nonquadratic) criteria, including mean *p*-power error (MPE) [200,201], maximum correntropy criterion [202–204], minimum error entropy criterion [205–207], and Huber criterion [208–210], among others. This book focuses on the MPE criterion which considers higher or lower order statistics and its application to adaptive filtering. Notably, the LS, MMSE, MAE, and MFE criteria can be viewed as special cases of the MPE.

1.2.1 MMPE Criterion

As a more general version of the MMSE approach, *lp*-norm minimization (also known as Minimum MPE or MMPE) has found widespread applications in various fields, including filter design, beamforming array, and deconvolution. In particular, when dealing with impulsive noise-contaminated signals, sinusoidal frequency estimation tends to favor *lp*-norm ($p = 1$) minimization [200,201,210]. Given the success of *lp*-norm minimization, there is growing interest in developing adaptive Finite Impulse Response (FIR) filter algorithms based on the MMPE criterion [184,194–198]. If we set $p = 2$, the generalized criterion becomes the conventional MSE criterion. However, for values of p other than 2, the MPE criterion may exhibit superior properties to the MSE criterion in certain circumstances. Notably, the MPE criterion reduces to the LAD criterion when $p = 1$, and the MFE criterion can be obtained by setting $p = 4$.

Pei and Tseng investigated the advantageous features of an adaptive FIR filter that utilizes the MMPE criterion [201]. Their findings demonstrated that the MMPE criterion outperforms the conventional MSE criterion in certain applications, provided that an appropriate value of p is selected. First, it is important to note that the optimum solution of the MPE function may outperform the Wiener solution of the MSE function. This is particularly relevant in system identification, where the MPE function may provide a solution that is closer to the true system parameters. Second, in cases where the optimal solution of the MPE function is the same as the Wiener solution of the MSE function for $p \neq 2$, the steepest descent algorithm based on the MPE criterion may exhibit superior performance, such as faster convergence speed, compared to the conventional Widrow-Hoff LMS algorithm. Third, when input signals or desired responses are corrupted by impulse noises, adaptive filters based on the MPE criterion with $p = 1$ may demonstrate stronger robustness than the LMS algorithm [156]. Furthermore, both analytical results and extensive simulations have shown that the new algorithms with $p = 3$ or $p = 4$ can perform better than the sign and LMS algorithms across a wide range of estimation scenarios.

1.2.2 MMPE Criterion based AFAs

As previously mentioned, the MMPE criterion serves as a useful cost function for designing various AFAs. This section will focus on summarizing MMPE-based AFAs, which can be broadly categorized into two types: linear and nonlinear. Figure 1.2 provides a detailed breakdown of this classification.

1.2.2.1 Linear AFAs under MMPE Criterion

1.2.2.1.1 Least Mean p-Power Error (LMP)

In [200,201], an adaptive FIR filter based on the MPE criterion is explored. This filter is a generalization of the instantaneous gradient descent algorithm for alpha-stable processes and is known as the least MPE (LMP) algorithm.

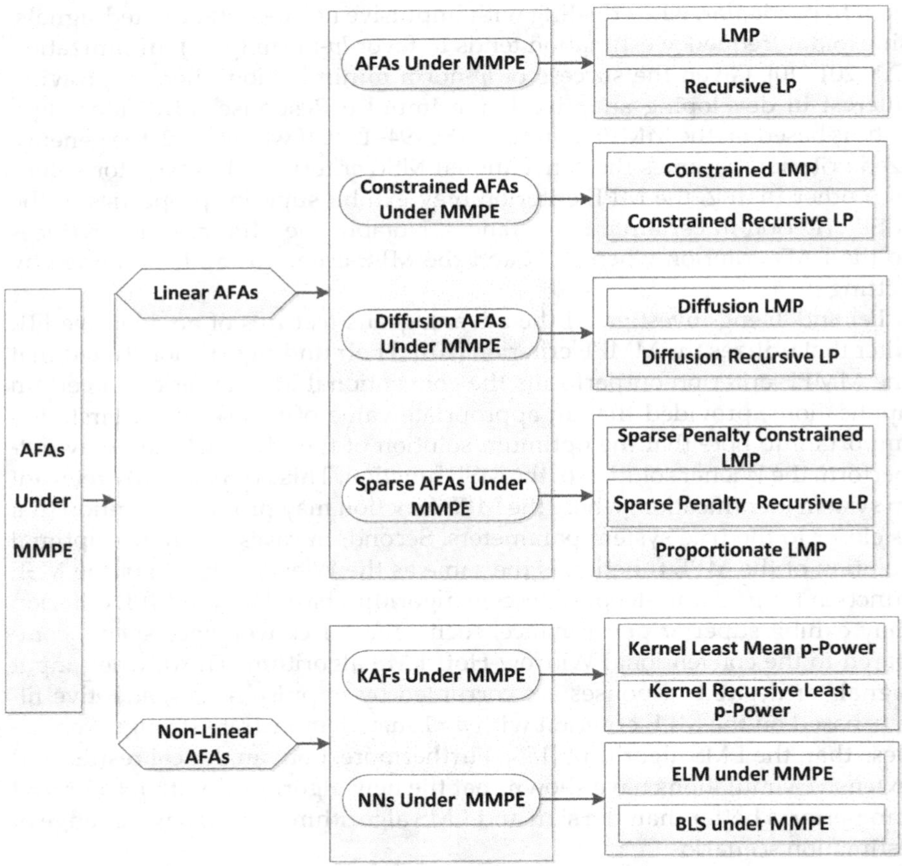

FIGURE 1.2
AFAs under MMPE.

When dealing with signals corrupted by impulsive noise, the LMP algorithm with $p=1$ is the preferred choice. However, when the signal is affected by noise or interference, the adaptive algorithm with an appropriate selection of p may be more suitable [201]. To tackle the challenge of identifying nonlinear systems in impulsive noise environments, researchers have turned to the LMP algorithm to identify the Volterra kernels [211]. Their findings demonstrate that the cost function is convex in relation to the filter weights for $p \geq 1$. Through an approximation analysis, they have determined the convergence range for the step size of the LMP algorithm. In addition, the authors have explored the impact of p on performance and have discovered that the optimal performance is achieved when p is closest to the characteristic factor α of the alpha-stable process. Using Taylor series expansion, the steady-state mean-square error (MSE) was analyzed for both real and complex LMP algorithms [211]. In [212], the authors provided closed-form analytical expressions for the steady-state MSE, along with the corresponding restrictive conditions for step size. Inspired by the NLMS algorithm, a normalized LMP algorithm (NLMP) was developed that utilizes a normalization by dividing the update term by p-norm of the input vector [213]. A normalized LMAD algorithm can be achieved by setting $p=1$ in the NLMP algorithm.

To enhance the robustness of the adaptive infinite impulse response (IIR) Notch filter (ANF), a new approach was proposed in [214], which utilizes the least MPE criterion. In addition, Maha [215] conducted a steady-state analysis of the constrained ANF with MPE. The findings indicate that the ANF with $p=1$ outperforms the LMS algorithm in canceling 60-Hz interference in electrocardiogram recordings. Furthermore, when the ANF with MPE is employed to estimate the frequency of a sinusoid embedded in white noise, it exhibits superior statistical accuracy compared to the LMS algorithm, particularly when p is set at 3. The success of MMPE has sparked interest in designing IIR filters based on the MMPE criterion [216,217]. Tseng [218] has proposed a digital IIR filter with MPE that uses the reweighted method, allowing for an arbitrarily prescribed frequency response. In addition, Xiao et al. [219] developed an adaptive algorithm based on the least MPE criterion for Fourier analysis in the presence of additive noise. Analytical results and extensive simulations have shown that the proposed algorithm for $p=3$ or $p=4$ generates improved discrete Fourier coefficient estimates in moderate to high SNR, with similar degrees of complexity. In [220], the filtered-x LMP algorithm (FxLMP) was proposed, which minimizes a fractional lower order moment (p-power of error) that is applicable to stable distributions. It has been demonstrated that the FxLMP algorithm with $p<a$ exhibits superior robustness to ANC of impulsive noise. To enhance the convergence performance of the FxLMP algorithm, two modified versions were proposed in [221]. The first algorithm aims to improve the robustness of the FxLMP algorithm by utilizing modified reference and error signals. The second algorithm, known as normalized FxLMP (NFxLMP), extends the concept of the NLMP to the FxLMP algorithm.

1.2.2.1.2 Constrained LMP

Constrained adaptive filters (AFAs) have a wide range of potential applications in signal-processing domain. The primary objective is to solve a constrained optimization problem explicitly. Typically, the MSE criterion is used in constrained adaptive filters, like CLMS [222], due to its attractive features of mathematical tractability, convexity, and low computational complexity. However, the CLMS is also susceptible to non-Gaussian noise interference. To address this issue, some robust constrained AFs have been developed based on the maximum correntropy criterion (MCC) and MPE criterion [223–225]. Peng et al. proposed a constrained LMP algorithm [225] by combining an equality constraint with the MPE criterion, which can achieve much better performance, especially in the presence of impulsive noises with a proper p value.

1.2.2.1.3 Diffusion LMP

The emergence of wireless sensor networks has spurred the development of distributed adaptive estimation schemes. Among these, the Diffusion LMS- [226] and RLS [100]-type algorithms have garnered significant attention. However, these algorithms rely on the MSE criterion and are therefore not well-suited for non-Gaussian noise environments. To address this issue, a diffusion LMP algorithm [227] has been proposed for distributed estimation in alpha-stable noise environments, which are commonly encountered in various settings. Despite its effectiveness, the DLMP algorithm suffers from a slow convergence rate. To overcome this limitation, a diffusion normalized LMP algorithm has been developed [228], inspired by the concept of normalized algorithms. To further enhance the performance of the DNLMP algorithm, a robust DNLMP algorithm has been introduced, which takes into account the error signal in the normalization factor and can effectively mitigate outliers' influence in impulsive noise environments. In addition, researchers have developed the diffusion LMF and LAD algorithms as special cases.

1.2.2.1.4 Sparsity-Aware LMP

Sparsity-aware AFAs have gained widespread popularity for sparse system identification. Most of these algorithms, including sparse LMS with l_0-norm constraint, proportionate LMS, and their variants, utilize the MMSE criterion as the cost function, which makes them well-suited for Gaussian noise environments. However, in practical applications, noise often exhibits non-Gaussian properties, and the MMSE criterion may result in poor performance, particularly when the noise is impulsive (e.g., alpha-stable noise). To address this issue, researchers have explored robust algorithms, such as those presented in references [162,229–232]. In [229], two frameworks, namely RLS-type and NG-type algorithms, were proposed for designing AFAs that exploit channel sparseness and achieve robust performance against impulsive noises. In addition, an improved proportionate affine projection sign algorithm (RIP-APSA) based on the p-norm of the error signal was introduced in [162]. The sparsity penalty terms play a crucial role in enabling the filters to fit well with the sparse structure of the system. Therefore, the adaptive filter and the sparsity penalty are the two main components of a sparse adaptive filter. However, finding the sparsest solution, which leads to an l_0-norm minimization problem, is a NP-hard combinatorial optimization problem. To tackle this challenging issue, the l_0-norm is often approximated by continuous functions. In recent years, the correntropy induced metric (CIM) has been proven to be an excellent approximation of the l_0-norm in [202,233], which can achieve arbitrarily close results to the 10-norm under certain conditions. To address sparse system identification in impulsive noise environments, several sparsity-aware LMP algorithms with different sparsity penalty terms (l_1-norm, reweighted l_1-norm, and CIM) have been developed in [234]. In addition, Zhang et al. proposed a proportionate LMP algorithm [235] based on the proportionate scheme, which utilizes an adaptive gain matrix to adjust the step size of each tap according to a specific rule.

1.2.2.1.5 Recursive LMP

The algorithms mentioned above that are designed to be LMP aware suffer from slow convergence when dealing with colored input signals due to the inconvenient stochastic gradient method. To address this issue and accelerate convergence in such conditions, RLS-type algorithms are typically preferred. In addition, various approaches have been proposed to improve the robustness of RLS to alpha-stable noise. For instance, a sliding window LMP algorithm has been introduced for filtering alpha-stable noise [236]. Unlike previous stochastic gradient-type algorithms, this algorithm precisely minimizes the MPE within a sliding window of fixed size, also known as the recursive LMP (RLMP) algorithm. Therefore, the RLMP algorithm exhibits similar convergence speed and computational complexity to the RLS algorithm, as opposed to stochastic gradient-based algorithms that behave like the LMS algorithm. The RLMP algorithm, proposed in [236], utilizes a reweighted least squares algorithm that converges to the minimum of reweighted MPE. While this approach benefits from a truly robust cost function, both schemes encounter practical issues. Specifically, the approach in [236] is not truly online, as it processes all samples in a window of past inputs to the filter in batch mode, requiring multiple iterations of the algorithm at every time instant. Consequently, this increases the memory and computational requirements of the filter. In this correspondence, a novel solution to the recursive least p-norm problem was proposed by utilizing a combination of adaptive filters [237]. The use of adaptive filter combinations has gained significant traction in recent times as a straightforward yet effective approach to address the various tradeoffs that impact the performance of adaptive filters. These tradeoffs include the steady-state error versus the convergence and tracking performance tradeoff. Zhang et al. proposed an enhanced RLMP algorithm [238] to further improve its tracking performance. This algorithm utilizes an adaptive gain factor in the cross-correlation vector and the input signal auto-correlation matrix. In addition, it employs the square of the estimated impulsive-free first moment of the error signal to control the updated gain factor. To address the limitations of the CLMP algorithm, a constrained AFA called the constrained recursive least *p*-power (CRLP) algorithm was proposed [239]. This algorithm incorporates a set of linear constraints into the MMPE criterion to directly solve a constrained optimization problem.

1.2.2.2 Nonlinear AFAs under MMPE Criterion

1.2.2.2.1 Kernel LMP

Most KAFs rely on the MSE criterion, which is chosen for mathematical simplicity and convenience. However, to enhance the performance of KAFs in the presence of non-Gaussian or impulsive noise with low probability but high amplitudes, some new KAFs based on the MMPE criterion have been developed. The kernel least mean p-power (KLMP) algorithm was designed to deal with alpha-stable distribution noise [240,241]. The KLMP algorithm is rooted in the conventional KAF framework, and it employs the MMPE criterion to mitigate the bad impact of impulsive noise on KAF. In addition, Ma [241] and Gao [242] have introduced two distinct kernel recursive least mean *p*-power (KRLP) algorithms that outperform KLMP in terms of convergence speed and steady-state accuracy. To enhance the convergence rate of the KRLP, a random Fourier features extended KRLP algorithm was developed [243]. This algorithm is designed to handle non-Gaussian impulsive noise and offers significant improvements in convergence rate, steady-state EMSE, and tracking ability in the presence of impulsive interference. In addition, it reduces computational complexity by replacing the calculation of kernel function with kernel approximation. Another approach to improving KAFs is the sparsified kernel adaptive filters (SKAF), which includes the projected kernel least mean p-power algorithm (PKLMP) based on the MPE criterion and vector projection method [244]. To utilize the information contained in the desired outputs, a modified PKLMP algorithm has been developed by smoothing the desired signal. In addition, Huang et al. have introduced a robust kernel conjugate gradient least mean p-power (KCGLMP) algorithm that combines the conjugate gradient optimization method with kernel trick, resulting in improved filtering accuracy and computational efficiency [245]. To address the challenges posed by nonlinear and non-Gaussian environments commonly encountered in real-world scenarios, a diffusion approximated KLMP algorithm has been developed for nonlinear distributed systems [246]. This algorithm approximates the property of shift-invariant kernel function using random Fourier features.

1.2.2.2.2 Neural Networks with Random Weights

Neural networks provide an important approach to construct nonlinear adaptive filters. As mentioned in Section 1.1.3, neural networks with random weights (NNRW) are a type of feedforward neural networks that utilize a non-iterative learning mechanism. The ELM and BLS, as prominent examples of NNRW, rely on the MSE loss function, which is susceptible to non-Gaussian noise or outliers in the training data. To improve the robustness of these models, researchers have developed robust ELM and BLS models that employ the MPE loss function.

a. *ELM under MPE:* The neural network is a nonlinear adaptive filter, and the ELM with MSE model has gained significant attention due to its simple structure and fast learning speed [247,248]. However, traditional ELM performance may deteriorate in non-Gaussian scenarios, leading to the development of robust ELMs under MPE criterion, as reported in [249–251]. For instance, Yang et al. [249] proposed a least mean *p*-power ELM, which maintains the computationally simple ELM architecture while utilizing the MPE criterion to sequentially update the output weights. Real industrial processes often involve measurement samples with different statistical characteristics and are obtained one by one, making it challenging to achieve optimal learning performance for systems affected by various types of noise. To address this issue, the authors proposed an online sequential learning algorithm, known as recursive LMPELM, which is capable of designing an online ELM [250] that can provide accurate predictions of variables even in the presence of non-Gaussian noise. This approach outperforms both ELM and online sequential ELM, making it a promising solution for industrial applications. Moreover, a novel online sparse RLMP-ELM approach is introduced, which incorporates a sparsity penalty constraint on the output weights as a cost function, in addition to the MPE criterion [251].
b. *BLS under MPE:* Several alternative optimization criteria have been proposed to improve the robustness of the original BLS. These criteria combine l_1-norm with different regularization terms, resulting in a class of robust BLS (RBLS) variants [188]. By using l_1-norm, which is less sensitive to outliers, the robustness of BLS has been significantly enhanced. In addition, Chu et al. [252] introduced the weighted BLS (WBLS), which has demonstrated good robustness in a nonlinear industrial process due to its well-designed weighted penalty factor. Another notable

approach to improving the robustness of BLS is the robust manifold BLS (RM-BLS) [253]. Zheng et al. proposed a robust BLS model that replaces the l_2-norm-based optimization model in BLS with a mixed-norm-based one. This model has been used to design a powerful classifier with strong generalization capability for brain computer interface (BCI) research [254]. Furthermore, Zheng [255] has developed a least p-norm-based BLS (LP-BLS) that utilizes the p-norm of the error vector as a cost function and incorporates a fixed-point iteration strategy. The LP-BLS approach allows for flexible adjustment of the value of p ($p \geq 1$) to effectively handle interferences from various types of noise, thereby improving the modeling of unknown data. To further enhance the robustness of BLS, Zheng has also incorporated the MCC [202] to train the output weights, resulting in a correntropy-based BLS (C-BLS). The proposed C-BLS is expected to exhibit superior robustness to outliers while maintaining the original performance of the standard BLS in Gaussian or noise-free environments [256].

1.2.2.3 AFAs under KMPE Criterion

The MPE with p-th absolute moment of the error is a powerful tool for handling non-Gaussian data when a suitable p value is chosen. It is generally robust to large outliers when $p < 2$. Chen et al. introduced a novel non-second order measure, called the kernel MPE (KMPE), which is essentially the MPE in kernel space [257]. When $p=2$, the KMPE reduces to the correntropy loss (C-Loss) [202], but with an appropriate p value, it can outperform the C-Loss when used as a cost function for robust learning. Drawing inspiration from KMPE, a novel measure called q-Gaussian kernel MPE (QKMPE) was proposed [258]. This measure is a generalization of the KMPE, defined with q-Gaussian kernel. In addition, a recursive kernel mean p-power is derived under the least QKMPE criterion for robust learning in noisy environments. This new algorithm has demonstrated superior performance against both Gaussian-type noise and non-Gaussian perturbations, particularly when the data contains large outliers. To further improve the performance of KMPE, a kernel mixture mean p-power error (KMP) criterion is proposed by combining the mixture of two Gaussian functions into the kernel function of KMPE [259]. The Nyström method is an efficient technique for controlling the growth of the network size of KAFs, and the recursive update form can enhance the tracking ability of KAFs. Finally, a recursive AFA is developed using KMPE with a forgetting factor as the cost function [260].

1.3 Outline of the Book

So far, numerous remarkable works have been accomplished on AFAs utilizing the MMPE criterion. Despite the existence of several books on AFAs designed under the MSE criterion, to our knowledge and investigation, there is still no book that presents AFAs under the MMPE criterion. Therefore, this book aims to provide a comprehensive treatment of AFAs under MMPE, with a focus on their properties, as well as linear and nonlinear AFAs. The remaining chapters of the book are organized as follows:

Chapter 2 delves into classical AFAs, such as the least mean square (LMS), recursive least square (RLS), and kernel adaptive filtering algorithms (e.g., kernel LMS and kernel RLS). This chapter serves as a foundation for readers to grasp the fundamental concepts that will be applied in subsequent chapters.

Chapter 3 presents a comprehensive overview of the minimum mean p-power error (MMPE) criterion, including its definition and properties. The chapter also delves into the relationship between MMPE and other conventional learning criteria such as MSE, MAE, and MFE. Furthermore, the chapter highlights several improved MMPE criteria, such as smoothed MMPE, adaptive MMPE, mixture MMPE, and kernel MMPE. This chapter is crucial for readers seeking to gain a deeper understanding of the fundamental principles underlying the MMPE criterion.

Chapter 4 focuses on various linear adaptive filtering algorithms that operate under the minimum mean p-power error criterion. These algorithms include the least mean p-power (LMP), adaptive LMP, smoothed LMP, sparsity-aware LMP, diffusion-aware LMP, and constrained LMP algorithms.

Chapter 5 mainly introduces the recursive AFAs under minimum mean p-power error criterion algorithms, such as recursive least mean p-power (RLP) algorithm, enhanced RLMP, sparsity RLP, diffusion RLP, and constrained RLP algorithm.

Chapter 6 presents nonlinear adaptive filtering algorithms that operate under the MMPE criterion. Specifically, we provide an overview of the kernel adaptive filtering and shallow neural network model under MMPE, including the kernel least mean p-power, kernel recursive least p-power, ELM under MMPE, and BLS under MMPE.

Chapter 7 primarily focuses on introducing the definition of the mixture MMPE criterion, along with various adaptive filtering algorithms that operate under this criterion. These include sparsity-aware AFAs, diffusion AFA, and kernel adaptive filters, all of which are designed to work effectively under mixture MMPE.

Chapter 8 provides a comprehensive overview of various adaptive filtering algorithms that are evaluated under the kernel mean p-power error criterion (KMPE). These algorithms include recursive KMPE, kernel adaptive filters (KAFs) that are based on KMPE family criteria (such as q-Gaussian KMPE and kernel mixture MPE-based KAFs), and ELM under KMPE.

2

Adaptive Filtering Algorithms under MMSE Criterion

Adaptive filtering algorithms (AFAs) have been extensively utilized in various practical applications, and the development of novel AFAs with distinct features remains a prominent research area in the field of signal processing. Nevertheless, the majority of new AFAs are based on traditional algorithms under minimum mean square error (MMSE) as their research foundation. To facilitate a better understanding of AFAs under minimum mean p-power error (MMPE), this chapter primarily focuses on reviewing some classical AFAs under MMSE, including least mean square (LMS), recursive least squares (RLS), kernel least mean square (KLMS), and kernel recursive least squares (KRLS).

2.1 LMS Algorithm

The traditional supervised adaptive filters rely on error-correction learning for their adaptive capability. To show the learning progress, the filtering structure depicted in Figure 2.1 is considered. The filter embodies a set of adjustable parameters (weights), which is denoted by the vector $\mathbf{w}(k-1)$, where k denotes discrete time instant, $\mathbf{u}(k) = [u(k), u(k-1), \ldots, u(k-M)]^T$ is an input signal vector applied to the filter at time k to produce the actual response $y(k) = \mathbf{u}^T(k)\mathbf{w}(k-1)$. This actual response is compared with an externally supplied desired response $d(k)$ to produce the error signal $e(k) = d(k) - y(k)$. This error signal is, in turn, used to produce an adjustment to the parameter vector $\mathbf{w}(k-1)$ of the filter by an incremental amount denoted by $\Delta\mathbf{w}(k)$. Accordingly, the updated parameter vector of the filter can be expressed by [261]

$$\mathbf{w}(k) = \mathbf{w}(k-1) - \Delta\mathbf{w}(k) \tag{2.1}$$

On the next iteration at time k, $\mathbf{w}(k)$ becomes the latest value of the parameter vector to be updated. The adaptive filtering process is continually repeated in this manner until the filter reaches a condition, whereafter the parameter

DOI: 10.1201/9781003176114-2

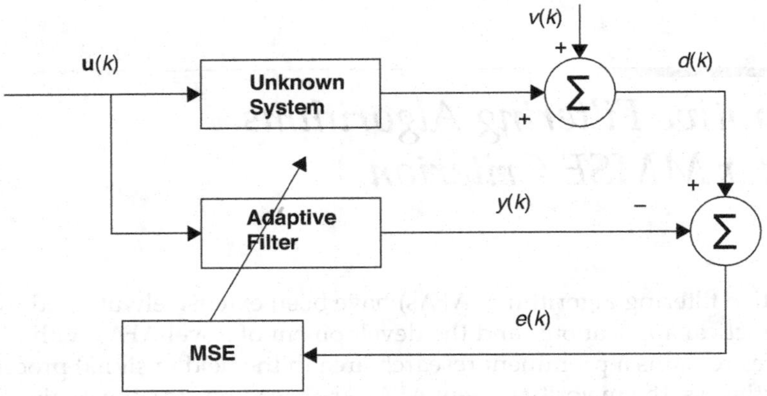

FIGURE 2.1
Basic structure of an adaptive filter.

adjustments become small enough to stop the adaptation. As is clear, the weights here embody the hypothesis in the definition of sequential learning.

Starting from some initial conditions denoted by $\mathbf{w}(0)$, the ensemble-averaged square error can be defined as:

$$J = E[e^2(k)], \qquad k = 1, 2, \ldots \tag{2.2}$$

where $E[]$ is the expectation operator. The (2.2) is carried out for an ensemble of different training sets, which can trace the learning curve of the adaptive filtering process. J is a quadratic function of the weight $\mathbf{w}$, i.e.

$$J = \mathbf{w}\mathbf{R}_{\mathbf{uu}}\mathbf{w} - 2\mathbf{P}_{\mathbf{u}d}^T\mathbf{w} + \sigma_d^2 \tag{2.3}$$

where $\mathbf{R}_{\mathbf{uu}} = E(\mathbf{u}(k)\mathbf{u}^T(k))$ is the correlation matrix, $\mathbf{P}_{\mathbf{u}d} = E(\mathbf{u}(k)d(k))$ denotes the cross-correlation vector, and $\sigma_d^2 = E(d^2(k))$ is the variance of desired signals. Hence, the following expression from gradient vector can be obtained

$$\nabla_2 = 2\mathbf{R}_{\mathbf{uu}}\mathbf{w} - 2\mathbf{P}_{\mathbf{u}d} \tag{2.4}$$

Let $\nabla_2 = 0$, we get the unique optimum weight vector as

$$\mathbf{w}_o = \mathbf{R}_{\mathbf{uu}}^{-1}\mathbf{P}_{\mathbf{u}d} \tag{2.5}$$

Equation (2.5) is the Wiener solution.

In the design of adaptive filters, a crucial consideration is to ensure that the learning curve converges as the number of iterations increases. This requires defining the speed of adaptation, such that the ensemble-averaged square error reaches a relatively stable value, indicating that the adaptive filter has converged in the mean square error (MSE) sense.

The LMS algorithm is the most widely used and straightforward form of AFAs. Essentially, it operates by minimizing the instantaneous MSE cost function as

$$J(k) = \frac{1}{2}e^2(k) \tag{2.6}$$

where the factor 1/2 is introduced to simplify the mathematical formulation. Given that the parameter vector of the filter is $\mathbf{w}(k-1)$, the error signal $e(k)$ is defined by

$$e(k) = d(k) - \mathbf{w}(k-1)^T \mathbf{u}(k) \tag{2.7}$$

Correspondingly, the instantaneous gradient vector can be calculated by

$$\frac{\partial}{\partial \mathbf{w}(k-1)} J(k) = -e(k)\mathbf{u}(k) \tag{2.8}$$

Following the instantaneous version of the method of gradient descent, the adjustment $\Delta\mathbf{w}(k)$ applied to the algorithm at time k is defined by

$$\Delta\mathbf{w}(k) = \mu e(k)\mathbf{u}(k) \tag{2.9}$$

where μ is the step size parameter which controls the convergence speed of the LMS algorithm. Thus, using Eq. (2.9) in Eq. (2.1) yields the following update rule for the filter's parameter vector:

$$\mathbf{w}(k) = \mathbf{w}(k-1) + \mu e(k)\mathbf{u}(k) \tag{2.10}$$

The LMS algorithm is also known as the stochastic gradient algorithm, and its simplicity is highlighted in Table 2.1. To initialize the algorithm, it is common practice to set the weight vector's initial value to zero.

TABLE 2.1

Least Mean Square Algorithm

Initialization: $\mathbf{w}(0)=0$, μ
For $k = 1, 2, \ldots$ **Do**
1. $y(k) = \mathbf{u}^T(k)\mathbf{w}(k-1)$
2. $e(k) = d(k) - y(k)$
3. $\mathbf{w}(k) = \mathbf{w}(k-1) + \mu e(k)\mathbf{u}(k)$
End

Upon examining the computations described above, it becomes clear that the LMS algorithm is fundamentally simple. Despite its simplicity, this algorithm can deliver effective performance, provided that the step size parameter μ is appropriately selected. One of the most significant advantages of the LMS algorithm is its model independence, as it imposes no structural restrictions on how the training data were generated. As a result, the LMS algorithm is renowned for its robustness. To achieve optimal performance, it is recommended to assign a relatively small value to the step size parameter μ. However, from a practical standpoint, this approach has a significant drawback: a small step size causes the LMS algorithm to converge slowly.

The benefits of utilizing the LMS can be succinctly summarized as follows: (i) it boasts a low computational complexity; (ii) it is straightforward to implement; (iii) it enables real-time operation; and (iv) it does not require any statistical knowledge of signals, such as $\mathbf{R_{uu}}$ and $\mathbf{P_{ud}}$.

The convergence of the LMS adaptive filter is dependent on the auto-correlation matrix $\mathbf{R_{uu}}$. To ensure that the system converges in the mean, two conditions must generally be met:

1. The auto-correlation matrix, $\mathbf{R_{uu}}$, is positive definite.
2. $0 < \mu < 1/\lambda_{max}$, where λ_{max} is the largest eigenvalue of $\mathbf{R_{uu}}$.

Here a brief analysis of the convergence condition presented in (2) is only performed. For ease of analysis, it is assumed that $\mathbf{w}(k)$ is independent of $\mathbf{u}(k)$. Taking expectation on both sides of (2.10), we have

$$\begin{aligned} E[\mathbf{w}(k)] &= E[\mathbf{w}(k-1)] + \mu E[e(k)\mathbf{u}(k)] \\ &= E[\mathbf{w}(k-1)] + \mu E[d(k)\mathbf{u}(k) - \mathbf{u}(k)(\mathbf{u}^T(k)\mathbf{w}(k-1))] \\ &= E[\mathbf{w}(k-1)] + \mu \mathbf{P}_{ud} - \mu \mathbf{R_{uu}} E[\mathbf{w}(k-1)] \\ &= (I - \mu \mathbf{R_{uu}}) E[\mathbf{w}(k-1)] + \mu \mathbf{P}_{ud} \end{aligned} \tag{2.11}$$

Following the previous derivation, it will converge to the Wiener filter weights in the mean sense if

$$\begin{bmatrix} \lim\limits_{n\to\infty}(1-\mu\lambda_1)^n & 0 & \cdots & 0 \\ 0 & \lim\limits_{n\to\infty}(1-\mu\lambda_2)^n & & \vdots \\ \vdots & & \ddots & 0 \\ 0 & \cdots & & \lim\limits_{n\to\infty}(1-\mu\lambda_L)^n \end{bmatrix} = 0$$

$$\Rightarrow |1-\mu\lambda_i| < 1, \qquad i = 1,2,\ldots,L$$

$$\Rightarrow 0 < \mu < \frac{1}{\lambda_{max}}$$

In addition, the rate of convergence is related to the eigenvalue spread. This is defined using the $\mathbf{R}_{uu}$ condition number of $\kappa = \lambda_{max} / \lambda_{min}$, where λ_{min} is the minimum eigenvalue of $\mathbf{R}_{uu}$. The fastest convergence of this system occurs when κ is close to 1, corresponding to white noise. This states that the fastest way to train a LMS adaptive system is to use white noise as the training input. As the noise becomes more and more colored, the speed of the training will decrease. The mean square convergence and tracking capability of the LMS under different conditions can be viewed in [57,56,261–264].

2.2 Recursive Least Square Algorithm

2.2.1 Original RLS Algorithm

To enhance the learning rate of the LMS algorithm, another classical AFA, called RLS algorithm, is an important method. The RLS algorithm shares a similar rationale with the LMS algorithm, as both are error-correction learning examples. However, the fundamental difference lies in their approach: while the LMS algorithm aims to minimize the instantaneous value of the squared estimation error $J(k)$ at each iteration k as in equation (2.12), the RLS algorithm aims to minimize the sum of squared estimation errors up to and including the current time k.

Mathematically, to introduce the RLS algorithm, a cost function is defined by

$$\mathbf{J}_{RLS}(k) = \sum_{i=0}^{k} (d(i) - \mathbf{w}^T \mathbf{u}(i))^2 \tag{2.12}$$

One can get the optimal parameter by solving the following optimization problem

$$\mathbf{w} = \arg\min_{\mathbf{w}} \mathbf{J}_{RLS}(k) \tag{2.13}$$

Setting the derivative of (2.12) with respect to $\mathbf{w}(k)$ to zero yields

$$\sum_{i=0}^{k} e(i)\mathbf{u}(i) = 0 \tag{2.14}$$

After some simple manipulations, we have

$$\sum_{i=0}^{k} \mathbf{u}(i)\mathbf{u}^T(i)\mathbf{w}(k) = \sum_{i=0}^{k} \mathbf{u}(i)d(i) \tag{2.15}$$

Let us define the following auto-correlation matrix and cross-correlation vector:

$$\Psi(k) = \sum_{i=0}^{k} \mathbf{u}(i)\mathbf{u}^T(i) \tag{2.16}$$

$$\Phi(k) = \sum_{i=0}^{k} \mathbf{u}(i)d(i) \tag{2.17}$$

The matrix form of equation (2.15) can then be represented by

$$\Psi(k)\mathbf{w}(k) = \Phi(k) \tag{2.18}$$

From equation (2.18), one can obtain an estimate of the optimal weight at each time k as

$$\mathbf{w}(k) = \Psi^{-1}(k)\Phi(k) \tag{2.19}$$

In general, $\Psi(k)$ is recursively expressed as shown in equation (2.20) to avoid the inverse operator of the auto-correlation matrix $\Psi(k)$ for each time instant.

$$\Psi(k) = \sum_{i=0}^{k-1} \mathbf{u}(i)\mathbf{u}^T(i) + \mathbf{u}(k)\mathbf{u}^T(k) \tag{2.20}$$

According to the definition of $\Psi(k)$, we know that the first term on the right-hand side of equation (2.20) is equal to $\Psi(k-1)$. As a result, the recursive form of equation (2.20) can be further expressed as

$$\Psi(k) = \Psi(k-1) + \mathbf{u}(k)\mathbf{u}^T(k) \tag{2.21}$$

Similarly, the recursive form of $\Phi(k)$ can be obtained as

$$\Phi(k) = \Phi(k-1) + \mathbf{u}(k)d(k) \tag{2.22}$$

Now the following development can be made by substituting (2.18) and (2.22) into (2.19), i.e.,

$$\begin{aligned}
\mathbf{w}(k) &= \mathbf{\Psi}^{-1}(k)\mathbf{\Phi}(k) \\
&= \mathbf{\Psi}^{-1}(k)[\mathbf{\Phi}(k-1) + \mathbf{u}(k)d(k)] \\
&= \mathbf{\Psi}^{-1}(k)[\mathbf{\Psi}(k-1)\mathbf{w}(k-1) + \mathbf{u}(k)d(k)] \\
&= \mathbf{\Psi}^{-1}(k)\left\{\mathbf{\Psi}(k-1)\mathbf{w}(k-1) + \mathbf{u}(k)d(k) + \mathbf{u}(k)\mathbf{u}(k)^T\mathbf{w}(k-1) - \mathbf{u}(k)\mathbf{u}(k)^T\mathbf{w}(k-1)\right\} \\
&= \mathbf{\Psi}^{-1}(k)[\mathbf{\Psi}(k)\mathbf{w}(k-1) + \mathbf{u}(k)(d(k) - \mathbf{u}(k)^T\mathbf{w}(k-1))] \\
&= \mathbf{\Psi}^{-1}(k)[\mathbf{\Psi}(k)\mathbf{w}(k-1) + \mathbf{u}(k)e(k)] \\
&= \mathbf{w}(k-1) + \mathbf{\Psi}^{-1}(k)\mathbf{u}(k)e(k)
\end{aligned} \tag{2.23}$$

To compute the inverse of $\mathbf{\Psi}(k)$, the following matrices are defined

$$\mathbf{A} = \mathbf{\Psi}(k-1),\ \ \mathbf{B} = \mathbf{u}(k),\ \ \mathbf{C} = \mathbf{I},\ \ \mathbf{D} = \mathbf{u}^T(k) \tag{2.24}$$

where $\mathbf{I}$ represents an identity matrix. Using the matrix inversion lemma (see Appendix A)stated below

$$[\mathbf{A} + \mathbf{BCD}]^{-1} = \mathbf{A}^{-1} - \mathbf{A}^{-1}\mathbf{B}\left(\mathbf{C}^{-1} + \mathbf{DA}^{-1}\mathbf{B}\right)^{-1}\mathbf{DA}^{-1} \tag{2.25}$$

we obtain the inverse of $\mathbf{\Psi}(k)$

$$\mathbf{\Psi}^{-1}(k) = \mathbf{\Psi}^{-1}(k-1) - \frac{\mathbf{\Psi}^{-1}(k-1)\mathbf{u}(k)\mathbf{u}^T(k)\mathbf{\Psi}^{-1}(k-1)}{1 + \mathbf{u}^T(k)\mathbf{\Psi}^{-1}(k-1)\mathbf{u}(k)} \tag{2.26}$$

For a simple description of (2.26), the following extended gain vectors $\Omega(k)$ and $\mathbf{K}(k)$ are introduced

$$\Omega(k) = \mathbf{\Psi}^{-1}(k) \tag{2.27}$$

$$\mathbf{K}(k) = \frac{\Omega(k-1)\mathbf{u}(k)}{1 + \mathbf{u}^T(k)\Omega(k-1)\mathbf{u}(k)} \tag{2.28}$$

Then (2.26) can be expressed as

$$\Omega(k) = \Omega(k-1) - \mathbf{K}(k)\mathbf{u}^T(k)\Omega(k-1) \tag{2.29}$$

Another equivalent form of (2.28) is

$$\mathbf{K}(k)\left[1 + \mathbf{u}^T(k)\Omega(k\text{-}1)\mathbf{u}(k)\right] = \Omega(k-1)\mathbf{u}(k) \tag{2.30}$$

According to (2.30) and using the result $\Omega(k)$ in (2.29), the gain vector $\mathbf{K}(k)$ can be further expressed in the following form

$$\begin{aligned}\mathbf{K}(k) &= \left[\Omega(k-1) - \mathbf{K}(k)\mathbf{u}^T(k)\Omega(k-1)\right]\mathbf{u}(k) \\ &= \Omega(k)\mathbf{u}(k)\end{aligned} \tag{2.31}$$

Inserting (2.31) into (2.23), we have

$$\mathbf{w}(k) = \mathbf{w}(k-1) + \mathbf{K}(k)e_p(k) \tag{2.32}$$

where $e_p(k) = d(k) - \mathbf{u}^T(k)\mathbf{w}(k-1)$.

To conclude the brief discussion on the LMS and RLS algorithms, it is summarized that each algorithm possesses distinct characteristics.

i. The computational complexity of the LMS algorithm increases linearly with the dimension of the parameter vector $\mathbf{w}$, while the computational complexity of the RLS algorithm grows exponentially with the dimension of the parameter vector.
ii. The LMS algorithm, being model independent, is generally considered to be more resilient than the RLS algorithm.
iii. The RLS algorithm typically achieves a convergence rate that is one order of magnitude faster than the LMS algorithm.
iv. Finally, it is worth noting that while the LMS algorithm propagates the estimation error from one iteration to the next, the RLS algorithm propagates the error covariance matrix. This distinction serves as a testament to the straightforwardness of the LMS algorithm. In addition, the mean square convergence of an adaptive RLS algorithm with stochastic excitation is demonstrated in [265].

2.2.2 Exponentially Weighted RLS Algorithm

In practice, the RLS algorithm often employs an exponentially weighted mechanism to give greater weight to recent data while reducing the influence of older data. As a result, the cost function for weighted least squares is defined as follows:

$$J_{\text{RLS}}(k) = \sum_{i=0}^{k}\left[\gamma^{k-i}(d(i) - \mathbf{w}(k)^T\mathbf{u}(i))^2\right] \tag{2.33}$$

where γ is usually called the forgetting factor.

Consequently, we differentiate (2.33) with respect to $\mathbf{w}(k)$ and set the derivatives to zero, yielding

$$\sum_{i=0}^{k}\gamma^{k-i}e(i)\mathbf{u}(i)=0, \tag{2.34}$$

After some manipulations in (2.34), we have

$$\sum_{i=0}^{k}\gamma^{k-i}\mathbf{u}(i)\mathbf{u}^T(i)\mathbf{w}(k)=\sum_{i=0}^{k}\gamma^{k-i}\mathbf{u}(i)d(i) \tag{2.35}$$

As a result, the recursive form of equation (2.35) can be further expressed as

$$\boldsymbol{\Psi}(k)=\gamma\boldsymbol{\Psi}(k-1)+\mathbf{u}(k)\mathbf{u}^T(k) \tag{2.36}$$

Similarly, we have the recursive form of $\boldsymbol{\Phi}(k)$ as

$$\boldsymbol{\Phi}(k)=\gamma\boldsymbol{\Phi}(k-1)+\mathbf{u}(k)d(k) \tag{2.37}$$

Then similar to the RLS derivation, the following extended gain vectors $\Omega(k)$ and $\mathbf{K}(k)$ are introduced

$$\Omega(k)=\boldsymbol{\Psi}^{-1}(k) \tag{2.38}$$

$$\mathbf{K}(k)=\frac{\Omega(k-1)\mathbf{u}(k)}{\gamma+\mathbf{u}^T(k)\Omega(k-1)\mathbf{u}(k)} \tag{2.39}$$

One can obtain the updated equation of the weighted RLS algorithm as

$$\mathbf{w}(k)=\mathbf{w}(k-1)+\mathbf{K}(k)e_p(k) \tag{2.40}$$

The weighted RLS algorithm is summarized in Table 2.2.

TABLE 2.2

Weighted RLS Algorithm

Initialization: $\mathbf{w}(0)=0,\ \Omega(0)=\gamma^{-1}\mathbf{I}$
For $k=1,2,\ldots$ **Do**
1. $y(k)=\mathbf{u}^T(k)\mathbf{w}(k-1)$
2. $e(k)=d(k)-y(k)$
3. $\mathbf{K}(k)=\dfrac{\Omega(k-1)\mathbf{u}(k)}{\gamma+\mathbf{u}^T(k)\Omega(k-1)\mathbf{u}(k)}$
4. $\mathbf{w}(k)=\mathbf{w}(k-1)+\mathbf{K}(k)e(k)$
5. $\Omega(k)=\gamma^{-1}[\mathbf{I}-\mathbf{K}(k)\mathbf{u}^T(k)]\Omega(k-1)$
End

2.3 KLMS Algorithm

2.3.1 Kernel Method

Consider the problem of learning a continuous input-output nonlinear mapping $f: \mathbf{U} \rightarrow \mathbf{R}$ with a sequence of input-output data set $\{\mathbf{u}(k), d(k)\}_{k=1}^{N}$. The relationship between the input and output is usually defined as $d = f(\mathbf{u}), \mathbf{u} \in \mathbf{U} \subseteq \mathbf{R}^M, d \in \mathbf{R}$, where $\mathbf{u}$ and d are the input and output, respectively. Here the goal is to estimate the mapping f using an adaptive filtering method.

Kernel adaptive filtering (KAF) is a powerful framework that facilitates the sequential learning of a nonlinear mapping through the use of the kernel method. The kernel method is a highly effective nonparametric modeling tool that is specifically designed to tackle the challenges posed by nonlinearity approximation problems. The fundamental concept behind this approach is to transform the input data (i.e., the input space) into a high-dimensional feature space using a specific nonlinear mapping, expressed as:

$$\boldsymbol{\varphi}: \mathbf{U} \rightarrow \mathbf{F} \tag{2.41}$$

To utilize the kernel method within a linear adaptive filtering framework, it is necessary to incorporate a kernel function $\kappa(\cdot,\cdot)$ that can perform inner product operations in feature space without requiring knowledge of precise nonlinear mapping. As per Mercer's theorem, every kernel generates a mapping from the input space $\mathbf{U}$ to a feature space, which allows for the transformation of data into a higher-dimensional space $\mathbf{F}$ where linear operations can be performed more effectively.

$$\kappa(\mathbf{u}, \mathbf{u}') = \boldsymbol{\varphi}(\mathbf{u})\boldsymbol{\varphi}^T(\mathbf{u}') \tag{2.42}$$

Equation (2.42) is called the kernel trick. The most commonly used kernel is the Gaussian kernel

$$\kappa(\mathbf{u}, \mathbf{u}') = \exp\left(-\frac{\|\mathbf{u}-\mathbf{u}'\|^2}{\sigma^2}\right) \tag{2.43}$$

with $\sigma > 0$ being the kernel bandwidth.

The unknown nonlinearity f can be estimated sequentially by a KAF algorithm such that f_k (the estimate at iteration k) is updated based on the last estimate f_{k-1} and the current example $\{\mathbf{u}(k), d(k)\}$. By the representer theorem, the output of a KAF filter can usually be expressed by

$$f_k(\mathbf{u}) = \sum_{j=1}^{k} \mathbf{a}(j)\langle \boldsymbol{\varphi}(\mathbf{u}), \boldsymbol{\varphi}(\mathbf{u}(j))\rangle \tag{2.44}$$

where $\mathbf{a}(j)$ denotes the coefficient of the *j-th* center, and $<\cdot,\cdot>$ stands for the inner product operator.

2.3.2 Kernel Least Mean Square

According to the kernel method, the kernel-induced mapping $\varphi(\cdot)$ is employed to transform the input $\mathbf{u}(k)$ into $\mathbf{F}$ as $\varphi(\mathbf{u}(k))$, usually denoted by $\varphi(k)=\varphi(\mathbf{u}(k))$ for simplicity. Then the weight vector $\boldsymbol{\omega}$ in $\mathbf{F}$ can be estimated by a stochastic gradient descent method. Applying the LMS algorithm on the example sequence $\{\varphi(k), d(k)\}$ in $\mathbf{F}$ yields

$$\begin{aligned}
&\boldsymbol{\omega}(0)=0\\
&e(k)=d(k)-\boldsymbol{\omega}^T(k-1)\varphi(k)\\
&\boldsymbol{\omega}(k)=\boldsymbol{\omega}(k-1)+\mu e(k)\varphi(k)
\end{aligned} \tag{2.45}$$

where $e(k)$ is the prediction error at iteration k, $\boldsymbol{\omega}(k)$ denotes the estimate of the weight vector in RKHS (see Appendix B for detailed description of RKHS), η denotes the step size. Thus, the output $f_k(\cdot)=\boldsymbol{\omega}^T(k)\varphi(\cdot)$. Since the dimensionality of $\varphi(\cdot)$ is high and it is only implicitly known, to perform the computation, the repeated application of the weight-update equation through iterations yields

$$\begin{aligned}
\boldsymbol{\omega}(k)&=\boldsymbol{\omega}(k-1)+\mu e(k)\varphi(k)\\
&=[\boldsymbol{\omega}(k-2)+\mu e(k-1)\varphi(k-1)]+\mu e(k)\varphi(k)\\
&=\boldsymbol{\omega}(k-2)+\mu[e(k-1)\varphi(k-1)+e(k)\varphi(k)]\\
&\cdots\\
&=\boldsymbol{\omega}(0)+\mu\sum_{j=1}^{k}e(j)\varphi(j)
\end{aligned} \tag{2.46}$$

If the initial $\boldsymbol{\omega}(0)=0$, we have $\boldsymbol{\omega}(k)=\mu\sum_{j=1}^{k}e(j)\varphi(j)$. Then, the output of the system to a new input $\mathbf{u}(k)$ can be expressed in terms of inner products between transformed inputs

$$<\boldsymbol{\omega}(k),\varphi(k)>=\left[\mu\sum_{j=1}^{k}(e(j)\varphi(j)\right]\varphi(n)=\mu\sum_{j=1}^{k}e(j)<\varphi(j),\varphi(k)> \tag{2.47}$$

By the kernel trick (2.42), the filter output can be computed by kernel evaluations as

$$<\omega(k),\varphi(k)>=\mu\sum_{j=1}^{k}|e(j)|^{p-2}\,e(j)\kappa(\mathbf{u}(j),\,\varphi(k)) \tag{2.48}$$

Then the following sequential learning rule in the original space is obtained

$$f_{k-1}(\cdot)=\mu\sum_{j=1}^{k-1}e(j)\kappa(\mathbf{u}(j),\cdot) \tag{2.49}$$

$$f_{k-1}(\mathbf{u}(k))=\mu\sum_{j=1}^{k-1}e(j)\kappa(\mathbf{u}(j),\mathbf{u}(k)) \tag{2.50}$$

$$e(k)=d(k)-f_{k-1}(\mathbf{u}(k)) \tag{2.51}$$

$$f_k(\cdot)=f_{k-1}(\cdot)+\mu e(k)\kappa(\mathbf{u}(k),\cdot) \tag{2.52}$$

Now the KLMS algorithm has been derived, which is the LMS in kernel space, and nonlinear filtering is done by kernel evaluation. KLMS allocates a new kernel unit for the new training data with input as the center and $\mu e(k)$ as the coefficient. The coefficients and the centers are stored in memory during training. Based on the description above, the KLMS is summarized in Table 2.3 and illustrated in Figure 2.2, where $\mathbf{a}(k)$ is the coefficient vector at iteration k, $\mathbf{a}_j(k)$ is its jth component, and $\mathbb{C}(k)$ is the corresponding set of centers. At iteration k, given a test input point, the output of the filter is

$$f_{k-1}(\mathbf{u}_*)=\mu\sum_{j=1}^{k}e(j)\kappa(\mathbf{u}(j),\mathbf{u}_*) \tag{2.53}$$

TABLE 2.3

Kernel Least Mean Square Algorithm

Initialization
$\mathbf{a}_1(1)=\mu d(1),\mathbb{C}(1)=\{\mathbf{u}(1)\},f_1=\mathbf{a}_1(1)k(\mathbf{u}(1),\cdot),\sigma=\sigma_0$
Computation
while $\{u(k),d(k)\}$ available do
$f_{k-1}(\mathbf{u}(k))=\sum_{j=1}^{k-1}\mathbf{a}_j(k-1)\kappa(\mathbf{u}(k),\mathbf{u}(j))$
$e(k)=d(k)-f_{k-1}(\mathbf{u}(k))$
%store the new center
$\mathbb{C}(k)=\{\mathbb{C}(k-1),\mathbf{u}(k)\}$
%Compute the coefficient
$\mathbf{a}_k(k)=\mu e(k)$
End while

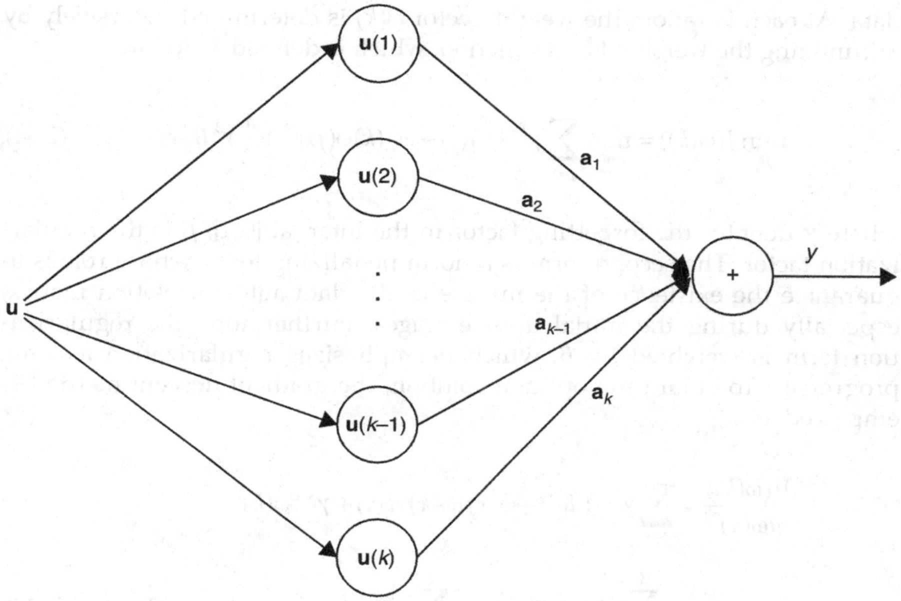

FIGURE 2.2
Network topology of KLMS at iteration k.

The KLMS topology bears resemblance to a radial-basis function (RBF) network, but with three key distinctions: first, the output weights represent the scaled prediction errors at each sample; second, it is a growing network where each new unit is placed over each new input; and third, the kernel function is not restricted to being a radial-basis function and can be any Mercer kernel (see Appendix C for a detailed description of Mercer kernel). Although KLMS is a straightforward algorithm that only requires a few operations per filter evaluation and weight update, there are still several unspecified aspects that require attention. First, the selection of the kernel function needs to be considered; second, the step size parameter must be chosen carefully; and finally, the growing memory and computation requirements for online operation must be addressed.

2.4 KRLS Algorithm

This section provides an overview of the exponentially weighted KRLS algorithm. The algorithm utilizes an exponentially weighted approach to give greater importance to recent data while reducing the significance of older data. At each iteration, the weight vector $\omega(k)$ is determined recursively by minimizing the weighted cost function, which is defined as follows:

$$\min_{\omega} J(\omega(k)) = \min_{\omega} \sum_{j=1}^{k} \gamma^{k-j} \, | d(j) - \omega^T(k)\varphi(j) |^p + \frac{1}{2}\gamma^k \beta \|\omega(k)\|^2 \qquad (2.54)$$

where γ denotes the forgetting factor in the interval [0 1], β is the regularization factor. The second term is a norm penalizing term, whose role is to guarantee the existence of the inverse of the data auto-correlation matrix, especially during the initial update stages. Furthermore, the regularization term is weighted by β, which deemphasizes regularization as time progresses. To obtain the optimal solution, the gradient descent method is employed as

$$\begin{aligned}\frac{\partial J(\omega(k))}{\partial \omega(k)} &= \sum_{j=1}^{k} \gamma^{k-j}(d(j) - \varphi^T(j)\omega(k))\varphi(j) + \gamma^k \beta \omega(k) \\ &= -\sum_{j=1}^{k} \gamma^{k-j} d(j)\varphi^T(j) + \sum_{j=1}^{k} \gamma^{k-j} \varphi^T(j)\varphi(j)\omega(k) + \gamma^k \beta \omega(k) \qquad (2.55) \\ &= -\sum_{j=1}^{k} \gamma^{k-j} d(j)\varphi^T(j) + \left(\sum_{j=1}^{k} \gamma^{k-j} \varphi^T(j)\varphi(j) + \gamma^k \beta \right) \omega(k)\end{aligned}$$

Setting the above gradient to zero, we obtain the solution

$$\omega = \left(\sum_{j=1}^{k} \gamma^{k-j} \varphi^T(j)\varphi(j) + \gamma^k \beta \right)^{-1} \sum_{j=1}^{k} \chi^{k-j} d(j)\varphi^T(j) \qquad (2.56)$$

To describe it concisely, we define

$$\mathbf{d}(k) = [d(1), d(2), \ldots, d(k)] \qquad (2.57)$$

$$\mathbf{\Phi}(k) = [\varphi(1), \varphi(2), \ldots, \varphi(k)] \qquad (2.58)$$

$$\mathbf{B}(k) = \mathrm{diag}[\gamma^{k-1}, \gamma^{k-2}, \ldots, 1] \qquad (2.59)$$

Then, the matrix form of the equation (2.56) at time k can be expressed as

$$\omega(k) = (\mathbf{\Phi}(k)\mathbf{B}(k)\mathbf{\Phi}^T(k) + \gamma^k \beta \mathbf{I})^{-1} \mathbf{\Phi}(k)\mathbf{B}(k)\mathbf{d}(k) \qquad (2.60)$$

Now, by using the matrix inversion lemma with the identifications $\beta\gamma^{i}\mathbf{I} \rightarrow \mathbf{A}$, $\mathbf{\Phi}(k) \rightarrow \mathbf{B}$, $\mathbf{B}(k) \rightarrow \mathbf{C}$, $\mathbf{\Phi}(k)^{T} \rightarrow \mathbf{D}$, one can obtain

$$(\mathbf{\Phi}(k)\mathbf{B}(k)\mathbf{\Phi}^{T}(k)+\gamma^{k}\beta\mathbf{I})^{-1}\mathbf{\Phi}(k)\mathbf{B}(k)=\mathbf{\Phi}(k)(\mathbf{\Phi}(k)\mathbf{\Phi}^{T}(k)+\gamma^{k}\beta\mathbf{B}(k)^{-1})^{-1} \quad (2.61)$$

Substituting (2.61) into (2.60) yields

$$\omega(k)=\mathbf{\Phi}(k)(\mathbf{\Phi}(k)\mathbf{\Phi}^{T}(k)+\gamma^{k}\beta\mathbf{B}(k)^{-1})^{-1}\mathbf{d}(k) \quad (2.62)$$

Now, the weight vector can be expressed explicitly as a linear combination of the input data in $\mathbf{F}$ as

$$\omega(k)=\mathbf{\Phi}(k)\mathbf{a}(k) \quad (2.63)$$

where $\mathbf{a}(k)$ denotes the computable expansion coefficients vector of the weight by the kernel trick, and it is defined by

$$\mathbf{a}(k)=(\mathbf{\Phi}(k)\mathbf{\Phi}^{T}(k)+\gamma^{k}\beta\mathbf{B}(k)^{-1})\mathbf{d}(k) \quad (2.64)$$

If we further denote

$$\mathbf{Q}(k)=(\mathbf{\Phi}(k)\mathbf{\Phi}^{T}(k)+\gamma^{k}\beta\mathbf{B}(k)^{-1})^{-1} \quad (2.65)$$

where $\mathbf{\Phi}(k)=\{\mathbf{\Phi}(k-1),\boldsymbol{\varphi}(k)\}$, it is easy to obtain that

$$\mathbf{Q}(k)^{-1}=\begin{bmatrix}\mathbf{Q}(k-1)^{-1} & \mathbf{h}(k)\\ \mathbf{h}(k)^{T} & \gamma^{k}\beta+\boldsymbol{\varphi}^{T}(k)\,\boldsymbol{\varphi}(k)\end{bmatrix} \quad (2.66)$$

where $\mathbf{h}(k)=\mathbf{\Phi}^{T}(k-1)\boldsymbol{\varphi}(k)$. By using the following block matrix inversion formula

$$\begin{bmatrix}\mathbf{A} & \mathbf{B}\\ \mathbf{C} & \mathbf{D}\end{bmatrix}^{-1}=\begin{bmatrix}(\mathbf{A}-\mathbf{B}\mathbf{D}^{-1}\mathbf{C})^{1} & -\mathbf{A}^{-1}\mathbf{B}(\mathbf{D}-\mathbf{C}\mathbf{A}^{-1}\mathbf{B})^{-1}\\ -\mathbf{D}^{-1}\mathbf{C}(\mathbf{A}-\mathbf{B}\mathbf{D}^{-1}\mathbf{C})^{-1} & (\mathbf{D}-\mathbf{C}\mathbf{A}^{-1}\mathbf{B})^{-1}\end{bmatrix} \quad (2.67)$$

we obtain the updated equation for the inverse of the growing matrix in (2.66) as

$$\mathbf{Q}(k)=r^{-1}(k)\begin{bmatrix}\mathbf{Q}(k-1)r(k)+z(k)z^{T}(k) & -z(k)\\ -z^{T}(k) & 1\end{bmatrix} \quad (2.68)$$

where $z(k)=\mathbf{Q}(k-1)\mathbf{h}(k)$, and $r(k)=\gamma^{k}\beta+\boldsymbol{\varphi}^{T}(k)\boldsymbol{\varphi}(k)-z^{T}(k)\mathbf{h}(k)$.

TABLE 2.4

Kernel Recursive Least Squares Algorithm

Initialization

$\mathbf{Q}(1) = (\lambda + \kappa(\mathbf{u}(1), \mathbf{u}(1)))^{-1}, \mathbf{a}(1) = \mathbf{Q}(1)d(1)$

Computation

Iterate for $k > 1$

$\mathbf{h}(k) = [\kappa(\mathbf{u}(k), \mathbf{u}(1)), \ldots, \kappa(\mathbf{u}(k), \mathbf{u}(k-1))]^T$

$e(k) = d(k) - \mathbf{h}(k)^T \mathbf{a}(k-1)$

$\mathbf{z}(k) = \mathbf{Q}(k-1)\mathbf{h}(k)$

$r(k) = \beta\gamma^k + \kappa(\mathbf{u}(k), \mathbf{u}(k)) - \mathbf{z}(k)^T \mathbf{h}(k)$

$$\mathbf{Q}(k) = r(k)^{-1} \begin{bmatrix} \mathbf{Q}(k-1)r(k) + \mathbf{z}(k)\mathbf{z}(k)^T & -\mathbf{z}(k) \\ -\mathbf{z}(k)^T & 1 \end{bmatrix}$$

$$\mathbf{a}(k) = \begin{bmatrix} \mathbf{a}(k-1) - \mathbf{z}(k) r(k)^{-1} e(k) \\ r(k)^{-1} e(k) \end{bmatrix}$$

Combining (2.63) and (2.68), we get

$$\mathbf{a}(k) = \mathbf{Q}(k)\mathbf{d}(k) = \begin{bmatrix} \mathbf{Q}(k-1) + \mathbf{z}(k)\mathbf{z}^T(k)r^{-1}(k) & -\mathbf{z}(k)r^{-1}(k) \\ -\mathbf{z}^T(k)r^{-1}(k) & r^{-1}(k) \end{bmatrix} \begin{bmatrix} \mathbf{d}(k-1) \\ d(k) \end{bmatrix}$$
$$= \begin{bmatrix} \mathbf{a}(k-1) - \mathbf{z}(k)r^{-1}(k)e(k) \\ r^{-1}(k)e(k) \end{bmatrix} \tag{2.69}$$

Then, the exponentially weighted kernel recursive least mean square (EWKRLS) algorithm has been derived, as summarized in Table 2.4. To summarize, the only difference between EWKRLS and KRLS is the regularization parameter in the later. The time and space complexities for them are both $O(k^2)$. With sparsification, the complexity of the EWKRLS algorithm will be reduced to $O(m_k^2)$, where m_k is the effective number of centers in the network at time k.

Appendix A: Block Matrix Inversion Lemma

$$\begin{bmatrix} \mathbf{A} & \mathbf{B} \\ \mathbf{C} & \mathbf{D} \end{bmatrix}^{-1} = \begin{bmatrix} (\mathbf{A}-\mathbf{B}\mathbf{D}^{-1}\mathbf{C})^{-1} & -\mathbf{A}^{-1}\mathbf{B}(\mathbf{D}-\mathbf{C}\mathbf{A}^{-1}\mathbf{B})^{-1} \\ -\mathbf{D}^{-1}\mathbf{C}(\mathbf{A}-\mathbf{B}\mathbf{D}^{-1}\mathbf{C})^{-1} & (\mathbf{D}-\mathbf{C}\mathbf{A}^{-1}\mathbf{B})^{-1} \end{bmatrix} \quad \text{(A.1)}$$

where **A** and **D** are arbitrary square matrix blocks.

Appendix B: Reproducing Kernel Hilbert Spaces

A pre-Hilbert space is an inner product space that has an orthonormal basis $\{\mathbf{x}_i\}_{i=1}^{\infty}$. Let H be the largest and most inclusive space of vectors for which the infinite set $\{\mathbf{x}_i\}_{i=1}^{\infty}$ is a basis. Then, vectors not necessarily lying in the original inner product space are represented in the form

$$\mathbf{x} = \sum_{i=1}^{\infty} a_i \mathbf{x}_i \quad \text{(B.1)}$$

are said to be spanned by the basis $\{\mathbf{x}_i\}_{i=1}^{\infty}$, and $\{a_i\}_{i=1}^{\infty}$ are the coefficients of these presentations. Define the new vector

$$\mathbf{y}_k = \sum_{i=1}^{k} a_i \mathbf{x}_i \quad \text{(B.2)}$$

Another vector $\mathbf{y}_j$ can be defined in a similar way. For $k > j$, we may express the squared Euclidean distance between the vectors $\mathbf{y}_k$ and $\mathbf{y}_j$ as

$$\begin{aligned} \left\|\mathbf{y}_k - \mathbf{y}_j\right\|^2 &= \left\| \sum_{i=1}^{k} a_i \mathbf{x}_i - \sum_{i=1}^{j} a_i \mathbf{x}_i \right\|^2 \\ &= \left\| \sum_{i=j+1}^{k} a_i \mathbf{x}_i \right\|^2 \\ &= \sum_{i=j+1}^{k} a_i^2 \end{aligned} \quad \text{(B.3)}$$

where, in the last line, we invoked the orthonormality condition. Therefore, to make the definition of **x** meaningful, we need the following conditions to hold:

i. $\sum_{i=j+1}^{k} a_i^2 \to 0$ as both $k, j \to \infty$.

ii. $\sum_{i=1}^{j} a_i^2 < \infty$.

In other words, a sequence of vectors $\{\mathbf{y}_i\}_{i=1}^{\infty}$ is a *Cauchy sequence*. Consequently, a vector **x** can be expanded on the basis $\{\mathbf{x}_i\}_{i=1}^{\infty}$ if, and only if, **x** is a linear combination of the basis vectors and the associated coefficients $\{a_i\}_{i=1}^{\infty}$ are square summable. From this discussion, it is apparent that the space H is more "complete" than the starting inner product space. We may therefore make the following important statement: *An inner product space* H *is complete if every Cauchy sequence of vectors taken from the space* H *converges to a limit in* H*; a complete inner product space is called a Hilbert space.*

Appendix C: Mercer Kernel

A *Mercer kernel* is a continuous, symmetric, positive-definite function $\kappa : \mathbf{U} \times \mathbf{U} \to \mathbb{R}$. U is the input domain, a subset of $\mathbb{R}^L$. The commonly used kernels include the Gaussian kernel [equation (B.4)] and the polynomial kernel:

$$\kappa(\mathbf{u}, \mathbf{u}') = \exp\left(-a \,||\mathbf{u} - \mathbf{u}'||^2\right) \tag{C.1}$$

$$\kappa(\mathbf{u}, \mathbf{u}') = \left(\mathbf{u}^T \mathbf{u}' + 1\right)^p \tag{C.2}$$

Let H be any vector space of all real-valued functions of **u** that are generated by the kernel $\kappa(\mathbf{u}, \cdot)$. Suppose now two functions $h(\cdot)$ and $g(\cdot)$ are picked from the space H that are respectively represented by

$$h = \sum_{i-1}^{l} a_i \kappa(\mathbf{c}_i, \cdot) \tag{C.3}$$

and

$$g = \sum_{j=1}^{m} b_j \kappa(\tilde{\mathbf{c}}_j, \cdot) \tag{C.4}$$

where the a_i and the b_j are expansion coefficients and both $\mathbf{c}_i$ and $\tilde{\mathbf{c}}_j$ belong to $\mathbf{U}$ for all i and j. The bilinear form is defined as

$$\langle h, g \rangle = \sum_{i-1}^{l} \sum_{j=1}^{m} a_i \kappa(\mathbf{c}_i, \tilde{\mathbf{c}}_j) b_j \tag{C.5}$$

which satisfies the following properties:

i. Symmetry

$$\langle h, g \rangle = \langle g, h \rangle \tag{C.6}$$

ii. Scaling and distributive property

$$\langle (cf + dg), h \rangle = c\langle f, h \rangle + d\langle g, h \rangle \tag{C.7}$$

iii. Squared norm

$$\|f\|^2 = \langle f, f \rangle \geq 0 \tag{C.8}$$

By virtue of these facts, the bilinear term h, g is indeed an inner product. There is one additional property that follows directly. Specifically, setting $g(\cdot) = \kappa(\mathbf{u}, \cdot)$, we obtain

$$\langle h, \kappa(\mathbf{u}, \cdot) \rangle = \sum_{i=1}^{l} a_i \kappa(\mathbf{c}_j, \mathbf{u}) = h(\mathbf{u}) \tag{C.9}$$

This property is known as the *reproducing property*. The kernel $\kappa(\mathbf{u}, \mathbf{u}')$, which represents a function of the two vectors $\mathbf{u}, \mathbf{u}' \in \mathbf{U}$, is called a reproducing kernel of the vector space $\mathbf{H}$ if it satisfies the following two conditions:

1. For every $\mathbf{u} \in \mathbf{U}$, $\kappa(\mathbf{u}, \mathbf{u}')$ as a function of the vector $\mathbf{u}'$ belongs to $\mathbf{H}$.
2. It satisfies the reproducing property.

The Mercer kernel satisfies both of these conditions, making it a reproducing kernel. When the reproducing kernel space is defined in a complete inner product space $\mathbf{H}$, it is referred to as a reproducing kernel Hilbert space (RKHS). The analytical strength of RKHS is demonstrated in the Mercer theorem, which states that any reproducing kernel $\kappa(\mathbf{u}, \mathbf{u}')$ can be expanded in the following manner:

$$\kappa(\mathbf{u}, \mathbf{u}') = \sum_{i=1}^{\infty} \varsigma_i \phi_i(\mathbf{u}) \phi_i(\mathbf{u}') \tag{C.10}$$

where ς_i and ϕ_i are the eigenvalues and the eigenfunctions, respectively. The eigenvalues are non-negative. Therefore, a mapping φ can be constructed as

$$\begin{aligned} &\varphi : \mathbf{U} \rightarrow \mathbf{F} \\ &\varphi(\mathbf{u}) = \left[\sqrt{\varsigma_1}\phi_1(\mathbf{u}), \sqrt{\varsigma_2}\phi_2(\mathbf{u}), \ldots \right] \end{aligned} \tag{C.11}$$

By construction, the dimension of **F** is determined by the number of strictly positive eigenvalues, which are infinite in the Gaussian kernel case. In the machine learning literature, $\boldsymbol{\phi}$ is usually treated as the feature mapping and $\boldsymbol{\phi}(\mathbf{u})$ is the transformed feature vector lying in the feature space **F**. By doing so, an important implication is

$$\varphi(\mathbf{u})^T \varphi(\mathbf{u}') = \kappa(\mathbf{u}, \mathbf{u}') \tag{C.12}$$

It is easy to check that **F** is essentially the same as the RKHS induced by the kernel by identifying $\varphi(\mathbf{u}) = \kappa(\mathbf{u}, \cdot)$, which are the bases of the two spaces, respectively. By slightly abusing the notation, we do not distinguish **F** and **H** in this book if no confusion is involved.

3

MMPE Family Criteria

As previously mentioned, the minimum mean *p-Power* error (MMPE) criterion can outperform the conventional MSE criterion when a suitable p value is selected [201]. There are three main scenarios to consider: First, if the optimal solution of the MMPE criterion differs from the Wiener solution of the MSE criterion, the former may be superior to the latter. In system identification, this implies that the optimal solution of the mean p-power error (MPE) cost function may be closer to the true system parameters than that of the MSE. Second, if the optimal solution of the MPE function is the same as the Wiener solution of the MSE function for a given $p=2$, then the steepest descent algorithm based on the MMPE criterion may exhibit better performance, such as faster convergence speed, than the conventional LMS algorithm. Third, when the input signal or desired process is corrupted by impulsive noise, an adaptive filter based on the MMPE criterion with $p=1$ may exhibit stronger robustness than the LMS algorithm. This chapter presents the definition and properties of the MMPE criterion, followed by an introduction to an adaptive MMPE criterion, which includes an online method for selecting a suitable p value to enhance the performance of the adaptive algorithm [266]. In addition, the smoothed MPE [267], the mixture MPE [268], and kernel MPE [257] criteria, which are extended versions of the MMPE criterion, will also be introduced.

3.1 Basic MMPE Criterion

3.1.1 Mean *p*-Power Error Criterion

Definition 3.1

Let $y(k)=\mathbf{w}^T(k)\mathbf{u}(k)$ and $e(k)=d(k)-y(k)$ denote the output and the prediction error of an FIR filter, respectively. Then, a cost function, called MPE is defined by

$$\begin{aligned} J_{\mathrm{MPE}}(k) &= E[\,|\,e(k)\,|^p\,] \\ &= E[\,|\,d(k)-y(k)\,|^p\,] \end{aligned} \tag{3.1}$$

DOI: 10.1201/9781003176114-3

where p is a positive constant. The filter parameters are adjusted so that the MPE is minimized.

However, $J_{\mathrm{MPE}}(k)$ is a function of degree p, whose value is unknown. Thus, it is possible that $J_{\mathrm{MPE}}(k)$ has local minima. In the next subsection, we use some results of convex function to prove that every minimum of performance function is a global minimum. In practice, the instantaneous MPE is usually used, i.e.,

$$\begin{aligned} J_{\mathrm{MPE}}(k) &= |e(k)|^p \\ &= |d(k) - y(k)|^p \end{aligned} \tag{3.2}$$

According to the definition of the MPE criterion in (3.1) and (3.2), one can see that it can reduce to the classical MSE, MAE, and MFE criterion when p is selected as 2.0, 1.0, and 4.0, respectively. Table 3.1 gives the MMPE with different p values.

The MSE cost function has a closed-form solution, namely, the Wiener solution. The solution to (3.2) can also be obtained by iterative procedures similar to LMS method. However, in alpha-stable noise environments, the error signal variance could be infinite, and therefore, the MSE cost is inappropriate. When there are anomalous points with large deviation in the data, the MSE-based algorithm may cause relatively large fluctuations after square. But the MPE with $p = 1$ (i.e., MAE) algorithm (otherwise known as the sign-error LMS or pilot LMS) only considers a power deviation and can overcome this defect well. Thus, anomalous points have little effect on it. Adaptive algorithms based on higher order moments of the error signal have been shown to perform better than MSE-based algorithms in some important applications. One of such algorithms is the least mean fourth (LMF), which has found application. The LMF algorithm seeks to minimize the mean fourth error, which is a convex function (and thus unimodal) of the weight vector. It has been shown that the LMF algorithm can outperform LMS for non-Gaussian additive noise, leading to considerably smaller excess mean square error (EMSE) for the same convergence speed.

TABLE 3.1

MMPE with Different p Values

Criterion	p	Cost function
MPE	p	$J_{\mathrm{MPE}} = E[\,|e(k)|^p\,]$
MSE	$p=2$	$J_{\mathrm{MSE}}(k) = E[\,|e(k)|^2\,]$
MAE	$p=1$	$J_{\mathrm{MAE}}(k) = E[\,|e(k)|\,]$
MFE	$p=4$	$J_{\mathrm{MFE}}(k) = E[\,|e(k)|^4\,]$

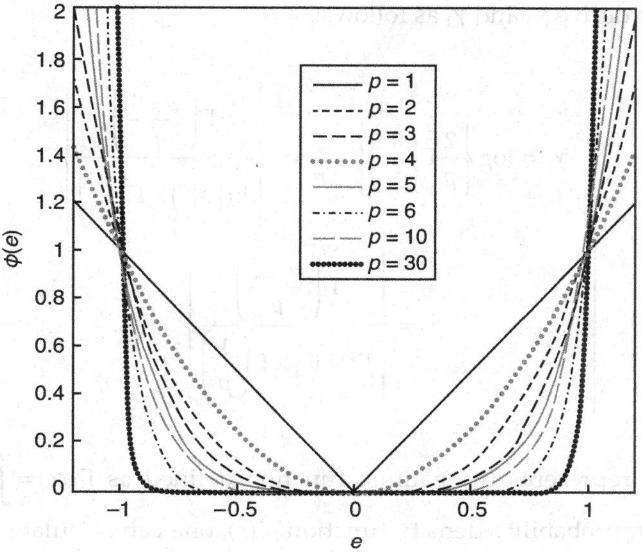

FIGURE 3.1
Cost functions of MPE with different p values [269].

The cost function $\phi(e) = |e|^p$ is a convex function defined on $\mathbf{R}^M$ for $p \geq 1$, and its values with different p are shown in Figure 3.1. One can observe that every minimum of the performance function $J_{\text{MPE}}(k)$ is a global minimum, which means that any optimum weight vector that an adaptive algorithm converges to will be a global optimal solution. Especially, there exists a fractional lower order moment (FLOM) for alpha-stable process when $0 \leq p \leq \alpha$.

Remark 3.1

For MPE criterion $E(|e|^p)$, from Theorem 1 in [269], one can know that there exists a worst-case probability density function according to Jaynes maximum entropy principle, such that $q(e) = \exp[-\gamma_0 - \gamma_1 |e|^p]$, where γ_0 and γ_1 are determined by

$$\begin{cases} \exp(\gamma_0) = \int_{-\infty}^{+\infty} \exp[\gamma_1 |e|^p] de \\ E(|e|^p)\exp(\gamma_0) = \int_{-\infty}^{+\infty} |e|^p \exp[\gamma_1 |e|^p] de \end{cases} \tag{3.3}$$

It is easy to derive γ_0 and γ_1 as follows:

$$\begin{cases} \gamma_0 = \log\left\{\dfrac{2}{p}\Gamma\left(\dfrac{1}{p}\right)\right\} - \dfrac{1}{p}\log\left\{\dfrac{\Gamma\left(\dfrac{p+1}{p}\right)}{E\left(|e|^p\right)\times\Gamma\left(\dfrac{1}{p}\right)}\right\} \\ \gamma_0 = \left\{\dfrac{\Gamma\left(\dfrac{p+1}{p}\right)}{E\left(|e|^p\right)\times\Gamma\left(\dfrac{1}{p}\right)}\right\} \end{cases} \tag{3.4}$$

where $\Gamma(.)$ represents the gamma function defined as $\Gamma(\alpha) = \int_0^\infty x^{\alpha-1}e^{-x}dx$ [270]. Given probability density function $p_e(e)$, one can calculate $E\left(|e|^p\right)$ and get the exact values of γ_0 and γ_1.

Remark 3.2

It is worth noting that the worst-case density $q(e) = \exp[-\gamma_0 - \gamma_1 |e|^p]$ is actually the generalized Gaussian density (GGD), in which p is called the shape parameter. The GGD model includes Laplace ($p = 1$) and Gaussian ($p = 2$) distributions as special cases and can be used to approximate a large number of distributions in the areas of image coding, speech recognition, blind source separation, and so on. For different p values, the worst-case densities $q(e)$ are depicted in Figure 3.2 (assuming $p_e(e) \sim N(0,1)$).

3.1.2 Properties of MPE Criterion

3.1.2.1 Convexity of MPE Criterion

On the convexity of the MPE criterion, the main results are summarized as follows [201].

Theorem 3.1

When $p \geq 1$, the performance function $J_{\text{MPE}}(\mathbf{w}) = E\left[|e(k)|^p\right]$ is a convex function defined on $\mathbb{R}^M$.

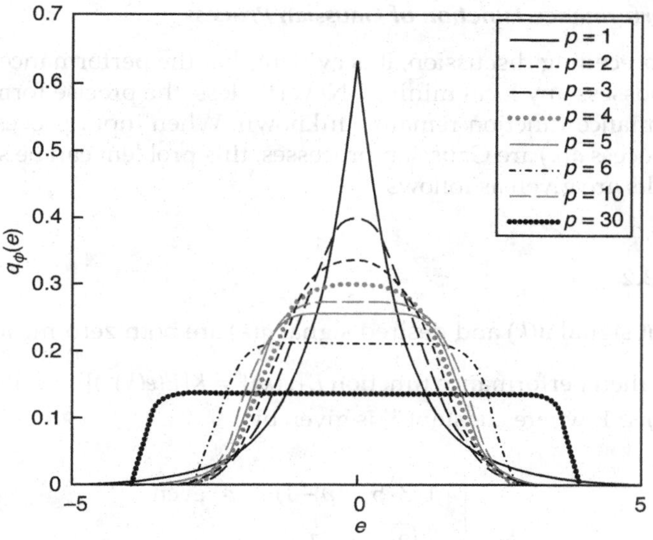

FIGURE 3.2
Worst-case densities $q(e)$ for different p values [269].

Proof: for every $\mathbf{w}_1$, $\mathbf{w}_2$ and $0 \le \alpha \le 1$, we have

$$\begin{aligned} & |d(k)-\mathbf{u}^T(k)[\alpha \mathbf{w}_1+(1-\alpha)\mathbf{w}_2]|^p \\ & =|d(k)-\mathbf{u}^T(k)[\alpha \mathbf{w}_1+(1-\alpha)\mathbf{w}_2]|^p \\ & =|\alpha[d(k)-\mathbf{u}^T(k)\mathbf{w}_1]+(1-\alpha)[d(k)-\mathbf{u}^T(k)\mathbf{w}_2]|^p \\ & \le \{\alpha\,|d(k)-\mathbf{u}^T(k)\mathbf{w}_1|^p+(1-\alpha)[d(k)-\mathbf{u}^T(k)\mathbf{w}_2]|^p\} \end{aligned} \tag{3.5}$$

Let $f\left(d(k),\mathbf{u}(k)\right)$ be the joint probability density function (p.d.f) of $d(k)$ and $\mathbf{u}(k)$. Multiply both sides of (3.5) by $f\left(d(k),\mathbf{u}(k)\right)$ and integrate with respect to variables $d(k)$ and $\mathbf{u}(k)$, i.e., taking the expectations of both sides of (3.3) yields

$$J_{\text{MPE}}(\alpha \mathbf{w}_1+(1-\alpha)\mathbf{w}_2) \le \alpha J_{\text{MPE}}(\mathbf{w}_1)+(1-\alpha)J_{\text{MPE}}(\mathbf{w}_2) \tag{3.6}$$

The proof is completed. Since J is a convex function, every minimum of the performance function J is a global minimum. This means that any optimum weight vector that the adaptive algorithm converges to is globally optimal.

3.1.2.2 Performance Function of Gaussian Process

After the preceding discussion, it is evident that the performance function J does not possess any local minima. Nevertheless, the precise formulation of the performance function remains unknown. When input process $\mathbf{u}(k)$ and desired process $d(k)$ are Gaussian processes, this problem can be solved. The main results are given as follows.

Theorem 3.2

If the input signal $\mathbf{u}(k)$ and desired signal $d(k)$ are both zero-mean Gaussian processes, then performance function $E|e(k)|^p = K[E(e(k)^2)]^{\frac{p}{2}}$ for all the choice of integer $p \geq 1$, where constant K is given by

$$K = \begin{cases} 1\cdot 3\cdot 5\ldots(p-1) & p: \text{even} \\ \sqrt{\dfrac{2^p}{\pi}}\left(\dfrac{p-1}{2}\right)! & p: \text{odd} \end{cases} \tag{3.7}$$

Proof: since $\mathbf{u}(k)$ and $d(k)$ are zero-mean Gaussian processes, the error $e(k)$ is also a zero-mean Gaussian process with probability density function as

$$f(e(k)) = \frac{1}{\sigma(k)\sqrt{2\pi}} \exp\left(\frac{-e(k)^2}{2\sigma(k)^2}\right) \tag{3.8}$$

where $\sigma(k)$ is given by

$$\sigma(k) = \left[E\left(e(k)^2\right)\right]^{1/2} \tag{3.9}$$

Thus, the following result can be obtained

$$\begin{aligned} E[|e(k)|^p] &= \frac{1}{\sigma(k)\sqrt{2\pi}} \int_{-\infty}^{\infty} |e(k)|^p \exp\left(\frac{-e(k)^2}{2\sigma(k)^2}\right) de(k) \\ &= \frac{2}{\sigma(k)\sqrt{2\pi}} \int_{0}^{\infty} |e(k)|^p \exp\left(\frac{-e(k)^2}{2\sigma(k)^2}\right) de(k) \\ &= \begin{cases} 1\cdot 3\cdot 5\ldots(p-1) & p: \text{even} \\ \sqrt{\dfrac{2^p}{\pi}}\left(\dfrac{p-1}{2}\right)! & p: \text{odd} \end{cases} \end{aligned} \tag{3.10}$$

Combining (3.9) and (3.10), one can get

$$E[|e(k)|^p] = K\left[E(e(k)^2)\right]^{p/2} \tag{3.11}$$

The proof is completed.

Remark 3.3

From (3.11), the gradient vector of $E(|e(k)|^p)$ can be obtained below

$$\nabla_p = S\nabla_2 \tag{3.12}$$

where the scaling factor S is given by

$$S = \frac{Kp}{2}\left[E(e(k)^2)\right]^{(p-2)/2} \tag{3.13}$$

When the steepest descent algorithm with constant step size is used to search the minimum of performance function, the larger the slope performance function has, the faster the convergence speed algorithm has. Thus, the larger scaling factor S is preferred. Moreover, from (3.12), it is clear that $E[|e(k)|^p]$ has the same optimum solution as $E[|e(k)|^2]$ for the choice of integer $p \geq 1$. Consequently, $E[|e(k)|^p]$ has the optimum solution at $\mathbf{w}_o = \mathbf{R}^{-1}\mathbf{P}$ for Gaussian processes. However, when $\mathbf{u}(k)$ and $d(k)$ are non-Gaussian processes, $E[|e(k)|^p]$ may not have the same optimum solution as $E[|e(k)|^2]$ for integer $p \neq 2$.

3.2 Adaptive MMPE Criterion

When applying the MMPE criterion, the challenge of selecting an appropriate p value to enhance the adaptive algorithm's performance often arises. A novel information-theoretic method has been employed to identify the optimal p value, as described in [269]

$$\begin{aligned} p_{\text{opt}} &= \arg\min_{p\in R_+}\left\{\min_{\tau\in R_+} D_{\text{KL}}(p_n(x)\,\|\,q(x))\right\} \\ &= \arg\min_{p\in R_+}\left\{\min_{\tau\in R_+}\int_R p_n(x)\log\left(\frac{p_n(x)}{q(x)}\right)dx\right\} \end{aligned} \tag{3.14}$$

where $D_{KL}(.\|\,.)$ denotes the KL-divergence [271], $p_n(.)$ represents the PDF of the additive noise $v(k)$, $q(x)=\exp[-\gamma_0-\gamma_1|x|^p]$ is the maximum entropy density (or worst-case density) under p-order absolute moment constraint, γ_0 and γ_1 are functions of τ ($\tau>0$), expressed as [269]

$$\begin{cases} \gamma_0=\log\left\{\dfrac{2}{p}\Gamma(1/p)\right\}-\dfrac{1}{p}\log\gamma_1 \\ \gamma_1=\dfrac{\Gamma((p+1)/p)}{\tau\Gamma(1/p)} \end{cases} \tag{3.15}$$

As $\Gamma((p+1)/p)=(1/p)\Gamma(1/p)$, the expression of (3.15) can be simplified as

$$\begin{cases} \gamma_0=\log\left\{\dfrac{2}{p}\Gamma(1/p)\right\}+\dfrac{1}{p}\log(\tau p) \\ \gamma_1=\dfrac{1}{\tau p} \end{cases} \tag{3.16}$$

The density $q(x)$ is actually the GGD with shape parameter p. In fact, any GGD density with zero-mean can be expressed as [270]

$$\begin{aligned} p_{GGD}(x,\tau,p)&=\frac{p}{2(p\tau)^{1/p}\Gamma(1/p)}\exp\left(-\frac{|x|^p}{p\tau}\right) \\ &=\exp[-\gamma_0-\gamma_1|x|^p] \end{aligned} \tag{3.17}$$

where $\tau=E[|x|^p]=\int|x|^p p_{GGD}(x,\tau,p)dx=\beta^p/p$, β denotes the dispersion parameter. The GGD family distributions include the Laplace ($p=1$), Gaussian ($p=2$) and Uniform ($p=\infty$) distributions as special cases.

The effectiveness of the optimization (3.14) has been verified by Monte Carlo simulation experiments [269]. However, there are two problems to the practical implementation of the optimum MMPE criterion: (1) the noise PDF $p_n(.)$ is usually unknown; (2) except for some special cases, there is no closed-form solution of the optimum p value, and hence a certain numerical optimization technique must be used, which increases the computational complexity dramatically.

In the following, a practicable online method is proposed to determine the p value. First, we give some further results on the optimization of (3.14).

Theorem 3.3

If the value of p value is fixed, function $D(\tau) \triangleq D_{KL}(p_n(x)\,||=\exp[-\gamma_0-\gamma_1|x|^p])$ attains its minimum at $\tau=\zeta$, where $\zeta = E[|x|^p] = \int_R |x|^p\, p_n(x)dx$.

Proof: Clearly, we have

$$\begin{aligned} D_{KL}(p_n(x) &= \exp[-\gamma_0-\gamma_1|x|^p]) \\ &= \int_R p_n(x)\log p_n(x)dx + \int_R p_n(x)\{\gamma_0+\gamma_1|x|^p\}dx - H(n)+\gamma_0+\gamma_1 E_n(|x|^p) \\ &= -H(n)+\gamma_0+\gamma_1\zeta \end{aligned} \tag{3.18}$$

where $H(n) = -\int_R p_n(x)\log p_n(x)dx$ is the differential entropy of noise. By expression (3.16), we get

$$D(\tau) = -H(n) + \left\{\log\left\{\frac{2}{p}\Gamma(1/p)\right\} + \frac{1}{p}\log(\tau p)\right\} + \left\{\frac{1}{\tau p}\right\}\zeta \tag{3.19}$$

It follows that

$$\frac{\partial}{\partial \tau}D(\tau) = 0 \Rightarrow \frac{1}{\tau p} - \left\{\frac{1}{\tau^2 p}\right\}\zeta = 0 \Rightarrow \tau = \zeta \tag{3.20}$$

which implies that $D(\tau)$ attains its minimum at $\tau=\zeta$.

Theorem 3.4

If the noise is GGD distributed with PDF $p_n(x) = p_{GGD}(x,\zeta,v)$, we have $p_{opt} = v$.

Proof: It is easy to derive

$$\begin{aligned} p_{opt} &= \arg\min_{p\in R_+}\left\{\min_{\tau\in R_+} D_{KL}(p_n(x)\,||\,p_{GGD}(x,\tau,p))\right\} \\ &\overset{(a)}{=} \arg\min_{p\in R_+} D_{KL}(p_{GGD}(x,\zeta,v)\,||\,p_{GGD}(x,\zeta,p)) \overset{(b)}{=} v \end{aligned} \tag{3.21}$$

where (a) follows from Theorem 3.1 and (b) follows from the fact that

$$D_{KL}(p_{GGD}(x,\zeta,v)\,||\,p_{GGD}(x,\zeta,p) \geq 0$$

with equality if and only if $p_{GGD}(x,\zeta,v) = p_{GGD}(x,\zeta,p)$ almost everywhere.

Theorem 3.2 indicates that when the noise PDF belongs to the generalized Gaussian distribution family, the optimum p value equals the shape parameter (v) of the noise distribution. To determine the value of p, we need to estimate the shape parameter of the noise, which is often unobservable in real-life scenarios. Therefore, we rely on error samples to estimate the shape parameter. This approach is reasonable because:

i. When the AFA is far from the optimum solution, the results can be obtained as $e(k) \approx e_a(k)$. Here the a priori error $e_a(k)$, as argued in [272], is usually of (approximate) Gaussian distribution. This implies that at the initial stage of adaptation, the estimated shape parameter (based on the error samples) will be approximately equal to 2 (i.e., the Gaussian shape parameter), and the adaptive algorithm is nearly the same as the LMS algorithm, which performs well in Gaussian environments.
ii. When the algorithm converges to the optimum solution, we have $W \approx W^*$ and $e(k) \approx n(k)$. This implies that at the final stage of adaptation, the estimated shape parameter (based on the error samples) will be approximately equal to the noise's shape parameter (v) which is the optimum p value.

    Several techniques have been suggested for the parametric estimation of the GGD distribution, with the most current advancements documented in [273]. One such approach, the moment matching method [273], enables real-time estimation of the shape parameter by utilizing error samples, as demonstrated by the following expression:

$$\hat{v}_k = K^{-1}\left(\frac{\left[(1/L)\sum_{i=k-L+1}^{k}(e(i))^2\right]}{\sqrt{(1/L)\sum_{i=k-L+1}^{k}(e(i))^4}}\right) \tag{3.22}$$

where $K^{-1}(.)$ is the inverse function of $K(x)=\Gamma(3/x)/\sqrt{\Gamma(5/x)\Gamma(1/x)}$, and L is the sliding data length. The curve of the inverse function $y = K^{-1}(x)$ is plotted in Figure 3.3.

Now an online method can be proposed to determine the p value, in which the p value at k iteration (denoted by p_k) is equal to $\hat{v}_k$ given by (3.22). In practical application, to achieve a better robust performance (avoid "very large gradient"), the p value is usually upper bound by a certain positive number. This yields the following adaptation cost:

$$J(k) = E[|e(k)|^{p_k}] = E\left[|e(k)|^{\min\{\hat{v}_k, p_{\text{upper}}\}}\right] \tag{3.23}$$

where $p_k = \min\{\hat{v}_k, p_{\text{upper}}\}$, and p_{upper} denotes the upper bound of p.

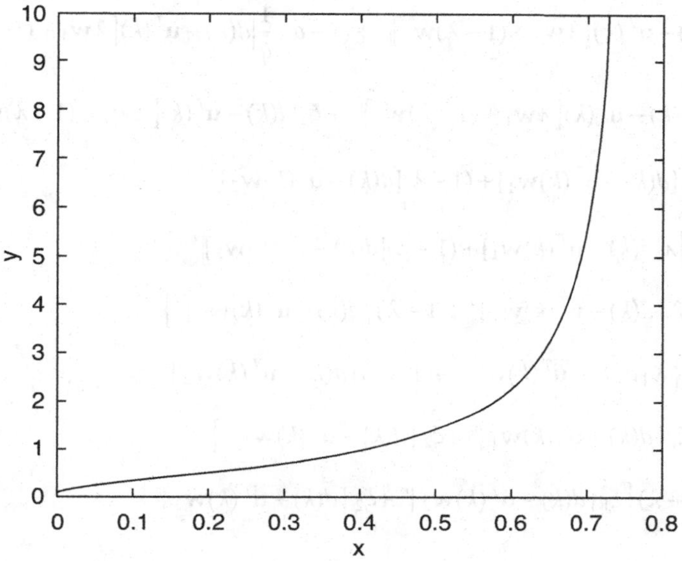

FIGURE 3.3
Curve of the inverse function $y = K^{-1}(x)$ [266].

3.3 Mixture MMPE Criterion

In certain complex conditions, relying solely on a single MPE criterion may lead to a decline in performance. To address this issue, we introduce the mixture MPE criterion, also known as GMN, which combines two MPE criteria. By utilizing a convex combination of two criteria, we define the mixture MPE criterion as [268]

$$J_{GMN}(\mathbf{w}(k)) = \omega \frac{1}{p} E[\,|e(k)|^p\,] + (1-\omega)\frac{1}{q} E[\,|e(k)|^q\,] \tag{3.24}$$

where $p>0$, $q>0$, and $\omega \in [0,1]$ is the mixing parameter.

The cost function $J_{GMN}(\mathbf{w}(k))$ is a convex function defined on $\mathbf{R}^L$ for $p>0$, $q>0$. This can be simply proved as follows. For every $\mathbf{w}_1$, $\mathbf{w}_2$, and $0 \le \lambda \le 1$, we have

$$
\begin{aligned}
&\omega\frac{1}{p}\left|d(k)-\mathbf{u}^T(k)[\lambda\mathbf{w}_1+(1-\lambda)\mathbf{w}_2]\right|^p+(1-\omega)\frac{1}{q}\left|d(k)-\mathbf{u}^T(k)[\lambda\mathbf{w}_1+(1-\lambda)\mathbf{w}_2]\right|^q \\
&=\xi_1\left|d(k)-\mathbf{u}^T(k)[\lambda\mathbf{w}_1+(1-\lambda)\mathbf{w}_2]\right|^p+\xi_2\left|d(k)-\mathbf{u}^T(k)[\lambda\mathbf{w}_1+(1-\lambda)\mathbf{w}_2]\right|^q \\
&=\xi_1\left|\lambda[d(k)-\mathbf{u}^T(k)\mathbf{w}_1]+(1-\lambda)[d(k)-\mathbf{u}^T(k)\mathbf{w}_2]\right|^p \\
&\quad+\xi_2\left|\lambda[d(k)-\mathbf{u}^T(k)\mathbf{w}_1]+(1-\lambda)[d(k)-\mathbf{u}^T(k)\mathbf{w}_2]\right|^q \\
&\le\xi_1\left\{\lambda\,|\,d(k)-\mathbf{u}^T(k)\mathbf{w}_1\,|^p+(1-\lambda)\,|\,d(k)-\mathbf{u}^T(k)\mathbf{w}_2\,|^p\right\} \\
&\quad+\xi_2\left\{\lambda\,|\,d(k)-\mathbf{u}^T(k)\mathbf{w}_1\,|^q+(1-\lambda)\,|\,d(k)-\mathbf{u}^T(k)\mathbf{w}_2\,|^q\right\} \\
&=\lambda\left\{\xi_1\,|\,d(k)-\mathbf{u}^T(k)\mathbf{w}_1\,|^p+\xi_2\,|\,d(k)-\mathbf{u}^T(k)\mathbf{w}_1\,|^q\right\} \\
&\quad+(1-\lambda)\left\{\xi_1\,|\,d(k)-\mathbf{u}^T(k)\mathbf{w}_2\,|^p+\xi_2\,|\,d(k)-\mathbf{u}^T(k)\mathbf{w}_2\,|^q\right\}
\end{aligned}
\tag{3.25}
$$

where $\xi_1=\omega\frac{1}{p},\xi_2=(1-\omega)\frac{1}{q}$. The joint probability density function (PDF) of $d(k)$and $\mathbf{u}(k)$ is defined as $P(d(k), \mathbf{u}(k))$. Using $P(d(k), \mathbf{u}(k))$to multiply both sides of (3.25) and taking expected value yields

$$
\mathrm{J}_{\mathrm{GMN}}(\lambda\mathbf{w}_1+(1-\lambda)\mathbf{w}_2)\le\lambda\,\mathrm{J}_{\mathrm{GMN}}(\mathbf{w}_1)+(1-\lambda)\,\mathrm{J}_{\mathrm{GMN}}(\mathbf{w}_2) \tag{3.26}
$$

The proof is completed.

Then the conclusion of $\mathrm{J}_{\mathrm{GMN}}(\mathbf{w}(k))$ as a convex function defined on convex set R^L is explained. This result means that any optimum weight vector that the adaptive algorithm convergences to is globally optimal. Taking advantage of the MMPE criterion, several novel adaptive filtering algorithms can be developed in the following section.

Remark 3.4

The mixture MPE criterion in (3.24) can be viewed as a generalized version of some existing criteria (such as MSE, MPE, LMMN [274], RMN [275], and l_2-l_p [276]). Table 3.2 summarizes the choices of p and q required to obtain different criteria mentioned.

TABLE 3.2
Suitable Choices for p and q Result in Different Criteria

	MSE	MPE	LMMN	RMN	$l_2 - l_p$
p	2	p	2	2	2
q	2	p	4	1	p

3.4 Smoothed MMPE Criterion

In this section, an interesting error criterion is further presented for adaptive filtering, namely the smoothed least mean p-power (SLMP) error criterion [267], which aims to minimize the mean p-power of the error plus an independent and scaled *smoothing variable.* If the distribution of the smoothing variable is symmetric and zero-mean, and p is an even number, i.e. $p=2k$ ($k \in$ N), then the SLMP error criterion will become a weighted sum of the even-order moments of the error, in which the weights are controlled by the *smoothing factor* (i.e., the scale factor). As the *smoothing factor* is large enough, this criterion will be approximately equivalent to the MSE criterion.

3.4.1 Smoothed MMPE Criterion

Definition 3.2

Let $e = d - g(U)$ be the estimation error, where $g(\cdot)$ stands for the collection of all measurable functions of U. Then the smoothed MPE cost is defined as

$$J_{\mathrm{SMPE}}(k) = E\left[| e(k) + h\xi |^p \right] \tag{3.27}$$

where ξ is a smoothing variable, which is a random variable that is independent of e, and $h \geq 0$ is a smoothing factor.

If the smoothing variable ξ has a continuous PDF $K(x)$, then the PDF of $h\xi$ can be expressed as

$$K_h(x) = \frac{1}{h} K\left(\frac{x}{h}\right) \tag{3.28}$$

Since ξ is independent of e, the PDF of the random variable $e + h\xi$ will be

$$p_{e+h\xi}(x) = (p_e * K_h)(x) = \int_{-\infty}^{\infty} p_e(\tau) K_h(x - \tau) d\tau \tag{3.29}$$

where * denotes the convolution operator, and p_e denotes the PDF of e. Then the following result can be got

$$
\begin{aligned}
J_{\mathrm{MPE}}(k) &= E\left[|\, e(k) + h\xi\,|^p\right] \\
&= \int_{-\infty}^{\infty} |\, x\,|^p \; p_{e+h\xi}(x)dx \\
&= \int_{-\infty}^{\infty} |\, x\,|^p \int_{-\infty}^{\infty} p_e(\tau)K_h(x-\tau)d\tau dx \\
&= \int_{-\infty}^{\infty} p_e(\tau)\Big(\int_{-\infty}^{\infty} |\, x\,|^p K_h(x-\tau)dx\Big)d\tau \\
&= E\left[\phi_h(e)\right]
\end{aligned}
\tag{3.30}
$$

where $\phi_h(e) = \int_{-\infty}^{\infty} |\, x\,|^p \; K_h(x-e)dx$. If the PDF function $K_h(x)$ is symmetric, then $\phi_h(x) = (\phi * K_h)(x)$, where $\phi(x) = |\, x\,|^p$. In this case, the cost function $\phi_h()$ of the SMPE is a "smoothed" function (by convolution) of the cost function MPE (hence called "smoothed LMP").

Given a distribution (usually with zero-mean and unit variance) of the smoothing variable ξ, one can easily derive the analytical expression of the smoothed cost function $\phi_h(e)$. Figure 3.4 shows the cost functions of the SLMP ($p=6$) criterion with different smoothing variable distributions (Gaussian, Uniform, and Binary) and smoothing factors ($h=0.5,1.0,1.5$), where the three smoothing variable distributions are

$$
\begin{cases}
\text{Binary} & K_h(x) = \dfrac{1}{2}\big(\delta(x+1) + \delta(x-1)\big) \\[2ex]
\text{Uniform} & K_h(x) = \begin{cases} \dfrac{1}{2\sqrt{3}} & \textit{if } -\sqrt{3} \le x \le \sqrt{3} \\ 0 & \text{otherwise} \end{cases} \\[3ex]
\text{Gaussian} & K_h(x) = \dfrac{1}{\sqrt{2\pi}} \exp\left(-\dfrac{x^2}{2}\right)
\end{cases}
$$

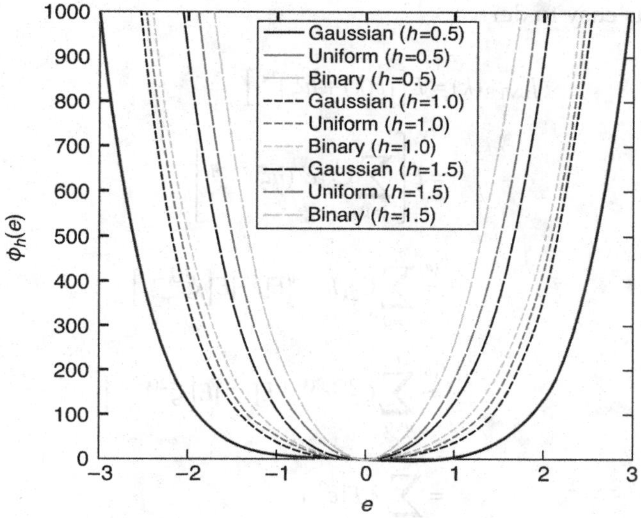

FIGURE 3.4
Cost functions of the SLMP (p=6) criterion with different smoothing variable distributions (Gaussian, Uniform, Binary) and smoothing factors (h=0.5, 1.0, 1.5) [267].

3.4.2 Properties of the SMPE Criterion

Property 3.1: As $h \to 0$, the SMPE criterion will be equivalent to the MMPE criterion.

Property 3.2: If the smoothing variable is zero-mean, then the SMPE criterion with $p=2$ is equivalent to the MSE criterion.

Proof: When the smoothing variable ξ is zero-mean and $p=2$, we have

$$\begin{aligned} J_{\text{SMPE}}(k) &= E\left[|\, e(k)+h\xi\,|^2\right] \\ &= E\left[e^2\right]+h^2E\left[\xi^2\right] \end{aligned} \tag{3.31}$$

Since the term $h^2E\left[\xi^2\right]$ is independent of the error e, we get $J_{\text{SMPE}} \sim J_{\text{MSE}} = E[e^2]$, where ~ denotes the equivalence relation.

Property 3.3: If the smoothing variable is symmetric and zero-mean, and p is an even number, $p=2k$, then the SLMP cost will be a weighted sum of the even-order moments of the error.

Proof. It is easy to derive

$$
\begin{aligned}
J_{\mathrm{SMPE}}(k) &= E\left[| e(k) + h\xi |^{2k} \right] \\
&= E\left[\sum_{q=0}^{2k} C_{2k}^{q} e^{q} (h\xi)^{2k-q} \right] \\
&\overset{(a)}{=} \sum_{q=0}^{2k} C_{2k}^{q} h^{2k-q} E[e^{q}] E\left[\xi^{2k-q} \right] \\
&\overset{(b)}{=} \sum_{l=0}^{k} C_{2k}^{2l} h^{2(k-1)} E[e^{2l}] E\left[\xi^{2(k-1)} \right] \\
&= \sum_{l=0}^{k} \lambda_l E[e^{2l}]
\end{aligned}
\tag{3.32}
$$

where $\lambda_l = C_{2k}^{2l} h^{2(k-1)} E\left[\xi^{2(k-1)} \right]$, (*a*) is independent of *e*, and (*b*) follows from the assumption that the PDF of $\xi$is symmetric and zero-mean.

Remark 3.5

According to Property 3.3, under certain conditions, the SMPE criterion will be equivalent to a linear combination of several even-order MPE criteria (or the LMF family criteria), where the combination weights are determined by the smoothing variable ξ and the smoothing factor h. Since the SMPE criterion can capture the information contained in more moments of the error, one can expect that algorithms developed under the SMPE criterion may behave more efficiently.

Property 3.4: If the smoothing variable ξ is symmetric and zero-mean, and p is an even number, then as h is large enough, the SMPE criterion will be approximately equivalent to the MSE criterion.

Proof: Since p is an even number, so $J_{\mathrm{SLMP}}(k) = E\left[(e(k) + h\xi)^{p} \right] = h^{p} E\left[\left(\frac{1}{h} e + \xi \right)^{p} \right]$, and $J_{\mathrm{SLMP}}(k) \sim E\left[(\tau e + h\xi)^{p} \right]$ where $\tau = 1/h$. As $h \to \infty$, we have $\tau \to 0$, and

$$\begin{aligned}
\psi(\tau) &= E\left[(\tau e + h\xi)^p\right] \\
&\approx \psi(0) + \psi'(0)\tau + \frac{1}{2}\psi''(0)\tau^2 \\
&= E[\xi^p] + pE[\xi^{p-1}e]\tau + \frac{p(p-1)}{2}E[\xi^{p-2}e^2]\tau^2 \\
&\overset{(c)}{=} E[\xi^p] + \frac{p(p-1)}{2}E[\xi^{p-2}]E[e^2] \\
&= E[\xi^p] + \lambda E[e^2]
\end{aligned} \tag{3.33}$$

where $\lambda = \frac{p(p-1)\tau^2}{2}E[\xi^{p-2}]$, and (*c*) is because that ξ is independent of e, and is symmetric and zero-mean. Thus as $h \to \infty$, we have $J_{\text{SMPE}} \sim E\left[e^2\right]$.

Property 3.5: If the smoothing variable ξ is symmetric and zero-mean, and p is an even number, then as e is small enough, the SMPE criterion will be approximately equivalent to the MSE criterion.

Proof: Since p is an even number, we have $J_{\text{SMPE}} = E\left[(e(k) + h\xi)^p\right]$, and

$$\begin{aligned}
J_{\text{SMPE}} &= E\left[(e(k) + h\xi)^p\right] = E_p\left[E_\xi\left(e + h\xi\right)^p\right] \\
&= E_e\left[E_\xi\left[\left(h\xi\right)^p + h\left(h\xi\right)^{p-1}e + \frac{1}{2}p(p-1)\left(h\xi\right)^{p-2}e^2 + o(e^2)\right]\right] \\
&\overset{(d)}{=} E_e\left[E_\xi\left[\left(h\xi\right)^p + \frac{1}{2}p(p-1)\left(h\xi\right)^{p-2}e^2 + o(e^2)\right]\right] \\
&\approx E_e\left[E_\xi\left[\left(h\xi\right)^p\right] + E_\xi\left[\frac{1}{2}p(p-1)\left(h\xi\right)^{p-2}\right]e^2\right] \\
&= c_1 + c_2E[e^2]
\end{aligned} \tag{3.34}$$

where (*d*) comes from the fact that ξ is independent of e, and is symmetric around zero, and $c_1 = E_\xi\left[\left(h\xi\right)^p\right], c_2 = E_\xi\left[+\frac{1}{2}p(p-1)\left(h\xi\right)^{p-2}\right]$. Therefore, as $e \to 0$, we have $J_{\text{SMPE}}(k) \sim E\left[e^2\right]$.

Property 3.6: The SLMP cost equals the asymptotic value of the empirical LMP cost estimated by the kernel density estimation (KDE) approach with a fixed kernel function $K_h(x)$, that is

$$J_{\mathrm{SMPE}} = \lim_{N\to\infty} \hat{J}_{\mathrm{SMPE}} = \lim_{N\to\infty} \int_{-\infty}^{\infty} |x|^p \hat{p}_e(x)dx \tag{3.35}$$

in which $\hat{p}_e(x)$ denotes the estimated PDF of the error:

$$\hat{p}_e(x) = \frac{1}{N}\sum_{i=1}^{N} K_h(x - e_i) \tag{3.36}$$

where $\{e_1,\ldots,e_N\}$ are N independent, identically distributed (i.i.d) error samples.

Proof: According to the theory of KDE, as sample number $N \to \infty$, the estimated PDF will uniformly converge (with probability 1) to the true PDF convolved with the kernel function.

$$\hat{p}_e(x) \overset{N\to\infty}{\to} = (p_e * K_h)(x) = p_{e+h\xi}(x) \tag{3.37}$$

It follows easily that

$$\begin{aligned}\lim_{N\to\infty} \int_{-\infty}^{\infty} |x|^p \hat{p}_e(x)dx &= \int_{-\infty}^{\infty} |x|^p \hat{p}_{e+h\xi}(x)dx = E\left[|e + h\xi|^p\right] \\ &= J_{\mathrm{SMPE}}\end{aligned} \tag{3.38}$$

Remark 3.6

The smoothed MMPE criterion is similar to the smoothed MEE (SMEE) criterion studied in [277]. The SMEE cost is defined as the entropy of the error plus an independent random variable (i.e. the smoothing variable), which is the limit (when sample size tends to infinity) of the empirical error entropy estimated by the KDE method with a fixed kernel function equal to the PDF of the smoothing variable [277]. The SMPE criterion is also inspired by the maximum correntropy criterion (MCC), which has been demonstrated to be equivalent to a smoothed maximum a posterior (MAP) criterion.

3.5 MMPE in Kernel Space

In recent years, information-theoretic learning [278] has introduced the concept of correntropy, which measures the similarity between two random variables over an observation window. Essentially, correntropy is a second-order statistical measure (or correlation) in kernel space, yielding a nonsecond order measure in original space. Other second-order statistical measures, such as MSE, can also be defined in kernel space. The MSE in kernel space is referred to as the correntropy loss (C-Loss) [279,280]. It has been shown that minimizing the C-Loss is equivalent to maximizing the correntropy. In this subchapter, we introduce a nonsecond order measure in kernel space called kernel MPE (KMPE) [257], which is the MPE in kernel space and of course, is also a nonsecond order measure in the original space. In addition, we introduce the q-Gaussian KMPE [258] and kernel mixture MPE (KMMPE) [259] as part of the KMPE family criteria.

3.5.1 Kernel Mean *p*-Power Error Criterion

3.5.1.1 Definition of KMPE

Non-second order statistical measures can be defined elegantly as a second-order measure in kernel space. For example, the correntropy between two random variables *X* and *Y*, is a correlation measure in kernel space, given by [278]

$$\begin{aligned} V(X,Y) &= E\left[\langle \varphi(X), \varphi(Y) \rangle_{\mathcal{H}}\right] \\ &= \int \langle \varphi(x), \varphi(y) \rangle_{\mathcal{H}} dF_{XY}(x,y) \end{aligned} \tag{3.39}$$

where $F_{XY}(x,y)$ stands for the joint distribution function, and $\varphi(x) = \kappa(x,.)$ is a nonlinear mapping induced by a Mercer kernel $\kappa(.,.)$, which transforms x from the original space to a functional Hilbert space (or kernel space) $\mathcal{H}$ equipped with an inner product $\langle .,. \rangle_{\mathcal{H}}$ satisfying $\langle \varphi(x), \varphi(y) \rangle_{\mathcal{H}} = \kappa(x,y)$. Obviously, we have $V(X,Y) = \mathbf{E}[\kappa(X,Y)]$. Without mentioned otherwise, the kernel function is the Gaussian kernel, given by

$$\kappa(x,y) = \kappa_\sigma(x-y) = \exp\left(-\frac{(x-y)^2}{2\sigma^2}\right) \tag{3.40}$$

with σ being the kernel bandwidth. Similarly, the C-Loss as MSE in kernel space can be defined by

$$\begin{aligned} C(X,Y) &= \frac{1}{2}E\left[\|\varphi(X)-\varphi(Y)\|_{\mathcal{H}}^2\right] \\ &= \frac{1}{2}E[\langle\varphi(X)-\varphi(Y),\varphi(X)-\varphi(Y)\rangle_{\mathcal{H}}] \\ &= \frac{1}{2}E[\langle\varphi(X),\varphi(X)\rangle_{\mathcal{H}}+\langle\varphi(Y),\varphi(Y)\rangle_{\mathcal{H}}-2\langle\varphi(X),\varphi(Y)\rangle_{\mathcal{H}}] \\ &= \frac{1}{2}E[2\kappa_\sigma(0)-2\kappa_\sigma(X-Y)] \\ &= E[1-\kappa_\sigma(X-Y)] \end{aligned} \tag{3.41}$$

where 1/2 is inserted to make the expression more convenient. It holds that $C(X, Y)=1-V(X, Y)$, hence minimizing the C-Loss will be equivalent to maximizing the correntropy. The MCC has drawn more and more attention due to its strong robustness to large outliers [281–285].

Here a new statistical measure in kernel space is defined in a nonsecond order manner. Specifically, we generalize the C-Loss to the case of arbitrary power and define the MPE in kernel space and call the new measure the KMPE. Given two random variables X and Y, the KMPE is defined by

$$\begin{aligned} C_p(X,Y) &= 2^{-p/2}E\left[\|\varphi(X)-\varphi(Y)\|_{\mathcal{H}}^p\right] \\ &= 2^{-p/2}E\left[\left(\|\varphi(X)-\varphi(Y)\|_{\mathcal{H}}^2\right)^{p/2}\right] \\ &= 2^{-p/2}E\left[(2-2\kappa_\sigma(X-Y))^{p/2}\right] \\ &= E\left[(1-\kappa_\sigma(X-Y))^{p/2}\right] \end{aligned} \tag{3.42}$$

where $p>0$ is the power parameter. Clearly, the KMPE includes the C-Loss as a special case (when $p=2$), but with a proper p value it can outperform the C-Loss when used as a cost function in robust learning. In addition, given N samples $\{x_i, y_i\}_{i=1}^N$, the empirical KMPE can be easily obtained as

$$\hat{C}_p(X,Y)=\frac{1}{N}\sum_{i=1}^{N}\left(1-\kappa_\sigma\left(x_i-y_i\right)\right)^{p/2} \tag{3.43}$$

Since $\hat{C}_p(X,Y)$ is a function of the sample vectors $\mathbf{X}=[x_1,x_2,\ldots,x_N]^T$ and $\mathbf{Y}=[y_1,y_2,\ldots,y_N]^T$, one can also denote $\hat{C}_p(X,Y)$ by $\hat{C}_p(\mathbf{X},\mathbf{Y})$ if no confusion arises. The cost function of KMPE for each sample can be expressed as $J(e)=(1-\kappa_\sigma(e))^{p/2}$, with $e=x-y$. Figure 3.5 shows the curves of $J(e)$ with different parameters.

3.5.1.2 Properties of KMPE

Here some important properties of the KMPE criterion are summarized.

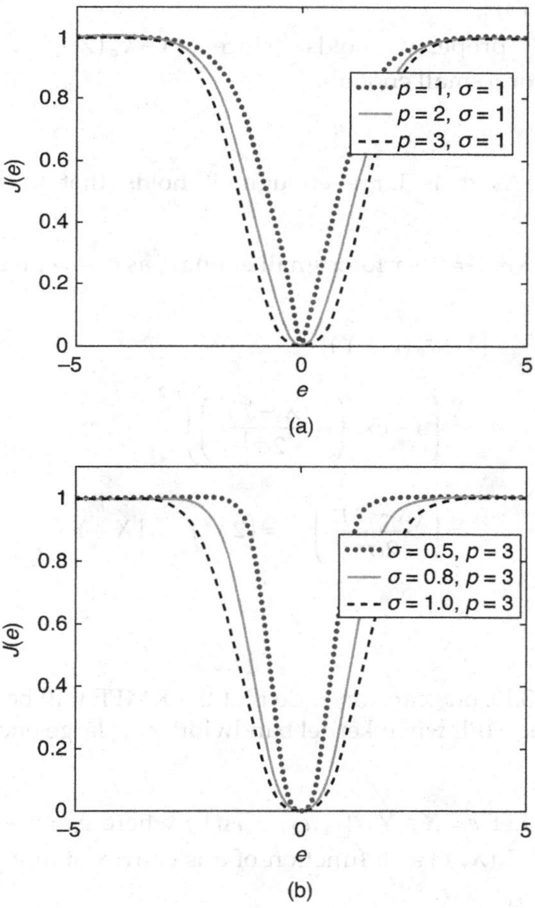

FIGURE 3.5
Cost function $J(e)$ with different parameters. (a) $p=1,2,3$ ($\sigma=1.0$); (b) $\sigma=0.5,0.8,1.0$ ($p=3$) [257].

Property 3.7: $C_p(X, Y)$ is symmetric, that is $C_p(X,Y)=C_p(Y,X)$.

Proof: The proof is straightforward, since $\kappa_\sigma(X-Y)=\kappa_\sigma(Y-X)$.

Property 3.8: $C_p(X, Y)$ is positive and bounded: $0 \le C_p(X,Y) < 1$, and it reaches its minimum if and only if $X=Y$.

Proof: The proof is straightforward, since $0<\kappa_\sigma(X-Y)\le 1$, with $\kappa_\sigma(X-Y)=1$ if and only if $X=Y$.

Property 3.9: As p is small enough, it holds that $C_p(X,Y)\approx 1+(p/2)E\left[\log\left(1-\kappa_\sigma(X-Y)\right)\right]$.

Proof: The property holds since $(1-\kappa_\sigma(X-Y))^{p/2} \approx 1+(p/2)\log(1-\kappa_\sigma(X-Y))$ for p small enough.

Property 3.10: As σ is large enough, it holds that $C_p(X,Y)\approx\left(2\sigma^2\right)^{-p/2}E\left[|X-Y|^p\right]$.

Proof: Since $\exp(x)\approx 1+x$ for x small enough, as $\sigma\to\infty$, we have

$$\begin{aligned}
&(1-\kappa_\sigma(X-Y))^{p/2}\\
&=\left(1-\exp\left(-\frac{(X-Y)^2}{2\sigma^2}\right)\right)^{p/2}\\
&\approx\left(\frac{(X-Y)^2}{2\sigma^2}\right)^{p/2}=\left(2\sigma^2\right)^{-p/2}|X-Y|^p
\end{aligned}\tag{3.44}$$

Remark 3.7

From Property 3.10, one can conclude that the KMPE will be, approximately, equivalent to the MPE when kernel bandwidth σ is large enough.

Property 3.11: Let $\boldsymbol{e}=\mathbf{X}-\mathbf{Y}=\left[e_1,e_2,\ldots,e_N\right]^T$, where $e_i=x_i-y_i$, if $p\ge 2$, the empirical KMP $\hat{C}_p(\mathbf{X},\mathbf{Y})$ as a function of $\boldsymbol{e}$ is convex at any point satisfying $\|\boldsymbol{e}\|_\infty=\max_{i=1,2,\ldots,N}|e_i|\le\sigma$.

Proof: Since $\hat{C}_p(\mathbf{X},\mathbf{Y})=(1/N)\sum_{i=1}^{N}\left(1-\kappa_\sigma(e_i)\right)^{p/2}$, the Hessian matrix of $\hat{C}_p(\mathbf{X},\mathbf{Y})$ with respect to e is

$$H_{\hat{C}_p(\mathbf{X},\mathbf{Y})}(e) = \left[\frac{\partial^2 \hat{C}_p(\mathbf{X},\mathbf{Y})}{\partial e_i\, \partial e_j}\right] = \operatorname{diag}\left[\xi_1, \xi_2, \ldots, \xi_N\right] \tag{3.45}$$

where

$$\begin{aligned} \xi_i &= \frac{p}{4N\sigma^4}\left(1-\kappa_\sigma(e_i)\right)^{(p-4)/2}\kappa_\sigma(e_i) \\ &\times\left\{(p-2)e_i^2\kappa_\sigma(e_i) - 2e_i^2\left(1-\kappa_\sigma(e_i)\right) + 2\sigma^2\left(1-\kappa_\sigma(e_i)\right)\right\} \end{aligned} \tag{3.46}$$

when $p \geq 2$, we have $\xi_i \geq 0$ if $|e_i| \leq \sigma$. Thus, for any point e with $\|e\|_\infty \leq \sigma$, we have $H_{\hat{C}_p(\mathbf{X},\mathbf{Y})}(e) \geq 0$.

Property 3.12: Given any point e with $\|e\|_\infty > \sigma$, the empirical KMPE $\hat{C}_p(\mathbf{X},\mathbf{Y})$ will be convex at e if p is larger than a certain value.

Proof: From (3.46), if $|e_i| \leq \sigma$ and $p \geq 2$, or if $|e_i| > \sigma$ and $p \geq \left[\left(2\left[e_i^2 - \sigma^2\right]\left(1-\kappa_\sigma(e_i)\right)\right) / e_i^2\kappa_\sigma(e_i)\right] + 2$, we have $\xi_i \geq 0$. So, it holds that $H_{\hat{C}_p(\mathbf{X},\mathbf{Y})}(e) \geq 0$ if

$$p \geq \max_{\substack{i=1,\ldots,N \\ |e_i|>\sigma}} \left\{\frac{2\left[e_i^2 - \sigma^2\right]\left(1-\kappa_\sigma(e_i)\right)}{e_i^2\kappa_\sigma(e_i)} + 2\right\} \tag{3.47}$$

This completes the proof.

Remark 3.8

According to Properties 5 and 6, the empirical KMPE as a function of e is convex at any point with $\|e\|_\infty \leq \sigma$ and it can also be convex at a point with $\|e\|_\infty > \sigma$ if the power parameter p is larger than a certain value.

Property 3.13: Let $\mathbf{0}$ be an N-dimensional zero vector. Then as $\sigma \to \infty$ $\left(\text{or } x_i \to 0, i = 1, \ldots, N\right)$, it holds that

$$\hat{C}_p(\mathbf{X},\mathbf{0}) \approx \frac{1}{N(\sqrt{2}\sigma)^p}\|\mathbf{X}\|_p^p \tag{3.48}$$

where $\|\mathbf{X}\|_p^p = \sum_{i=1}^{N}|x_i|^p$.

Proof: As σ is large enough, we have

$$\begin{aligned}\hat{C}_p(\mathbf{X},\mathbf{0}) &= \frac{1}{N}\sum_{i=1}^{N}\left(1-\kappa_\sigma(x_i)\right)^{p/2} \\ &\overset{(a)}{\approx} \frac{1}{N}\sum_{i=1}^{N}\left(1-\left(1-\frac{x_i^2}{2\sigma^2}\right)\right)^{p/2} \\ &= \frac{1}{N}\sum_{i=1}^{N}\left(\frac{x_i^2}{2\sigma^2}\right)^{p/2} \\ &= \frac{1}{N(\sqrt{2}\sigma)^p}\sum_{i=1}^{N}|x_i|^p\end{aligned} \tag{3.49}$$

Property 3.14: Assume that $|x_i| > \delta,\ \forall i: x_i \neq 0$, where δ is a small positive number. As $\sigma \to 0+$, minimizing the empirical KMPE $\hat{C}_p(\mathbf{X},\mathbf{0})$ will be, approximately, equivalent to minimizing the l_0 -norm of $\mathbf{X}$, that is

$$\min_{\mathbf{X}\in\Omega}\hat{C}_p(\mathbf{X},\mathbf{0}) \sim \min_{\mathbf{X}\in\Omega}\|\mathbf{X}\|_0, \quad \text{as } \sigma \to 0+ \tag{3.50}$$

where Ω denotes a feasible set of $\mathbf{X}$.

Proof: Let $\mathbf{X}_0$ be the solution obtained by minimizing $\|\mathbf{X}\|_0$ over Ω and $\mathbf{X}_C$ the solution achieved by minimizing $\hat{C}_p(\mathbf{X},\mathbf{0})$.Then $\hat{C}_p(\mathbf{X}_C,\mathbf{0}) \leq \hat{C}_p(\mathbf{X}_0,\mathbf{0})$, and

$$\sum_{i=1}^{N}\left[\left(1-\kappa_\sigma\left((\mathbf{X}_C)_i\right)\right)^{p/2}-1\right] \leq \sum_{i=1}^{N}\left[\left(1-\kappa_\sigma\left((\mathbf{X}_0)_i\right)\right)^{p/2}-1\right] \tag{3.51}$$

where $(\mathbf{X}_C)_i$ denotes the ith component of $\mathbf{X}_C$. It follows that:

$$\begin{aligned}&\|\mathbf{X}_C\|_0 - N + \sum_{i=1,(\mathbf{X}_C)_i\neq 0}^{N}\left[\left(1-\kappa_\sigma\left((\mathbf{X}_C)_i\right)\right)^{p/2}-1\right] \\ &\leq \|\mathbf{X}_0\|_0 - N + \sum_{i=1,(\mathbf{X}_0)_i\neq 0}^{N}\left[\left(1-\kappa_\sigma\left((\mathbf{X}_0)_i\right)\right)^{p/2}-1\right]\end{aligned} \tag{3.52}$$

Hence, we have

$$\|\mathbf{X}_C\|_0 - \|\mathbf{X}_0\|_0 \le \sum_{i=1,(\mathbf{X}_0)_i \neq 0}^{N} \left[\left(1 - \kappa_\sigma\left((\mathbf{X}_0)_i\right)\right)^{p/2} - 1 \right] - \sum_{i=1,(\mathbf{X}_C)_i \neq 0} \left[\left(1 - \kappa_\sigma\left((\mathbf{X}_C)_i\right)\right)^{p/2} - 1 \right] \tag{3.53}$$

Since $|x_i| > \delta,\ \forall i : x_i \neq 0$, as $\sigma \to 0+$, the right side of (3.53) will approach zero. Thus, if σ is small enough, it holds that

$$\|\mathbf{X}_0\|_0 \le \|\mathbf{X}_C\|_0 \le \|\mathbf{X}_0\|_0 + \varepsilon \tag{3.54}$$

where ε is a small positive number arbitrarily close to zero. This completes the proof.

Remark 3.9

From Properties 3.7 and 3.8, one can see that the empirical KMPE $\hat{C}_p(\mathbf{X},\mathbf{0})$ behaves like an l_p-norm of $\mathbf{X}$ when kernel bandwidth σ is very large, and like an l_p-norm of $\mathbf{X}$ when σ is very small.

3.5.2 *q*-Gaussian Kernel Mean *p*-Power Error

As mentioned before, the KMPE can be used to build a loss function for robust learning with outliers [257]. The default kernel in KMPE is the Gaussian kernel, but it does not always obtain the best performance. In this part, a q-Gaussian KMPE criterion [258] is introduced. The q-Gaussian distribution, which arises from the maximization of the Tsallis entropy under appropriate constraints [286], is given by

$$f(x) = \frac{\sqrt{\beta}}{C_q} e_q\left(-\beta x^2\right) \tag{3.55}$$

where

$$e_q(x) = \begin{cases} \left[1 + (1-q)x\right]^{\frac{1}{1-q}}, & 1 + (1-q)x \ge 0 \\ 0, & \text{otherwise} \end{cases} \tag{3.56}$$

and $\beta = \left[(3-q)\sigma_v^2\right]^{-1}$ is the scale parameter, where σ_v^2 denotes the variance for random variables subject to q-Gaussian distribution. Besides, the normalization factor C_q is characterized by

$$C_q = \begin{cases} \dfrac{\sqrt{\pi}\Gamma\left(\dfrac{2-q}{1-q}\right)}{\sqrt{1-q}\Gamma\left(\dfrac{5-3q}{2(1-q)}\right)}, & -\infty < q < 1 \\ \sqrt{\pi}, & q = 1 \\ \dfrac{\sqrt{\pi}\Gamma\left(\dfrac{3-q}{2(q-1)}\right)}{\sqrt{q-1}\Gamma\left(\dfrac{1}{q-1}\right)} & 1 < q < 3 \end{cases} \tag{3.57}$$

where $\Gamma(\cdot)$ is the gamma function. When $q<1$, the q-Gaussian distribution denotes the probability distribution function of a bounded random variable $x \in \left[-\dfrac{\sqrt{3-q}\sigma}{\sqrt{1-q}}, \dfrac{\sqrt{3-q}\sigma}{\sqrt{1-q}}\right]$. For $q \to 1$, the q-Gaussian distribution is gradually closed to the Gaussian distribution. When $q \in (1,3)$, the q-Gaussian distribution is more favorable than the Gaussian distribution for its heavy tails, and then the q-Gaussian kernel can be defined under Mercer theorem as

$$\kappa_{q,\sigma}(X-Y) = \left[1 + \left(\frac{q-1}{3-q}\right)\frac{\|X-Y\|^2}{\sigma^2}\right]^{\frac{1}{1-q}} \tag{3.58}$$

where $q \in (1,3)$ controls the shape of the kernel.

Remark 3.10

It is noted that the q-Gaussian kernel contains different types of kernels as its special cases. When $q=1$, the q-Gaussian kernel is equivalent to the Gaussian kernel. When $q=2$, the q-Gaussian kernel reduces to the Cauchy kernel [287], and for $q \equiv \dfrac{v+1}{v+3}$, the q-Gaussian kernel is recovered by the scaled Student's-t kernel [288] with v degrees of freedom.

The q-Gaussian density function, which can adaptively change the shape of the distribution by setting a tuning parameter q, is more suitable than Gaussian density to model the effect of external data randomness [289]. When the q-Gaussian density is taken as the activation of neurons, the approximation capability and generalization performance of the neural network can be enhanced [290,291].

One can employ the q-Gaussian kernel to construct the KMPE. Given a sample vector $\boldsymbol{e}=[e_1,e_2,\ldots,e_N]$, the q-Gaussian KMPE can be approximated by the sample estimator as follows:

$$\hat{J}_{\text{QKMPE}}(\boldsymbol{e})=\frac{1}{N}\sum_{i=1}^{N}\left(1-\kappa_{q,\sigma}(e_i)\right)^{\frac{p}{2}} \tag{3.59}$$

The QKMPE measure in (3.59) reflects the similarity of two random variables in high-dimensional feature space induced by the q-Gaussian kernel. Based on the properties of q-Gaussian kernel, the QKMPE will reduce to KMPE when $q\rightarrow 1$. Clearly, one can apply the QKMPE to develop new adaptive filtering algorithms. With a proper shape parameter q, the adaptive filters under QKMPE may obtain better performance than those under KMPE.

3.5.3 Kernel Mixture Mean p-Power Error Criterion

Inspired by the mixture correntropy (MC) [204] and MPE, a kernel mixture mean p-power error (KMP) criterion [259] was proposed by applying a mixed Gaussian function to MPE. The MC measure can be viewed as a special case of the KMP with $p=2$. The KMP with an appropriate p can achieve better accuracy than MC for robust learning.

3.5.3.1 Definition

Given two random variables X and Y with $X,Y\in\mathbb{R}$, the KMP using a mixture of two Gaussian functions can be defined by

$$\begin{aligned} L_p(X,Y)&=E\left[\left(1-\left(\alpha\kappa_{\sigma_1}(e)+(1-\alpha)\kappa_{\sigma_2}(e)\right)\right)^{p/2}\right] \\ &=\int\left(1-\left(\alpha\kappa_{\sigma_1}(e)+(1-\alpha)\kappa_{\sigma_2}(e)\right)\right)^{p/2}dF_{X,Y}(x,y) \end{aligned} \tag{3.60}$$

where p is a power parameter; σ_1 and $\sigma_2$are the kernel bandwidths of the Gaussian functions $\kappa_{\sigma 1}(\cdot)$ and $\kappa_{\sigma 2}(\cdot)$, respectively; $0\leq\alpha\leq 1$ is the mixture coefficient; $e=X-Y$ is the error variable. Without loss of generality, we assume $\sigma_1\leq\sigma_2$ here. Note that the MC is also a special case of KMP with $p=2$. Therefore, KMP is a more universal similarity measure than MC, leading to performance improvement for robust learning with a proper p.

Given N available data $\boldsymbol{x} = [x(1), x(2), \dots, x(N)]^T$, $\boldsymbol{y} = [y(1), y(2), \dots, y(N)]^T$, the empirical KMP can be obtained as

$$\hat{L}_p(\boldsymbol{x}, \boldsymbol{y}) = \frac{1}{N} \sum_{i=1}^{N} \left(1 - \left(S(i) + F(i)\right)\right)^{p/2} \tag{3.61}$$

where $S(i) = \alpha \kappa_{\sigma 1}(e(i))$, $F(i) = (1-\alpha)\kappa_{\sigma 1}(e(i))$ and $e(i) = x(i) - y(i)$.

3.5.3.2 Properties

Some important properties of the KMP are presented as follows:

Property 3.15: As σ_1 is large enough, it holds that $L_p(X,Y) \approx \left(\frac{\alpha\sigma_2^2 + (1-\alpha)\sigma_1^2}{2\sigma_1^2\sigma_2^2}\right)^{p/2} E\left[|e|^p\right]$.

Proof: Since exp(x) is approximated by $1+x$ for small enough x, on the condition of large enough σ_1, we have

$$\begin{aligned} L_p(X,Y) &= E\left[\left(1 - \left(\alpha\kappa_{\sigma 1}(e) + (1-\alpha)\kappa_{\sigma 2}(e)\right)\right)^{p/2}\right] \\ &\approx E\left[\left(\frac{\left(\alpha\sigma_2^2 + (1-\alpha)\sigma_1^2\right)}{2\sigma_1^2\sigma_2^2} e^2\right)^{p/2}\right] \\ &= \left(\frac{\alpha\sigma_2^2 + (1-\alpha)\sigma_1^2}{2\sigma_1^2\sigma_2^2}\right)^{p/2} E\left[|e|^p\right] \end{aligned} \tag{3.62}$$

Remark 3.11

According to Property 3.15, KMP is approximately equivalent to MPE, when σ_1 is large enough. Thus, MPE can be viewed as an extreme case of KMP.

Property 3.16: Let $\boldsymbol{e} = \boldsymbol{x} - \boldsymbol{y} = [e(1), e(2), \dots, e(N)]^T$, where $e(i) = x(i) - y(i)$, $i = 1, 2, \dots, N$. The empirical KMP $\hat{L}_p(\boldsymbol{x}, \boldsymbol{y})$ as a function of $_e$ is convex at any point satisfying $\|e\|_\infty = \max|e(i)| \le \sigma_1$, $i = 1, 2, \dots, N$, and $p \ge 2$..

Proof: The Hessian matrix of $\hat{L}_p(\boldsymbol{x}, \boldsymbol{y})$ regarding to $\boldsymbol{e}$ is

$$\mathbf{H}_{\hat{L}_{p(x,y)}}(e) = \left[\frac{\partial \hat{L}_{p(x,y)}}{\partial e(i)\partial e(j)}\right] = \mathrm{diag}[\gamma_1, \gamma_2, \ldots, \gamma_N] \tag{3.63}$$

where diag[·] denotes a diagonal matrix with diagonal entries $\gamma_1, \gamma_2, \ldots, \gamma_N$ and

$$\begin{aligned}\gamma_i = \zeta_i \Big(&\left(\sigma_1^2 S(i) - e^2(i) S(i)\right)/\sigma_1^4 + \left(\sigma_2^2 F(i) - e^2(i) F(i)\right)/\sigma_2^4 \\ &+ \left((p-2)/2\right)\left(1 - \left(S(i) + F(i)\right)\right)^{-1} \times \left(S(i)e(i)/\sigma_1^2 + F(i)e(i)/\sigma_2^2\right)^2 \Big)\end{aligned} \tag{3.64}$$

with $\zeta_i = \frac{p}{2N}\left(1 - \left(S(i) + F(i)\right)\right)^{(p-2)/2} > 0, \quad i = 1, 2, \ldots, N$. When $\max|e(i)| \leq \sigma_1$, $i = 1, 2, \ldots N$ and $p \geq 2$ are satisfied, from (3.64) we have $\mathbf{H}_{\hat{L}_p(\mathbf{x},\mathbf{y})}(\boldsymbol{e}) \geq 0$.

4

Adaptive Filtering Algorithms under MMPE

4.1 The Original Least Mean p-Power Algorithm

4.1.1 Derivation of the LMP Algorithm

Figure 4.1 shows a general scheme of adaptive filtering under the MPE criterion. The desired signal $d(k)$ is written as

$$d(k) = \mathbf{w}_o^T \mathbf{u}(k) + v(k) \tag{4.1}$$

in which $\mathbf{w}_o = [w_{1,o}, w_{2,o}, \ldots, w_{o,L}]^T$ denotes the weight (parameter) vector of the unknown system (L is the size of channel memory) that we wish to estimate, and $\mathbf{u}(k) = [u(k-L+1), \ldots, u(k-1), u(k)]^T$ is the input vector at time k. $v(k)$ accounts for both measurement noise and modeling errors. Then the instantaneous error can be calculated as

$$e(k) = d(k) - y(k) \tag{4.2}$$

where the filter output signal can be expressed as

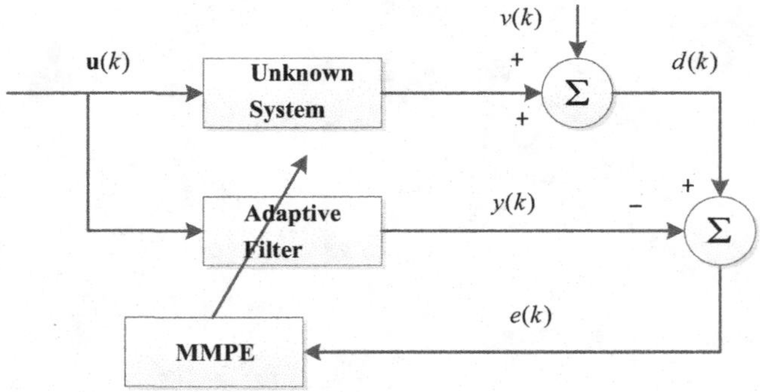

FIGURE 4.1
Adaptive filtering under MMPE criterion.

DOI: 10.1201/9781003176114-4

$$y(k) = \mathbf{w}^T(k)\mathbf{u}(k) \tag{4.3}$$

where $\mathbf{w}(k)$ denotes the weight vector of the linear filter. According to the MMPE criterion, an adaptive algorithm can be derived by using the steepest descent method to adjust filter parameters. When the instantaneous value of J is used to replace its ensemble averaged value, we have [212]

$$\mathbf{w}(k+1) = \mathbf{w}(k) - \eta \frac{\partial |e(k)|^p}{\partial \mathbf{w}(k)} \tag{4.4}$$

where

$$|e(k)|^p = \begin{cases} (e(k))^p & p\text{:even} \\ \text{sign}(e(k))e(k)^p & p\text{:odd} \end{cases} \tag{4.5}$$

with sign(·) being a sign operator, defined by

$$\text{sign}(x) = \begin{cases} x/|x| & x \neq 0 \\ 0 & x = 0 \end{cases} \tag{4.6}$$

Then the following expression can be obtained

$$\begin{aligned} \frac{\partial |e(k)|^p}{\partial \mathbf{w}(k)} &= \begin{cases} p \cdot (e(k))^{p-1} \dfrac{\partial e(k)}{\partial \mathbf{w}(k)} & p\text{:even} \\ p \cdot \text{sign}(e(k)) e(k)^{p-1} \dfrac{\partial e(k)}{\partial \mathbf{w}(k)} & p\text{:odd} \end{cases} \\ &= p \cdot |e(k)|^{p-2}\, e(k) \frac{\partial e(k)}{\partial \mathbf{w}(k)} \end{aligned} \tag{4.7}$$

Thus, (4.4) becomes

$$\begin{aligned} \mathbf{w}(k+1) &= \mathbf{w}(k) - \eta \frac{\partial J_{\text{MPE}}(k)}{\partial \mathbf{w}(k)} \\ &= \mathbf{w}(k) - \eta \left[p\,|e(k)|^{p-2} \frac{\partial e(k)}{\partial \mathbf{w}(k)} \right] \end{aligned} \tag{4.8}$$

Moreover, the gradient vector is

$$\frac{\partial e(k)}{\partial \mathbf{w}(k)} = -\mathbf{u}(k) \tag{4.9}$$

Therefore, the LMP algorithm can be obtained as

$$\begin{aligned}\mathbf{w}(k+1) &= \mathbf{w}(k)+\eta p\,|\,e(k)\,|^{p-2}\, e(k)\mathbf{u}(k)\\ &= \mathbf{w}(k)+\mu\,|\,e(k)\,|^{p-2}\, e(k)\mathbf{u}(k)\end{aligned} \tag{4.10}$$

where $\mu = \eta p$ is the step size. It is clear that when $p = 1$ or $p = 2$, the LMP algorithm will become the sign-LMS or LMS, respectively.

Remark 4.1

The LMP algorithm can be expressed as

$$\mathbf{w}(k+1) = \mathbf{w}(k)+2\frac{\mu\,|\,e(k)\,|^{p-2}}{2}e(k)\mathbf{u}(k) \tag{4.11}$$

For $p = 2$, the LMP algorithm will reduce to the LMS algorithm

$$\mathbf{w}(k+1)=\mathbf{w}(k)+\mu e(k)\mathbf{u}(k) \tag{4.12}$$

Comparing (4.11) and (4.12), the LMP can be viewed as the LMS with time-varying step size

$$\mu(k) = \frac{\mu\,|\,e(k)\,|^{p-2}}{2} \tag{4.13}$$

For the LMS algorithm, it is well known that the convergence speed and misadjustment are both proportional to the step size. A better adaptive algorithm will be the one that has faster speed of adaption and smaller misadjustment. Thus, we hope that the step size $\mu(k)$ is larger at the beginning of adaption for fast convergence and is smaller at the end of adaption for smaller misadjustment. Fortunately, for $p > 2$, the step size in (4.13) fulfills these two requirements due to the following observations: (1) When the LMP algorithm has not converged, the error $|\,e(k)\,|$ is large. This makes the step size $\mu(k)$ be large, and the LMP will have fast convergence speed. (2) When the LMP algorithm has converged, the error $|\,e(k)\,|$ is small. This makes both step size and misadjustment small.

When the signals are of non-Gaussian light-tailed distributions, the steepest descent algorithm based on MMPE criterion with $p > 2$ (e.g., $p = 4$) may achieve better convergence performance (say, achieve either faster convergence speed or lower misadjustment). On the other hand, the LMP algorithm with $p < 2$ (e.g. $p = 1$) will be robust to non-Gaussian heavy-tailed noises.

Remark 4.2

Although the LMP algorithm can be viewed as an LMS algorithm with time-varying step size, the point that the LMP algorithm converges to may be different for various p. If the LMP algorithm is convergent in the mean, then $E(\mathbf{w}(k+1)) = E(\mathbf{w}(k))$. As a result, the point of convergence must obey the equation

$$E(|e(k)|^{p-2} e(k)\mathbf{u}(k)) = 0 \tag{4.14}$$

From (4.7), this equation can be rewritten as

$$\frac{\partial |e(k)|^p}{\partial \mathbf{w}(k)} = 0 \tag{4.15}$$

One can know that the solution satisfying (4.15) may be different for various p. Moreover, the term $\mathrm{H}(k) = |e(k)|^{p-2}$ in (4.10) is a crucial factor for the adaptive algorithm based on MPE, which is robust to the outliers modeled by alpha-stable distribution when $0 < p \le \alpha$ (See Appendix D for a detailed description of alpha-stable distribution). Figure 4.2

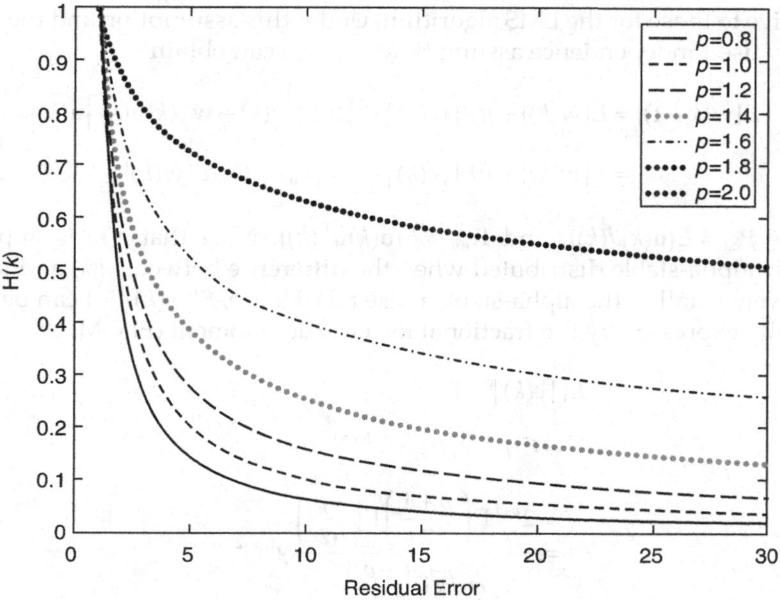

FIGURE 4.2
Illustration of H(k) with different values of p.

illustrates H(k) as the function of the error with different p. As the residual error increases, the H(k) falls to zero. Consequently, the adaptive filters will be insensitive to large outliers.

4.1.2 Convergence Analysis

The convergence analysis for the LMP algorithm is in general a difficult problem, but the convergence range of the step size μ in (4.10) can be determined through an approximation analysis [211]. Here the updating equation (4.10) is represented as [212]

$$\begin{aligned} \mathbf{w}(k+1) &= \mathbf{w}(k) + \mu |e(k)|^{p-2} e(k)\mathbf{u}(k) \\ &= \mathbf{w}(k) + \mu |e(k)|^{p-2} \left[d(k) - \mathbf{w}^T(k)\mathbf{u}(k)\right]\mathbf{u}(k) \end{aligned} \tag{4.16}$$

Taking the expected value on both sides of (4.16) yields

$$E\{\mathbf{w}(k+1)\} = E\{\mathbf{w}(k)\} + \mu E\{|e(k)|^{p-2}\left[d(k) - \mathbf{w}^T(k)\mathbf{u}(k)\right]\mathbf{u}(k)\} \tag{4.17}$$

We may further assume that the term $|e(k)|^{p-2}$ is uncorrelated with $\left[d(k) - \mathbf{w}^T(k)\mathbf{u}(k)\right]\mathbf{u}(k)$, although such approximation is rather rough. This approximation not only facilitates the analysis but also yields results comparative to those for the LMS algorithm. Under this assumption and the commonly used independence assumptions [5], one can obtain

$$\begin{aligned} E\{\mathbf{w}(k+1)\} &= E\{\mathbf{w}(k)\} + \mu E\{|e(k)|^{p-2}\} \times E\{\left[d(k) - \mathbf{w}^T(k)\mathbf{u}(k)\right]\mathbf{u}(k)\} \\ &= E\{\mathbf{w}(k)\} + \mu E\{|e(k)|^{p-2}\} \times \left[\mathbf{P}_{\mathbf{u}d} - \mathbf{R}_{\mathbf{uu}} E\{\mathbf{w}(k)\}\right] \end{aligned} \tag{4.18}$$

where $\mathbf{P}_{\mathbf{u}d} = E\{\mathbf{u}(k)d(k)\}$, and $\mathbf{R}_{\mathbf{uu}} = E\{\mathbf{u}(k)\mathbf{u}^T(k)\}$. Note that $e(k)$ is approximately alpha-stable distributed when the difference between $d(k)$ and $y(k)$ is relatively small to the alpha-stable noise $v(k)$. Hence, $E\{|e(k)|^{p-2}\}$ can be analytically expressed by the fractional lower-order moment (FLOM) as

$$\begin{aligned} & E\{|e(k)|^{p-2}\} \\ &= C(p-2,\alpha)\gamma^{((p-2)/\alpha)} \\ &= \frac{2^{p+1}\Gamma\left(\frac{p+1}{2}\right)\Gamma\left(\frac{-p}{\alpha}\right)}{\alpha\sqrt{\pi}\Gamma\left(\frac{-p}{2}\right)}\gamma^{((p-2)/\alpha)} \\ &:= D(p,\alpha,\gamma) \end{aligned} \tag{4.19}$$

The mean weight error vector is further defined as $\tilde{\mathbf{w}} := E\{\mathbf{w}(k) - \mathbf{w}_o\}$. Substracting $E\{\mathbf{w}_o\}$ from both sides of (4.18), the following stochastic difference equation can be obtained

$$\tilde{\mathbf{w}}(k+1) = \{\mathbf{I} - \mu D(p,\alpha,\gamma)\mathbf{R}_{\mathbf{uu}}\}\tilde{\mathbf{w}}(k) + \mu D(p,\alpha,\gamma)[\mathbf{P}_{\mathbf{u}d} - \mathbf{R}_{\mathbf{uu}}\mathbf{w}_o] \tag{4.20}$$

It is known that the solution to the difference equation converges if and only if the magnitudes of all the eigenvalues of $\mathbf{I} - \mu D(p,\alpha,\gamma)\mathbf{R}_{\mathbf{uu}}$ are less than one. As a result, for convergence, we have

$$|I - \mu D(p,\alpha,\gamma)\lambda_{\mathbf{R}_{\mathbf{uu}}}| < 1 \Rightarrow 0 < \mu < \frac{2}{D(p,\alpha,\gamma)\lambda_{\mathbf{R}_{\mathbf{uu}}}} \tag{4.21}$$

A more conservative yet more convenient condition is given by [211]

$$0 < \mu < \frac{2}{D(p,\alpha,\gamma)\mathrm{Tr}(\mathbf{R}_{\mathbf{uu}})} \tag{4.22}$$

where Tr($\mathbf{A}$) denotes the trace of the matrix $\mathbf{A}$.

From an examination of (4.22), one can know that the convergence range of the step size depends on the eigenvalues of the correlation matrix of the inputs and the noise characteristics. For a chosen p and fixed noise parameters, the convergence range is determined by the maximum eigenvalue of the correlation matrix. This result is similar to that of the LMS algorithm. Indeed, if p=2 and $\alpha = 2$ in (4.19), which makes the LMP equivalent to the LMS in the Gaussian noise case, and we have $D(p,\alpha,\gamma) = 1$. It follows that the convergence range of LMP reduces to the established convergence range for the LMS algorithm:

$$0 < \mu < \frac{2}{\lambda_{\max}} \tag{4.23}$$

This result gives us insight into the similarity between the LMP and LMS algorithms. It is also noted that since the auto-correlation matrix $\mathbf{R}_{\mathbf{uu}}$ has larger eigenvalue spread in the Volterra filter case than in the linear FIR filter case, more caution should be taken in the selection of μ. Using a time-varying step size normalized by the norm of the input vector is an option to reduce the eigenvalue spread of the correlation matrix. While the derivations presented here are through approximation, they give some quantitative measures of the convergence property.

4.1.3 Steady-State MSE Analysis

Due to the complexity of analyzing the nonlinear estimation error signal, the steady-state mean squared error (MSE) analysis has been done only for the special cases of the LMP algorithm (e.g., LMS algorithm $p = 2$ and least-mean fourth (LMF) algorithm $p = 4$). For general LMP algorithm with an arbitrary choice of parameter p, it has largely been left undone. In this section, the steady-state MSE for real and complex LMP algorithm with an arbitrary choice of parameter p is investigated based on series expansion [293, 294], and the corresponding restrictive conditions for step size are given. Moreover, in Gaussian noise environments, its steady-state performance is also investigated.

To perform the steady-state MSE analysis, we rewrite the update equation of the LMP algorithm as

$$\mathbf{w}(k+1) = \mathbf{w}(k) + \mu \mathbf{u}(k) f(e(k)) \tag{4.24}$$

where $f(e(k))$ is the estimation error signal expressed as

$$f(e(k)) = |e(k)|^{p-2}\, e(k) \tag{4.25}$$

Define so-called a priori estimation error to be $e_a(k) = \mathbf{u}^T(k)\tilde{\mathbf{w}}(k)$, and $\tilde{\mathbf{w}}_k = \mathbf{w}_o - \mathbf{w}_k$ to be the weight error vector. From (4.2), the relationship between $e(k)$ and $e_a(k)$ can be expressed by

$$e(k) = e_a(k) + v(k) \tag{4.26}$$

The steady-state MSE can be written as $\zeta_{\text{MSE}} = \lim_{k\to\infty} E|e(k)|^2$. To get ζ_{MSE}, the following two popular assumptions are used [261,295,296].

Assumption 4.1: The noise sequence $\{v(k)\}$ with zero-mean and variance σ_v^2 is independent of identical distribution (i.i.d.), and statistically independent of the regressor sequence $\{\mathbf{u}(k)\}$.

Assumption 4.2: The *a priori* estimation error $e_a(k)$ with zero-mean is independent of $v(k)$. Also, for complex-valued cases, it satisfies the circularity condition, namely, $E\left[e_a^2(k)\right] = 0$.

Under Assumptions 4.1 and 4.2, and (4.26), the steady-state MSE can be rewritten as $\zeta_{\text{MSE}} = \sigma_v^2 + \zeta_{\text{EMSE}}$ [296], where ζ_{EMSE} is the excess MSE (EMSE), defined by

$$\zeta_{\text{EMSE}} = \lim_{k\to\infty} E|e_a(k)|^2 \tag{4.27}$$

Using the definition of $\tilde{\mathbf{w}}(k)$ and (4.24), we have

$$\tilde{\mathbf{w}}(k+1) = \tilde{\mathbf{w}}(k) - \mu \mathbf{u}(k) f(e(k)) \tag{4.28}$$

Hence, we obtain

$$\begin{aligned}\|\tilde{\mathbf{w}}(k+1)\|^2 &= \|\tilde{\mathbf{w}}(k)\|^2 - \mu(\mathbf{u}(k)\tilde{\mathbf{w}}(k))^* f(e(k)) - \mu \mathbf{u}(k)\tilde{\mathbf{w}}(k) f^*(e(k)) \\ &\quad + \mu^2 \|\mathbf{u}(k)\|^2 |f(e(k))|^2\end{aligned} \tag{4.29}$$

Using the definition of $e_a(k)$ and taking expectations on both sides of (4.29), we have

$$E\left[\|\tilde{\mathbf{w}}(k+1)\|^2\right]=E\left[\|\tilde{\mathbf{w}}(k)\|^2\right]-\mu 2\,\mathrm{Re}\,E\left[e_a^*(k)f\left(e(k)\right)\right]+\mu^2 E\left[\|\mathbf{u}(k)\|^2\left|f\left(e(k)\right)\right|^2\right] \tag{4.30}$$

At steady state, the adaptive filters hold $E\|\tilde{\mathbf{w}}(k+1)\|^2=E\|\tilde{\mathbf{w}}(k)\|^2$ [58], and the time index "k" will be omitted for the easiness of reading, i.e.,

$$e_a(k)\to e_a,\mathbf{u}(k)\to\mathbf{u},v(k)\to v,\tilde{\mathbf{w}}(k)\to\tilde{\mathbf{w}},e(k)\to e\quad k\to\infty$$

Then, (4.30) can be rewritten compactly as

$$2\mathrm{Re}E[e_a^*f(e)]=\mu\mathrm{Tr}(\mathbf{R}_{\mathbf{uu}})E;f(e)|^2 \tag{4.31}$$

where $\mathrm{Tr}(\mathbf{R}_{\mathbf{uu}})=E\left[\|\mathbf{u}\|^2\right]$ denotes the trace of the correlation matrix $\mathbf{R}_{\mathbf{uu}}$, and $\mathrm{Re}(\cdot)$ denotes the real part of complex-valued data.

4.1.3.1 EMSE Analysis for Complex LMP Algorithm

The series expansion of the estimation error signal $f(e)$ with respect to (e_a,e_a^*) around (v,v^*) can be written as [293,294]

$$\begin{aligned}f(e)&=f(v,v^*)+f_e^{(1)}(v,v^*)e_a+f_{e^*}^{(1)}(v,v^*)e_a^*+\\&\frac{1}{2}\left[f_{e,e}^{(2)}(v,v^*)e_a^2+f_{e^*,e^*}^{(2)}(v,v^*)(e_a^*)^2+2f_{e,e^*}^{(2)}(v,v^*)|e_a|^2\right]+\mathrm{O}(e_a,e_a^*)\end{aligned} \tag{4.32}$$

where $f_x^{(1)}(a,b)$ denotes the first-order partial derivative of $f(x,y)$ with respect to x at the value (a,b), $f_{x,y}^{(2)}(a,b)$ denotes the second-order partial derivative of $f(x,y)$ with respect to x and y at the value (a,b), and $\mathrm{O}(e_a,e_a^*)$ denotes third and higher-power terms of e_a or e_a^*. Substituting (4.32) into the left-hand side of (4.31) yields

$$2\,\mathrm{Re}\,E[e_a^*f(e)]=2\,\mathrm{Re}\,E\left[f(v,v^*)e_a^*+f_e^{(1)}(v,v^*)|e_a|^2+f_{e^*}^{(1)}(v,v^*)(e_a^*)^2+\mathrm{O}(e_a,e_a^*)\right] \tag{4.33}$$

Using Assumptions 4.1 and 4.2 and ignoring $E[\mathrm{O}(e_a,e_a^*)]$, one can obtain

$$2\,\mathrm{Re}\,E[e_a^*f(e)]=2\,\mathrm{Re}\,E\left[f_e^{(1)}(v,v^*)\right]\zeta_{\mathrm{EMSE}} \tag{4.34}$$

Substituting (4.32) into the right-hand side of (4.31) yields

$$\mu \operatorname{Tr}(\mathbf{R}_{\mathbf{uu}})E[|f(e)|^2] = \mu \operatorname{Tr}(\mathbf{R}_{\mathbf{uu}})E\Big[|f(v,v^*)|^2 + B_0 e_a + B_1 e_a^* + B_2 |e_a|^2 + B_3 e_a^2 + B_4 (e_a^*)^2 + \mathrm{O}(e_a, e_a^*)\Big] \tag{4.35}$$

$$\begin{aligned}
B_0 &= 2\operatorname{Re}\Big[f(v,v^*)^* f_e^{(1)}(v,v^*)\Big] \\
B_1 &= 2\operatorname{Re}\Big[f(v,v^*) f_e^{(1)}(v,v^*)^*\Big] \\
B_2 &= \Big|f_e^{(1)}(v,v^*)\Big|^2 + \Big|f_{e^*}^{(1)}(v,v^*)\Big|^2 + 2\operatorname{Re}\Big[f(v,v^*)^* f_{e,e^*}^{(2)}(v,v^*)\Big] \\
B_3 &= 2\operatorname{Re}\Big[f_e^{(1)}(v,v^*) f_{e^*}^{(1)}(v,v^*)^*\Big] + \operatorname{Re}\Big[f(v,v^*)^* f_{e,e}^{(2)}(v,v^*)\Big] \\
B_4 &= 2\operatorname{Re}\Big[f_e^{(1)}(v,v^*)^* f_{e^*}^{(1)}(v,v^*)\Big] + \operatorname{Re}\Big[f(v,v^*) f_{e,e}^{(2)}(v,v^*)^*\Big]
\end{aligned} \tag{4.36}$$

Similarly, using Assumptions 4.1 and 4.2 and ignoring $\mathrm{E}\big[\mathrm{O}(e_a, e_a^*)\big]$, we obtain

$$\mu \operatorname{Tr}(\mathbf{R}_{\mathbf{uu}})E[|f(e)|^2] = \mu \operatorname{Tr}(\mathbf{R}_{\mathbf{uu}})\Big[E|f(v,v^*)|^2 + E[B_2]\zeta_{\mathrm{EMSE}}\Big] \tag{4.37}$$

Using (4.25), the following results can be obtained

$$\begin{aligned}
E[|f(v,v^*)|^2] &= E[|v|^{2p-2}] \\
E[f_e^{(1)}(v,v^*)] &= \frac{p}{2}E[|v|^{p-2}] \\
E[B_2] &= (p-1)^2 E[|v|^{2p-4}]
\end{aligned} \tag{4.38}$$

Then substituting (4.38) into (4.34) and (4.37), respectively, and using (4.31), we get

$$\Big[p\xi_v^{p-2} - \mu \operatorname{Tr}(\mathbf{R}_{\mathbf{uu}})(p-1)^2 \xi_v^{2p-4}\Big]\zeta_{\mathrm{EMSE}} = \mu \operatorname{Tr}(\mathbf{R}_{\mathbf{uu}})\xi_v^{2p-2} \tag{4.39}$$

where $\xi_v^k = E|v|^k$. Since $\mu > 0$, $\zeta_{\mathrm{EMSE}} \geq 0$ and $\mu \operatorname{Tr}(\mathbf{R}_{\mathbf{uu}})\xi_v^{2p-2} \geq 0$, if the step size μ satisfies

$$0 < \mu < \frac{p\xi_v^{p-2}}{(p-1)^2 \operatorname{Tr}(\mathbf{R}_{\mathbf{uu}})\xi_v^{2(p-2)}} \tag{4.40}$$

the steady-state EMSE for complex LMP algorithm can be written as

$$\zeta_{\mathrm{EMSE}} = \frac{\mu \operatorname{Tr}(\mathbf{R}_{\mathbf{uu}})\xi_v^{2p-2}}{p\xi_v^{p-2} - (p-1)^2 \mu \operatorname{Tr}(\mathbf{R}_{\mathbf{uu}})\xi_v^{2(p-2)}} \tag{4.41}$$

Remark 4.3

i. For the complex LMS and LMF algorithms, substituting $p=2$ and $p=4$ into (4.41), respectively, yields the same results in [261].
ii. Here we only focus on the steady-state analysis, the inequality of (4.40), determining whether (4.41) is suitable to achieve the steady-state EMSE for complex LMP algorithm, cannot be directly acted as the strict convergence condition for adaptive filters.
iii. In view of the small step size μ, (4.41) can be simplified to

$$\zeta_{\text{EMSE}} = \frac{\mu \text{Tr}(\mathbf{R}_{\mathbf{uu}})\xi_v^{2p-2}}{p\xi_v^{p-2}} \tag{4.42}$$

iv. In complex Gaussian noise environments, based on the following formula summarized from [297]

$$\xi_v^k = E|v|^k = \begin{cases} \dfrac{k}{2}!\sigma_v^k, & k: \text{ even} \\ 0, & k: \text{ odd} \end{cases} \tag{4.43}$$

the equality (4.41) can be rewritten as

$$\zeta_{\text{EMSE}} = \frac{\mu \text{Tr}(\mathbf{R}_{\mathbf{uu}})\sigma_v^p}{p\left(\dfrac{p-2}{2}\right)! - \mu \text{Tr}(\mathbf{R}_{\mathbf{uu}})(p-1)(p-1)!\sigma_v^{p-2}} \tag{4.44}$$

for even parameter p. However, for odd parameter p, from (4.43), we can see $\xi_v^{p-2}=0$, which leads to the restrictive condition for step size (4.40) being not satisfied. So the expression (4.41) cannot be used directly to get the steady-state EMSE.

4.1.3.2 EMSE Analysis for Real LMP Algorithm

Since $\{\mathbf{u}, v, e_a, e\}$ are real-valued data, the (4.31) can be simplified to

$$2E[e_a f(e)] = \mu \text{Tr}(\mathbf{R}_{\mathbf{uu}})E|f(e)|^2 \tag{4.45}$$

and the Taylor series expansion for $f(e)$ with respect to e around v can be written as

$$f(e) = f(v) + f_e^{(1)}(v)e_a + \frac{1}{2}f_{e,e}^{(2)}(v)e_a^2 + \text{O}(e_a) \tag{4.46}$$

where $\text{O}(e_a)$ denotes third and higher-power terms of e_a. Using (4.25), we can obtain

$$\begin{aligned} f_e^{(1)}(e) &= (p-1)|e|^{p-2} \\ f_{e,e}^{(2)}(e) &= (p-1)(p-2)|e|^{p-4}\, e \end{aligned} \tag{4.47}$$

Substituting (4.46) into the left-hand side of (4.45) yields

$$2E[e_a f(e)] = 2E\left[f(v)e_a + f_e^{(1)}(v)e_a^2 + O(e_a)\right] \tag{4.48}$$

Using Assumptions 4.1 and 4.2, and neglecting $E\left[O(e_a)\right]$, we obtain

$$2E[e_a f(e)] = 2(p-1)\xi_v^{p-2}\zeta_{\text{EMSE}} \tag{4.49}$$

Similarly, by substituting (4.46) into the right-hand side of (4.45), and using (4.47), we get

$$\mu \text{Tr}(\mathbf{R_{uu}})E|f(e)|^2 = \mu \text{Tr}(\mathbf{R_{uu}})\xi_v^{2p-2} + \mu \text{Tr}(\mathbf{R_{uu}})(p-1)(2p-3)\xi_v^{2p-4}\zeta_{\text{EMSE}} \tag{4.50}$$

Then, substituting (4.49) and (4.50) into (4.45), we obtain

$$\left[2(p-1)\xi_v^{p-2} - \mu \text{Tr}(\mathbf{R_{uu}})(p-1)(2p-3)\xi_v^{2p-4}\right] \times \zeta_{\text{EMSE}} = \mu \text{Tr}(\mathbf{R_{uu}})\xi_v^{2p-2} \tag{4.51}$$

Since $\mu > 0$, $\mu \text{Tr}(\mathbf{R_{uu}})\xi_v^{2p-2} \geq 0$ and $\zeta_{\text{EMSE}} \geq 0$, if the step size μ satisfies

$$0 < \mu < \frac{2\xi_v^{p-2}}{(2p-3)\text{Tr}(\mathbf{R_{uu}})\xi_v^{2(p-2)}} \tag{4.52}$$

the steady-state EMSE for real LMP algorithm can be written as

$$\zeta_{\text{EMSE}} = \frac{\mu \text{Tr}(\mathbf{R_{uu}})\xi_v^{2p-2}}{2(p-1)\xi_v^{p-2} - (p-1)(2p-3)\mu \text{Tr}(\mathbf{R_{uu}})\xi_v^{2(p-2)}} \tag{4.53}$$

Remark 4.4

i. Substituting $p = 2$ and $p = 4$ into (4.53), respectively, we obtain the same results for real LMS and LMF algorithms in [261].
ii. For cases of step size μ being small enough, (4.53) can be simplified to

$$\zeta_{\text{EMSE}} = \frac{\mu \text{Tr}(\mathbf{R_{uu}})\xi_v^{2p-2}}{2(p-1)\xi_v^{p-2}} \tag{4.54}$$

iii. In Gaussian noise environments, based on the following formula

$$\xi_v^k = \begin{cases} (k-1)!!\sigma_v^k, & k: \text{even} \\ \sqrt{\frac{2^k}{\pi}}(\frac{k-1}{2})!\sigma_v^k, & k: \text{odd} \end{cases} \tag{4.55}$$

where $(k-1)!! = 1\cdot 3\cdot 5\ldots(k-1)$, (4.53) can be rewritten as (4.56). Here, we use $0!! = 1$ and $(-1)!! = 1$.

$$\zeta_{\text{EMSE}} = \begin{cases} \dfrac{\mu \text{Tr}(\mathbf{R}_{\mathbf{uu}})(2p-3)!!\xi_v^p}{2(p-1)(p-3)!! - \mu \text{Tr}(\mathbf{R}_{\mathbf{uu}})(p-1)(2p-3)!!\xi_v^{p-2}} & p: \text{even} \\ \dfrac{\mu \text{Tr}(\mathbf{R}_{\mathbf{uu}})(2p-3)!!\xi_v^p}{\sqrt{\frac{2p}{\pi}}(p-1)\left(\frac{p-3}{2}\right)! - \mu \text{Tr}(\mathbf{R}_{\mathbf{uu}})(p-1)(2p-3)!!\xi_v^{p-2}} & p: \text{odd} \end{cases} \tag{4.56}$$

4.1.4 Simulation Results

The values of the steady-state MSE of an 11-tap real LMP filter with odd parameter $p = 3$ for different choices of the step size μ (from 1×10^{-4} to 7×10^{-2}) are shown in Figure 4.3. The data are generated in Gaussian noise environments with variance $\sigma_v^2 = 0.001$ by feeding correlated data into a tapped delay time. The theoretical values are obtained by using the expressions (4.56). For each step size, the experimental value is obtained by running the LMP algorithm for 4×10^{-5} iterations and averaging the squares-error curve over 100 experiments in order to generate the ensemble-average curve. The average of the last 4×10^4 entries of the ensemble-average curve is then used as the

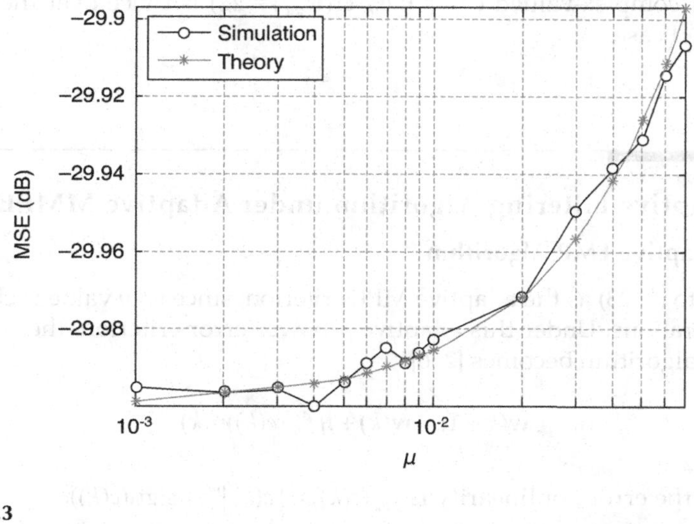

FIGURE 4.3
Theoretical and simulated MSE curves for LMP algorithm with $p = 3$ versus μ [212].

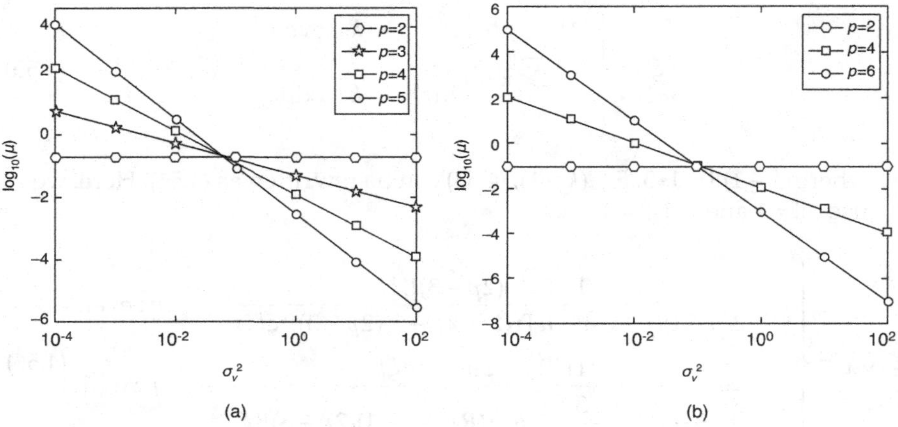

FIGURE 4.4
Logarithmic maximum values of step size for different choices of the parameter p of LMP algorithm in Gaussian noise environments. (a) Real-valued cases; (b) Complex-valued cases [212].

experimental value for the MSE. From Figure 4.3, one can observe that the theoretical and simulated MSE curves are matched reasonably well.

In Gaussian noise environments, Figure 4.4(a) and (b) show the logarithmic maximum-value curves of step size for different choices of parameter p (from 2 to 6) for real and complex LMP algorithms, respectively. Here, the range of the variance σ_v^2 is set from 1×10^{-4} to 1×10^{2}. From these results, one can see that the allowable maximum value of the step size decreases with σ_v^2 increasing, and it does not relate to σ_v^2 while $p=2$, i.e., $\mu_{\max}=1/11$ for real LMS algorithm, and $\mu_{\max}=1/22$ for complex LMS algorithm. Here, the trace of $\mathbf{R}_{\mathbf{uu}}$ for complex-valued cases (i.e., $\mathrm{Tr}(\mathbf{R}_{\mathbf{uu}})=22$) is twice over the one for real-valued cases.

4.2 Adaptive Filtering Algorithm under Adaptive MMPE

4.2.1 Adaptive LMP Algorithm

We refer to (3–23) as the adaptive MPE criterion, since its p value is changing across iterations. Under this adaptive p-power error criterion, the stochastic gradient algorithm becomes [266]

$$\mathbf{w}(k+1)=\mathbf{w}(k)+\eta f_{p_k}(e(k))\mathbf{u}(k) \tag{4.57}$$

in which the error nonlinearity is $f_{p_k}(e(k))=|e(k)|^{p_k-1}\,\mathrm{sign}(e(k))$.

The adaptive p-power error criterion (3.23) was originally proposed assuming that the additive noise follows a GGD distribution. However, this criterion remains reasonable even when the noise PDF does not belong to the GGD family. This is because GGD densities belong to a family of maximum entropy densities [296,297] under moment constraints, which can approximate a wide range of different signals, including non-GGD distributions [298–300]. As demonstrated in the simulation section, the algorithm (4.57) also performs well in situations involving non-GGD noise.

4.2.2 Mean-Square Convergence Analysis

This part presents a mean-square convergence analysis for the adaptive LMP (ALMP) algorithm in (4.57). The analysis of mean-square convergence has been extensively studied for adaptive algorithms that involve error nonlinearities in update equations. In this regard, the works of Sayed and his research group [295,301,302] are noteworthy, as their approaches are based on the energy conservation relation. However, to the best of our knowledge, Sayed et al. did not consider the case where the error nonlinearity changes with iterations. In this section, we demonstrate that the energy conservation relation still holds even when the error nonlinearity is time-varying. We consider the general algorithm as [266]

$$\mathbf{w}(k+1) = \mathbf{w}(k) + \eta f_{p_k}(e(k))\mathbf{u}(k) \tag{4.58}$$

in which $f_{pk}(.)$ is the general error nonlinearity at k iteration. Subtracting $\mathbf{w}_o$ from both sides of (4.58), we get

$$\tilde{\mathbf{w}}(k+1) = \tilde{\mathbf{w}}(k) - \eta f_{p_k}(e(k))\mathbf{u}(k) \tag{4.59}$$

By defining the *a posteriori* error $e_p(k) \triangleq \tilde{\mathbf{w}}^T(k+1)\mathbf{u}(k)$, we have

$$e_p(k) = e_a(k) + \mathbf{u}^T(k)(\tilde{\mathbf{w}}(k+1) - \tilde{\mathbf{w}}(k)) \tag{4.60}$$

By incorporating (4.59),

$$e_p(k) = e_a(k) - \eta \|\mathbf{u}(k)\|^2 f_{pk}(e(k)) \tag{4.61}$$

where $\|\mathbf{u}(k)\|^2 \triangleq \mathbf{u}^T(k)\mathbf{u}(k)$.

Combining (4.59) and (4.61) so as to eliminate the nonlinearity $f_{pk}(.)$, we have

$$\tilde{\mathbf{w}}(k+1) = \tilde{\mathbf{w}}(k) + (e_p(k) - e_a(k))\frac{\mathbf{u}(k)}{\|\mathbf{u}(k)\|^2} \tag{4.62}$$

Squaring both sides of (4.62) and after some straightforward manipulations, we obtain the energy conservation relation:

$$\|\tilde{\mathbf{w}}(k+1)\|^2 + \frac{(e_a(k))^2}{\|\mathbf{u}(k)\|^2} = \|\tilde{\mathbf{w}}(k)\|^2 + \frac{(e_p(k))^2}{\|\mathbf{u}(k)\|^2} \tag{4.63}$$

where $\|\tilde{\mathbf{w}}(k)\|^2 \triangleq \tilde{\mathbf{w}}^T(k)\tilde{\mathbf{w}}(k)$ is the weight error power (WEP). By taking expectations of both sides, the energy conservation relation becomes

$$E\left[\|\tilde{\mathbf{w}}(k+1)\|^2\right] + E\left[\frac{(e_a(k))^2}{\|\mathbf{u}(k)\|^2}\right] = E\left[\|\tilde{\mathbf{w}}(k)\|^2\right] + E\left[\frac{(e_p(k))^2}{\|\mathbf{u}(k)\|^2}\right] \tag{4.64}$$

Note that the energy conservation relation is actually irrelevant to the error nonlinearity $f_{pk}(.)$ since it has been eliminated.

The energy conservation relation (4.64) can be used to analyze the mean-square convergence of the algorithm in (4.57). By substituting $e_p(k) = e_a(k) - \eta\|\mathbf{u}(k)\|^2 f_{pk}(e(k))$ into (4.64), we obtain

$$E\left[\|\tilde{\mathbf{w}}(k+1)\|^2\right] = E\left[\|\tilde{\mathbf{w}}(k)\|^2\right] - 2\eta E\left[e_a(k) f_{pk}(e(k))\right] + \eta^2 E\left[\|\mathbf{u}(k)\|^2 f_{pk}^2(e(k))\right] \tag{4.65}$$

From (4.65), it is easy to observe that

$$\begin{aligned} &E\left[\|\tilde{\mathbf{w}}(k+1)\|^2\right] \le E\left[\|\tilde{\mathbf{w}}(k)\|^2\right] \\ &\Leftrightarrow -2\eta E\left[e_a(k) f_{pk}(e(k))\right] + \eta^2 E\left[\|\mathbf{u}(k)\|^2 f_{pk}^2(e(k))\right] \le 0 \end{aligned} \tag{4.66}$$

Therefore, if we choose η such that for all k

$$\eta \le \frac{2E\left[e_a(k) f_{pk}(e(k))\right]}{E\left[\|\mathbf{u}(k)\|^2 f_{pk}^2(e(k))\right]} \tag{4.67}$$

then the sequence of WEP will be decreasing, and hence convergent. Thus, a sufficient condition for the mean-square stability of the algorithm (4.57) would be

$$\eta \le \inf_{k \ge 0} \frac{2E\left[e_a(k) f_{pk}(e(k))\right]}{E\left[\|\mathbf{u}(k)\|^2 f_{pk}^2(e(k))\right]} \tag{4.68}$$

To further evaluate the upper bound of the step size, we use the following three assumptions:

Assumption 4.3: The noise $\{v(k)\}$ is independent, identically distributed, and is independent of the stationary input $\{\mathbf{u}(k)\}$;

Assumption 4.4: The *a priori* error $e_a(k)$ is Gaussian distributed;

Assumption 4.5: The filter is long enough such that the random variables $\|\mathbf{u}(k)\|^2$ and $f_{p_k}^2(e(k))$ are uncorrelated.

Under these assumptions, (4.68) becomes

$$\begin{aligned}\eta &\le \inf_{k\ge 0}\frac{2E\left[e_a(k)f_{p_k}(e(k))\right]}{E\left[\|\mathbf{u}(k)\|^2\right]E\left[f_{p_k}^2(e(k))\right]}\\ &= \frac{1}{E\left[\|\mathbf{u}(k)\|^2\right]}\inf_{k\ge 0}\frac{2E\left[e_a(k)f_{p_k}(e(k))\right]}{E\left[f_{p_k}^2(e(k))\right]}\\ &= \frac{1}{E\left[\|\mathbf{u}(k)\|^2\right]}\inf_{k\ge 0}\frac{2E[e_a^2(k)]h_G^{(p_k)}(E[e_a^2(k)])}{h_U^{(p_k)}(E[e_a^2(k)])}\end{aligned} \tag{4.69}$$

where

$$\begin{cases} h_G^{(p_k)}(E[e_a^2(k)]) \triangleq \dfrac{E\left[e_a(k)f_{p_k}(e(k))\right]}{E[e_a^2(k)]} \\ h_U^{(p_k)}(E[e_a^2(k)]) \triangleq E\left[f_{p_k}^2(e(k))\right] \end{cases} \tag{4.70}$$

Note that given a specific p value, the above two functions depend only on the second moment $E[e_a^2(k)]$. This is because $e_a(k)$ is Gaussian and independent of the noise $v(k)$.

At present, it is difficult to find the general closed-form expressions of functions $h_G^{(p)}(.)$ and $h_U^{(p)}(.)$. However, for the case in which $p=2l$ $(l=1,2,3,\cdots)$, one can easily derive the expressions of the two functions (see Appendix E).

Because all terms in the minimization are functions of p_k and $E[e_a^2(k)]$, equation (4.69) can be expressed as

$$\eta \le \frac{1}{E\left[\|\mathbf{u}(k)\|^2\right]}\inf_{p\in\Theta_p,\sigma_{e_a}^2\in\Theta_{e_a}}\frac{2\sigma_{e_a}^2 h_G^{(p)}\left(\sigma_{e_a}^2\right)}{h_U^{(p)}\left(\sigma_{e_a}^2\right)} \tag{4.71}$$

where $\sigma_{e_a}^2 \triangleq \mathbf{E}[e_a^2(k)]$, Θ_p and Θ_{e_a} denote the feasible set of p_k and $\sigma_{e_a}^2$, respectively. As $p_k = \min\{\hat{v}_k, p_{\text{upper}}\}$, the feasible set of p_k can be written as $\Theta_p = \{p : 0 < p \le p_{\text{upper}}\}$. Further in literature [272], a feasible set of $\sigma_{e_a}^2$ is given by

$$\Theta_{e_a} = \left\{\sigma_{e_a}^2 : \lambda \le \sigma_{e_a}^2 \le \frac{1}{4}\text{Tr}(\mathbf{R}_{\mathbf{uu}})E\left[\|\tilde{\mathbf{w}}_0\|^2\right]\right\} \tag{4.72}$$

where $\mathbf{R}_{\mathbf{uu}} = E[\mathbf{u}(k)\mathbf{u}^T(k)]$, λ denotes the Cramer-Rao lower bound (CRLB) associated with the problem of estimating $\mathbf{w}_o^T\mathbf{u}(k)$ by using $\mathbf{w}^T(k)\mathbf{u}(k)$.

4.2.3 Simulation Results

Here some simulation results are given to demonstrate the satisfactory performance of the ALMP algorithm (4.57) in comparison with the LMP algorithm with p=1, 2 and 4, i.e., the LAD, LMS, and LMF algorithms, respectively. In the calculation of p_k, the sliding data length L is set to 50 and the upper bound $p_{\text{upper}} = 4$. Consider the system identification case, and the unknown weight vector is $\mathbf{w}_o = [0.1, 0.2, 0.3, 0.4, 0.5, 0.4, 0.3, 0.2, 0.1]^T$. The input signal $\{\mathbf{u}(k)\}$ is white Gaussian noise process with unit-power. The initial weights of the adaptive filter are set to zero. In the experiments, the following noise distributions are considered:

$$
\begin{cases}
\text{(a) Laplace: } p_v(x) = \dfrac{\sqrt{2}}{2}\exp(-\sqrt{2}\,|x|) \\[2mm]
\text{(b) Gaussian: } p_v(x) = \dfrac{1}{\sqrt{2\pi}}\exp(-0.5x^2) \\[2mm]
\text{(c) Uniform: } p_v(x) = \begin{cases} 0.5 & -1 \le x \le 1 \\ 0 & \text{otherwise} \end{cases} \\[2mm]
\text{(d) Mixed Gaussian: } p_v(x) = \dfrac{1}{\sqrt{2\pi}}\{\exp(-2(x-0.5)^2) + \exp(-2(x+0.5)^2)\} \\[2mm]
\text{(e) Binary: } \begin{cases} \Pr(v(k) = 1) = 0.5 \\ \Pr(v(k) = -1) = 0.5 \end{cases} \\[2mm]
\text{(f) Symmetric } \alpha \text{ stable } (S\alpha S): \phi_{\gamma,\alpha}(\omega) = \exp(-\gamma\,|\omega|^{\alpha}) \text{ with } \gamma = 1, \alpha = 1.8
\end{cases}
$$

where the noise PDFs (a)–(c) belong to the GGD family (correspond to v=1, 2 and ∞), $\phi_{\gamma,\alpha}(\omega)$ denotes the characteristic function of the symmetric alpha-stable (SαS) distribution with the dispersion parameter γ and the characteristic exponent α. Simulation results are presented in Figure 4.5 and Table 4.1, among which Figure 4.5 illustrates the average convergence curves for each p value and each noise distribution over 100 independent Monte Carlo runs. Figure 4.6 plots the averaged adaptive p values (p_k) across iterations when disturbed by GGD family noises (Laplace, Gaussian and Uniform). Table 4.1 gives the sample mean and standard deviation of the WEP at the final stage of adaptation. From the simulation results, one can obtain the following observations:

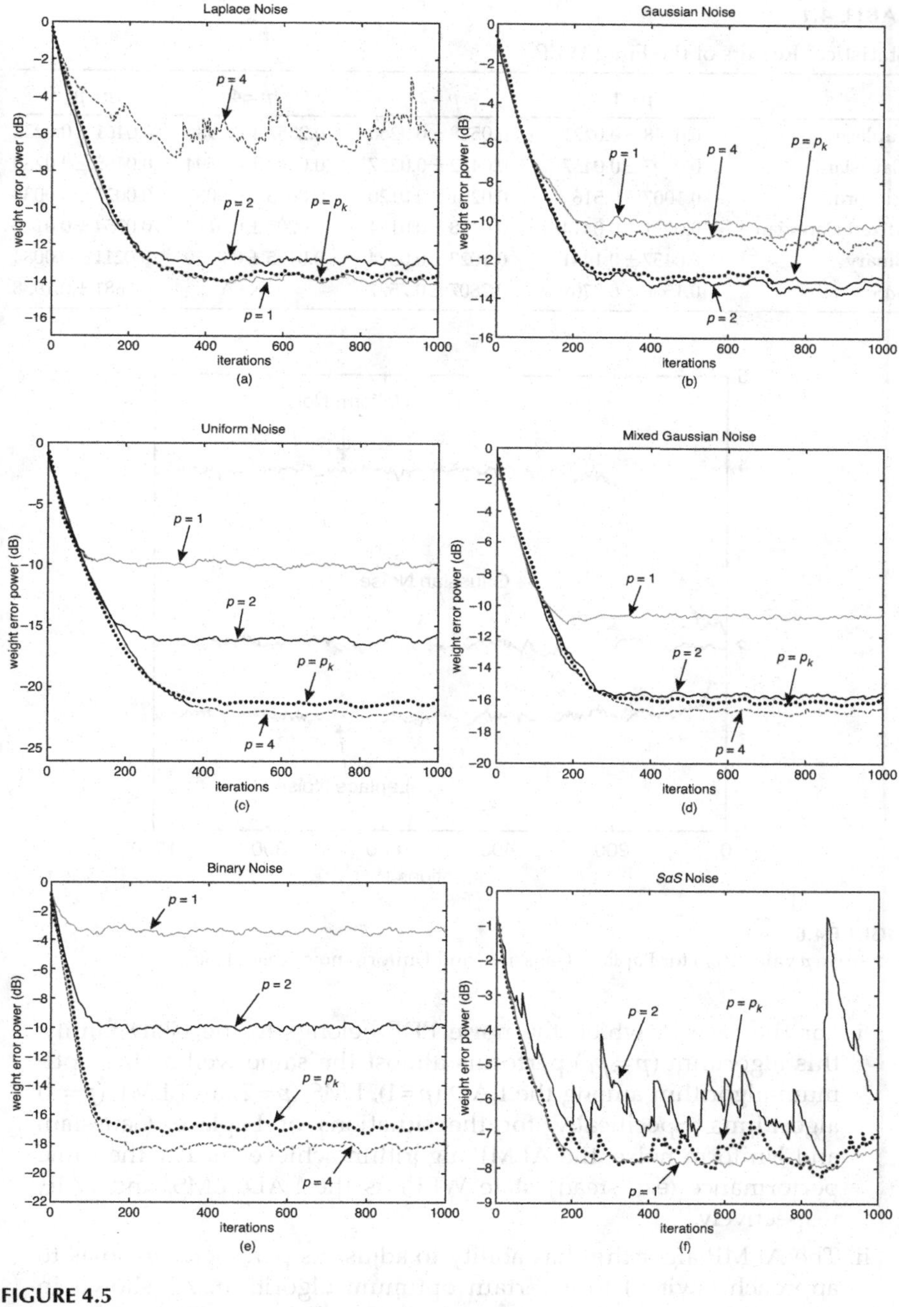

FIGURE 4.5
Average convergence curves for each p value and each noise distribution [266].

TABLE 4.1
Statistical Results of the Final WEP

	$p=1$	$p=2$	$p=4$	$p=p\_k$
Laplace	0.0448 ± 0.0221	0.0528 ± 0.0266	0.2552 ± 0.2986	0.0461 ± 0.0232
Gaussian	0.0997 ± 0.0457	0.0472 ± 0.0227	0.0809 ± 0.0594	0.0485 ± 0.0230
Uniform	0.1007 ± 0.516	0.0251 ± 0.0120	0.0063 ± 0.0031	0.0077 ± 0.0038
Mixed Gaussian	0.0931 ± 0.0414	0.0278 ± 0.0134	0.0207 ± 0.0117	0.0254 ± 0.0125
Binary	0.0457 ± 0.1441	0.1023 ± 0.0374	0.0155 ± 0.0060	0.0211 ± 0.0084
SαS	0.1586 ± 0.0708	0.2807 ± 0.7527	N/A	0.1681 ± 0.1938

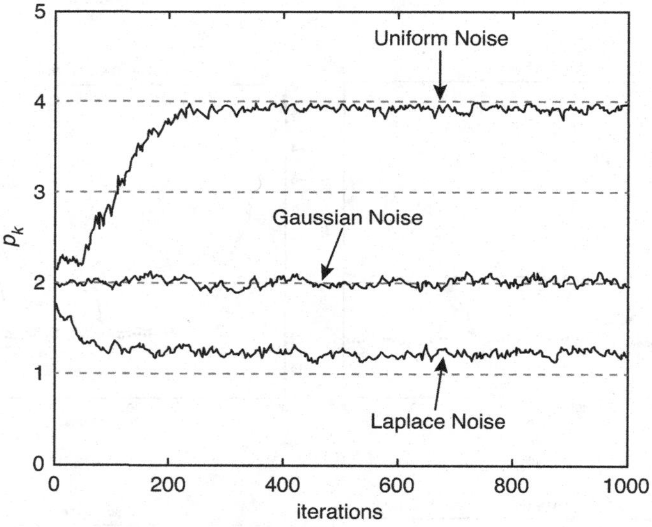

FIGURE 4.6
Adaptive p values (p_k) for Laplace, Gaussian, and Uniform noise cases [266].

i. For the case in which the noise PDF belongs to the GGD family, this algorithm ($p = p_k$) performs almost the same well as the optimum algorithm among the LAD ($p=1$), LMS ($p=2$) and LMF ($p=4$) algorithms. Specifically, for the situations of Laplace, Gaussian, and Uniform noise, the ALMP algorithm achieves nearly the same performance (e.g. steady-state WEP) as the LAD, LMS, and LMF, respectively.

ii. The ALMP algorithm has ability to adjust its parameter p_k so as to approach (switch) to a certain optimum algorithm. As shown in Figure 4.6, the adaptive parameter p_k starts from an initial value, i.e., 2, and converges approximately to 1, 2, and 4 (p_k is artificially limited to $p_k \leq 4$) when disturbed by Laplace, Gaussian, and Uniform noise. This agrees with the discussion in 4.2.1.

iii. For the non-GGD noise cases (Mixed Gaussian, Binary, and $S\alpha S$), the ALMP algorithm still performs well and achieves the second-best performance. This confirms the previous claim that this algorithm is applicable to a wide range of noise distributions.

4.3 Adaptive Filtering Algorithm with Variable Normalization under MMPE

4.3.1 Variable Normalization LMP Algorithm

Inspired by the LMS-based algorithm of which the shape of the error nonlinearity is related to the power of the estimation error, a variable normalization is designed for adaptive change of the shape of the error nonlinearity in LMP to combat both Gaussian and non-Gaussian noises efficiently [292]. First, the normalized term of LMP in the weight update is considered using the product of the qth moment of the norm of the input and the $(p\text{-}q)^{\text{th}}$ moment of the error, i.e.

$$\mathbf{w}(k+1)=\mathbf{w}(k)+\mu\frac{|e(k)|^{p-1}\operatorname{sign}(e(k))}{\varepsilon+|\,|\mathbf{u}(k)|\,|^{q}|e(k)|^{p-q}}\mathbf{u}(k) \tag{4.73}$$

where ε is a small positive constant. It has been proved that the error nonlinear function is optimal regarding the steady-state mean square error when the order of the error in the denominator of the weight update is one order larger than that of the numerator [272]. Therefore, $q \in [0, 2]$ is chosen to ensure the stability and filtering accuracy, which is also explained in [303,304]. According to (4.73), the weight update in (4.73) with error nonlinear function can be expressed as [292]

$$\mathbf{w}(k+1)=\mathbf{w}(k)+\mu\mathbf{u}(k)e(k)f(e(k)) \tag{4.74}$$

where

$$f(e(k))=\frac{|e(k)|^{p-2}}{\varepsilon+\|\mathbf{u}(k)\|^{q}\,|e(k)|^{p-q}} \tag{4.75}$$

The input signal is assumed to be a stationary sequence of independent zero-mean Gaussian random variable with a finite variance $\sigma_{\mathbf{u}}^2$. By setting $\sigma_{\mathbf{u}}^2=0.01$, $N=32$, $p=4$, and $\alpha=1$, the error surface of (4.75) can be plotted in Figure 4.7. As one can see from Figure 4.7, equation (4.75) is a V-shaped algorithm when $q=2$ and an M-shaped algorithm otherwise.

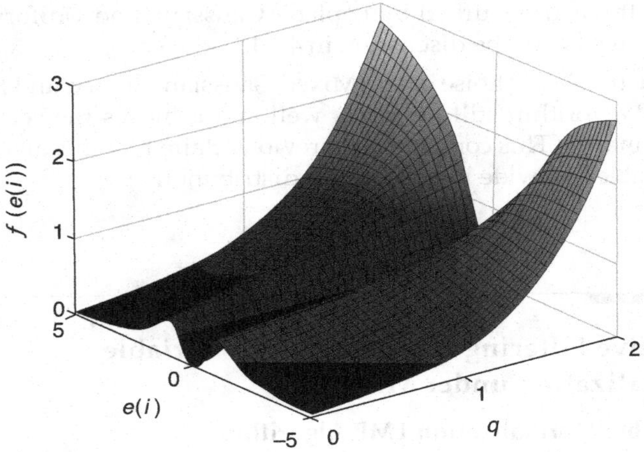

FIGURE 4.7
Surface of the error nonlinear function [292].

It is worth mentioning that (4.75) with a fixed q can only be used in a specific noise environment. Thus, a variable order of error in (4.75) is used to combine the characteristics of both V-shaped and M-shaped algorithms. Generally, different noises require different shapes of error nonlinear functions. For example, the V-shaped algorithms used in the presence of Gaussian and sub-Gaussian noises cannot combat impulsive noise, and the M-shaped algorithms used in the presence of impulsive noise cannot improve filtering accuracy for Gaussian and sub-Gaussian noises. Therefore, $q \in [0, 2]$ in (4.75) is required for different noises. In Figure 4.7, a large error requires a small q for guaranteeing the convergence of the algorithm, and a small error requires a large q to improve the steady-state performance. Thus to obtain variable $q(i)$, a monotone decreasing function with respect to $e(k)$ with upper and lower bounds of 2 and 0 is designed. Before proceeding, the sigmoidal function is introduced, and it can be expressed by

$$\text{sgm}\,(e(k)) = \frac{1}{1+\exp(-\beta\,|\,e(k)\,|^{p})} \tag{4.76}$$

where parameter $p > 0$, and $\beta > 0$ is the steepness parameter which controls the steepness of the sigmoidal function curve. The sigmoid function (4.76) is a symmetric function regarding the origin and has the maximum of 1 and only a global minimum of sgm[0] = 0.5. Thus, using the sigmoidal function (4.76), the following variable method for $q(k)$ can be designed:

$$q(k) = 4(1-\text{sgm}(e(k))) = 4 - \frac{4}{1+\exp(-\beta\,|\,e(k)\,|^{p})} \in [0,2] \tag{4.77}$$

In (4.77), a larger β results in a steeper sigmoid curve, which can be used in impulsive noise, while equation (4.75) requires a smaller q to combat such noise. On the contrary, a smaller β can be used in nonimpulsive noises to smooth the steady-state performance. In addition, $q(k)$ with a larger p has a steeper curve at moderate errors and a smoother curve at smaller errors. Otherwise, $q(k)$ has a contrary curve. Therefore, $q(k)$ with different values of p can be applied in different noise environments. Finally, replacing q in (4.73) with $q(k)$ in (4.77) gives the robust VNLMP algorithm, i.e.,

$$\mathbf{w}(k+1) = \mathbf{w}(k) + \mu \frac{|e(k)|^{p-1}\,\text{sign}(e(k))}{\varepsilon + \|\mathbf{u}(k)\|^{q}\,|e(k)|^{p-q(k)}}\mathbf{u}(k) \tag{4.78}$$

Remark 4.5

According to (4.77) used in the normalization, VNLMP can automatically switch between the V-shaped and M-shaped algorithms. When $e(k)$ is very large, i.e., a large $|e(k)|$, we have that $q(k) \to 0$ from (4.77) and VNLMP in (4.78) can be viewed as an M-shaped algorithm. This means that VNLMP can combat various outliers including impulsive noise efficiently. When $e(k)$ is very small, i.e., $|e(k)| \to 0$, $q(k) \to 2$ is obtained from (4.77) and VNLMP reduces to a V-shaped algorithm. Therefore, VNLMP can improve the steady-state performance in the presence of both Gaussian and non-Gaussian noises simultaneously.

Remark 4.6

In the absence of impulsive noise, the error shown in (4.77) is usually small, and thus a small $q(k)$ is required in (4.78) to improve the convergence performance. Note that $q(k)$ with p=2 has a smooth curve when the error is not large enough, which leads to the desirable performance. Thus, in the non-impulsive noise environments, we reasonably choose p=2 in (4.77).

Remark 4.7

In the VNLMP algorithm shown in (4.78), there exist four parameters, i.e., error power p, positive constant α, steepness parameter β, and step size μ. We generally choose p as a positive integer for different applications [303,305]. As α is used to avoid the normalization of VNLMP being 0, we can set it to a small positive number. According to Figure 4.7, we see that β can affect the switching rate of VNLMP between the V-shaped and M-shaped algorithms. Specifically, a lager β leads to a more probable M-shaped algorithm, whereas a smaller β leads to a more probable V-shaped algorithm. Therefore, steepness parameter β together with step size μ can achieve a trade-off between the transient and steady-state filtering performance in different noise environments. Because the V-shaped algorithm can smooth a small error and the M-shaped algorithm can combat a large error, a relatively small β is set for non-impulsive noise and a relatively large β for impulsive noise generally.

4.3.2 Convergence Analysis

By using the weight error vector $\tilde{\mathbf{w}}(k) = \mathbf{w}(k) - \mathbf{w}_o$, the instantaneous mean square deviation (MSD) is defined by the following expectation of the squared 2-norm based on the weight error vector:

$$y_{\text{MSD}}(k) = E[\|\tilde{\mathbf{w}}(k)\|^2] \tag{4.79}$$

where $\|\tilde{\mathbf{w}}(k)\|$ is the Euclidean norm of $\tilde{\mathbf{w}}(k)$. The excess mean-square error (EMSE) is therefore defined as:

$$\xi(k) = E[e_a^2(k)] \tag{4.80}$$

where the *a priori* estimation error is $e_a(k) = \tilde{\mathbf{w}}^T(k)\mathbf{u}(k)$. In general, the EMSE is also used to evaluate the steady-state performance and tracking ability of adaptive filtering algorithms.

The steady-state performance analysis of the VNLMP algorithm will be performed in the following. To make the analysis mathematically tractable, the following assumptions are considered [136,303].

Assumption 4.6: $\mathbf{u}(k)$ is a stationary sequence of zero-mean independently and identically distributed (i.i.d.) Gaussian random variable with a finite variance $\sigma_{\mathbf{u}}^2$ and a positive-definite covariance matrix $\mathbf{R}_{\mathbf{uu}} = E[\mathbf{u}(k)\mathbf{u}^{\mathrm{T}}(k)]$.

Assumption 4.7: $v(k)$ is a stationary sequence of zero-mean i.i.d. random variable with a finite variance σ_v^2.

Assumption 4.8: The sequences $\mathbf{u}(k)$ and $v(k)$ are mutually independent and jointly Gaussian.

Assumption 4.9: The *a priori* estimation error $e_a(k)$ has a Gaussian distribution and is independent of the noise.

From the viewpoint of the adaptive filters with the error nonlinearity, the weight update form of VNLMP can be written as

$$\mathbf{w}(k+1) = \mathbf{w}(k) + \mu g(e(k))\mathbf{u}(k) \tag{4.81}$$

where $g(e(k))$ is the error nonlinear function, which can be given from the weight update form of VNLMP by

$$g(e(k)) = \frac{|e(k)|^{p-1}\operatorname{sign}[e(k)]}{\alpha + \|\mathbf{u}(k)\|^{q(k)}|e(k)|^{p-q(k)}} \tag{4.82}$$

According to the definition of the weight error vector and (4.81), we have

$$\tilde{\mathbf{w}}(k+1) = \tilde{\mathbf{w}}(k) - \mu g(e(k))\mathbf{u}(k) \tag{4.83}$$

Pre-multiplying both sides of (4.83) by their transposes and taking the expected value, we have

$$E\left[\|\tilde{\mathbf{w}}(k+1)\|^2\right] = E\left[\|\tilde{\mathbf{w}}(k)\|^2\right] - 2\mu E\left[\tilde{\mathbf{w}}^T(k)\mathbf{u}(k)g(e(k))\right] + \mu^2 E\left[\|\mathbf{u}(k)\|^2 g^2(e(k))\right] \tag{4.84}$$

According to Assumption 4.9 and using Price's theorem [237,251,268,274], the second expectation on the right side of (4.84) can be derived as $E[\tilde{\mathbf{w}}^{\mathrm{T}}(k)\mathbf{u}(k)g(e(k))] = E\{e(k)g(e(k))\} / E[e^2(k)]$. Then, from Assumptions 4.6–4.9 and using (4.79), we can obtain the recursive MSD relation of VNLMP as follows:

$$y_{\mathrm{MSD}}(k+1) = \{1 - 2\mu\sigma_u^2 h_G[e(k)]\}y_{\mathrm{MSD}}(k) + \mu^2 E[\|\mathbf{u}(k)\|^2]h_U[e(k)] \tag{4.85}$$

where

$$h_G[e(k)] = \frac{E\{e(k)g(e(k))\}}{E[e^2(k)]} \tag{4.86}$$

$$h_U[e(k)] = E\{g^2(e(k))\} \tag{4.87}$$

Substituting (4.82) into (4.86) and (4.87), we can obtain $h_G[e(k)]$ and $h_U[e(k)]$ of VNLMP as follows:

$$h_G[e(k)] = \frac{E\left\{\dfrac{|e(k)|^p}{\alpha + \|\mathbf{u}(k)\|^{q(k)}|e(k)|^{p-q(k)}}\right\}}{E[e^2(k)]} \tag{4.88}$$

$$h_U[e(k)] = E\left\{\frac{|e(k)|^{2p-2}}{\left[\alpha + \|\mathbf{u}(k)\|^{q(k)}|e(k)|^{p-q(k)}\right]^2}\right\} \tag{4.89}$$

When the VNLMP approaches the steady state, $y_{\mathrm{MSD}}(k+1) = y_{\mathrm{MSD}}(k)$ in (4.85) is obtained. Thus, the steady-state MSD can be obtained as follows:

$$y_{\mathrm{MSD}}(\infty)=\frac{1}{2}\mu\frac{\lim\limits_{k\to\infty}E[\|\mathbf{u}(i)\|^2]h_U[e(k)]}{\sigma_u^2\lim\limits_{k\to\infty}h_G[e(k)]}\approx\frac{1}{2}\mu N\frac{\lim\limits_{k\to\infty}h_U[e(k)]}{\lim\limits_{k\to\infty}h_G[e(k)]}\tag{4.90}$$

where $\|\mathbf{u}(k)\|^2=\mathbf{u}^{\mathrm{T}}(k)\mathbf{u}(k)\approx N\sigma_u^2$ is used, which is valid when N is assumed to be sufficiently large under Assumption 4.6 [301,302]. This is also an ergodic assumption, which means that the time average over the taps is equal to its ensemble average. Under Assumptions 4.6–4.9, we have that $e(k)$ is a zero-mean Gaussian variable. Using (4.79) and (4.80), we have

$$\begin{aligned}\sigma_e^2&=E[e^2(k)]\\&=E[v(k)-\tilde{\mathbf{w}}^{\mathrm{T}}(k)\mathbf{u}(k)]\\&=\sigma_v^2+\xi(k)\\&=\sigma_v^2+\sigma_u^2y_{MSD}(k)\end{aligned}\tag{4.91}$$

Then, from (4.80), (4.90), and (4.91), we obtain the steady-state EMSE of VNLMP as follows:

$$\xi(\infty)=\sigma_u^2y_{\mathrm{MSD}}(\infty)=\frac{1}{2}\mu\frac{\lim\limits_{k\to\infty}E[\|\mathbf{u}(k)\|^2]h_U[e(k)]}{\lim\limits_{k\to\infty}h_G[e(k)]}=\frac{1}{2}\mu\mathrm{Tr}[\mathbf{R_{uu}}]\frac{\lim\limits_{k\to\infty}h_U[e(k)]}{\lim\limits_{k\to\infty}h_G[e(k)]}\tag{4.92}$$

where $\mathrm{Tr}[\mathbf{R_{uu}}]$ denotes the trace of the matrix $\mathbf{R_{uu}}$.

When the VNLMP approaches the steady state, i.e., the step size is sufficiently small with $\lim\limits_{k\to\infty}q(k)\to 2$, one can get

$$\begin{aligned}\lim_{k\to\infty}h_G[e(k)]&=\lim_{k\to\infty}\frac{E\left\{\dfrac{|e(k)|^p}{\alpha+\|\mathbf{u}(k)\|^{q(i)}|e(k)|^{p-q(k)}}\right\}}{E[e^2(k)]}\\&\approx\frac{\lim\limits_{i\to\infty}E\left\{\dfrac{|e(k)|^p}{\alpha+\|\mathbf{u}(k)\|^2|e(k)|^{p-2}}\right\}}{\lim\limits_{i\to\infty}E[e^2(k)]}\\&=\frac{1}{\sigma_e^2}\lim_{k\to\infty}E\left\{\frac{|e(k)|^p}{\alpha+\|\mathbf{u}(k)\|^2|e(k)|^{p-2}}\right\}\end{aligned}\tag{4.93}$$

$$\lim_{k\to\infty} h_U[e(k)] = \lim_{i\to\infty} E\left\{\frac{|e(k)|^{2p-2}}{\left[\alpha + \|\mathbf{u}(k)\|^{q(k)} |e(k)|^{p-q(k)}\right]^2}\right\}$$
$$\approx \lim_{k\to\infty} E\left\{\frac{|e(k)|^{2p-2}}{\left[\alpha + \|\mathbf{u}(k)\|^{2} |e(k)|^{p-2}\right]^2}\right\} \tag{4.94}$$

where $\sigma_e^2 = E[e^2(k)]$.

Obtaining the derivations of (4.92) and (4.94) with different p is a trivial but rather tedious task, since only the second order of error (i.e., $p=2$) is optimal under the Gaussian assumption [261]. Thus, we only discuss the case of $p=2$ in (4.93) and (4.94) in this book. In addition, α generally is set as a very small number. Therefore, substituting $p=2$ into (4.93) and (4.94) gives

$$\lim_{k\to\infty} h_G[e(k)] = \frac{1}{\sigma_e^2} \lim_{k\to\infty} E\left\{\frac{e^2(k)}{\alpha + \|\mathbf{u}(k)\|^2}\right\} = E\left\{\frac{1}{\|\mathbf{u}(k)\|^2}\right\} \tag{4.95}$$

$$\lim_{k\to\infty} h_U[e(k)] = \lim_{k\to\infty} E\left\{\frac{e^2(k)}{\left[\alpha + \|\mathbf{u}(k)\|^2\right]^2}\right\} = \lim_{k\to\infty} E\left\{\frac{e^2(k)}{\|\mathbf{u}(k)\|^4}\right\} = \sigma_e^2 E\left\{\frac{1}{\|\mathbf{u}(k)\|^4}\right\} \tag{4.96}$$

Then combining (4.90), (4.91), (4.92), (4.95), and (4.96), one can obtain the steady-state EMSE and MSD of the VNLMP with p=2 as follows:

$$\xi(\infty) = \frac{\mu\sigma_v^2 \mathrm{Tr}[\mathbf{R_{uu}}]}{2E\left\{\frac{1}{\|\mathbf{u}(k)\|^2}\right\} - \mu \mathrm{Tr}[\mathbf{R_{uu}}] E\left\{\frac{1}{\|\mathbf{u}(k)\|^4}\right\}} E\left\{\frac{1}{\|\mathbf{u}(k)\|^4}\right\} \tag{4.97}$$

$$y_{\mathrm{MSD}}(\infty) = \frac{\mu N\sigma_v^2}{2E\left\{\frac{1}{\|\mathbf{u}(k)\|^2}\right\} - \mu \mathrm{Tr}[\mathbf{R_{uu}}] E\left\{\frac{1}{\|\mathbf{u}(k)\|^4}\right\}} E\left\{\frac{1}{\|\mathbf{u}(k)\|^4}\right\} \tag{4.98}$$

Under Assumption 4.6 and using the approximation $\|\mathbf{u}(k)\|^2 \approx N\sigma_{\mathbf{u}}^2$, the steady-state EMSE and MSD of the VNLMP with p=2 can be rewritten as

$$\xi(\infty) = \frac{\mu\sigma_v^2 \mathrm{Tr}[\mathbf{R_{uu}}]}{2N\sigma_u^2 - \mu \mathrm{Tr}[\mathbf{R_{uu}}]} \tag{4.99}$$

$$y_{MSD}(\infty) = \frac{\mu N\sigma_v^2}{2N\sigma_u^2 - \mu \mathrm{Tr}[\mathbf{R_{uu}}]} \tag{4.100}$$

Remark 4.8

When $p=2$, the steady-state EMSE and MSD of VNLMP are the same as those of NLMS. Note that (4.99) and (4.100) are only valid for small μ and α. In addition, to guarantee $q(k) \to 2$ in VNLMP at steady-state, β needs to be set as a small value. When α is small, as long as μ or β is small enough, the validity of the derived EMSE and MSD can be guaranteed. The steady-state performance of the VNLMP with different p also can be performed under different noise distributions using a Taylor expansion method [285].

4.3.3 Simulation Results

This section gives some simulation results to show the obtained theoretical results and the MSD performance of the VNLMP algorithm. The simulations are done in the example of adaptive FIR system identification. The theoretical result validation and performance comparison are performed by the simulations as follows.

4.3.3.1 Theoretical Validation

To verify the steady-state behavior of VNLMP, the theoretical MSD in (4.84) is compared with the simulated one under different values of μ and β. The simulations are performed with the zero-mean Gaussian input and noise. Figure 4.8 shows the evolution of the theoretical and simulated steady-state MSDs of the VNLMP with $\alpha = 10^{-4}$, $\beta = 10^{-4}$, and $\mu = 0.004$ to 0.1. As one can

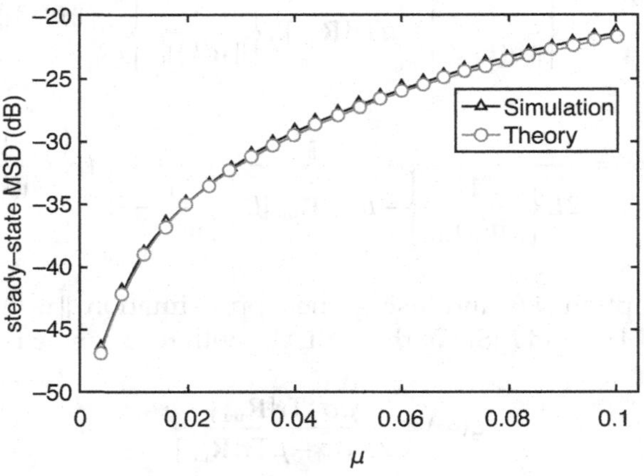

FIGURE 4.8
Theoretical and simulated steady-state MSDs of VNLMP ($p = 2$) with different values of μ [292].

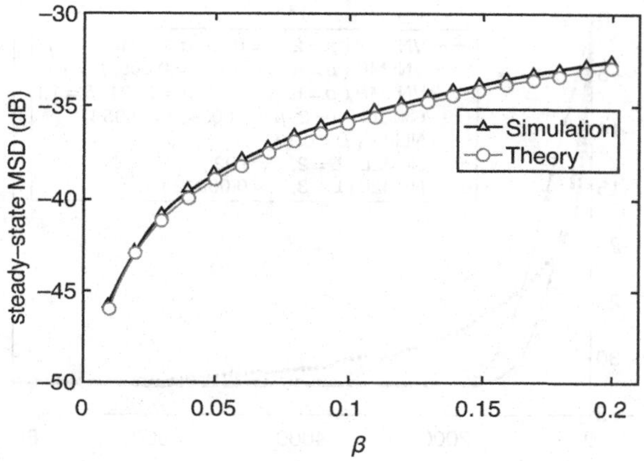

FIGURE 4.9
Theoretical and simulated steady-state MSDs of VNLMP ($p = 2$) with different values of β [292].

see from Figure 4.8, the simulated steady-state MSDs strictly match well with the theoretical ones when μ and β are small. Figure 4.9 shows the theoretical and simulated MSDs of the VNLMP. It can be seen from Figure 4.9 that the theoretical MSDs match well with the simulated ones when μ and β are small. Therefore, from Figures 4.8 and 4.9, we obtain that the derived steady-state MSD of VNLMP shown in (4.84) is valid for small step size and steepness parameter.

4.3.3.2 Performance Comparison

In this section, the MSD performance of VNLMP is compared with other algorithms in different noise environments. In stable LMS-based algorithms, NLMS is typically used for comparison in the Gaussian noise case, and NLM2L [303] is used in the sub-Gaussian noise case due to its superior filtering performance. Therefore, the NLMS and NLM2L are selected for comparisons in the Gaussian and uniform noise environments, respectively. In the impulsive noise environment, the generalized MCC (GMCC) [283] and MCC with variable center (MVC) [306] methods are used for comparisons, as they provide better performance than other robust filters.

We begin by comparing the MSDs of VNLMP, GMCC, NLMS, and NLM2L in Gaussian and uniform noise environments with zero-mean and variance 0.01. Figures 4.10 and 4.11 display the compared MSDs. From Figure 4.10, it is evident that VNLMP has a faster convergence rate than NLMS, GMCC, and NLM2L for Gaussian noise. Similarly, from Figure 4.11, one can observe that VNLMP and NLM2L have comparable convergence rates while being faster than NLMS and GMCC for uniform noise. It is worth noting that in Gaussian and uniform noise environments, GMCC can suffer from instability

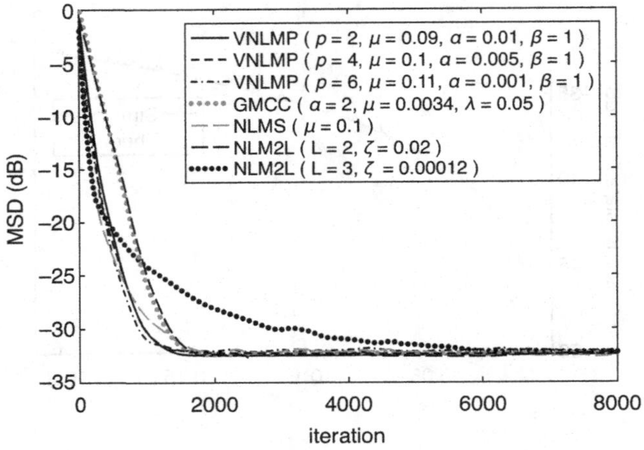

FIGURE 4.10
Comparison of the MSDs of VNLMP, NLMS, GMCC, and NLM2L in Gaussian noise [292].

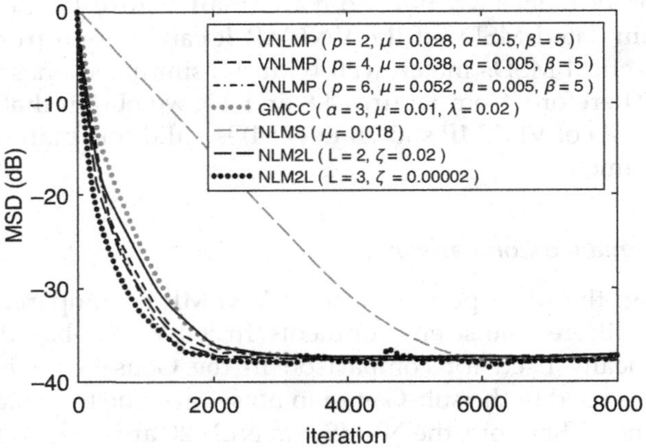

FIGURE 4.11
Comparison of the MSDs of VNLMP, NLMS, GMCC, and NLM2L in uniform noise [292].

when the same steady-state performance as other algorithms is required. Therefore, the filtering performance of VNLMP is superior to that of GMCC for both Gaussian and uniform noise. Thus, in Gaussian and sub-Gaussian noise environments, the VNLMP algorithm can provide a faster convergence rate when the same steady-state MSD is required.

Second, an impulsive noise environment is considered that the noise model is $v(i) = v_o(i) + b(i)v_i(i)$, where $v_o(i)$ is the ordinary noise and $v_i(i)$ is the impulsive noise to represent large outliers, $b(i)$ is generated using a Bernoulli random process. In this case, as a large outlier, the α-stable noise is chosen as $v_i(i)$. The distribution of $v_o(i)$ is considered as: (1) zero-mean Gaussian with variance $\sigma_{v_o}^2 = 1$; (2) Binary distribution over {1,–1} with probability mass $\Pr\{x=1\} = \Pr\{x=-1\} = 0.5$. First, Figures 4.12 and 4.13 show the compared MSDs of all algorithms for different distributions of $v_o(i)$ in the mixed noises environments. As one can see from Figures 4.12 and 4.13, VNLMP can combat impulsive noises effectively and provides better filtering performance than MVC and GMCC in different impulsive noises. Specifically, for the Gaussian $v_o(i)$, VNLMP with $p=2$ has the best filtering performance and

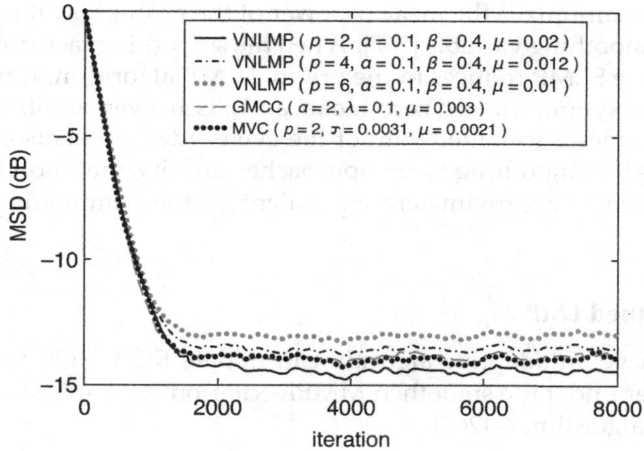

FIGURE 4.12
Comparison of the MSDs of VNLMP, MVC, and GMCC with Gaussian $v(k)$ [292].

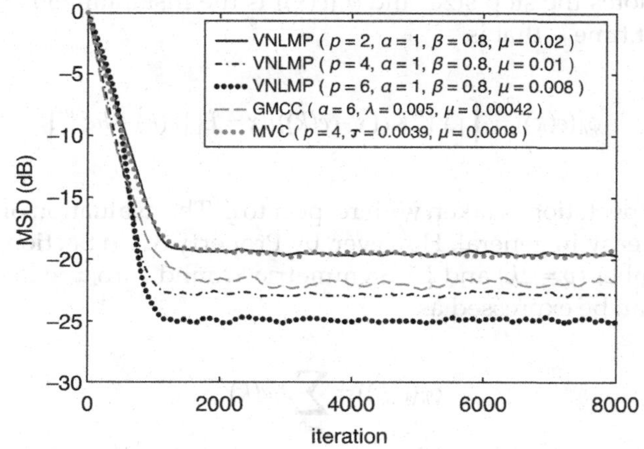

FIGURE 4.13
Comparison of the MSDs of VNLMP, MVC, and GMCC with Binary $v(k)$ [292].

slightly outperforms MVC and GMCC. For binary $v_o(i)$, VNLMP with $p = 6$ has the best performance and is dramatically superior to MVC and GMCC.

4.4 Adaptive Filtering Algorithm under Smoothed MMPE

This subchapter introduces an adaptive filtering algorithms (AFA) based on the smoothed MPE criterion, known as the smoothed LMP (SLMP) algorithm, which minimizes the mean p-power of the error plus an independent and scaled smoothing variable [267]. When the smoothing factor (or scale factor) is zero, the SLMP reduces to the original LMP algorithm. If the smoothing variable is symmetric and zero-mean, and p is an even number, the SMPE criterion becomes a weighted sum of the even-order moments of the error. Moreover, as the smoothing factor approaches infinity, the smoothed MMPE criterion becomes approximately equivalent to the commonly used MSE criterion.

4.4.1 Smoothed LMP Algorithm

One can derive an adaptive algorithm to update the weight vector of the adaptive filter under the smoothed MMPE criterion. A simple stochastic gradient-based algorithm is [267]

$$\mathbf{w}(k+1) = \mathbf{w}(k) - \eta \frac{\partial}{\partial \mathbf{w}(k)} \phi_h(e(k)) \tag{4.101}$$

where η denotes the step size, and $\phi_h(e(k))$ is the instantaneous value of the SMPE cost at time k, that is

$$\phi_h(e(k)) = \int_{-\infty}^{\infty} |x|^p K_h(x - e(k))dx = E[|e(k) + h\xi|^p] \tag{4.102}$$

where the expectation is taken with respect to ξ. The evaluation of this expectation is not easy in general. However, by Property 3.3 in Section 3.4.2, if p is an even number ($p = 2k$) and ξ is symmetric around zero, the instantaneous SMPE cost can be expressed as

$$\phi_h(e(k)) = \sum_{l=0}^{m} \lambda_l e(k)^{2l} \tag{4.103}$$

where $\lambda_l = C_{2k}^{2l} h^{2(m-l)} E[\xi^{2(m-l)}]$. In this case, the adaptive algorithm (4.101) can be derived as

$$\begin{aligned} \mathbf{w}(k+1) &= \mathbf{w}(k) - \eta \frac{\partial}{\partial \mathbf{w}(k)} \left\{ \sum_{l=0}^{m} \lambda_l e(k)^{2l} \right\} \\ &= \mathbf{w}(k) + \eta \left\{ \sum_{l=1}^{m} 2l\lambda_l e(k)^{2l-1} \right\} \mathbf{u}(k) \end{aligned} \tag{4.104}$$

which is a mixture of the LMF family adaptive algorithms, and the mixture weights are controlled by the smoothing variable ξ and the smoothing factor h. Here for simplicity, we always assume that p is an even number and ξ is symmetric around zero. Here we refer to (4.104) as the SLMP algorithm.

Remark 4.9

Given a smoothing variable ξ, the performance of the SLMP algorithm is influenced by the smoothing factor h. As $h \to 0$, the SLMP algorithm reduces to the original LMP algorithm, while when $h \to \infty$, the SLMP algorithm converges to the popular LMS algorithm (see **Property 3.4 in** Section 3.4.2). Thus, the smoothing factor h provides a nice bridge between LMP and LMS. The distribution of the smoothing variable ξ also has influence on the performance of the SLMP algorithm. However, as shown in simulation results, this influence is much smaller than the influence of h.

4.4.2 Convergence Performance

The SLMP algorithm (4.104) can be written in a general form as

$$\mathbf{w}(k) = \mathbf{w}(k-1) + \eta f(e(k))\mathbf{u}(k) \tag{4.105}$$

where

$$f(e(k)) = \sum_{l=1}^{m} 2l\lambda_l e(k)^{2l-1} \tag{4.106}$$

Here the desired signal $d(k)$, the unknown weight vector $\mathbf{w}_o$, the disturbance noise $v(k)$ with variance σ_v^2, and the weighted error vector $\tilde{\mathbf{w}}(k-1) = \mathbf{w}_0 - \mathbf{w}(k-1)$ at iteration $k-1$ are also the same as the definition given above, and the relationship of them follows that

$$e(k) = \tilde{\mathbf{w}}(k-1)^T \mathbf{u}(k) + v(k) = e_a(k) + v(k) \tag{4.107}$$

where $e_a(k) = \tilde{\mathbf{w}}(k-1)^T \mathbf{u}(k)$ is referred to as the *a priori* error.

First, the mean convergence performance of the SLMP is performed by using the similar method used in [307]. According to [307], one can immediately obtain the following results:

Theorem 4.1

The SLMP algorithm is convergent in mean if the step size η satisfies

$$0 < \eta < \frac{2}{f(\lambda_{\max}) + \frac{1}{2} f^{(2)}(\lambda_{\max})\sigma_v^2} \tag{4.108}$$

where $f^{(2)}$ denotes the second-order derivative of the function $f(x) = \sum_{l=1}^{m} 2l\lambda_l x^{2l-1}$, and $\lambda_{\max}$ is the largest eigenvalue of the input correlation matrix $\mathbf{R}_{\mathbf{uu}} = E[\mathbf{u}(k)\mathbf{u}(k)^T]$. Further, if one unit of time corresponds to one iteration, the time constant corresponding to the ith eigenvalue of $\mathbf{R}_{\mathbf{uu}}$ is

$$\tau_i \approx \frac{1}{\eta(f(\lambda_i) + \frac{1}{2} f^{(2)}(\lambda_i)\sigma_v^2)} \tag{4.109}$$

Next, the mean square convergence performance is further analyzed. According to the energy conservation relation [261,272,302], the following equality holds:

$$E[\|\tilde{\mathbf{w}}(k)\|^2] = E[\|\tilde{\mathbf{w}}(k-1)\|^2] - 2\eta E[e_a(k) f(e(k))] + \eta^2 E[\|\mathbf{u}(k)\|^2 f^2(e(k))] \tag{4.110}$$

Therefore, if the step size η is chosen such that for all k

$$\eta \le \frac{2E[e_a(k) f(e(k))]}{E[\|\mathbf{u}(k)\|^2 f^2(e(k))]} = \frac{2E[\sum_{l=1}^{m} 2l\lambda_l e_a(k) e(k)^{2l-1}]}{E[\|\mathbf{u}(k)\|^2 (\sum_{l=1}^{m} 2l\lambda_l e(k)^{2l-1})^2]} \tag{4.111}$$

then the sequence of WEP $E[\|\tilde{\mathbf{w}}(k)\|^2]$ will be decreasing and converging. One can further analyze the mean square transient behaviors of the algorithm, but this is not an easy task since we have to evaluate the expectations $E[\sum_{l=1}^{m} 2l\lambda_l e_a(k)e(k)^{2l-1}]$ and $E\left[\|\mathbf{u}(k)\|^2\left(\sum_{l=1}^{m} 2l\lambda_l e(k)^{2l-1}\right)^2\right]$. In the following, we present only some analysis results about the steady-state mean square performance. The analysis method is based on the Taylor expansion method [212,285]. When the filter reaches the steady-state, the distributions of $e_a(k)$ and $e(k)$ are independent of k, and one can simply write e_a and e by omitting the time index. The following lemma holds [272]:

Lemma 4.1

If the filter is stable and reaches the steady-state, then

$$2E[e_a f(e)] = \eta Tr(\mathbf{R}_{\mathbf{uu}})E[f^2(e)] \tag{4.112}$$

provided that $\|\mathbf{u}(k)\|^2$ is asymptotically uncorrelated with $f^2(e(k))$.

Remark 4.10

The rationality of the assumption that $\|\mathbf{u}(k)\|^2$ and $f^2(e(k))$ are asymptotically uncorrelated has been discussed in [272].

Let $S = \lim_{k\to\infty} E[e_a^2(k)] = E[e_a^2]$ be the steady-state EMSE. Below we will derive an approximate analytical expression of S. First, we give two assumptions:

Assumption 4.10: The noise $v(k)$ is zero-mean, independent, identically distributed, and is independent of the input.

Assumption 4.11: The *a priori* error $e_a(k)$ is zero-mean and independent of the noise.

Taking the Taylor expansion of $f(e)$ with respect to e_a around v, we have

$$f(e) = f(e_a + v) = f(v) + f'(v)e_a + \frac{1}{2}f''(v)e_a^2 + o(e_a^2) \tag{4.113}$$

where $o(e_a^2)$ denotes the third and higher-order terms, and

$$\begin{aligned} f'(v) &= \sum_{l=1}^{m} 2l(2l-1)\lambda_l v^{2l-2} \\ f''(v) &= \sum_{l=2}^{m} 2l(2l-1)(2l-2)\lambda_l v^{2l-3} \end{aligned} \tag{4.114}$$

If $E[o(e_a^2)]$ is small enough, then based on Assumptions 4.10 and 4.11, we have

$$E[e_a f(e)] = E[e_a f(v) + f'(v)e_a^2 + o(e_a^2)] \approx E[f'(v)]S \tag{4.115}$$

$$E\left[f^2(e)\right] \approx E\left[f^2(v)\right] + E\left[f(v)f''(v) + \left|f'(v)\right|^2\right]S \tag{4.116}$$

Substituting (4.115) and (4.116) into (4.112), we obtain

$$S = \frac{\eta \mathrm{Tr}(\mathbf{R_{uu}})E\left[f^2(v)\right]}{2E\left[f'(v)\right] - \eta \mathrm{Tr}(\mathbf{R_{uu}})E\left[f(v)f''(v) + \left|f'(v)\right|^2\right]} \tag{4.117}$$

Further, substituting (4.106) and (4.114) into (4.117) yields

$$S = \left(\eta \mathrm{Tr}(\mathbf{R_{uu}})E\left[\left(\sum_{l=1}^{k} 2l\lambda_l v^{2l-1}\right)^2\right]\right) \Bigg/ \left(2E\left[\sum_{l=1}^{k} 2l(2l-1)\lambda_l v^{2l-2}\right] - \eta \mathrm{Tr}(\mathbf{R_{uu}})E\left[\left(\sum_{l=1}^{k} 2l\lambda_l v^{2l-1}\right)\right.\right.$$
$$\left.\left.\times\left(\sum_{l=2}^{k} 2l(2l-1)(2l-2)\lambda_l v^{2l-3}\right) + \left(\sum_{l=1}^{k} 2l(2l-1)\lambda_l v^{2l-2}\right)^2\right]\right) \tag{4.118}$$

When the step size is small enough, (4.118) can be simplified to

$$S = \frac{\eta \mathrm{Tr}(\mathbf{R_{uu}})E\left[\left(\sum_{l=1}^{m} 2l\lambda_l v^{2l-1}\right)^2\right]}{2E\left[\sum_{l=1}^{m} 2l(2l-1)\lambda_l v^{2l-2}\right]} \tag{4.119}$$

Remark 4.11

Given a distribution of the noise v, one can evaluate the expectations in (4.118) and obtain a theoretical value of the steady-state EMSE. We should emphasize that the steady-state EMSE of (4.118) is derived under the assumption that the steady-state *a priori* error e_a is small such that its third and higher-order terms are negligible. When the step size or noise power is too large, the *a priori* error will also be large. In this situation, the derived EMSE value will not characterize the performance accurately enough.

4.4.3 Simulation Results

In this subsection, some simulation results are presented to confirm the theoretical analyses and demonstrate the performance of the SLMP algorithm.

4.4.3.1 Theoretical and Simulated Steady-State Performance

Here the filter length is 20, and the input signal is a zero-mean white Gaussian process with unit variance. The disturbance noise is assumed to be zero-mean and uniform distributed, and the smoothing factor is set to 1.5. The steady-state EMSEs with different noise variances and step sizes are plotted in Figure 4.14. One can observe that: (1) the steady-state EMSEs are increasing with both noise variance and step size; (2) when the noise variance or step size is small, the steady-state EMSEs computed by simulations match very well the theoretical values computed by (4.99); (3) when the noise variance or step size becomes larger, however, the experimental results may gradually differ from the theoretical values, and this confirms the theoretical prediction.

4.4.3.2 Stability Problems

Here some simulation results are given to test the probability of divergence of the SLMP algorithm. The input signal and the disturbance noise are both zero-mean and unit-power Gaussian. The probabilities of divergence for different smoothing factors are illustrated in Figure 4.15. For comparison

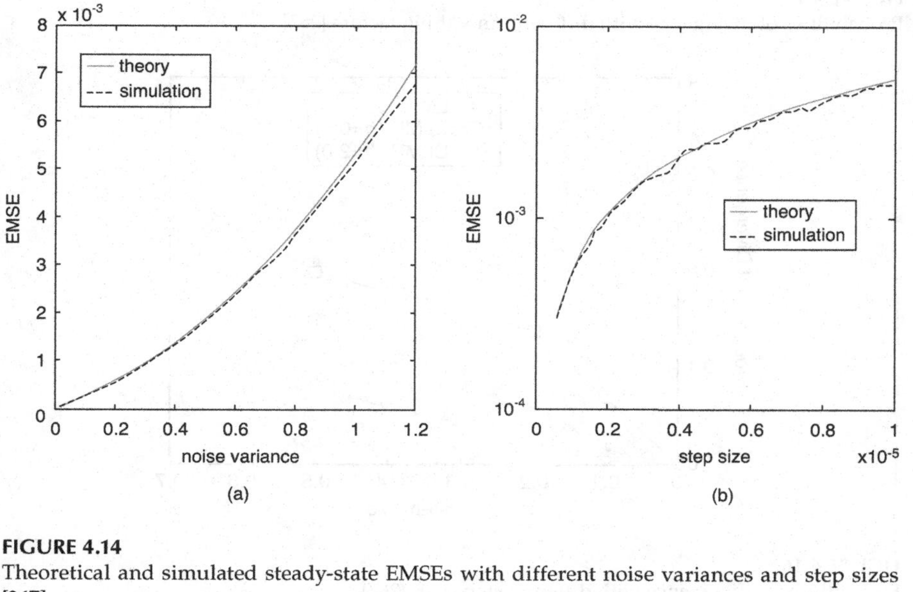

FIGURE 4.14
Theoretical and simulated steady-state EMSEs with different noise variances and step sizes [267].

purpose, the probability of divergence of the LMF algorithm is also plotted in Figure 4.15. One can observe in Figure 4.15 that: (1) when the smoothing factor is very small, the probability of divergence of the SLMP can be larger than that of the LMF; (2) when the smoothing factor becomes larger, the probability of divergence of the SLMP can be much smaller than that of the LMF and will approach to zero as the smoothing factor becomes very large. The probabilities of divergence for different step sizes are given in Figure 4.16. Obviously, a smaller step size leads to a more stable algorithm.

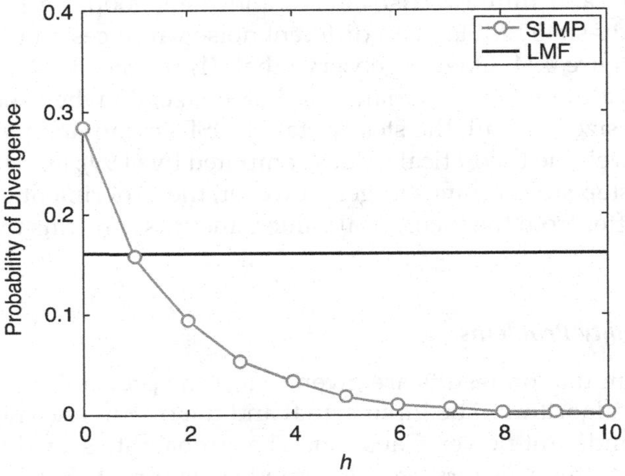

FIGURE 4.15
Probabilities of divergence with different smoothing factors [267].

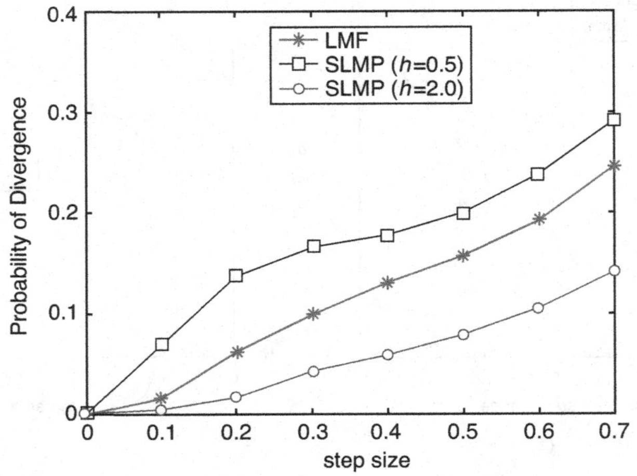

FIGURE 4.16
Probabilities of divergence with different step sizes [267].

4.4.3.3 Performance Comparison

In this case, the normalized versions of the SLMP, LMP, and LMMN (denoted by NSLMP, NLMP, and NLMMN) are used, since in general, the normalized algorithms are more robust and faster when compared with the original versions. Here, four noise distributions are considered: Binary, Gaussian, MixGaussian, and Laplace. The convergence curves are shown in Figure 4.17. The smoothing factor in SLMP is experimentally chosen such that the algorithm achieves a desirable performance. From the simulation results we observe that: (1) when the noise is Binary, the NSLMP (with a smaller h) and the NLMP with $p=6$ may achieve a much lower misadjustment than others; (2) when the noise is Gaussian, the NSLMP (with a larger h) can perform almost as well as the NLMS $p=6$ and NLMMN; (3) when the noise is MixGaussian, the NSLMP may perform best; (4) when the noise is Laplacian, the NLMP with $p \geq 4$ performs very poorly, particularly the NLMP with $p=6$ will diverge and its convergence curve cannot be plotted, but in this case the NSLMP, NLMMN, and NLMP with $p=1$ or $p=2$ can still work well.

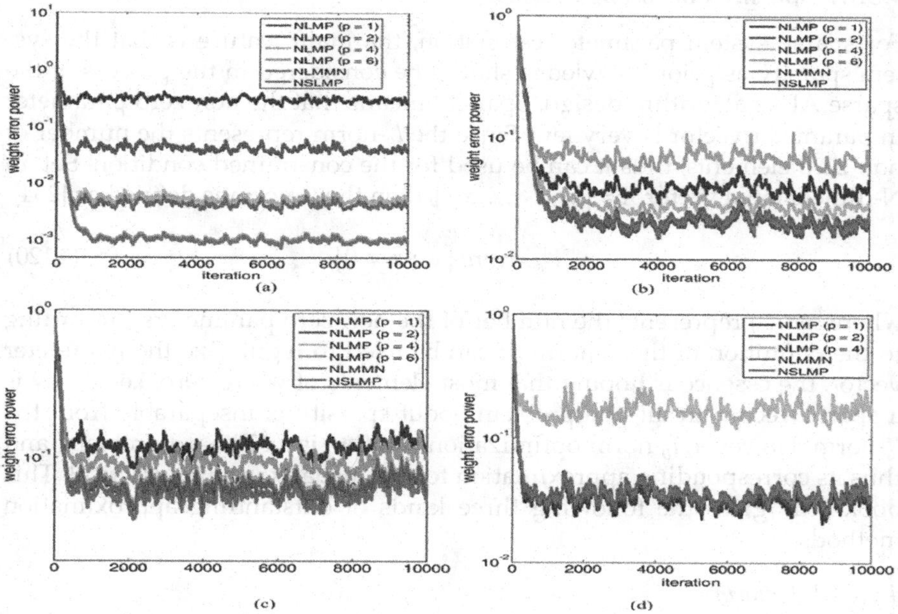

FIGURE 4.17
Convergence curves in different noises: (a) Binary; (b) Gaussian; (c) MixGaussian; (d) Laplace [267].

4.5 Sparsity-Aware Adaptive Filtering Algorithm under MMPE

The AFAs under the MMPE criterion are robust to outliers when $p<2$. In its standard form, however, the standard form of the MMPE does not take into account any special characteristics of the system model, such as in sparse channel estimation which may have a sparse impulsive (or frequency) response. In this subchapter, some sparsity-aware LMP algorithms will be presented [234,235]. In detail, several sparse LMP algorithms based on different sparsity penalty terms (l_1-norm, reweighted l_1-norm, and correntropy induced metric (CIM)) are given to address the SCPE in impulsive noise environments. The corresponding algorithms are denoted as ZALMP, RZALMP, and CIMLMP, respectively. In addition, the proportionate least mean p-power (PLMP) algorithm is further presented, which can achieve much better performance in terms of the MSD.

4.5.1 LMP Algorithm with Sparsity Constrained Term

4.5.1.1 Sparsity Constrained Terms

For sparse system parameter estimation, the main feature is that the system sparsity as prior knowledge should be considered in the process of the sparse AFAs algorithm design. Sparsity means that the non-zero parameter in parameter vector is very few. Since the l_0-norm represents the number of non-zero elements, thus it can be used for the constrained condition. Set an N-dimensional vector $X=[x_1,x_2,\ldots,x_N]$, then the l_0-norm is defined as [234]

$$\|X\|_0 = card\{x_i : x_i \neq 0\} \tag{4.120}$$

where $card(\cdot)$ represents the number of the non-zero parameters. According to the definition of the l_0-norm, it can be used to regularize the parameter vector, the essence is hoping that most elements of $\mathbf{w}$ are zero, i.e., the $\mathbf{w}$ is a sparse vector. In all, the problem about sparsity is inseparable from the l_0-norm. However, l_0-norm optimization solution is a NP-hard problem, and thus its corresponding approximation form is usually used in practical. This book only gives the following three kinds of outstanding approximation methods.

4.5.1.1.1 l_1-norm

l_1-norm is the sum of absolute values of the given vector, i.e., "Sparse regularization operator", which is defined by

$$\|X\|_1 = \sum_{i=1}^{N} |x_i| \tag{4.121}$$

l_1-norm as an optimal convex approximation is easily optimization and solving solution compared with the l_0-norm. Moreover, the l_0-norm will be equivalent to l_1-norm when it meets certain conditions, and thus the l_1-norm has been used in sparsity studies.

4.5.1.1.2 Reweighted l1-norm

As another important approximation of the l_0-norm, the reweighted l_1-norm is defined as:

$$\|X\|_0 \sim \sum_{i=1}^{N} \log\left(1+\frac{|x_i|}{\varepsilon'}\right) \tag{4.122}$$

where ε' is a positive constant. The reweighted l_1-norm has excellent approximation performance for l_0-norm and is better than l_1-norm. Let X_0 is the optimal solution obtained by minimizing l_0-norm, and the X_l is the solution from the (4.103) minimization, then the following result can be obtained [308]:

$$\|X_l\|_0 < \|X_0\|_0 + O\left(\frac{1}{\text{In } \varepsilon'}\right) \tag{4.123}$$

The (4.123) means that the formula (4.122) can commendably approach l_0-norm when the absolute of the non-zero parameter in vector X_l is less than a small positive number, i.e. ε' infinitely approaches to zero.

4.5.1.1.3 Correntropy Induced Metric

In practice, the data distribution is usually unknown, and only a finite number of samples $\{x_i, y_i\}$ are available. Therefore, the correntropy defined in [202] can be estimated by

$$\hat{V}(\mathbf{x},\mathbf{y}) = \frac{1}{N}\sum_{i=1}^{N} \kappa(x_i, y_i) \tag{4.124}$$

Given two vectors in the sample space: $\mathbf{x} = [x_1, \ldots, x_N]^T$, and $\mathbf{y} = [y_1, \ldots, y_N]^T$, the CIM in the sample space is

$$\text{CIM}(\mathbf{x},\mathbf{y}) = \left(\kappa(0) - \hat{V}(\mathbf{x},\mathbf{y})\right)^{1/2} \tag{4.125}$$

where the kernel is assumed to be a Gaussian kernel with $\kappa(0) = 1/\sigma\sqrt{2\pi}$. The CIM provides a nice approximation for the l_0-norm [233]. Given a vector $\mathbf{x} = [x_1, \ldots, x_N]^T$, the l_0-norm can be approximated by

$$\|\mathbf{x}\|_0 \sim \mathrm{CIM}^2(\mathbf{x},0) = \frac{\kappa(0)}{N}\sum_{i=1}^{N}\left(1-\exp\left(-\frac{x_i^2}{2\sigma^2}\right)\right) \tag{4.126}$$

For the Gaussian kernel, it has been observed that CIM behaves like an l_2-norm when the two vectors are close, like an l_1- norm outside the l_2- norm zone, and like an l_0-norm as they go farther apart [233]. It can be shown that if $|x_i| > \delta, \forall x_i \neq 0$, then as $\sigma \to 0$, one can get a solution arbitrarily close to that of the l_0 -norm, where δ is a small positive number. The extent of the space over which the CIM acts as l_2 norm or l_0 norm is directly related to the kernel width σ. As an approximation of the l_0-norm, the CIM favors sparsity and can be used as a sparsity penalty term (see Appendix F).

Remark 4.12

Laplace function is another approximation of the l_0-norm, and it is defined as

$$\|X\|_0 \approx \sum_i 1-\exp(-\beta|x_i|)$$

One can see that it is very similar to CIM from the form, but the difference also exists between them. First, CIM is a smooth function while the Laplace function is not smooth at 0 points. Second, by selecting different kernel width, The CIM can approach to the norm from l_0 to l_2, while the Laplace function only approaches l_0-norm to l_1-norm. Furthermore, it is worth noting that the definition of CIM does not restrict the kernel to Gaussian kernel, if the Laplacian kernel is used, i.e., $\kappa(x,y) = \frac{\beta}{2}\exp(-\beta|x-y|)$, then the CIM will be identical to the Laplacian function.

4.5.1.2 Sparsity-Aware LMP Algorithms

In this section, a sparse channel model as (4.127) is considered to introduce the sparsity-aware LMP algorithms. The input vector $\mathbf{u}(k) = [u_k, u_{k-1}, \ldots u_{k-L+1})]^T$ is sent over the FIR channel with parameter vector $\mathbf{w}_o = [w_{o,1}, w_{o,2}, \ldots, w_{o,L}]^T$, where L is the size of the channel memory. It is assumed that the channel parameters are real-valued, and most of them are zero. The received signal $d(k)$ is modeled as:

$$d(k) = \mathbf{w}_o^T\mathbf{u}(k) + v(k) \tag{4.127}$$

where $v(k)$ denotes the interference noise at the end of receiver. In many practical situations, the noise $v(k)$ is non-Gaussian due to the impulsive nature of man-made electromagnetic interference as well as nature noises. To robustly estimate such sparse system in impulsive noise environments, we will develop the following sparse LMP algorithms to address this problem.

4.5.1.2.1 *Sparse LMP with Zero-attracting (l_1-norm) Penalty Term*

To develop a sparse LMP algorithm with zero-attracting (l_1-norm) penalty term, we introduce the cost function as [234]

$$\begin{aligned} J_{\mathrm{ZALMP}}(k) &= J_{\mathrm{LMP}}(k) + \lambda J_{\mathrm{ZA}}(k) \\ &= |e(k)|^p + \lambda \|\mathbf{w}(k)\|_1 \end{aligned} \tag{4.128}$$

where $J_{\mathrm{ZA}}(k) = \|\mathbf{w}(k)\|_1$ denotes the l_1-norm of the estimated parameter vector. In (4.128), the LMP term is robust to impulsive noise, and the ZA penalty term is a sparsity-inducing term, and the two terms are balanced by a weight factor $\lambda \geq 0$. Based on the cost function (4.128), the following algorithm is derived:

$$\begin{aligned} \mathbf{w}(k+1) &= \mathbf{w}(k) - \eta \frac{\partial J_{\mathrm{ZALMP}}(k)}{\partial \mathbf{w}(k)} \\ &= \mathbf{w}(k) - \eta\left[-p|e(k)|^{p-1}\operatorname{sign}(e(k))\mathbf{u}(k) + \lambda \operatorname{sign}(\mathbf{w}(k))\right] \\ &= \mathbf{w}(k) + \mu|e(k)|^{p-1}\operatorname{sign}(e(k))\mathbf{u}(k) - \rho \operatorname{sign}(\mathbf{w}(k)) \end{aligned} \tag{4.129}$$

where $\mu = \eta p$ is the step size, $\rho = \eta\lambda$ is the zero-attractor control factor. This algorithm is referred as the ZALMP algorithm. Clearly, the ZALMP algorithm will reduce to the ZALMS algorithm when $p=2$.

4.5.1.2.2 *Sparse LMP with Logarithmic Penalty Term*

Here the following cost function is further defined to derive sparse LMP algorithm with logarithmic penalty term [234]:

$$\begin{aligned} J_{\mathrm{RZALMP}}(k) &= J_{\mathrm{LMP}}(k) + \gamma J_{\mathrm{RZA}}(k) \\ &= |e(k)|^p + \lambda \sum_{i=1}^{L} \log\left(1 + |w_i(k)/\delta|\right) \end{aligned} \tag{4.130}$$

where the log-sum penalty $\sum_{i=1}^{L}\log\left(1+|w_i|/\delta\right)$ is introduced as it behaves more similarly to the l_0-norm than $\|\mathbf{w}\|_1$, δ is a positive number. Then a gradient-based adaptive algorithm can be easily derived as

$$\begin{aligned} w_i(k+1) &= w_i(k)-\eta\frac{\partial J_{\text{ZALMP}}(k)}{\partial w_i(k)} \\ &= w_i(k)-\eta\left[-p|e(k)|^{p-1}\operatorname{sign}(e(k))u_i(k)+\lambda\frac{\operatorname{sign}(w_i(k))}{1+\delta'|w_i(k)|}\right] \\ &= w_i(k)+\mu|e(k)|^{p-1}\operatorname{sign}(e(k))u_i(k)-\rho\frac{\operatorname{sign}(w_i(k))}{1+\delta'|w_i(k)|} \end{aligned} \tag{4.131}$$

Or equivalently, in vector form

$$\mathbf{w}(k+1)=\mathbf{w}(k)+\mu|e(k)|^{p-1}\operatorname{sign}(e(k))\mathbf{u}(k)-\rho\frac{\operatorname{sign}(\mathbf{w}(k))}{1+\delta'|\mathbf{w}(k)|} \tag{4.132}$$

where μ and ρ are the same as in (4.129), and $\delta'=1/\delta$. This algorithm is referred to as the RZALMP algorithm. Obviously, the RZALMP will become the RZALMS algorithm when $p=2$.

4.5.1.2.3 Sparse LMP with CIM Penalty Term

Now we derive a sparse LMP algorithm based on CIM by using the following cost function [234]:

$$\begin{aligned} J_{\text{CIMLMP}}(k) &= J_{\text{LMP}}(k)+\lambda J_{\text{CIM}}(k) \\ &= |e(k)|^p+\lambda\frac{1}{L\sigma\sqrt{2\pi}}\sum_{i=1}^{L}\left(1-\exp\left(-w_i(k)^2/2\sigma^2\right)\right) \end{aligned} \tag{4.133}$$

Where σ denotes the kernel width in CIM. The first term in (4.133) is robust to impulsive noises, and the CIM in the second term with a smaller kernel width is a sparsity-inducing term.

With the cost function (4.133), a gradient-based adaptive algorithm can be derived as

$$
\begin{aligned}
w_i(k+1) &= w_i(k) - \eta \frac{\partial J_{\text{CIMLMP}}(k)}{\partial w_i(k)} \\
&= w_i(k) - \eta\left[-p\left|e(k)\right|^{p-1}\operatorname{sign}(e(k))u_i(k) + \lambda\frac{1}{L\sigma^3\sqrt{2\pi}}w_i(k)\exp\left(-\frac{w_i^2(k)}{2\sigma^2}\right)\right] \\
&= w_i(k) - \eta\left[-p\left|e(k)\right|^{p-1}\operatorname{sign}(e(k))u_i(k) + \lambda\frac{1}{L\sigma^3\sqrt{2\pi}}w_i(k)\exp\left(-\frac{w_i^2(k)}{2\sigma^2}\right)\right] \\
&= w_i(k) + \mu\left|e(k)\right|^{p-1}\operatorname{sign}(e(k))u_i(k) - \rho\frac{1}{L\sigma^3\sqrt{2\pi}}w_i(k)\exp\left(-\frac{w_i^2(k)}{2\sigma^2}\right)
\end{aligned}
\tag{4.134}
$$

or equivalently, in vector form

$$
\mathbf{w}(k+1) = \mathbf{w}(k) + \mu\left|e(k)\right|^{p-1}\operatorname{sign}(e(k))\mathbf{u}(k) - \rho\frac{1}{L\sigma^3\sqrt{2\pi}}\mathbf{w}(k).*\exp\left(-\frac{\mathbf{w}^2(k)}{2\sigma^2}\right)
\tag{4.135}
$$

The above algorithm is referred as the CIMLMP algorithm. The kernel width is a free parameter in the penalty term. A proper kernel width will make CIM as a good approximator for the l_0-norm [233]. As an approximation of the l_0-norm, the CIM favors sparsity and can improve the steady-state performance.

4.5.1.3 Convergence Analysis

To perform the mean and the mean square convergence analysis of the sparse LMP algorithms by using energy conservation through an approximation approach, we give a unifying form of the sparsity-aware algorithms presented above as [234]

$$
\mathbf{w}(k+1) = \mathbf{w}(k) + \mu f(e(k))e(k)\mathbf{u}(k) + \rho g(\mathbf{w}(k))
\tag{4.136}
$$

where $f(e(k)) = \left|e(k)\right|^{p-2}$, and the vector $g(\mathbf{w}(k)) = \left[g(w_1(k)), g(w_2(k)), \cdots, g(w_L(k))\right]$, where $g(x)$ are $\operatorname{sign}(x)$, $-\operatorname{sign}(x)/(1+\delta'|x|)$ and $-1\left(L\sigma^3\sqrt{2\pi}\right)x\exp\left(-x^2/2\sigma^2\right)$ for, respectively, ZALMP, RZALMP, and CIMLMP. To simplify the analysis, we denote $g(\mathbf{w}(k)) = g(k)$, and the following statistical assumptions are given:

Assumption 4.12: The input signal $\{\mathbf{u}(k)\}$ is independent and identically distributed (i.i.d.) with zero-mean Gaussian distribution.

Assumption 4.13: The noise signal $\{v(k)\}$ is i.i.d. with zero-mean and variance σ_v^2, and is independent of $\{\mathbf{u}(k)\}$.

Assumption 4.14: The error nonlinearity $f(e(k)) = |e(k)|^{p-2}$ is independent of the input signal $\mathbf{u}(k)$ and $\mathbf{w}(k)$.

Assumption 4.15: The $\mathbf{w}(k)$ and $g(k)$ are independent of the $\mathbf{u}(k)$.

Assumption 4.16: The expectation $E[f(e(\infty))]$ is upper bounded.

Remark 4.13

Assumptions 4.12 and 4.13 are commonly used in the Refs. [70,309–311]. Assumption 4.14 is valid when the weight vector $\mathbf{w}(k)$ lies in the neighborhood of the optimal solution.

4.5.1.3.1 Mean Performance

Here we define the filter misalignment vector $\tilde{\mathbf{w}}(k) = \mathbf{w}_o - \mathbf{w}(k)$, and then the mean and auto-correlation matrix of $\tilde{\mathbf{w}}(k)$ are denoted by

$$\delta(k) = E[\tilde{\mathbf{w}}(k)] \tag{4.137}$$

$$S(k) = E[\Delta(k)\Delta^T(k)] \tag{4.138}$$

where $\Delta(k)$ is defined as

$$\Delta(k) = \tilde{\mathbf{w}}(k) - \delta(k) = \tilde{\mathbf{w}}(k) - E[\tilde{\mathbf{w}}(k)] \tag{4.139}$$

Combining (4.127) and (4.136), we obtain

$$\begin{aligned}
\tilde{\mathbf{w}}(k+1) &= \tilde{\mathbf{w}}(k) - \mu f(e(k))e(k)\mathbf{u}(k) - \rho g(k) \\
&= \tilde{\mathbf{w}}(k) - \mu f(e(k))[d(k) - \mathbf{w}^T(k)\mathbf{u}(k)]\mathbf{u}(k) - \rho g(k) \\
&= \tilde{\mathbf{w}}(k) - \mu f(e(k))[\mathbf{w}_o^T(k)\mathbf{u}(k) + v(k) - \mathbf{w}^T(k)\mathbf{u}(k)]\mathbf{u}(k) - \rho g(k) \\
&= \tilde{\mathbf{w}}(k) - \mu f(e(k))\mathbf{u}(k)\mathbf{u}^T(k)\tilde{\mathbf{w}}(k) - \mu f(e(k))v(k)\mathbf{u}(k) - \rho g(k) \\
&= [\mathbf{I} - \mu f(e(k))\mathbf{u}(k)\mathbf{u}^T(k)\tilde{\mathbf{w}}(k)]\tilde{\mathbf{w}}(k) - \mu f(e(k))v(k)\mathbf{u}(k) - \rho g(k) \\
&= \mathbf{A}(k)\tilde{\mathbf{w}}(k) - \mu f(e(k))v(k)\mathbf{u}(k) - \rho g(k)
\end{aligned} \tag{4.140}$$

where $\mathbf{A}(k) = \mathbf{I} - \mu f(e(k))\mathbf{u}(k)\mathbf{u}^T(k)\tilde{\mathbf{w}}(k)$.

Taking the expectation of (4.140) and using the independence Assumptions 4.12–4.14, we get

$$\delta(k+1) = E\left[\tilde{\mathbf{w}}(k+1)\right] = \left[1-\mu E\left[f\left(e(k)\right)\right]\sigma_u^2\right]\delta(k) - \rho E\left[g(k)\right] \quad (4.141)$$

Where σ_u^2 denotes the variance of $\mathbf{u}(k)$.
From (4.141), one can easily obtain

$$\delta(\infty) = E\left[\tilde{\mathbf{w}}(\infty)\right] = -\frac{\rho}{\mu E\left[f\left(e(\infty)\right)\right]\sigma_u^2}E\left[g(\infty)\right] \quad (4.142)$$

Combining (4.137) and (4.142), one can get

$$E\left[\mathbf{w}(\infty)\right] = \mathbf{w}_o = -\frac{\rho}{\mu E\left[f\left(e(\infty)\right)\right]\sigma_u^2}E\left[g(\infty)\right] \quad (4.143)$$

From the definition of $g(k)$ for ZALMP and RZALMP, we note that the vector $E\left[g(\infty)\right]$ is bounded between **−1** and **1**, and similar demonstration can be read in [312]. For the CIMLMP, we show that $E\left[g(\infty)\right]$ is still limited (See Appendix G). Then under Assumption 4.16, $E\left[\mathbf{w}(\infty)\right]$ is bounded, and $E\left[\mathbf{w}(k)\right]$ will converge to a limiting vector as shown in (4.143).

4.5.1.3.2 Mean Square Performance

Subtracting (4.141) from (4.140) and applying (4.139) yields

$$\begin{aligned}
\Delta(k+1) &= [\mathbf{I} - \mu f(e(k))\mathbf{u}(k)\mathbf{u}^T(k)]\tilde{\mathbf{w}}(k) - \mu f(e(k))v(k)\mathbf{u}(k) - \rho g(k) \\
&\quad -\left\{[1-\mu E[f(e(k)]\sigma_{\mathbf{u}}^2]\delta(k) - \rho E[g(k)]\right\} \\
&= [\mathbf{I} - \mu f(e(k))\tilde{\mathbf{w}}(k)]\Delta(k) + \mu[E[f(e(k)]\sigma_{\mathbf{u}}^2 \\
&\quad - f(e(k))\mathbf{u}(k)\mathbf{u}^T(k)]\delta(k) \\
&\quad - \mu f(e(k))v(k)\mathbf{u}(k) - \rho(g(k) - E[g(k)]) \\
&= \mathbf{A}(k)\Delta(k) + \mu B(k)\delta(k) - \mu f(e(k))v(k)\mathbf{u}(k) - \rho C(k)
\end{aligned} \quad (4.145)$$

where $B(k) = E[f(e(k)]\sigma_{\mathbf{u}}^2 - f(e(k))\mathbf{u}(k)\mathbf{u}^T(k), C(k) = g(k) - E\left[g(k)\right]$. Under Assumptions 4.12, 4.13, 4.14, and 4.15, it is straightforward to verify that $B(k)$ and $C(k)$ are zero-mean, and $\tilde{\mathbf{w}}(k)$, $\mathbf{u}(k)$ and $v(k)$ are mutually independent. So, substituting (4.145) into (4.138) and after some tedious calculations, we have

$$
\begin{aligned}
&S(k+1)\\
&= E[\Delta(k+1)\Delta^T(k+1)]\\
&= E\begin{bmatrix}(\mathrm{A}(k)\Delta(k)+\mu B(k)\delta(k)-\mu f(e(k))v(k)u(k)-\gamma C(k))\times\\ (\mathrm{A}(k)\Delta(k)+\mu B(k)\delta(k)-\mu f(e(k))v(k)u(k)-\gamma C(k))^T\end{bmatrix} \qquad (4.146)\\
&= E[\mathrm{A}(k)\Delta(k)\Delta^T(k)A^T(k)]-\rho E[C(k)\Delta^T(k)\mathrm{A}^T(k)]\\
&\quad +\mu^2 E[B(k)\delta(k)\delta^T(k)B^T(k)]+\mu^2 E[f^2(e(k))]\sigma_u^2\sigma_v^2\\
&\quad -\rho E[\mathrm{A}(k)\Delta(k)C^T(k)]+\rho^2 E[C(k)C^T(k)]
\end{aligned}
$$

Using the facts mentioned in [70,309–311] that the fourth-order moment of a Gaussian variable is three times the variance squared and that $S(k)$ is symmetric, and under Assumption 4.14, we get

$$
\begin{aligned}
&E[\mathbf{A}(k)\Delta(k)\Delta(k)\mathbf{A}^T(n)]\\
&= (1-2\mu E[f(e(k))]\sigma_{\mathbf{u}}^2+2\mu^2 E[f^2(e(k))]\sigma_{\mathbf{u}}^4)S(k) \qquad (4.147)\\
&\quad +\mu^2 E[f^2(e(k))]\sigma_{\mathbf{u}}^4 \mathrm{Tr}[S(k)]\mathbf{I}
\end{aligned}
$$

$$
\begin{aligned}
&E[B(k)\delta(k)\delta^T(k)B^T(k)]\\
&= \mu^2 E[f^2(e(k))]\sigma_{\mathbf{u}}^4(\varepsilon(k)\varepsilon^T(k)+\mathrm{Tr}[\varepsilon(k)\varepsilon^T(k)]\mathbf{I}) \qquad (4.148)
\end{aligned}
$$

Also using (4.139) and the definition of $C(k)$, we obtain

$$
\begin{aligned}
&E[\mathbf{A}(k)\Delta(k)C^T(k)]\\
&= E^T[C(k)\Delta^T(k)\mathbf{A}^T(k)] \qquad (4.149)\\
&= (1-\mu E[f(k)]\sigma_{\mathbf{u}}^2)E[\mathbf{w}(k)C^T(k)]
\end{aligned}
$$

Using (4.147), (4.148) and (4.149), one can derive

$$
\begin{aligned}
&S(k+1)\\
&= (1-2\mu E[f(e(k))]\sigma_{\mathbf{u}}^2+2\mu^2 E[f^2(e(k))]\sigma_{\mathbf{u}}^4)\mathrm{S}(k)\\
&\quad +\mu^2 E[f^2(e(k))]\sigma_{\mathbf{u}}^4\mathrm{Tr}[S(k)]\mathbf{I}\\
&\quad -2\rho(1-\mu E[f(e(k))]\sigma_{\mathbf{u}}^2)E[\mathbf{w}(k)C^T(k)] \qquad (4.150)\\
&\quad +\mu^2 E[f^2(e(k))]\sigma_{\mathbf{u}}^4(\delta(k)\delta^T(k)+\mathrm{Tr}[\delta(k)\delta^T(k)]\mathbf{I}\\
&\quad +\mu^2 E[f^2(e(k))]\sigma_{\mathbf{u}}^2\sigma_v^2+\rho^2 E[C(k)C^T(k)]
\end{aligned}
$$

Further, the trace of (4.150) is written as

$$\begin{aligned}
&\mathrm{Tr}[S(k+1)]\\
&= (1-2\mu E[f(e(k))]\sigma_u^2\\
&+(L+2)\mu^2 E[f^2(e(k))]\sigma_u^4)\mathrm{Tr}[S(k)]\\
&+(L+1)\mu^2 E[f^2(e(k))]\sigma_u^4\delta(k)\delta^T(k)\\
&+L\mu^2 E[f^2(e(k))]\sigma_v^2\sigma_u^2\\
&+\rho^2 E[C(k)C^T(k)]\\
&-2\rho(1-\mu E[f(e(k))]\sigma_u^2)E[\mathbf{w}(k)C^T(k)]
\end{aligned} \tag{4.151}$$

In equation (4.151), $E[\mathbf{w}(k)]$ converges (see in (4.143)) and $C(k)$ is bounded, and hence, $E[\mathbf{w}(k)C^T(k)]$ converges. This conclusion is the same as that in [309–311] when the sparsity penalty term $g(k)$ is chosen as the l_1-norm or logarithmic penalty term. While CIM penalty term is also bounded because the negative exponential term can reach its maximum value. Therefore, the adaptive algorithm is stable if and only if

$$|(1-2\mu E[f(e(k))]\sigma_u^2+(L+2)\mu^2 E[f^2(e(k))]\sigma_u^4)|<1 \tag{4.152}$$

Hence, the equation (4.152) simplifies to

$$0<\mu<\frac{2E[f(e(k))]}{(L+2)E[f^2(e(k))]\sigma_u^2} \tag{4.153}$$

It is worth noting that the condition of (4.153) is equal to the stability condition of the standard ZALMS when $f(e(k))=1$. This result shows that if the step size satisfies (4.153), the convergence of the sparsity-aware LMP algorithms is guaranteed.

4.5.1.4 Simulation Results

This part presents simulation results to demonstrate the performance of the ZALMP, RZALMP, and CIMLMP, compared with several other algorithms including LMP, ZALMS, and RZALMS. The time-varying channel and sparse echo cancellation cases are used to verify the performance of the sparsity-aware LMP algorithms under impulsive noise environments. The alpha-stable random variables do not have finite variance and have only finite pth-order moments for $p<\alpha$, and the FLOM of the alpha-stable random variable X is defined as

$$E\left(|X|^p\right) = C(p,\alpha)\gamma^{p/\alpha} \qquad 0 < p < \alpha \tag{4.154}$$

where $C(p,\alpha) = 2^{p+1}\Gamma\left(\frac{p+1}{2}\right)\Gamma\left(-\frac{p}{\alpha}\right)\Big/\alpha\sqrt{\pi}\Gamma\left(-\frac{p}{2}\right)$ is a constant that depends only on p and α. In addition, $\Gamma(\cdot)$ denotes the gamma function. Since the cost function of LMP is convex for $p \geq 1$ and the FLOM only exists for $0 < p < \alpha$ for alpha-stable process, we limit the discussion to the case where $1 < \alpha < 2$. Note that $e(k)$ is of approximately alpha-stable distribution when the difference between $d(k)$ and $y(k)$ is relatively small with respect to the alpha-stable noise. Hence, $E\left[f^2(e(k))\right]$ and $E\left[f(e(k))\right]$ in (4.153) can be analytically expressed by (4.154) as:

$$E\left[f(e(k))\right] = E\left(|e(k)|^{p-2}\right) = C(p-2,\alpha)\gamma^{(p-2)/\alpha} \tag{4.155}$$

$$E\left[f^2(e(k))\right] = E\left(|e(k)|^{2(p-2)}\right) = C\left(2(p-2),\alpha\right)\gamma^{2(p-2)/\alpha} \tag{4.156}$$

Here, we note that the condition of the FLOM should be satisfied in (4.155) and (4.156). Combining (4.153), (4.155), and (4.156), we obtain a conservative upper bound of the step size that ensures the convergence of the sparse LMP algorithms:

$$0 < \mu < \mu_b = \frac{2C(p-2,\alpha)}{(L+2)C\left(2(p-2),\alpha\right)\gamma^{(p-2)/\alpha}\sigma_{\mathbf{u}}^2} \tag{4.157}$$

4.5.1.4.1 Time-Varying Channel Estimation

In this case, the input signal is assumed to be a white Gaussian process with zero-mean and unit variance. The channel memory size is set to 20.

First, the convergence curves in terms of the average estimate of the MSD are given in Figure 4.18. As one can see, when the system is very sparse (before the 2000th iteration), the ZALMP, RZALMP, and CIMLMP yield faster convergence rate and better steady-state performances than the LMP algorithm, whereas ZALMS and RZALMS cannot work stably due to their sensitivities to large outliers caused by the impulsive noises. Note that the CIMLMP achieves the best performance among all the algorithms. After the 2,000th iteration, as the number of non-zero taps increases to ten, the performances of the sparse LMP algorithms (especially the ZALMP) will deteriorate, whereas CIMLMP still maintains the best performance. After the 3,000th iteration, the RZALMP and CIMLMP will perform comparably with the standard LMP even though the system is now completely non-sparse.

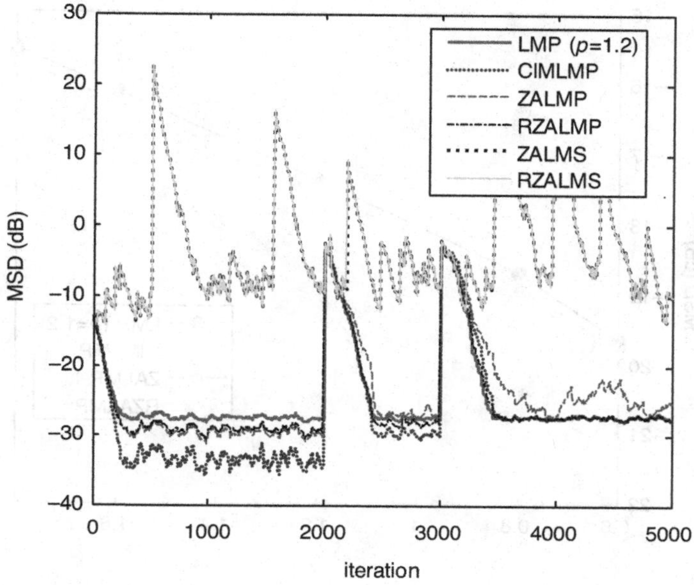

FIGURE 4.18
Tracking and steady-state behaviors [234].

Second, the robust property of the sparse LMP algorithms is tested. The dispersion parameter γ and the characteristic parameter α are the main factors for the impulsive property of the alpha-stable distribution. So, we examine how the two parameters (γ and α) affect the performance. Figure 4.19 illustrates the steady-state MSD versus different γ, where the parameter α is fixed at 1.2. Figure 4.20 shows the ssMSD versus different α, where the dispersion parameter $\gamma=0.2$. Simulation results confirm that the sparse LMP algorithms, especially the CIMLMP, may achieve a better accuracy than the standard LMP.

Third, the steady-state MSD of the CIMLMP algorithm as a function of the order p for different α is investigated. The steady-state MSD results are shown in Figure 4.21. One can observe that for different p, a more impulsive noise (smaller α) leads to a larger MSD. Further, the lowest MSD is obtained when p is close to α. In general, there should be satisfied with $0<p<\alpha$ between p and α for alpha-stable noise. Moreover, one can see that the steady-state MSD is worse when $\alpha=0.5$ and 1. This conclusion can be verified by the next simulation.

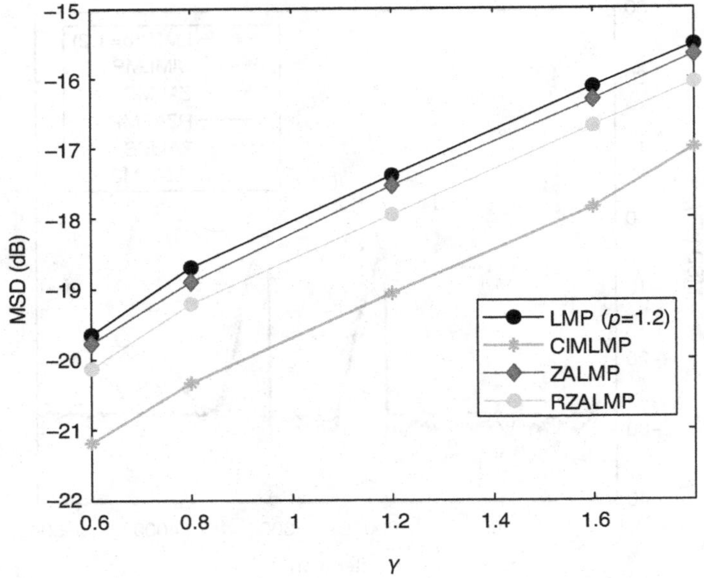

FIGURE 4.19
Steady-state MSD of the channel estimates versus γ [234].

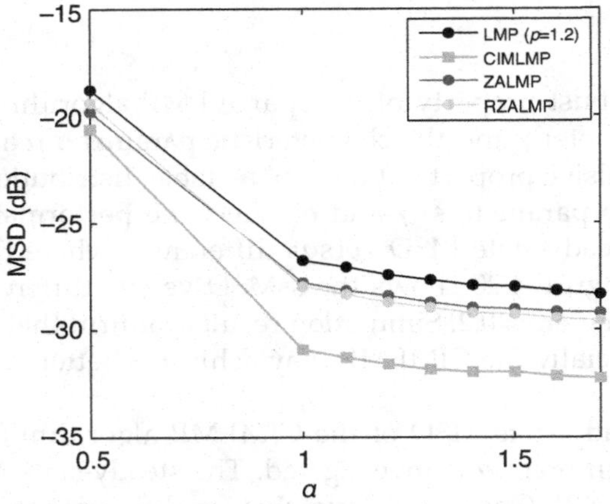

FIGURE 4.20
Steady-state MSD of the channel estimates versus α [234].

Finally, the effects of the step size on the performance of the CIMLMP algorithm is tested. As derived before, the step size μ can be chosen in the range of (4.157). Then, by (4.157) we calculate a conservative upper bound on the step size at μ_b=0.0609. The convergence curves with different step sizes are shown in Figure 4.22. One can see that the satisfactory convergence

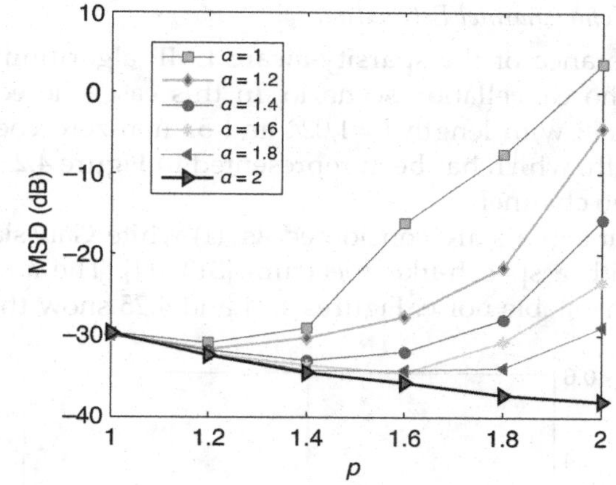

FIGURE 4.21
Steady-state MSD of CIMLMP as a function of order p for different α [234].

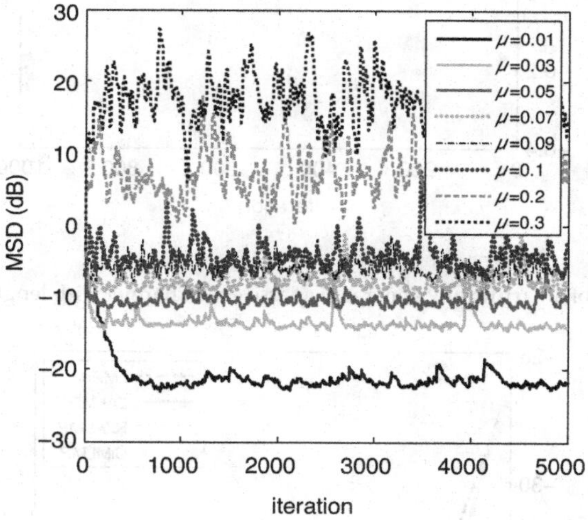

FIGURE 4.22
Convergence of the CIMLMP with different μ [234].

performance can be achieved when $\mu = 0.01 < \mu_b$. Since μ_b is a conservative upper bound, the algorithm still shows the convergence behavior when the step size is slightly larger than μ_b (e.g. $\mu = 0.07$), although the performance may become much poorer. When the step size is too large (say $\mu > 0.1 > \mu_b$), the algorithm shows no obvious convergence.

4.5.1.4.2 Sparse Echo Channel Estimation

Here the performance of the sparsity-aware LMP algorithms is evaluated under sparse echo cancellation scenario. In this case, the echo path commutes to a channel with length $L = 1{,}024$ and 52 non-zero coefficients. This artificial echo path, which has been represented in Figure 4.23, is referred to as the sparse echo channel.

First, two input signals are considered as: (1) white Gaussian process; (2) USASI signal with a speech-like spectrum [313,314]. The additive noise at output is an alpha-stable noise. Figures 4.24 and 4.25 show the convergence

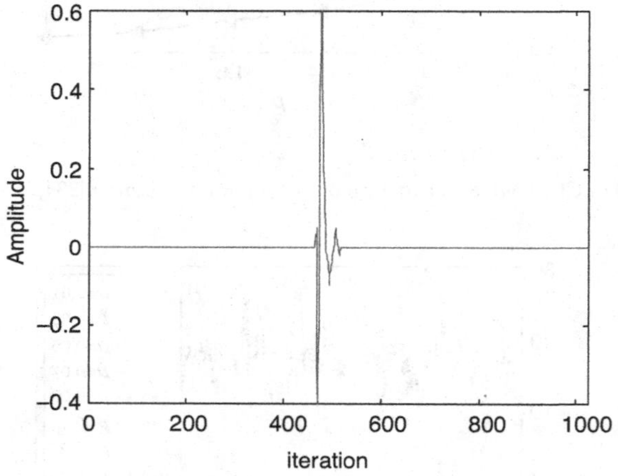

FIGURE 4.23
Impulse response of an artificially generated sparse echo channel with length $M = 1024$ [234].

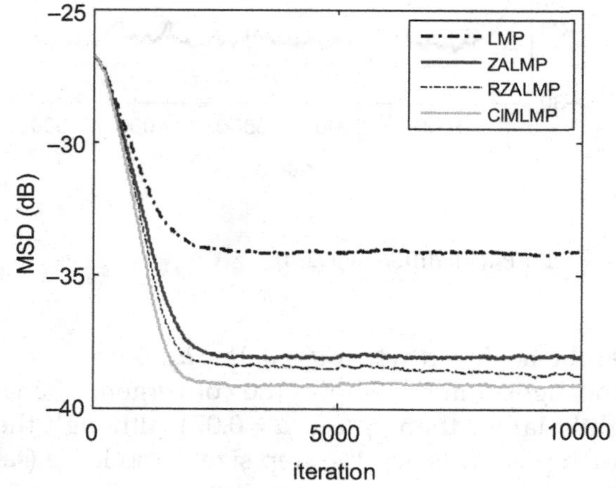

FIGURE 4.24
Convergence behaviors of sparse echo response with Gaussian input [234].

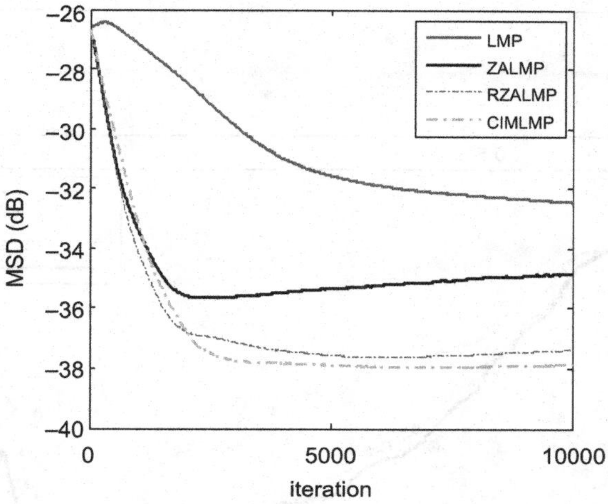

FIGURE 4.25
Convergence behaviors of sparse echo response with USASI input [234].

curves. As expected, all the sparse LMP methods can converge quickly and reach at a steady state with high accuracy, and particularly the CIMLMP method exhibits the best performance.

Second, the input signal is considered as a fragment of 2*s* of real speech, sampled at 8*k* Hz, and the strongly sparse echo paths depicted in Figure 4.26 is used, and the output noise is still the alpha-stable noise as before. The convergence behaviors of the sparsity-aware LMP algorithms are shown in Figure 4.26. One can see that these algorithms perform well for the real speech under impulsive noises.

4.5.2 Proportionate LMP Algorithm

Proportionate AFAs, such as the proportionate normalized least mean square (PNLMS) [79], μ-law PNLMS (MPNLMS) [82], improved PNLMS (IPNLMS) [81], and individual-activation-factor PNLMS (IAF-PNLMS) [315], have been shown to outperform conventional adaptive algorithms like the LMS and NLMS algorithms in sparse system identification problems, in terms of convergence speed or tracking capability. In this part, we introduce a novel PLMP algorithm [316], which utilizes the MPE as the cost function and exhibits superior performance in sparse system identification, particularly in the presence of non-Gaussian noise.

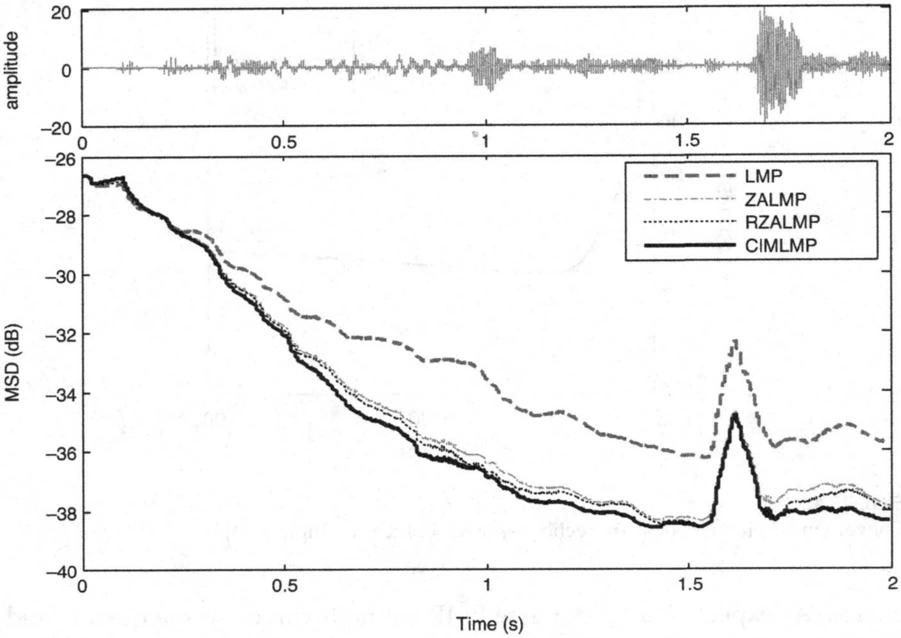

FIGURE 4.26
Convergence behaviors of sparse echo response with speech signal input [234].

4.5.2.1 Formulation of PLMP Algorithm

The weight update equation of the PLMP algorithm, in an intuitively evident fashion, can be expressed as [235]:

$$\begin{aligned}\mathbf{w}(k+1) &= \mathbf{w}(k) - \mu G(k)\frac{\partial J_{LMP}(k)}{\partial \mathbf{w}(k)} \\ &= \mathbf{w}(k) - \mu G(k)\left[-p\,|\,e(k)\,|^{p-1}\ \mathrm{sign}(e(k))\mathbf{u}(k)\right] \\ &= \mathbf{w}(k) + \eta G(k)\,|\,e(k)\,|^{p-1}\ e(k)\mathbf{u}(k)\end{aligned} \tag{4.158}$$

where $\eta = \mu p$, μ denotes the step size control parameter, $G(k) = \mathrm{diag}(g_1(k), g_2(k), \ldots, g_L(k))$ is a diagonal gain distribution matrix that modifies the step size of each tap according to a specific rule, $g_l(k)$ is calculated from the following procedure [83]:

$$g_l(k) = \frac{\chi_l(k)}{\sum_{i=1}^{M} \chi_i(k)}, \quad l = 1, 2, \ldots, L \tag{4.159}$$

where

$$\chi_l(k) = \max\left[\varepsilon \max\left\{\phi, |w_1(k)|, \ldots, |w_M(k)|\right\}, |w_l(k)|\right] \tag{4.160}$$

The parameter ϕ, together with the parameter ε, is a constant parameter, which is used to prevent the very small coefficients of $\mathbf{w}(k)$ from stalling. $\chi_l(k)$ is the bigger value between l^{th} element of the weight vector $\mathbf{w}(k)$ at instant k and others, such as shown in (4.159). Here we summarize the PLMP algorithm in Table 4.2 [235].

4.5.2.2 Convergence Analysis

This section performs the convergence analysis of the PLMP algorithm. First, we rewrite (4.158) in the following form [235]:

$$\mathbf{w}(k+1) = \mathbf{w}(k) + \eta f(e(k))e(k)G(k)\mathbf{u}(k) \tag{4.161}$$

where $f(e(k)) = |e(k)|^{p-2}$. Furthermore, the mean and auto-covariance matrix of $\tilde{\mathbf{w}}(k)$ are defined as

$$\xi(k) = E\left[\tilde{\mathbf{w}}(k)\right] \tag{4.162}$$

and

$$S(k) = E\left[\vartheta(k)\vartheta^{\mathrm{T}}(k)\right] \tag{4.163}$$

TABLE 4.2
PLMP Algorithm

Proportionate Least Mean p-Power Error Algorithm.
Initialization: $\mathbf{w}(0) = \vec{0}, \mu, p, \phi, \varepsilon, \eta = \mu p$
Beginning Computation:
Iterate for $k>1$:
$e(k) = d(k) - \mathbf{w}^T(k)\mathbf{u}(k);$
$\chi_l(n) = \max\left[\varepsilon \max\left\{\phi, \lvert w_1(k)\rvert, \ldots, \lvert w_L(k)\rvert\right\}, \lvert w_l(k)\rvert\right];$
$g_l(k) = \dfrac{\chi_l(k)}{\sum_{i=1}^{L} \chi_i(k)};$
$G(k) = \mathrm{diag}(g_1(k), g_2(k), \ldots, g_L(k));$
$\mathbf{w}(k+1) = \mathbf{w}(k) + \eta \lvert e(k)^{p-2}\rvert e(k)G(k)\mathbf{u}(k);$
End

where $\vartheta(k)$ is defined as

$$\vartheta(k) = \tilde{\mathbf{w}}(k) - \xi(k) \tag{4.164}$$

For simplification of the following analysis, we denote $G=G(k)$. Furthermore, the following assumptions should be given.

Assumption 4.17: The input signal $\mathbf{u}(k)$ is a zero-mean stationary Gaussian process and the noise $v(k)$ is a zero-mean process with variance σ_v^2 and uncorrelated with any other signals in the system.

Assumption 4.18: The weight vector $\mathbf{w}(k)$ and $\mathbf{u}(k)$ are statistically independent.

Assumption 4.19: The error nonlinearity $f(e(k))$ is independent of the $\mathbf{u}(k)$.

Assumption 4.20: The matrix G is almost time-invariant.

Remark 4.14

Assumptions 4.17 and 4.18 have been widely used in [234,285]. Assumption 4.19 is rational when the weight vector $\mathbf{w}(k)$ is close to the optimal solution. The validation of Assumption 4.20 has been explained in [317,318].

4.5.2.2.1 Mean Performance

Combining (4.161) and weight error vector, we have

$$\begin{aligned}
\tilde{\mathbf{w}}(k+1) &= \tilde{\mathbf{w}}(k) - \eta f(e(k))e(k)G\mathbf{u}(k) \\
&= \tilde{\mathbf{w}}(k) - \eta f(e(k))G\left(\mathbf{w}_o\mathbf{u}(k) + v(k) - \mathbf{w}^T(k)\mathbf{u}(k)\right)\mathbf{u}(k) \\
&= \left[I - \eta f(e(k))G\mathbf{u}(k)\mathbf{u}^T(k)\right]\tilde{\mathbf{w}}(k) - \eta f(e(k))v(k)G\mathbf{u}(k) \\
&= A(k)\tilde{\mathbf{w}}(k) - \eta f(e(k))v(k)G\mathbf{u}(k)
\end{aligned} \tag{4.165}$$

where $A(k) = I - \eta f(e(k))G\mathbf{R}(k)$ and $\mathbf{R}(k) = \mathbf{u}(k)\mathbf{u}^T(k)$.

Taking the expectation of (4.165) and using Assumptions 4.17-4.20, we can obtain

$$\xi(k+1) = \left[1 - \eta\sigma_u^2 E\left[f(e(k))\right]G\right]\xi(k) \tag{4.166}$$

In order to obtain the steady-state error $\xi(\infty)$, we set $\xi(k)$ and $\xi(k+1)$ as $\xi(\infty)$ in (4.166), then one can obtain

$$-\eta\sigma_x^2 E\left[f(e(k))G\right]\xi(\infty) = \mathbf{0} \tag{4.167}$$

From (4.167), we have

$$\xi(\infty) = \mathbf{w}_o - E\left[\mathbf{w}(\infty)\right] = 0 \tag{4.168}$$

Then we have

$$\mathbf{w}_o = E\left[\mathbf{w}(\infty)\right] \tag{4.169}$$

4.5.2.2.2 Mean Square Performance

Combining (4.165) and (4.166) yields

$$\begin{aligned}\vartheta(k+1) = &\left(I - \eta f\left(e(k)G\mathbf{R}(k)\right)\right)\vartheta(k) - \eta G\left(f\left(e(k)\right)\right)\mathbf{R}(k) - \sigma_u^2 E\left[f\left(e(k)\right)\right]\xi(n) \\ &- \eta f\left(e(k)\right)G\mathbf{u}(k)v(k)\end{aligned} \tag{4.170}$$

Substituting (4.170) into (4.163), one can obtain

$$\begin{aligned}S(k+1) = &S(k) - \eta E\left[f\left(e(k)\right)\right]\sigma_x^2 GS(k) - \eta E\left[f\left(e(k)\right)\right]\sigma_u^2 S(k)G \\ &+ \eta^2 E\left[f^2\left(e(k)\right)\right]GE\left[\mathbf{R}(k)S(k)\mathbf{R}^T(k)\right]G \\ &+ \eta^2 E\left[f^2\left(e(k)\right)\right]GE\left[\mathbf{R}(k)\xi(k)\xi^T(k)\mathbf{R}(k)\right]G \\ &- \eta^2 E\left[f^2\left(e(k)\right)\right]\sigma_u^2 G\xi(k)\xi^T(k)G \\ &+ \eta^2 E\left[f^2\left(e(k)\right)\right]\sigma_u^2\sigma_v^2 G^2\end{aligned} \tag{4.171}$$

As mentioned in [79,286], the fourth moment of a Gaussian variable is three times the variance squared, and $S(k)$ is symmetric. Therefore, one can get

$$E\left[\mathbf{R}(k)S(k)\mathbf{R}^T(k)\right] = 2S(k) + \Phi(k)\mathbf{I} \tag{4.172}$$

$$E\left[\mathbf{R}(k)\xi(k)\xi^T(k)\mathbf{R}^T(k)\right] = 2\xi(k)\xi^T(k) + \left\|\xi(k)\right\|^2\mathbf{I} \tag{4.173}$$

where $\Phi(k)$ denotes the trace of the matrix $S(k)$, which can also be interpreted as misalignment noise power. Substituting (4.172) and (4.173) into (4.171), one can observe

$$\begin{aligned} S(k+1) = S(k) &- \eta E\left[f(e(k))\right]\sigma_u^2 GS(k) \\ &- \eta E\left[f(e(k))\right]\sigma_u^2 S(k)G \\ &+ 2\eta^2 E\left[f^2(e(k))\right]GS(k)G \\ &+ \eta^2 E\left[f^2(e(k))\right](2-\sigma_u^4)G\xi(k)\xi^{\mathrm{T}}(n)G \\ &+ \eta^2 E\left[f^2(e(k))\right]\left(\Phi(k)+\|\xi(k)\|^2+\sigma_u^2\sigma_v^2\right)G^2 \end{aligned} \tag{4.174}$$

In order to obtain the steady-state error, we take $S(k)$ and $S(k+1)$ as $S(\infty)$ in (4.174), and also set $\xi(k)$ to zero. Then we obtain

$$\begin{aligned} &E\left[f(e(\infty))\right]\sigma_u^2 GS(\infty) + E\left[f(e(\infty))\right]\sigma_v^2 GS(\infty) \\ &-2\eta E\left[f^2(e(\infty))\right]GS(\infty)G \\ &= \eta E\left[f^2(e(\infty))\right]\left(\Phi(\infty)+\sigma_u^2\sigma_v^2\right)G^2 \end{aligned} \tag{4.175}$$

Under the assumption that the scalar equations for the diagonal elements of the matrix $S(\infty)$ are uncoupled, then we have

$$S_{kk} = \eta E[f^2(e(\infty))](\Phi(\infty)+\sigma_u^2\sigma_v^2)G_{kk} \,/\, (2E[f(e(\infty))]\sigma_u^2 - 2\eta E[f^2(e(\infty))]G_{kk}) \tag{4.176}$$

where S_{kk} is the kth of these diagonal elements of $S(\infty)$. Due to $\sum_{k=1}^{M} S_{kk}(\infty) = \Phi(\infty)$, summing all M of these equations yields

$$\Phi(\infty) = \sum_{k=1}^{M} \frac{(\Phi(\infty)+\sigma_u^2\sigma_v^2)\eta E[f^2(e(\infty))]G_{kk}}{2E[f(e(\infty))]\sigma_u^2 - 2\eta E[f^2(e(\infty))]G_{kk}} \tag{4.177}$$

Therefore, we have

$$\Phi(\infty) = \sigma_u^2\sigma_v^2 \frac{\varpi}{1-\varpi} \tag{4.178}$$

where

$$\varpi = \sum_{k=1}^{M} \frac{\eta E[f^2(e(\infty))]G_{kk}}{2E[f(e(\infty))]\sigma_u^2 - 2\eta E[f^2(e(\infty))]G_{kk}}$$

Similar to [37], the element of $\xi(k)$ can be removed, and we have

$$\begin{aligned}S(k+1) = &-\eta E\left[f\left(e(k)\right)\right]\sigma_u^2 GS(k) - \eta E\left[f\left(e(k)\right)\right]\sigma_u^2 S(k)G \\ &+ 2\eta^2 E\left[f^2\left(e(k)\right)\right]GS(k)G + \eta^2 E\left[f\left(e(k)\right)\right] \\ &\left(2-\sigma_u^4\right)G\xi(k)\xi^{\mathrm{T}}(k)G + \eta^2 E\left[f^2\left(e(k)\right)\right] \\ &\left(\Phi(k) + \left\|\xi(k)\right\|^2 + \sigma_u^2\sigma_v^2 G^2\right) + S(k)\end{aligned} \tag{4.179}$$

Let $\varphi(k)$ be the vector with the elements equal to the diagonal elements of the matrix $S(k)$, then we have

$$\begin{aligned}\varphi(k+1) &= \varphi(k) - 2\eta E\left[f\left(e(k)\right)\right]\sigma_u^2 G\varphi(k) \\ &2\eta^2 E\left[f\left(e(k)\right)\right]G^2\varphi(k) \\ &+\eta^2 E\left[f\left(e(k)\right)\right]\left(\Phi(k) + \sigma_u^2\sigma_v^2\right)\Gamma \\ &= B\varphi(k) + \eta^2 E\left[f^2\left(e(k)\right)\right]\left(\Phi(k) + \sigma_u^2\sigma_v^2\right)\Gamma\end{aligned} \tag{4.180}$$

where the vector Γ is the nth element of G_{kk}^2, and $B = 1 - 2\eta E\left[f\left(e(k)\right)\right]\sigma_u^2 G + 2\eta^2 E\left[f^2\left(e(k)\right)\right]\ E\left[f^2\left(e(k)\right)\right]G^2$, in which $B_{kk} = 1 - 2\eta E[f^2(e(k))]\sigma_u^2 G_{kk} + 2\eta^2 E[f^2(e(k))]G_{kk}^2$. After vector manipulation, we obtain

$$\varphi(k+1) = \eta^2 E\left[f^2\left(e(k)\right)\right]\sum_{m=0}^{k} B^m\Gamma\left(\Phi(k-m) + \sigma_u^2\sigma_v^2\right) \tag{4.181}$$

Note that $\Phi(k-m)$ is a function of $\varphi(k-m)$, then (4.181) is satisfied

$$\begin{aligned}\Phi(k+1) &= \eta^2 E\left[f^2\left(e(k)\right)\right]\sum_{m=0}^{k} 1^T B^m\Gamma\left(\Phi(k-m) + \sigma_u^2\sigma_v^2\right) \\ &= \sum_{m=0}^{k} \Lambda(m)\left(\Phi(k-m) + \sigma_u^2\sigma_v^2\right)\end{aligned} \tag{4.182}$$

where

$$\begin{aligned}\Lambda(m) &= \eta^2 E[f^2(e(k))]1^T B^m\Gamma \\ &= \eta^2 E[f^2(e(k))]\sum_{k=1}^{M} G_{kk}(1 - 2\eta E[f(e(k))]\sigma_u^2 G_{kk} + 2\eta^2 E[f^2(e(k))]G_{kk}^2)^m\end{aligned} \tag{4.183}$$

and the 1^T denotes a vector of all ones. Therefore, the condition for the mean square stability is

$$\sum_{m=1}^{\infty} \Lambda(m) < 1 \tag{4.184}$$

Therefore, substituting (4.183) into (4.184), we can derive sufficient conditions for the convergence of the PLMP algorithm, which are given by

$$\begin{cases} \eta > 0 \\ \eta \sum_{k=1}^{m} \dfrac{G_{kk}}{2E[f(e(k))]\sigma_u^2 - 2\eta E[f^2(e(k))]G_{kk}} < 1 \end{cases} \tag{4.185}$$

4.5.2.3 Simulation Results

In this part, two experiments are presented to verify the effectiveness of the PLMP algorithm for the sparse system identification. The convergence can be evaluated by the MSD. The alpha-stable distribution has been provided a good model for non-Gaussian impulsive noises. The ZALMP, RZALMP, and CIMLMP algorithms have been proposed in [234].

4.5.2.3.1 Sparse System Identification

Here impulsive responses for four unknown channels are considered as shown in Figure 4.27. In this case, the input signal is considered as a Gaussian process with zero-mean and unit variance. The convergence curves for different channels are shown in Figure 4.28. The results show that when the channel is extremely sparse (e.g., channel 1 and channel 2), the PLMP algorithm can yield the lowest MSD. When the system is less sparse (e.g., channel 3), the performances deteriorate, while the PLMP still maintains the best performance. The performance of PLMP will be approximately equivalent to the CIMLMP and RZALMP since the sparsity ratio is 1/2 (e.g., channel 4). One can also note that the PNLMS performs worst in all cases. The reason is that it is very sensitive to impulsive noises.

4.5.2.3.2 Sparse Echo Channel Estimation

Here another sparse echo channel estimation issue is further considered to demonstrate the performance of the PLMP algorithm, in which the input signal is a fragment of 2*s* of the real speed, sampled at 8*k* Hz. An acoustic echo path of a 1024-tap system with 52 non-zero coefficients is considered to be very sparse, which is shown in Figure 4.29, and used in the simulation. Figure 4.30 shows the simulation results. Obviously, the PLMP achieves a smaller MSD compared with the ZALMP, RZALMP, CIMLMP, and PNLMS.

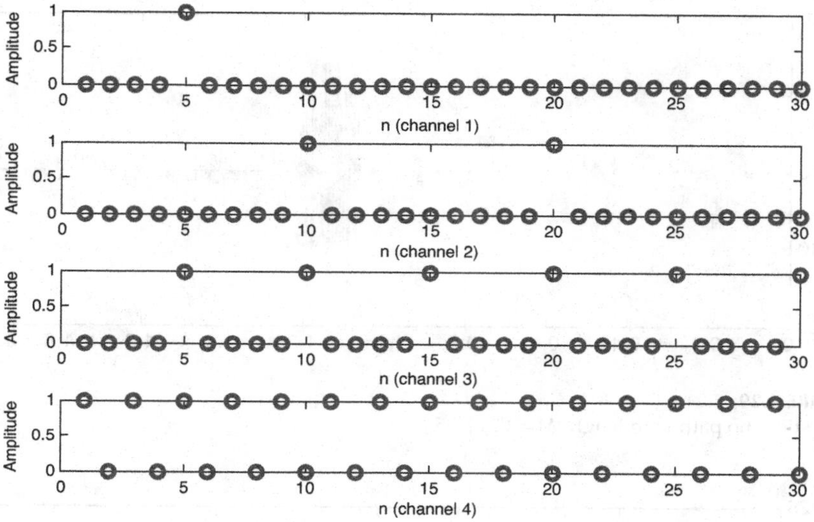

FIGURE 4.27
The impulse responses for four different channels [235]

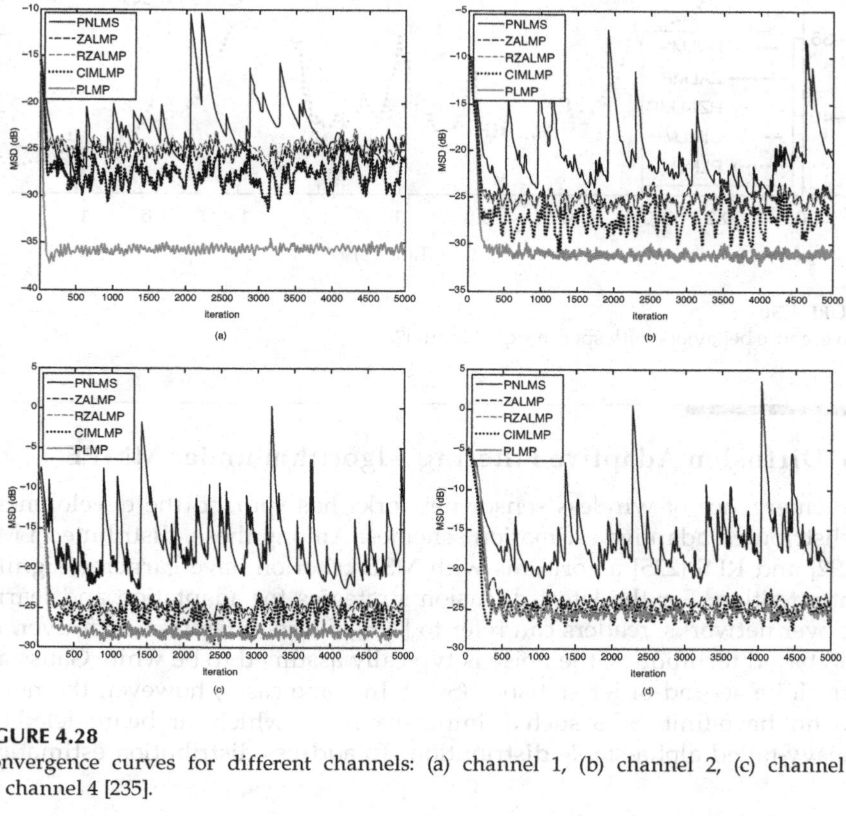

FIGURE 4.28
Convergence curves for different channels: (a) channel 1, (b) channel 2, (c) channel 3, (d) channel 4 [235].

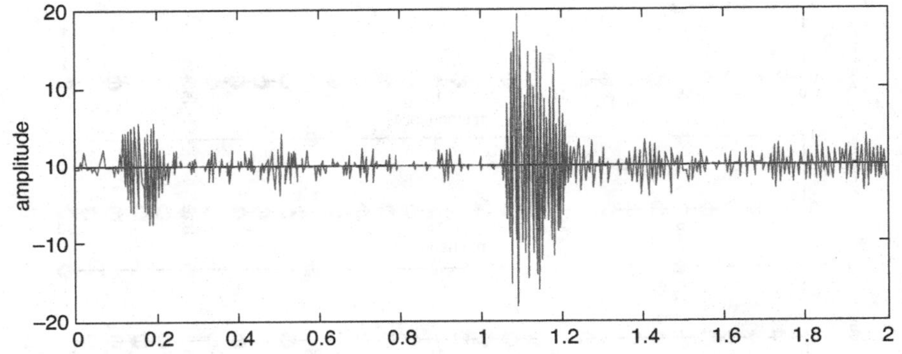

FIGURE 4.29
Acoustic echo path with length $M=1024$ [235].

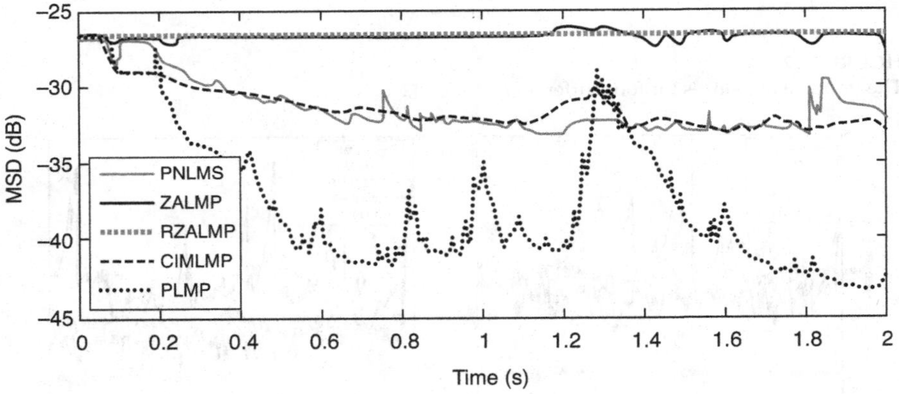

FIGURE 4.30
Convergence behaviors with speech signal input [235].

4.6 Diffusion Adaptive Filtering Algorithm under MMPE

The emergence of wireless sensor networks has spurred the development of distributed adaptive estimation schemes. Among these, distributed LMS [91,92] and RLS [225] algorithms with MSE criterion have garnered significant attention. For the latest diffusion strategies for adaptation and learning over networks, readers can refer to [226] and its references. However, in MSE-based techniques, the noise is typically assumed to be white Gaussian with finite second-order statistics (SOS). In some cases, however, the noise may not have finite SOS, such as impulsive noise, which can be modeled by a heavy-tailed alpha-stable distribution. To address distribution estimation under non-Gaussian (or impulsive) noise, this chapter presents diffusion LMP methods [227,228] for distributed estimation.

4.6.1 Diffusion LMP Algorithm

4.6.1.1 General Diffusion LMP Algorithm

Here a network composed of N nodes distributed over a geographic area is considered to estimate an unknown vector $\mathbf{w}_o$ of size $M\times 1$ from measurements collected at N nodes. At each time instant i (i=1, 2, ... I), each node k has access to the realization of a scalar measurement d_k and a regression vector $\mathbf{u}_k(i)=\left[u_{k-1}(i),u_{k-2}(i),\cdots,u_{k-M}(i)\right]^T$ of size $M\times 1$, related as

$$d_k(i)=\mathbf{w}_o^T\mathbf{u}_k(i)+v_k(i) \tag{4.186}$$

where $v_k(i)$ denotes the measurement noise, and superscript T denotes transposition. The objective is for every node in the network to use the data $\{d_k(i),\mathbf{u}_k(i)\}$ to estimate the unknown column vector. We assume that all the signals are real, and extension to the complex case is straightforward. For the global LMP, the parameter $\mathbf{w}_o$ can be estimated by minimizing the following global cost function [227]

$$J_k^{\text{global}}(\mathbf{w})=\sum_{l=1}^{N}E\left\{\left|d_l(i)-\mathbf{w}^T\mathbf{u}_l(i)\right|^p\right\} \tag{4.187}$$

The unknown parameters are updated along the steepest descent of the cost function in (4.187), which is given by

$$\begin{aligned}\mathbf{w}(i)&=\mathbf{w}(i-1)+\mu_k\nabla J_k^{\text{global}}(\mathbf{w}_k)\\&=\mathbf{w}(i-1)+\sum_{l=1}^{N}\mu_l\left|e_l(i)\right|^{p-2}e_l(i)\mathbf{u}_l(i)\end{aligned} \tag{4.188}$$

where $e_l(i)=d_l(i)-\mathbf{w}_k^T\mathbf{u}_l(i)$ is the error signal.

For local estimation and for each node k, the DLMP seeks to estimate $\mathbf{w}_o$ by minimizing a linear combination of the local MPE within the node k' s neighbor N_k, and the following cost function for each node can be expressed

$$J_k^{\text{local}}(\mathbf{w}_k)=\sum_{l\in N_k}\alpha_{l,k}E\left\{\left|d_l(i)-\mathbf{w}_k^T\mathbf{u}_l(i)\right|^p\right\}=\sum_{l\in N_k}\alpha_{l,k}E\left\{\left|e_l(i)\right|^p\right\} \tag{4.189}$$

where $\{\alpha_{l,k}\}$ are some non-negative combination coefficients satisfying $\sum_{l\in N_k}\alpha_{l,k}=1$, and $\alpha_{l,k}=0$ if $l\notin N_k$, and it determines which nodes $l\notin N_k$ should share their measurements $\{d_l(i), \mathbf{u}_l(i)\}$ with node k. Taking the derivative of (4.189), and a gradient-based algorithm for estimating $\mathbf{w}_o$ at node k can thus be derived as

$$\begin{aligned} \mathbf{w}_k(i) &= \mathbf{w}_k(i-1) + \mu_k \nabla J_k^{\text{local}}(\mathbf{w}_k) \\ &= \mathbf{w}_k(i-1) + \mu_k \sum_{l \in N_k} \left[\alpha_{l,k} |e_l(i)|^{p-2} e_l(i) \mathbf{u}_l(i) \right] \end{aligned} \tag{4.190}$$

where $\mathbf{w}_k(i)$ stands for the estimate of $\mathbf{w}_o$ at time instant i, and μ_k is the step size for node k. There are mainly two different schemes (including the ATC scheme and the CTA scheme) for the diffusion estimation. The ATC scheme first updates the local estimates using the adaptive algorithm and then the estimates of the neighbors are fused together, while the CTA scheme performs the operations of the ATC scheme in a reverse order. In the next section, we will give these two versions of DLMP algorithms. For each node, we calculate the intermediate estimates by

$$\varphi_k(i-1) = \sum_{l \in N_k} \beta_{l,k} \mathbf{w}_l(i-1) \tag{4.191}$$

where $\varphi_k(i-1)$ denotes an intermediate estimate offered by node k at instant $i-1$, and $\beta_{l,k}$ denotes a weight with which a node should share its intermediate estimate $w_l(i-1)$ with node k. With all the intermediate estimates, the nodes update their estimates by

$$\phi_k(i) = \varphi_k(i-1) + \mu_k \sum_{l \in N_k} \alpha_{l,k} |e_l(i)|^{p-2} e_l(i) \mathbf{u}_l(i) \tag{4.192}$$

The above iteration in (4.192) is referenced as incremental step. The coefficients $\{\alpha_{l,k}\}$ determine which nodes should share their measurements $\{d_l(i), u_l(i)\}$ with node k. The combination is then performed as

$$\mathbf{w}_k(i) = \sum_{l \in N_k} \delta_{l,k} \phi_l(i) \tag{4.193}$$

This result in (4.193) represents a convex combination of estimates from incremental step (4.192) fed by spatially distinct data $\{d_k(i), \mathbf{u}_k(i)\}$, and it is referenced as a diffusion step. The coefficients in $\{\delta_{l,k}\}$ determine which nodes should share their intermediate estimates $\phi_l(i)$ with node k.

According to the above analysis, one can obtain the following general diffusion LMP by combining (4.191), (4.192), and (4.193)

$$\left\{\begin{array}{l} \varphi_k(i-1)=\sum_{l\in N_k}\beta_{l,k}\mathbf{w}_l(i-1) \qquad \textbf{diffusion} \quad I \\ \phi_k(i)=\varphi_k(i-1)+\eta_k\sum_{l\in N_k}\left[\alpha_{l,k}\,|(d_l(i)-\mathbf{u}_l(i)\varphi_k(i-1))|^{p-2}\,(d_l(\mathrm{i})-\mathbf{u}_l(i)\varphi_k(i-1))\mathbf{u}_l(i)\right] \quad \textbf{incremental} \\ \mathbf{w}_k(i)=\sum_{l\in N_k}\delta_{l,k}\phi_l(i) \qquad \textbf{diffusion} \quad II \end{array}\right. \tag{4.194}$$

Details on the selection of the weights $\beta_{l,k},\alpha_{l,k}$, and $\delta_{l,k}$ can be found in [95]. The diffusion LMP (DLMP) algorithm is summarized in Table 4.3 [227].

4.6.1.2 ATC and CTA Diffusion LMP

The non-negative real coefficients $\{\beta_{l,k}\},\{\alpha_{l,k}\}$, and $\{\delta_{l,k}\}$ in (4.194) are corresponding to the $\{l,k\}$ entries of matrices $\mathbf{P}_1,\mathbf{P}_2$, and $\mathbf{P}_3$, respectively, and satisfy

$$\mathbf{1}^T\mathbf{P}_1=\mathbf{1}^T,\mathbf{1}^T\mathbf{P}_2=\mathbf{1},\mathbf{1}^T\mathbf{P}_3=\mathbf{1}^T$$

where **1** denotes the vector with unit entries. Below we give the ATC and CTA diffusion LMP algorithms.

4.6.1.2.1 ATC Diffusion LMP

When $\mathbf{P}_1=\mathrm{I},\mathbf{P}_2=\mathrm{I}$, the algorithm in (4.174) will reduce to the uncomplicated ATC diffusion LMP (ATCDLMP) version as

TABLE 4.3

Diffusion LMP Algorithm

Initialization: Given non-negative real coefficients

Start with $w_l(-1)=0$ for all l

1. for i=0 to $i$max do
2. for k=1 to N do

$$\varphi_k(i-1)=\sum_{l\in N_k}\beta_{l,k}\mathbf{w}_l(i-1)$$

3. $\phi_k(i)=\varphi_k(i-1)+\eta_k\sum_{l\in N_k}\left[\alpha_{l,k}\,|(d_l(i)-\mathbf{u}_l(i)\varphi_k(i\text{-}1))|^{p-2}\,(d_l(\mathrm{i})-u_l(i)\varphi_k(i-1))\mathbf{u}_l(i)\right]$

$$\mathbf{w}_k(i)=\sum_{l\in N_k}\delta_{l,k}\phi_l(i)$$

4. end
5. end

$$\begin{cases} \phi_k(i) = \mathbf{w}_k(i-1) + \eta_k\left[\left|d_k(i) - \mathbf{u}_k(i)\mathbf{w}_k(i-1)\right|^{p-2}\left(d_k(i) - \mathbf{u}_k(i)\mathbf{w}_k(i-1)\right)\mathbf{u}_k(i)\right] \\ \mathbf{w}_k(i) = \sum_{l \in N_k} \delta_{l,k}\phi_l(i) \end{cases} \tag{4.195}$$

4.6.1.2.2 CTA Diffusion LMP

Similar to the ATC version, one can get a simple CTA diffusion LMP (CTADLMP) algorithm by choosing $\mathbf{P}_2$=I and $\mathbf{P}_3$=I

$$\begin{cases} \varphi_k(i-1) = \sum_{l \in N_k} \beta_{l,k}\mathbf{w}_l(i-1) \\ \mathbf{w}_k(i) = \varphi_k(i-1) = \eta_k\left[\left|d_k(i) - \mathbf{u}_k(i)\varphi_k(i-1)\right|^{p-2}\left(d_k(i) - \mathbf{u}_k(i)\varphi_k(i-1)\right)\mathbf{u}_k(i)\right] \end{cases} \tag{4.196}$$

Equations (4.195) and (4.196) are similar to the ATC diffusion LMS (ATCLMS) [92] and the CTA diffusion LMS (CTALMS) [91], respectively. Clearly, the ATCDLMP and CTADLMP will reduce to the ATCDLMS and CTADLMS when p=2, where $e_k(i)$ is $d_k(i) - \mathbf{u}_k(i)\mathbf{w}_k(i-1)$ and $d_k(i) - \mathbf{u}_k(i)\varphi_k(i-1)$ for ATC and CTA versions, respectively. In addition, no exchange of data is needed during the adaptation of the step size, which makes the communication cost relatively low.

4.6.1.3 Simulation Results

This section considers a connected ad hoc wireless sensor network composed of N=20 nodes, and it is generated as a realization of the random geometric graph model on the unit square, with communication range $r = 0.5$. The topology of the network is shown in Figure 4.31. The desired signal is assumed to be white Gaussian process in the following simulations, and the unknown parameter is drawn randomly from the standard uniform distribution.

Since the variance of the alpha-stable process is infinite, instead of the signal-to-noise ratio (SNR), we use generalized SNR (GSNR), which is defined in [319]. The transient and steady network MSD of the diffusion LMP as a function of order p are given in Figures 4.32 and 4.33, respectively. From Figure 4.31, one can observe that, due to the impulsive noise, the LMP algorithm is non-convergent for order p=2 or p close to 2. Meanwhile, the algorithm is converged when p=1 or p close to 1. One can see from Figure 4.32 that the more impulsive the noise (smaller α), the larger the MSD for p=2. The impulsive noise has a great influence on the diffusion LMS algorithms. Furthermore, lower MSD is obtained when p is close to α.

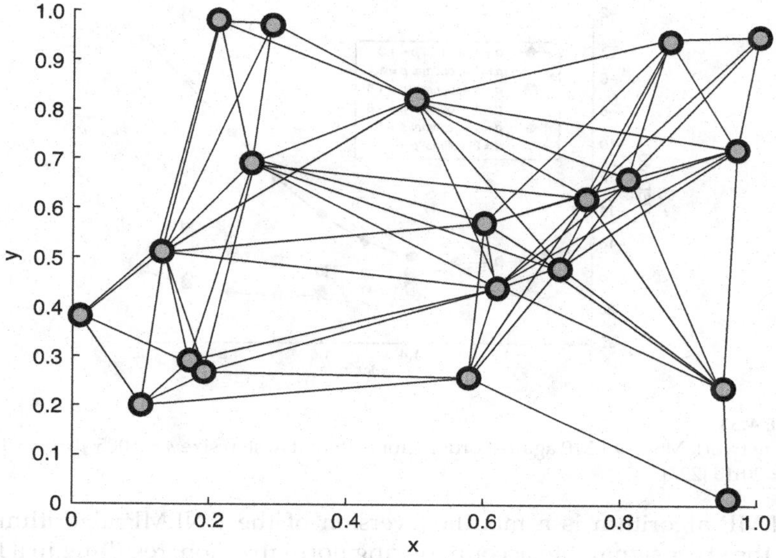

FIGURE 4.31
Ad hoc wireless sensor network with $N = 20$ sensors, generated as realization of random geometric graph model [227].

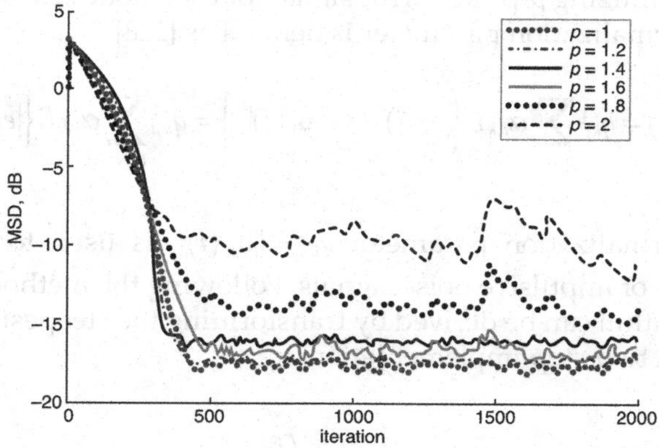

FIGURE 4.32
Transient network MSD of LMP for different p, step size $\mu = 0.005$, $\alpha = 1.2$, $\beta = 0$, $\gamma = 1$ and $\delta = 0$. GSNR = 20 dB [227].

4.6.2 Robust Diffusion NLMP Algorithm

This section introduces the diffusion normalized LMP (DNLMP) algorithm and the robust DNLMP (RDNLMP) algorithm. The DNLMP algorithm is obtained by incorporating normalization into adaptation, while the

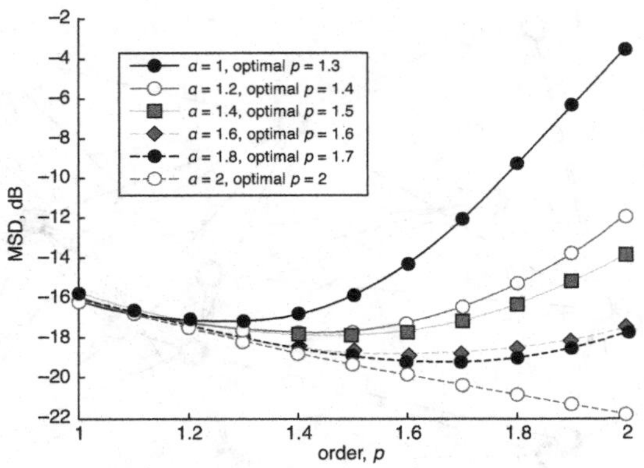

FIGURE 4.33
Steady network MSD of LMP against order p for different α, step size $\mu = 0.005$, $\beta = 0$, $\gamma = 1$, $\delta = 0$. GSNR=20 dB [227].

RDNLMP algorithm is a modified version of the DNLMP algorithm that takes the error signal into account during normalization, resulting in a faster convergence rate.

4.6.2.1 Diffusion NLMP Algorithm

Instead of minimizing p-power error signal for each node k, a local cost function with a normalization parameter is defined as [228]

$$J_k^{\text{local}}(\mathbf{w}) = q_{k,i}^{-1} \sum_{l \in N_k} \alpha_{l,k} E\left\{ \left| d_l(i) - \mathbf{w}_k^T \mathbf{u}_l(i) \right|^p \right\} = q_{k,i}^{-1} \sum_{l \in N_k} \alpha_{l,k} E\left\{ \left| e_l(i) \right|^p \right\} \quad (4.197)$$

where the normalization parameter $q_{k,i} = \left\| \mathbf{u}_k(i) \right\|_p^p$ is used to mitigate the adverse effects of impulsive noise signals. Following the method in [292], the DNLMP algorithm can be derived by transforming the steepest descent type iteration into a two-step implementation:

$$\begin{cases} \phi_k(i) = \mathbf{w}_k(i-1) - \eta q_{k,i}^{-1} \dfrac{\partial E\left| e_l(i)^p \right|}{\partial \mathbf{w}} \Big|_{\mathbf{w}_k(i-1)} (\text{adaptation}) \\ \mathbf{w}_k(i) = \sum_{l \in N_k} \delta_{l,k} \phi_l(i) (\text{combination}) \end{cases} \quad (4.198)$$

In adaptation step of (4.198), the local estimate $\phi_k(i-1)$ is replaced by linear combination $\mathbf{w}_k(i-1)$. Such substitution is reasonable because the linear combination contains more data information from neighbor nodes than $\phi_k(i-1)$

[320]. In general, the ATC-based algorithm outperforms the CTA-based one, and thus the DNLMP algorithm is only extended to ATC forms leading to the following adaptation.

4.6.2.1.1 *ATC DNLMP Algorithm*

$$\begin{cases} \phi_k(i) = \mathbf{w}_k(i-1) + \eta \dfrac{\left|d_k(i) - \mathbf{u}_k^T(i)\mathbf{w}_k(i-1)\right|^{p-2}\left(d_k(i) - \mathbf{u}_k^T(i)\mathbf{w}_k(i-1)\right)\mathbf{u}_k(i)}{\left\|\mathbf{u}_k(i)\right\|_p^p + \varepsilon} \\ \mathbf{w}_k(i) = \sum_{l \in N_k} \delta_{l,k}\phi_l(i) \end{cases} \tag{4.199}$$

where $\varepsilon = 0.01$ is the regularization parameter.

4.6.2.2 Robust DNLMP Algorithm

Although the DNLMP algorithm presented above exhibits outstanding performance for distributed estimation, it encounters issues when the noise signal is impulsive, i.e., has a large peak. In such cases, the energy of the normalization in the DNLMP algorithm increases, causing the step size to decrease to ensure convergence to the global minimum. One method to overcome this problem is the AT-based method [321], which clips error signals whose magnitudes exceed a certain threshold set by signal statistics. However, this method requires offline estimation of statistics and defining suitable thresholds is an ill-posed problem due to the high sensitivity of the bounds on percentile levels. Therefore, it is clear that there is a need for a more effective diffusion algorithm that satisfies three requirements: faster convergence, smaller misadjustment, and easier implementation. In addition, during the initial convergence stage, the error signals at each node are also peaky in nature and their effect on adaptation must be taken into account to achieve an improved convergence rate. To this end, the following modified normalized step size is designed for the DNLMP algorithm [228]:

$$\eta_k(i) = \frac{\eta}{\left\|\mathbf{u}_k(i)\right\|_p^p + \sigma_{e,k}^p(i) + \varepsilon} \tag{4.200}$$

where $\sigma_{e,k}^p(i)$ denotes the low-pass filtered estimate of the p-power error signal at node k and time i, which can be calculated by $\sigma_{e,k}^p(i) = \beta\sigma_{e,k}^p(i-1) + (1+\beta)\left|e_k(i)\right|^p$, where $\beta = 0.9$ is the forgetting factor.

Based on (4.179) and (4.180), the RDNLMP algorithm can be given as:

$$\begin{cases} \phi_k(i) = \mathbf{w}_k(i-1) + \eta_{k,i}\left|d_k(i) - \mathbf{u}_k^T(i)\mathbf{w}_k(i-1)\right|^{p-2}\left(d_k(i) - \mathbf{u}_k^T(i)\mathbf{w}_k(i-1)\right)\mathbf{u}_k(i) \\ \mathbf{w}_k(i) = \sum_{l\in N_k} \delta_{l,k}\phi_l(i) \end{cases} \tag{4.201}$$

The adaptation step of (4.181) can be expressed in a generalized form as $\phi_k(i) = \mathbf{w}_k(i-1) + f(e_k(i))(d_k(i) - \mathbf{u}_k(i)\mathbf{w}_k(i-1)\mathbf{u}_k(i))$, where $f(e_k(i)) = \frac{\left|e_k(i)^{p-2}\right|}{\|\mathbf{u}_k(i)\|_p^p + \sigma_{e,k}^p(i) + \varepsilon}$ stands for the step size normalized function (nonlinear term), and the constant p is blended into the step size. From Figure 4.34, it can be observed that the step size function of the DNLMP is a soft limiter for $\|\mathbf{u}_k(i)\|_p^p = 10$, and has no limiting performance in low-input variance ($\|\mathbf{u}_k(i)\|_p^p = 1$). The step size normalized function of the RDNLMP algorithm is similar to the probability density function of Gaussian distribution. The larger $e_k(i)$ is, the smaller the filter coefficient adaptation size is. Therefore, the RDNLMP algorithm is more stable and robustness against impulsive noises.

4.6.2.3 Performance Analysis

In this section, the steady-state performance of the RDNLMP algorithm is theoretically studied. For tractable analysis, the following assumptions should be first given:

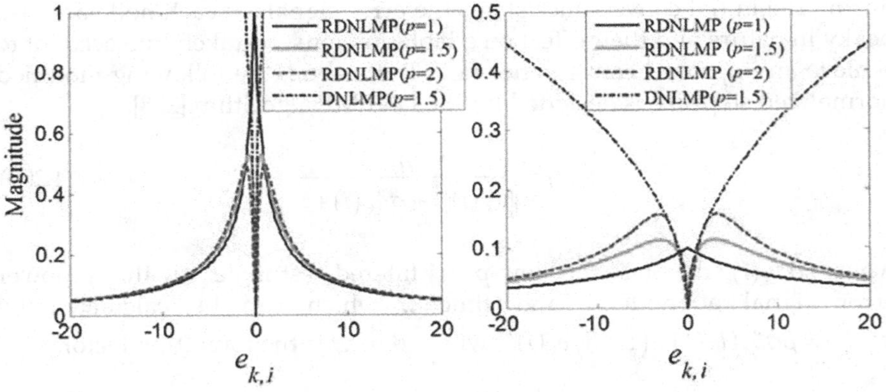

FIGURE 4.34
Step size function $f(e_k(i))$ for the algorithms. (From left to right: $\|\mathbf{u}_k(i)\|_p^p = 1$, $\|\mathbf{u}_k(i)\|_p^p = 10$.)

Assumption 4.21: All regressors $\{\mathbf{u}_k(i)\}$ arise from Gaussian sources with zero-mean and spatially and temporally independent.

Assumption 4.22: The noise signal $v_k(i)$ at each node k is assumed to be a Gaussian noise with zero-mean and is independent of any other signals.

Assumption 4.23: The error nonlinearity $f(e_k(i))$ is independent of the input regressors $\mathbf{u}_k(i)$, the error components $e_k(i)$ and the noise signals $v_k(i)$.

Assumption 4.24: At the steady state, the error signal $e_k(i)$ is assumed to be approximately equal to the noise component $v_k(i)$.

The network weight error vector and intermediate network weight error vector of the RDNLMP for node k can be defined as

$$\tilde{\mathbf{w}}_k(i) \triangleq \mathbf{w}_o - \mathbf{w}_k(i), \tilde{\varphi}_k(i) \triangleq \mathbf{w}_o - \varphi_k(i) \tag{4.202}$$

Then, we can introduce the following global quantities of the network weight error vector and intermediate network weight error vector:

$$\tilde{\mathbf{w}}_k(i) \triangleq col\{\tilde{\mathbf{w}}_1(i), \tilde{\mathbf{w}}_2(i), \ldots \tilde{\mathbf{w}}_N(i)\}, \tilde{\varphi}(i) \triangleq col\{\tilde{\varphi}_1(i), \tilde{\varphi}_2(i) \ldots \tilde{\varphi}_N(i)\} \tag{4.203}$$

The diagonal matrix of the step size is defined as

$$\mathbf{S}_e \triangleq \mathbf{M}\Theta \otimes \mathbf{I}_M \tag{4.204}$$

where $\mathbf{M} = \mathrm{diag}\{\mu, \mu, \ldots \mu\}$ is the $N \times N$ matrix, and $\mathbf{S}_e$ is the matrix with the size of $MN \times MN$. The nonlinearity error is defined as

$$\mathbf{F}(i) = \begin{cases} \dfrac{|e_1(i)|^{p-2}}{\|\mathbf{u}_1(i)\|_p^p + \sigma_{e,N}^p(i) + \varepsilon}, \dfrac{|e_2(i)|^{p-2}}{\|\mathbf{u}_2(i)\|_p^p + \sigma_{e,2}^p(i) + \varepsilon}, \\ \dfrac{|e_N(i)|^{p-2}}{\|\mathbf{u}_N(i)\|_p^p + \sigma_{e,N}^p(i) + \varepsilon} \end{cases} \tag{4.205}$$

Subtracting $\mathbf{w}_o$ from both sides of (4.201), a matrix form of (4.201) can be expressed as

$$\begin{cases} \tilde{\varphi}(i) = \tilde{\mathbf{w}}(i-1) - \mathbf{S}_e\mathbf{F}(i)\left[\mathbf{D}(i)\tilde{\mathbf{w}}(i-1) + \mathbf{V}(i)\right] \\ \tilde{\mathbf{w}}(i) = \mathbf{B}^T\tilde{\varphi}(i) \end{cases} \tag{4.206}$$

where $\mathbf{B} = \alpha \otimes \mathbf{I}_M$ is the extended weighting matrix, $\mathbf{D}(i) = \mathrm{diag}\{\mathbf{u}_1(i)\mathbf{u}_1(i)^T, \mathbf{u}_2(i)\mathbf{u}_2(i)^T, \ldots \mathbf{u}_N(i)\mathbf{u}_N(i)^T\}$, and $\mathbf{V}(i) = \mathrm{diag}\{\mathbf{u}_1(i)v_1(i), \mathbf{u}_2(i)v_2(i), \ldots \mathbf{u}_N(i)v_N(i)\}$. Combining the above two formulas into one, we have

$$\begin{aligned}\tilde{\mathbf{w}}(i) &= \mathbf{B}^T\tilde{\mathbf{w}}(i-1) - \mathbf{B}^T\mathbf{S}_e\mathbf{F}(i)\left[\mathbf{D}(i)\tilde{\mathbf{w}}(i-1) + \mathbf{V}(i)\right] \\ &= \mathbf{B}^T\left[\mathbf{I}_{MN} - \mathbf{S}_e\mathbf{F}(i)\mathbf{D}(i)\right]\tilde{\mathbf{w}}(i-1) - \mathbf{B}\mathbf{S}_e\mathbf{F}(i)\mathrm{V}(i)\end{aligned} \tag{4.207}$$

To make the following mathematical expressions concise, the following qualities will be introduced

$$\begin{aligned}\bar{\mathbf{D}} &\triangleq E\{\mathbf{D}(i)\} = \mathrm{diag}\left\{E\left\{\mathbf{u}_1(i)\mathbf{u}_1(i)^T\right\}, E\left\{\mathbf{u}_2(i)\mathbf{u}_2(i)^T\right\}, \ldots E\left\{\mathbf{u}_N(i)\mathbf{u}_N(i)^T\right\}\right\} \\ &= \mathrm{diag}\left\{\mathbf{R}_{\mathbf{u}_1}, \mathbf{R}_{\mathbf{u}2}, \ldots \mathbf{R}_{\mathbf{u}_N}\right\}\end{aligned} \tag{4.208}$$

$$\bar{\mathbf{V}} = E\left\{\mathbf{V}(i)\mathbf{V}(i)^T\right\} = \mathrm{diag}\left\{\sigma_{v_1}\mathbf{R}_{\mathbf{u}1}, \sigma_{v_2}\mathbf{R}_{\mathbf{u}2}, \ldots \sigma_{v_N}\mathbf{R}_{\mathbf{u}N}\right\} \tag{4.209}$$

where $\mathbf{R}_{\mathbf{u}_k} = E\left\{\mathbf{u}_k(i)\mathbf{u}_k(i)^T\right\}$ is the auto-correlation matrix of the input signal at node k, which is assumed to be positive-definite. Taking the weighted Euclidean norms of both sides of (4.207), the fundamental weighted energy conservation relation can be shown as:

$$\begin{aligned}&E\left[\|\tilde{\mathbf{w}}(i)\|_\Sigma^2\right] \\ &= E\left[\|\tilde{\mathbf{w}}(i-1)\|^2_{\left[\mathbf{I}-\mathbf{S}_e\mathbf{F}(i)\mathbf{D}(i)\mathbf{B}^T\Sigma\mathbf{B}\left[\mathbf{I}-\mathbf{S}_e\mathbf{F}(i)\mathbf{D}(i)\right]\right]}\right] + E\left[\mathbf{V}^T(i)\mathbf{F}(i)\mathbf{S}_e(i)\mathbf{B}^T\Sigma\mathbf{B}\mathbf{F}(i)\mathbf{D}(i)\right]\end{aligned} \tag{4.210}$$

where $\boldsymbol{\Sigma}$ is any Hermitian positive-definite matrix, and $\|x\|_\Sigma^2$ denotes the weighted squared norm $x^T\Sigma x$. Under Assumption 4.21, the following approximation can be obtained

$$\begin{aligned}&E\left[\|\tilde{\mathbf{w}}(i)\|_\Sigma^2\right] \\ &= E\left[\|\tilde{\mathbf{w}}(i-1)\|_\Theta^2\right] + \mathrm{Tr}\left[\Sigma\mathbf{B}^T\mathbf{S}_e\mathbf{F}(i)\bar{\mathbf{V}}\mathbf{S}_e\mathbf{F}(i)\mathbf{B}\right]\end{aligned} \tag{4.211}$$

where

$$\begin{aligned}\Theta = &\mathbf{B}^T\Sigma\mathbf{B} - \bar{\mathbf{F}}(i)\bar{\mathbf{D}}\mathbf{S}_e\mathbf{B}^T\Sigma\mathbf{B} - \mathbf{B}^T\Sigma\mathbf{B}\mathbf{S}_e\bar{\mathbf{F}}(i)\bar{\mathbf{D}} \\ &+ E\left[\mathbf{D}(i)\mathbf{S}_e\mathbf{F}(i)\mathbf{S}_e\mathbf{B}^T\Sigma\mathbf{B}\mathbf{S}_e\mathbf{F}(i)\mathbf{D}(i)\right]\end{aligned} \tag{4.212}$$

and $\bar{\mathbf{F}}(i) = E\{\mathbf{F}(i)\}$. Now, let us define

$$\sigma \triangleq vec(\Sigma), \Sigma \triangleq vec^{-1}(\sigma) \tag{4.213}$$

Herein, we use the notation $\|\tilde{\mathbf{w}}\|_{\sigma}^{2}$ to denote $\|\tilde{\mathbf{w}}\|_{\Sigma}^{2}$. Applying the property of the Kronecker product, thus (4.212) can be rewritten as:

$$vec(\Theta) = \mathbf{Q}_i\sigma \tag{4.214}$$

Where $\mathbf{Q}_i$ is a matrix and can be expressed as:

$$\mathbf{Q}_i = \left[\mathbf{I} - \mathbf{I}\otimes\left(\bar{\mathbf{F}}(i)\bar{\mathbf{D}}\mathbf{S}_e\right) - \left(\bar{\mathbf{F}}(i)\bar{\mathbf{D}}\mathbf{S}_e\right)\otimes\mathbf{I} + E\left[\mathbf{D}(i)^T\mathbf{F}(i)\mathbf{S}_e \otimes \mathbf{D}(i)^T\mathbf{F}(i)\mathbf{S}_e\right]\right](\mathbf{B}\otimes\mathbf{B}) \tag{4.215}$$

Using the property $Tr(\Sigma\mathbf{X}) = \text{vec}\left(\mathbf{X}^T\right)^T\sigma$, (4.211) can be rewritten as

$$E\left[\|\tilde{\mathbf{w}}(i)\|_{\sigma}^{2}\right] = E\left[\|\tilde{\mathbf{w}}(i-1)\|_{\mathbf{Q}_i\sigma}^{2} + \mathbf{r}_i\sigma\right] \tag{4.216}$$

where $\mathbf{r}_i = \text{vec}\left[\mathbf{B}^T\mathbf{S}_e\bar{\mathbf{F}}(i)\bar{\mathbf{V}}^T\mathbf{S}_e\bar{\mathbf{F}}(i)\mathbf{B}^T\right]$. The MSD at node k can be defined as

$$\text{MSF}_k(i) \triangleq E\left[\|\tilde{\mathbf{w}}_k(i)\|_{m_k}^{2}\right] = E\left[\|\mathbf{w}_o - \mathbf{w}_k(i)\|^2\right] \tag{4.217}$$

where $m_k \triangleq \text{vec}\left(\text{diag}(\kappa_k)\otimes\mathbf{I}_M\right)$ and κ_k is the column vectors with a unit entry at position k and zeros elsewhere. The network-averaged MSD is defined across all nodes in the network as

$$\mathbf{MSD}^{\text{network}}(i) = \frac{1}{N}\sum_{k=1}^{N}\text{MSD}_k(i) \tag{4.218}$$

To obtain the theoretical value for (4.218), $\boldsymbol{Q}_i$ and $\boldsymbol{F}_i$ should be calculated explicitly. Supposing the independence in nonlinearity error, we have

$$E\left[\frac{|e_1(i)|^{p-2}}{\|\mathbf{u}_1(i)\|_p^p + |e_k(i)|^p + \varepsilon}\right] \approx \frac{E\left\{|e_1(i)|^{p-2}\right\}}{E\left\{\|\mathbf{u}_1(i)\|_p^p\right\} + E\left\{|e_k(i)|^p + \varepsilon\right\}} \tag{4.219}$$

Owing to the small fluctuation of $e_k(i)$ at the steady state, $E\left\{|e_k(i)|^p\right\}$ can be approximated as $E\left\{|e_k(i)|^{\frac{p}{2}}\right\}$. Though this assumption is not true in general, it simplifies the analysis of the RDNLMP. Therefore, (4.219) can be expressed as

$$\frac{E\left\{|e_1(i)|^{p-2}\right\}}{E\left\{\|\mathbf{u}_1(i)\|_p^p\right\} + E\left\{|e_k(i)|^p\right\} + \varepsilon} \approx \frac{E\left\{\sigma_{ek(i)}^2\right\}^{\frac{p-2}{2}}}{E\left\{\|\mathbf{u}_1(i)\|_p^p\right\} + E\left\{\sigma_{ek(i)}^2\right\}^{\frac{p}{2}} + \varepsilon} \tag{4.220}$$

where $\sigma^2_{e_k(i)} = \sigma^2_{\varepsilon_k(i)} + \sigma^2_{v_k}$, and $\sigma^2_{\varepsilon_k(i)} = E\left|\mathbf{u}_k(i)\tilde{\mathbf{w}}(i-1)\right|^2$. Now, to evaluate (4.220) we need to compute $\left\|\mathbf{u}_k(i)\right\|_p$. According to [322], $\left\|\mathbf{u}_k(i)\right\|_p$ can be written as

$$\left\|\mathbf{u}_k(i)\right\|_p = \left(\sum_{n=0}^{M-1}\left|\mathbf{u}_{k,n}(i)\right|^p\right)^{\frac{1}{p}} \tag{4.221}$$

where $\mathbf{u}_{k,n}(i)$ denotes the element of the input vector $\mathbf{u}_k(i)$. Without loss of generality, $\left|\mathbf{u}_{k,n}(i)\right|^p$ can be approximated by the soft parameter function [323]

$$S\left(\mathbf{u}_{k,n}(i)\right) = \left(1+\beta^{-1}\right)\left(1-\exp\left(-\beta\left|\mathbf{u}_{k,n}(i)\right|\right)\right) \tag{4.222}$$

where β is the constant, which can be regarded as a hyper parameter superseding for p to control the sparsity prior. For p=1 or 0, (4.222) is equal to $\left|\mathbf{u}_{k,n}(i)\right|^p$ with $\beta=\infty$ and $\beta=0$. According to Assumptions 4.21 and 4.22, $\left|\mathbf{u}_{k,n}(i)\right|^p$ can be derived from the standard Gaussian distribution with variance $\sigma^2_{u,k}$ given by (4.223), i.e.,

$$E\left\{S\left(\mathbf{u}_{k,n}(i)\right)\right\} = \frac{1}{\sqrt{2\pi\sigma_{u,k}}}\int_{-\infty}^{\infty} S\left(\mathbf{u}_{k,n}(i)\right)\exp\left(\frac{\mathbf{u}^2_{k,n}(i)}{2\sigma^2_{u,k}}\right)d\mathbf{u}_{k,n}(i) \tag{4.223}$$

After conducting the integral operation, we have

$$E\left\{S\left(\mathbf{u}_{k,n}(i)\right)\right\} = \frac{1}{\sqrt{2\pi\sigma_{u,k}}}\begin{cases}\dfrac{(\beta+1)\sqrt{2\pi\sigma^2_{u,k}}\left[\exp\left(\dfrac{\beta^2\sigma^2_{u,k}}{2}\right)\operatorname{erf}\left(\dfrac{\sqrt{2}\beta\sigma_{u,k}}{2}\right)-\exp\left(\dfrac{\beta^2\sigma^2_{u,k}}{2}\right)+1\right]}{\beta} \\ ,\text{for} \quad \sigma_{u,k}>0\end{cases} \tag{4.224}$$

Supposing the RDNLMP algorithm converges and the matrix $\mathbf{I}-\mathbf{Q}_i$ is invertible. Therefore, in steady-state $i\to\infty$, (4.216) will reduce to

$$E\left[\left\|\tilde{\mathbf{w}}(\infty)\right\|^2_{(\mathbf{I}-\mathbf{Q}(\infty))\sigma}\right] = \left\{\operatorname{vec}\left[\mathbf{B}^T\mathbf{S}_e\mathbf{F}(\infty)\bar{\mathbf{V}}\mathbf{S}_e\mathbf{F}(\infty)\mathbf{B}\right]\right\}^T\sigma \tag{4.225}$$

where $\mathbf{Q}(\infty) = \left[\mathbf{I}-\mathbf{I}\otimes\left(\bar{\mathbf{F}}(\infty)\bar{\mathbf{D}}\mathbf{S}_e\right)-\left(\bar{\mathbf{D}}\bar{\mathbf{F}}(\infty)\mathbf{S}_e\right)\otimes\mathbf{I}\right](\mathbf{B}\otimes\mathbf{B})$

Using the above equations, the steady-state MSD at node k and the network MSD can be respectively expressed as

$$\mathbf{MSD}_k(\infty) = E\left[\left\|\tilde{\mathbf{w}}(\infty)\right\|^2_{\kappa_k}\right] = \left\{\operatorname{vec}\left[\mathbf{B}^T\mathbf{S}_e\mathbf{F}(\infty)\bar{\mathbf{V}}\mathbf{S}_e\mathbf{F}(\infty)\mathbf{B}\right]\right\}^T\left(\mathbf{I}-\mathbf{Q}(\infty)\right)^{-1}\kappa_k \tag{4.226}$$

$$\mathbf{MSD}^{\text{network}}(\infty) \triangleq \frac{1}{N}\sum_{k=1}^{N}\mathbf{MSD}_k(\infty) = \frac{1}{N}E\left[\left\|\tilde{\mathbf{w}}_k(\infty)\right\|_2^2\right] \tag{4.227}$$

To obtain $\overline{\mathbf{F}}(\infty)$, $E\{\mathbf{F}(\infty)\}$ needs to be calculated. This term cannot be easily replaced by a closed form and is difficult to analyze exactly. Instead of using a numerical method to derive an exact solution, we employ Assumptions 4.24 and (4.220) to obtain a closed-form solution which is more intuitive than that of the numerical method, that is

$$E\{f(e_k(\infty))\} \approx \frac{E\{\sigma_{v,k}^2\}^{\frac{p-2}{2}}}{E\left\{\left\|\mathbf{u}_1(i)\right\|_p^p\right\} + E\{\sigma_{v,k}^2\}^{\frac{p}{2}} + \varepsilon} \tag{4.228}$$

Therefore, the network MSD in the steady-state is obtained as:

$$\mathbf{MSD}^{\text{network}}(\infty) = \frac{1}{N}E\left\|\tilde{\mathbf{w}}_k(\infty)\right\|_2^2 \approx \frac{1}{N}\sum_{j=0}^{\infty}\mathrm{Tr}\left(\Omega^j\Phi\left(\Omega^T\right)^j\right) \tag{4.229}$$

where $\Omega = \mathbf{B}^T\left(\mathbf{I} - \mathbf{S}_e\mathbf{F}(i)\mathbf{D}\right)$ and $\Phi = \mathbf{B}^T\mathbf{S}_e\mathbf{F}(i)\overline{\mathbf{V}}\mathbf{F}(i)\mathbf{S}_e\mathbf{B}$. Then, we can attempt to rewrite (4.229) in a general nonlinear form:

$$\mathbf{MSD}^{\text{network}}(\infty) \approx \frac{1}{N}\sum_{j=0}^{Y}\mathrm{Tr}\left(\Omega^j\Phi\left(\Omega^T\right)^j\right) \tag{4.230}$$

where Y is a positive constant, which is usually chosen between 10 and 50.

4.6.2.4 Simulation Results

This section gives the simulation results to show the performance of the DNLMP and RDNLMP algorithms. The metropolis weights [92] are used for the adaptation matrix C, and every pair of nodes has a 10% probability of being connected. The network mean square deviation (NMSD), is used to evaluate the performance.

4.6.2.4.1 Verification of the Analysis

In this case, the network is composed of four nodes, and the length of the unknown parameter vector is 5. The white Gaussian noise (WGN) with zero mean and unit variance is used as the input signal, and the noise signal is WGN with different SNRs. Figure 4.35 shows the theoretical and simulated steady-state NMSD of the RDNLMP algorithm in Gaussian noise

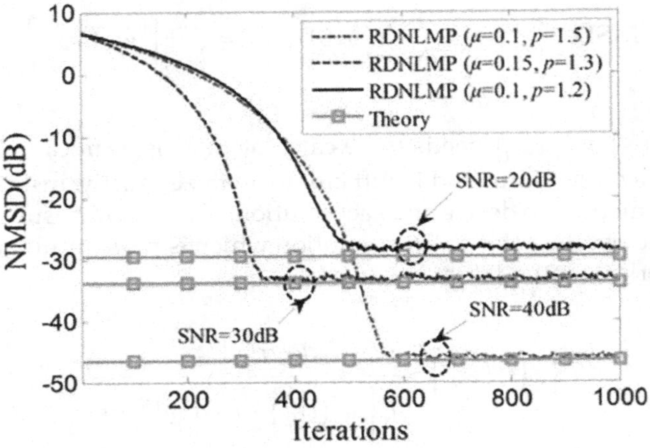

FIGURE 4.35
A comparison of the theoretical NMSD with simulation results [228].

environments with different SNRs. It can be seen from Figure 4.35 that the simulated steady-state NMSDs match well with the theoretical calculation.

4.6.2.4.2 Impulsive Noise Environments

The performance of the RDNLMP algorithm is evaluated in a network with 10 nodes under impulsive noise environments. First, the effect of p and step size for the RDNLMP algorithm is investigated. Figure 4.36 shows that a large step size results in a faster convergence rate, and RDNLMP with $p=1.5$ and $\mu=0.5$ achieves the fastest convergence rate. In addition, the performance

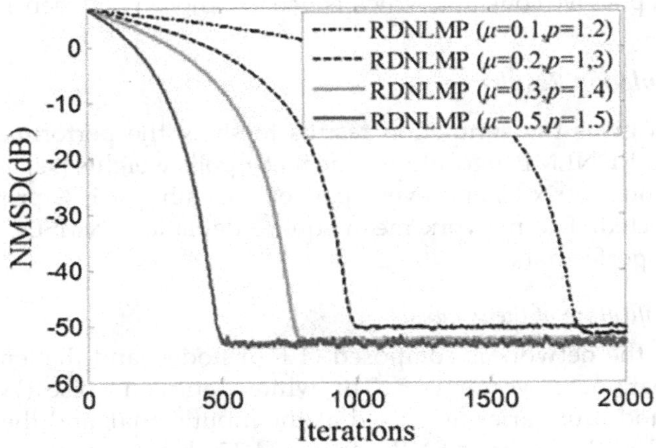

FIGURE 4.36
NMSD curves of the RDNLMP algorithm for uniform input with impulsive noises [228].

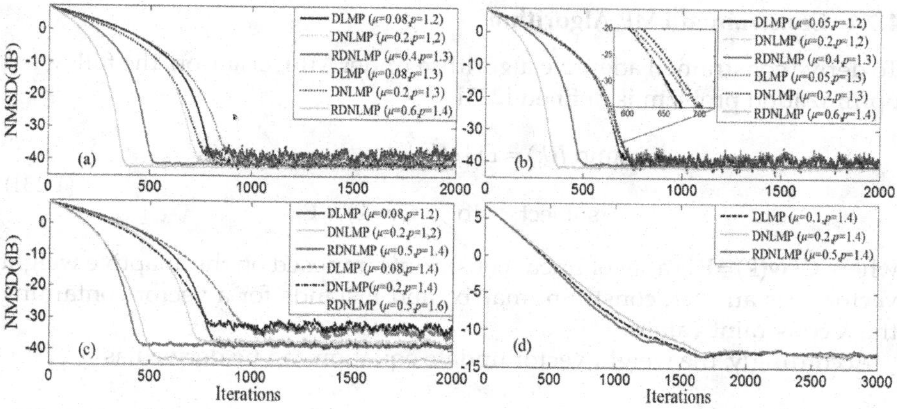

FIGURE 4.37
NMSD curves of the algorithms for different impulsive environments: (a) Uniform input with impulsive noises, (b) WGN input with impulsive noises, (c) Uniform input with mixed noises (impulsive noises and WGN with SNR=40dB), (d) WGN input with symmetric alpha-stable noise (α=1.5) [228].

of the DNLMP and RDNLMP algorithms is compared with the DLMP algorithm under four different impulsive environments, as shown in Figure 4.37. These cases demonstrate the effect of the input signal and additive noise on the algorithms. As can be seen, the DNLMP algorithm achieves a faster convergence rate than the DLMP algorithm for cases (a) and (c), whereas the RDNLMP algorithm significantly enhances the convergence rate compared to other algorithms in all cases.

4.7 Constrained Adaptive Filtering Algorithm under MMPE

Constrained adaptive filters (AFs) have a wide range of potential applications in signal-processing domains. The primary objective is to explicitly solve a constrained optimization problem. The MSE criterion has been extensively used in most existing constrained-type AFs, such as the constrained LMS (CLMS) [111–115], due to its attractive features of mathematical tractability, convexity, and low computational complexity. However, the MSE criterion only considers the SOS of data and assumes optimality under Gaussian assumption, which is not always valid in practical applications. To address this issue, a constrained AF algorithm, called the constrained least mean p-power error (CLMP) algorithm [223], is introduced in this subchapter by combining an equality constraint with the minimum p-power error (MPE) criterion.

4.7.1 Constrained LMP Algorithm

To derive constrained adaptive algorithm under MPE criterion, the following optimization problem is defined [225]

$$\begin{aligned} &\min J(k) = E[\,|e(k)|^p] \\ &\text{subject to} \quad \mathbf{C}\mathbf{w}(k) = \mathbf{F} \end{aligned} \tag{4.231}$$

where $\mathbf{C}^T\mathbf{w}(k)=\mathbf{F}$ is a set of linear constraints imposed on the adaptive weight vector, $\mathbf{C}$ is an $L \times K$ constraint matrix, and $\mathbf{F}$ stands for a vector containing the K constraint values.

Accordingly, the weight vector update equation can be derived as

$$\mathbf{w}(k) = P\left[\mathbf{w}(k-1) + \mu h(e(k))\left(d(k) - \mathbf{w}^T(k-1)\mathbf{u}(k)\right)\mathbf{u}(k)\right] + Q \tag{4.232}$$

where $h(e(k))$ is a nonlinear function of $e(k)$, given by

$$h(e(k)) = p\left|e(k)^{p-2}\right| \tag{4.233}$$

μ stands for the step-size parameter, $P = \mathbf{I}_L - \mathbf{C}\left(\mathbf{C}^T\mathbf{C}\right)^{-1}\mathbf{C}$, and $Q = \mathbf{C}\left(\mathbf{C}^T\mathbf{C}\right)^{-1}\mathbf{F}$. The above algorithm is referred to as the CLMP algorithm, which includes the CLMS algorithm as a special case (when $p=2$). The CLMP algorithm is summarized in Table 4.4.

4.7.2 Convergence Analysis

This section gives the convergence analysis of the CLMP algorithm, and the following independent assumptions are first given [225]:

Assumption 4.25: The input vector $\mathbf{u}(k)$ is independent with zero-mean and diagonal auto-correlation matrix $\mathbf{R}$.

Assumption 4.26: The noise $v(k)$ is a zero-mean Gaussian random process and uncorrelated with any other signals.

TABLE 4.4

CLMP Algorithm

Parameters: μ, $\mathbf{C}$, and $\mathbf{F}$
Initialization: $P = \mathbf{I}_L - \mathbf{C}\left(\mathbf{C}^T\mathbf{C}\right)^{-1}\mathbf{C}$
$Q = \mathbf{C}\left(\mathbf{C}^T\mathbf{C}\right)^{-1}\mathbf{F}$
$\mathbf{w}(0)=Q$
Update:
$y(k) = \mathbf{w}^T(k)\mathbf{u}(k)$
$e(k) = d(k) - y(k)$
$\mathbf{w}(k) = P\left[\mathbf{w}(k-1) + \mu h(e(k))\left(d(k) - \mathbf{w}^T(k-1)\mathbf{u}(k)\right)\mathbf{u}(k)\right] + Q$

Under the above assumptions, one can derive the optimal solution of the constrained optimization problem. Combining the constrain function (4.231) and a $K\times 1$ vector undetermined Lagrange multipliers α, we have

$$J_{\mathrm{LMP}}(k)=|e(k)|^{p}+\alpha^{T}\left(\mathbf{C}^{T}\mathbf{w}(k)-\mathbf{F}\right) \tag{4.234}$$

By setting $\frac{J_{\mathrm{LMP}}(k)}{\mathbf{w}(k)}=\vec{0}_L$, an optimal weight vector can be obtained by the following method:

$$\begin{aligned}&\frac{J_{\mathrm{LMP}}(k)}{\mathbf{w}(k)}=h(e(k))\left(d(k)-\mathbf{w}^{T}(k)\mathbf{u}(k)\right)\mathbf{u}(k)+\mathbf{C}\alpha(k)=\vec{0}_L\\&\Leftrightarrow \mathbf{w}_o=d(k)\mathbf{R}^{-1}\mathbf{u}(k)+\frac{\mathbf{R}^{-1}\mathbf{C}\alpha(k)}{h(e(k))}\end{aligned} \tag{4.235}$$

Since

$$\begin{aligned}&\mathbf{C}^{T}\mathbf{w}_o=\mathbf{F}\\&\Leftrightarrow \alpha(k)=h(e(k))\left(\mathbf{C}^{T}\mathbf{R}^{-1}\mathbf{C}\right)^{-1}\left(\mathbf{F}-d(k)\mathbf{C}^{T}\mathbf{R}^{-1}\mathbf{u}(k)\right)\end{aligned} \tag{4.236}$$

we obtain

$$\mathbf{w}_o=d(n)\mathbf{R}^{-1}\mathbf{u}(k)+\mathbf{R}^{-1}\mathbf{C}\left(\mathbf{C}^{T}\mathbf{R}^{-1}\mathbf{C}\right)^{-1}\left(\mathbf{F}-d(k)\mathbf{C}^{T}\mathbf{R}^{-1}\mathbf{u}(k)\right) \tag{4.237}$$

Using the relationship between the desired and input signal in the FIR model, and after some straightforward vector manipulations, we have

$$d(k)\mathbf{R}^{-1}\mathbf{u}(k)=\mathbf{w}^{*} \tag{4.238}$$

Therefore, we can rewrite (4.237) as

$$\mathbf{w}_o=\mathbf{w}^{*}+\mathbf{R}^{-1}\mathbf{C}\left(\mathbf{C}^{T}\mathbf{R}^{-1}\mathbf{C}\right)^{-1}\left(\mathbf{F}-\mathbf{C}^{T}\mathbf{w}^{*}\right) \tag{4.239}$$

Combining (4.232) and (4.239), the weight error vector ss $\tilde{\mathbf{w}}(k)$ can be expressed as

$$\tilde{\mathbf{w}}(k)=\left[\begin{array}{c}P\left(\mathbf{I}_L-\mu h(e(k))P\mathbf{u}(k)\mathbf{u}^{T}(k)\right)\tilde{\mathbf{w}}(k-1)+\mu h(e(k))v(k)P\mathbf{u}(k)\\+\mu h(e(k))P\mathbf{u}(k)\mathbf{u}^{T}(k)\ell+P\mathbf{w}_o-\mathbf{w}+Q\end{array}\right] \tag{4.240}$$

where $\ell = \mathbf{w}^* - \mathbf{w}_o$. Since $P\mathbf{w}_o - \mathbf{w}_o + Q = \vec{0}_L$, one can rewrite (4.240) as

$$\tilde{\mathbf{w}}(k) = \left[\begin{array}{c} P\left(\mathbf{I}_L - \mu h(e(k)) P\mathbf{u}(k)\mathbf{u}^T(k)\right)\tilde{\mathbf{w}}(k-1) + \mu h(e(k)) v(k) P\mathbf{u}(k) \\ +\mu h(e(k)) P\mathbf{u}(k)\mathbf{u}^T(k)\ell \end{array} \right] \tag{4.241}$$

Combining (4.241) and the fact that matrix P is idempotent, i.e., $P = P^2$ and $P = P^T$, which can be easily verified, we derive

$$\tilde{\mathbf{w}}(k) = \left[\begin{array}{c} \left(\mathbf{I}_L - \mu h(e(k)) P\mathbf{u}(k)\mathbf{u}^T(k)P\right)\tilde{\mathbf{w}}(k-1) + \mu h(e(k)) v(k) P\mathbf{u}(k) \\ +\mu h(e(k)) P\mathbf{u}(k)\mathbf{u}^T(k)\ell \end{array} \right] \tag{4.242}$$

Assuming that $\xi(k) = E\left[\tilde{\mathbf{w}}(k)\right]$, taking the expectation of (4.242) and using the above assumptions yields

$$\xi(k) = \left(\mathbf{I}_L - \mu E\left[h(e(k))\right] P\mathbf{R}P\right)\xi(k-1) + \mu E\left[h(e(k))\right] P\mathbf{R}\ell \tag{4.243}$$

Since $P\mathbf{R}\ell = \vec{0}_L$, we have

$$\xi(k) = \left(\mathbf{I}_L - \mu E\left[h(e(k))\right] P\mathbf{R}P\right)\xi(k-1) \tag{4.244}$$

Assuming the filter achieves the steady state, i.e. $\lim_{k\to\infty}\xi(k) = \lim_{k\to\infty}\xi(k-1)$, we have

$$\mu E\left[h(e(k))\right] P\mathbf{R}P\xi(\infty) = \vec{0}_L \tag{4.245}$$

From (4.245), we get

$$\xi(\infty) = E\left[\mathbf{w}(\infty)\right] - \mathbf{w}^* = \vec{0}_L \tag{4.246}$$

Then we obtain

$$E\left[\mathbf{w}(\infty)\right] = \mathbf{w}^* \tag{4.247}$$

4.7.3 Simulation Results

Here some simulation results are presented to illustrate the performance of the CLMP algorithm compared with the CLMS algorithm in non-Gaussian situations. The performance measure is the MSD. Consider a problem of the constrained system identification where the system order is set to 7, and the number of the equality constraint is 3, accordingly. We set the system parameter, the constraint parameters, and the input covariance matrix with the conditions that $\mathbf{w}^*$ has unit energy, $\mathbf{C}$ is full-rank, and $\mathbf{R}$ is symmetric positive-definite [113]. The input signal is a zero-mean multivariate Gaussian distribution. First, the convergence behaviors of CLMP and CLMS in non-Gaussian noises are plotted in Figure 4.38. As one can see from the simulation results, the CLMP has much better steady-state performance than the CLMS.

Second, the simulation results with different parameters p are given to demonstrate the performance of the CLMP algorithm. The convergence curves with different are given in Figure 4.39. It is clear that when $p = 1.4$, the CLMP plays the best performance in this case. If p is too small or too large, the convergence performance will become worse. Therefore, the parameter p has significant influence on the CLMP algorithm, and how to select the best p is an interesting research in the future.

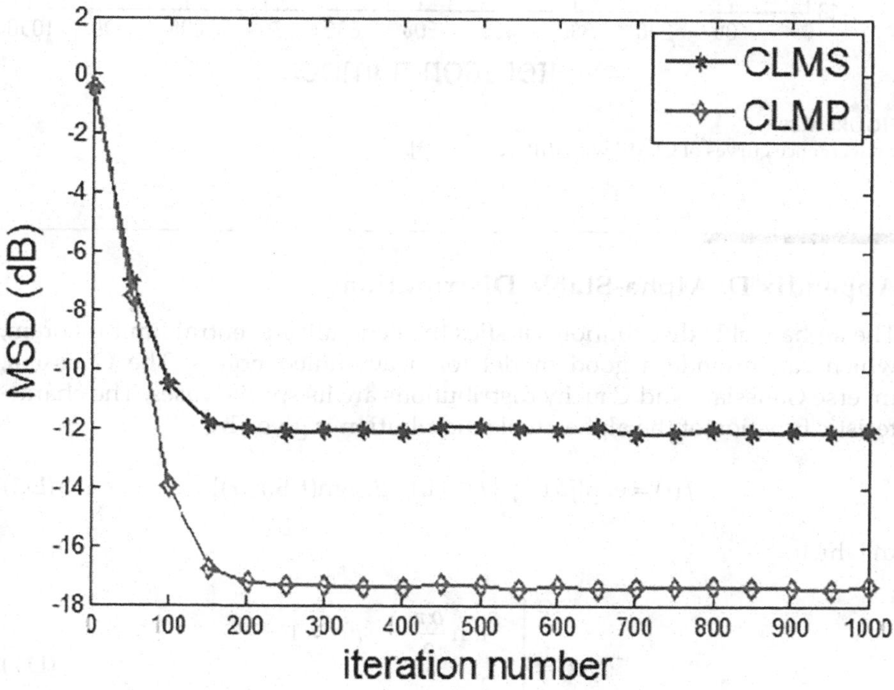

FIGURE 4.38
Convergence curves of CLMP and CLMS [225]

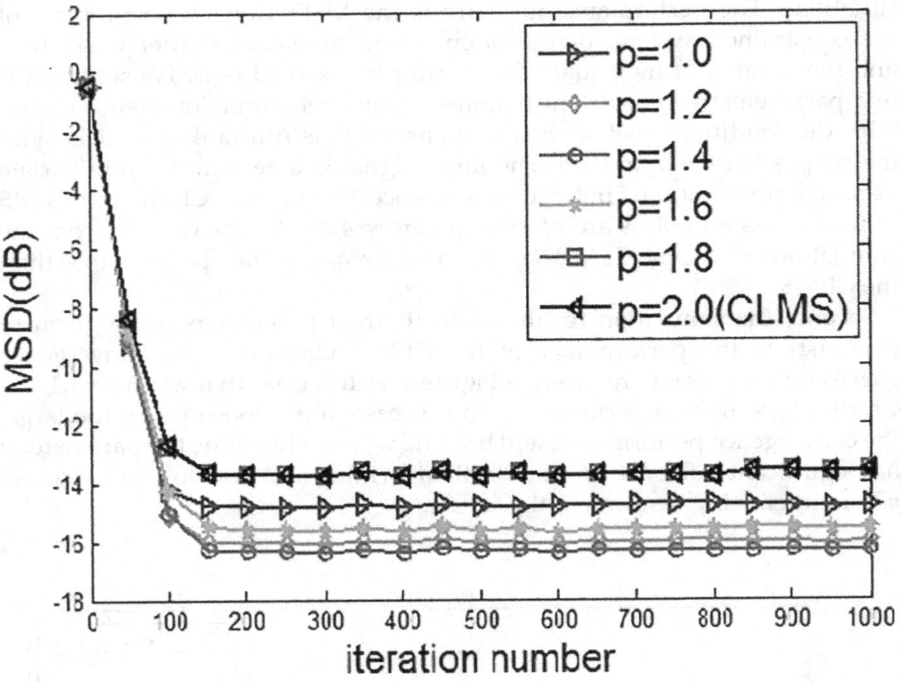

FIGURE 4.39
Convergence curves of CLMP with different p [225].

Appendix D: Alpha-Stable Distribution

The alpha-stable distribution satisfies the generalized central limit theorem, which can provide a good model for heavy-tailed noises. The Gaussian, inverse Gaussian, and Cauchy distributions are its special cases. The characteristic function of the alpha-stable distribution is given by

$$f(t) = \exp\{j\delta t - \gamma |t|^{\alpha} [1 + j\beta \operatorname{sign}(t) S(t,\alpha)]\} \tag{D.1}$$

in which

$$S(t,\alpha) = \begin{cases} \tan\dfrac{\alpha\pi}{2} & if \alpha \neq 1 \\ \dfrac{2}{\pi}\log|t| & if \alpha = 1 \end{cases} \tag{D.2}$$

where $\alpha \in (0,2]$ is the characteristic factor, $-\infty < \delta < +\infty$ is the location parameter, $\beta \in [-1,1]$ is the symmetry parameter, and $\gamma > 0$ is the dispersion parameter. The characteristic factor α measures the tail heaviness of the distribution. The smaller α is, the heavier the tail is. In addition, γ measures the dispersion of the distribution, which plays a similar role as the variance of Gaussian distribution. The distribution is symmetric about its location δ when $\beta = 0$, called a symmetric alpha-stable distribution ($S\alpha S$). The Gaussian and Cauchy distributions are alpha-stable distributions with $\alpha = 2$ and $\alpha = 1$, respectively.

When $\alpha < 2$, the tail attenuation of alpha-stable distribution is slower than that of Gaussian distribution, which can be used to describe the outlier data or impulsive noises. In this book, the impulsive interference noise is sometimes assumed to be drawn from alpha-stable distribution, and the parameters vector of the noise model is denoted by $V = (\alpha, \beta, \gamma, \delta)$.

Appendix E

For the case in which $p = 2l$, the functions $h_G^{(p)}(\cdot)$ and $h_U^{(p)}(\cdot)$ can be derived as follows:

$$
\begin{aligned}
h_G^{(p)}\left(\sigma_{e_a}^2\right) &= \frac{1}{\sigma_{e_a}^2} E\left[e_a(k) f_P\left(e_a(k)+v(k)\right)\right] \\
&= \frac{1}{\sigma_{e_a}^2} E\left[e_a(k)\left|e_a(k)+v(k)\right|^{p-1} sign\left(e_a(k)+v(k)\right)\right] \\
&\overset{p=2l}{=} \frac{1}{\sigma_{e_a}^2} E\left[e_a(k)\left\{e_a(k)+v(k)\right\}^{2l-1}\right] \\
&= \frac{1}{\sigma_e^2} E\left[e_a(k)\left\{\sum_{j=0}^{2l-1}\binom{2l-1}{j}\left(e_a(k)\right)^j + \left(v(k)\right)^{2l-1-j}\right\}\right] \qquad \text{(E.1)} \\
&= \frac{1}{\sigma_{e_a}^2}\left\{\sum_{j=1}^{l}\left[\binom{2l-1}{2j-1} E\left[\left(e_a(k)\right)^{2j}\right] E\left[\left(v(k)\right)^{2(l-j)}\right]\right]\right\} \\
&= \frac{1}{\sigma_{e_a}^2}\left\{\sum_{j=1}^{l}\left[\binom{2l-1}{2j-1}\frac{(2j)!}{2^j j!}\sigma_{e_a}^{2j} E\left(v(k)\right)^{2(l-j)}\right]\right\}
\end{aligned}
$$

$$\begin{aligned}
h_U^{(p)}\left(\sigma_{e_a}^2\right) &= E\left[f_p^2\left(e_a(k)+v(k)\right)\right] \\
&= E\left[\left\{e_a(k)+n(k)\right\}^{2(p-1)}\right] \\
&= E\left[\left\{\sum_{j=0}^{2(p-1)}\binom{2(p-1)}{j}\left(e_a(k)\right)^j\left(v(k)\right)^{2(p-1)-j}\right\}\right] \\
&= \sum_{j=0}^{p-1}\left\{\binom{2(p-1)}{2j}E\left[\left(e_a(k)\right)^{2j}\right]E\left[\left(v(k)\right)^{2(p-j-1)}\right]\right\} \\
&= \sum_{j=0}^{p-1}\left\{\binom{2(p-1)}{2j}\frac{(2j)!}{2^j j!}\sigma_{e_a}^{2j}E\left[\left(v(k)\right)^{2(p-j-1)}\right]\right\} \\
&\overset{p=2l}{=} \sum_{j=0}^{2l-1}\left\{\binom{4l-2}{2j}\frac{(2j)!}{2^j j!}\sigma_{e_a}^{2j}E\left[\left(v(k)\right)^{2(2l-j-1)}\right]\right\}
\end{aligned} \tag{E.2}$$

Appendix F

Theorem 4.2

Let the X_0 be the solution we get by minimizing the l_0-norm and X_l be the solution we achieve by minimizing CIM. Then we have

$$\mathrm{CIM}(X_l,0) < \mathrm{CIM}(X_0,0) \tag{F.1}$$

Proof: According to the definition of the CIM

$$\Rightarrow \sum_{i=1}^{N} \exp\left(-\frac{(X_l)_i^2}{2\sigma^2}\right) \geq \sum_{i=1}^{N} \exp\left(-\frac{(X_0)_i^2}{2\sigma^2}\right)$$

$$\Rightarrow \sum_{i=1,(X_l)^2=0}^{N} \exp\left(-\frac{(X_l)_i^2}{2\sigma^2}\right) + \sum_{i=1,(X_l)^2\neq 0}^{N} \exp\left(-\frac{(X_l)_i^2}{2\sigma^2}\right) \geq \sum_{i=1,(X_0)^2=0}^{N} \exp\left(-\frac{(X_0)_i^2}{2\sigma^2}\right)$$

$$+ \sum_{i=1,(X_0)^2\neq 0}^{N} \exp\left(-\frac{(X_0)_i^2}{2\sigma^2}\right)$$

$$\Rightarrow \left(N - \|X_l\|_0\right) + \sum_{i=1(X_l)^2\neq 0}^{N} \exp\left(-\frac{(X_l)_i^2}{2\sigma^2}\right) \geq \left(N - \|X_0\|_0\right) + \sum_{i=1(X_0)^2\neq 0}^{N} \exp\left(-\frac{(X_0)_i^2}{2\sigma^2}\right)$$

$$\Rightarrow \|X_l\|_0 - \sum_{i=1(X_l)^2\neq 0}^{N} \exp\left(-\frac{(X_l)_i^2}{2\sigma^2}\right) \geq \|X_0\|_0 - \sum_{i=1(X_0)^2\neq 0}^{N} \exp\left(-\frac{(X_0)_i^2}{2\sigma^2}\right)$$

$$\Rightarrow \|X_l\|_0 - \|X_0\|_0 \leq \sum_{i=1(X_l)^2\neq 0}^{N} \exp\left(-\frac{(X_l)_i^2}{2\sigma^2}\right) - \sum_{i=1(X_0)^2\neq 0}^{N} \exp\left(-\frac{(X_0)_i^2}{2\sigma^2}\right)$$

Now if we assume that $\left|(X_0)_{j,(X_0)_j\neq 0}\right|, \left|(X_l)_{j,(X_l)_j\neq 0}\right|_0 > \delta$, where δ is a small positive number, then we can choose a suitable kernel width to make the right-hand side of the equation arbitrarily close to zero. Therefore, we arrive at the condition given by $\|X_0\|_0 \leq \|X_l\|_0 \leq \|X_0\|_0 + \tau$, where τ is a small positive number.

Appendix G

Defining a function as follows:

$$\bar{g}(x) = \tau x \exp\left(-\frac{x^2}{2\sigma^2}\right) \tag{G.1}$$

where τ is a constant. Obviously, one can get $\bar{g}(x) = g_{\text{CIM}}(w_i(k)) = -\frac{1}{L\sigma^3\sqrt{2\pi}}$ when $\tau = -\frac{1}{L\sigma^3\sqrt{2\pi}}$ and $x = w_i(k)$. Further, we compute the limit of the function $\bar{g}(x)$ with $x \to \infty$:

$$\lim_{x\to\infty}\bar{g}(x)=\lim_{x\to\infty}\tau x\exp\left(-\frac{x^2}{2\sigma^2}\right)=\tau\lim_{x\to\infty}\frac{x}{\frac{x^2}{e^{2\sigma^2}}}=\tau\lim_{x\to\infty}\frac{\sigma^2}{\frac{x^2}{xe^{2\sigma^2}}}=0 \tag{G.2}$$

The estimated weight $w_i(k)$ is limited in general, and even if the channel parameter $w_i(k)$ tends to infinity, the vector $g_{CIM}(w_i(k))$ is still limited.

5

Recursive Adaptive Filtering Algorithms under MMPE

In Chapter 4, the LMP-aware algorithms have been presented. A major drawback of LMP-aware algorithms with stochastic gradient method is the slow convergence in the presence of colored input signals. To accelerate the convergence speed in such conditions, RLS typed algorithms are usually preferred. As a result, some recursive LMP (RLMP) algorithms are introduced in this chapter. A sliding window adaptation algorithm, called RLMP, is proposed [236,324], which is similar to the RLS algorithm both in terms of derivation and convergence characteristics. The RLMP algorithm provides much increased convergence rate at the expense of increased computational complexity. However, the RLMP algorithm with a large forgetting factor yields a reduced steady-state misalignment at the expense of a poor tracking capability and with a small forgetting factor they offer an improved tracking capability at the cost of an increased steady-state misalignment [237]. Thus, a convex adaptive combination of two RLpN filters (AC-RLpN) was proposed in [237], which adopts a small forgetting factor to improve the tracking performance. Furthermore, a method is proposed to enhance the tracking performance of the RLpN (named ET-RLpN algorithm) [238], which employs the square of the estimated impulsive-free first moment of the error signal to control the gain factor update. Similar to the RLS algorithm that is used for sparse parameter estimation, and distributed estimation, the RLMP with sparsity penalty term and the diffusion RLMP algorithms are further presented. In addition, we also introduce a constrained recursive least *p*-power (CRLP) algorithm [239], which is designed by incorporating a linear constraint into the MMPE criterion.

5.1 Recursive AFAs under MMPE

In this part, the RLMP algorithm, aiming at recursively minimizing the MMPE (or l_p-norm), is first introduced. To improve the tracking ability of the RLMP, an enhanced tracking RLMP algorithm is further presented.

DOI: 10.1201/9781003176114-5

5.1.1 Recursive Least Mean *p*-Power Error Algorithm

The primary goal of an adaptation algorithm is to reduce the average error at the filter output by adjusting the filter coefficients. When estimating a time-varying finite impulse response system adaptively, the filtering memory is typically restricted. In this case, the adaptation of filter coefficients is achieved through the use of a true finite memory or a sliding window approach. Specifically, the error's average p-norm is minimized within a window of size L as

$$J = \sum_{i=1}^{k} \gamma^{k-i} \left|e_i(k)\right|^p \tag{5.1}$$

where $0 \ll \gamma < 1, 0 < p < \alpha$ and $e_i(k)$ is defined as

$$e_i(k) = d(i) - \mathbf{w}^T(k)\mathbf{u}(i) \tag{5.2}$$

The optimal solution for minimizing J can be obtained by differentiating (5.1) with respect to $\mathbf{w}(k)$ and setting the derivatives to zero. After some straightforward manipulations, we obtain the following normal equation:

$$\mathbf{\Psi}_{\text{MPE}}(k)\mathbf{w}\ (k) = \mathbf{\Phi}_{\text{MPE}}(k) \tag{5.3}$$

where

$$\mathbf{\Psi}_{\text{MPE}}(k) = \sum_{i=1}^{k} \gamma^{k-i} s\left(d(i) - \mathbf{w}^T(k)\mathbf{u}(i)\right)\mathbf{u}(i)\mathbf{u}^T(i) \tag{5.4}$$

$$\mathbf{\Phi}_{\text{MPE}}(k) = \sum_{i=1}^{k} \gamma^{k-i} s\left(d(i) - \mathbf{w}^T(k)\mathbf{u}(i)\right)d(i)\mathbf{u}(i) \tag{5.5}$$

where the weighted window function $s(\cdot)$ of RLP is $s(x) = |x|^{p-2}$.

To derive a truly online algorithm, the following approximations are used:

$$\mathbf{\Psi}_{\text{MPE}}(k) \approx \gamma\ \mathbf{\Psi}_{\text{MPE}}(k-1) + s\left(d(k) - \mathbf{w}^T(k)\mathbf{u}(k)\right)\mathbf{u}(k)\mathbf{u}^T(k) \tag{5.6}$$

$$\mathbf{\Phi}_{\text{MPE}}(k) \approx \gamma\ \mathbf{\Phi}_{\text{MPE}}(k-1) + s\left(d(k) - \mathbf{w}^T(k)\mathbf{u}(k)\right)d(k)\mathbf{u}(k) \tag{5.7}$$

Using matrix inversion lemma, the following update equations for the RLP algorithm can be obtained

$$\mathbf{K}_{\mathrm{MPE}}(k) = \frac{s\left(d(k) - \mathbf{w}^T(k)\mathbf{u}(k)\right)\mathbf{\Omega}_{\mathrm{MPE}}(k-1)\mathbf{u}(k)}{\gamma + s\left(d(k) - \mathbf{w}^T(k)\mathbf{u}(k)\right)\mathbf{u}^T(k)\mathbf{\Omega}_{\mathrm{MPE}}(k-1)\mathbf{u}(k)} \tag{5.8}$$

$$\mathbf{w}(k) = \mathbf{w}(k) + \mathbf{K}_{\mathrm{MPE}}(k)\left(d(k) - \mathbf{w}(k-1)\mathbf{u}(k)\right) \tag{5.9}$$

$$\mathbf{\Omega}_{\mathrm{MPE}}(k) = \gamma^{-1}\left(\mathbf{I} - \mathbf{K}_{\mathrm{MPE}}(k)\mathbf{u}^T(k)\right)\mathbf{\Omega}_{\mathrm{MPE}}(k-1) \tag{5.10}$$

where $\mathbf{\Omega}_{\mathrm{MPE}}(k) = \mathbf{\Psi}^T_{\mathrm{MPE}}(k)$ is inverse covariance-like matrix, $\mathbf{K}_{\mathrm{MPE}}(k)$ is the gain vector and the symbol $\mathbf{I}$ denotes an identity matrix of an appropriate dimension.

We know that the RLP algorithm, aiming at recursively minimizing the MPE, offers a more stable and robust solution than adaptive filtering schemes based on the minimization of the MSE. However, since the RLMP solution cannot be obtained in closed form for $p \neq 2$, it is necessary to introduce some approximations that critically affect the filter behavior. The main observed drawback is a poor convergence rate in nonstationary scenarios, especially in the presence of abrupt changes in the model. In this correspondence, this problem can be overcome by relying on convex combinations of two RLP filters with long and short memories [237].

5.1.2 Enhanced Recursive Mean *p*-Power Error Algorithm

The RLMP algorithm's primary limitation is its subpar tracking performance when faced with sudden changes in the model. Therefore, an enhanced tracking RLMP (ET-RLpN) is further presented to enhance tracking capability of the RLMP (ET-RLpN) algorithm in this section, which uses the adaptive gain factor in the cross-correlation vector and the input-signal auto-correlation matrix to enhance tracking capability. The RLP algorithm forces an iterative approximation to the solution (5.3). Inserting (5.6) and (5.7) into (5.3), we get [238]

$$\begin{aligned}\mathbf{w}(k) = &\left(\gamma\mathbf{\Psi}_{\mathrm{MPE}}(k-1) + s\left(d(k) - \mathbf{w}^T(k)\mathbf{u}(k)\right)\mathbf{u}(k)\mathbf{u}^T(k)\right)^{-1} \times \\ &\left(\gamma\mathbf{\Phi}(k-1) + s\left(d(k) - \mathbf{w}^T(k)\mathbf{u}(k)\right)d(k)\mathbf{u}(k)\right)\end{aligned} \tag{5.11}$$

To illustrate slow tracking performance, Figure 5.1(a) shows the function $s(\cdot)$. For a sudden system disturbance, $|e(k)|$ would be very large for large filter length L, in which case the weighted window function $s(e(k))$ would be a small value, and therefore, $\mathbf{w}(k) \approx \mathbf{\Psi}^{-1}_{\mathrm{MPE}}(k)\mathbf{\Phi}_{\mathrm{MPE}}(k)$, the tracking capability of the RLpN would be degraded.

To reduce the negative effects of the function $s(\cdot)$ for a sudden system disturbance, (5.6) and (5.7) can be modified by

$$\mathbf{\Psi}_{\mathrm{MPE}}(k) \approx \gamma \mathbf{\Psi}_{\mathrm{MPE}}(k-1) + T(k) \times s\left(d(k) - \mathbf{w}^T(k)\mathbf{u}(k)\right)\mathbf{u}(k)\mathbf{u}^T(k) \tag{5.12}$$

$$\mathbf{\Phi}_{\mathrm{MPE}}(k) \approx \gamma \mathbf{\Phi}_{\mathrm{MPE}}(k-1) + T(k) \times s\left(d(k) - \mathbf{w}^T(k)\mathbf{u}(k)\right)d(k)\mathbf{u}(k) \tag{5.13}$$

where $T(k) = \left(E\left[|e(k)|\right]\right)^2$ is an adaptive gain factor to enhance the tracking performance. Exploiting the MAD in [325], the impulsive-free estimation of $E[|e(k)|]$ in time instant n is estimated as

$$\hat{\sigma}(k) = \beta\hat{\sigma}(k-1) + 1.483(1+5/(\mathrm{N}-1))(1-\beta)\,\mathrm{median}(\psi(k)) \tag{5.14}$$

where $0 \ll \beta < 1$, $\psi(k) = \left[|e(k)|, |e(k-1)|, \cdots |e(k-N+1)|\right]^T$, $1.483(1+5/(N-1))$ is a finite sample correction factor [326], and N is the length of the estimation window, which is usually chosen between 5 and 9 [326]. The medium operation helps to remove the effect of additive impulsive noise in estimating $E\left[|e(k)|\right]$. Applying the matrix inversion lemma in (5.12), the ET-RLpN algorithm can be derived as follows:

$$\mathbf{K}_{\mathrm{MPE}}(k) = \frac{T(k) \times s\left(d(k) - \mathbf{w}^T(k)\mathbf{u}(k)\right)\mathbf{\Omega}_{\mathrm{MPE}}(k-1)\mathbf{u}(k)}{\gamma + T(k) \times s\left(d(k) - \mathbf{w}^T(k)\mathbf{u}(k)\right)\mathbf{u}^T(k)\mathbf{\Omega}_{\mathrm{MPE}}(k-1)\mathbf{u}(k)} \tag{5.15}$$

$$\mathbf{w}(k) = \mathbf{w}(k-1) + \mathbf{K}_{\mathrm{MPE}}(k)\mathrm{e}(k) \tag{5.16}$$

$$\mathbf{\Omega}_{\mathrm{MPE}}(k) = \gamma^{-1}\left(I - \mathbf{K}_{\mathrm{MPE}}(k)\mathbf{u}^T(k)\right)\mathbf{\Omega}_{\mathrm{MPE}}(k-1) \tag{5.17}$$

Remark 5.1

In (5.12) and (5.13), the weighted window function $T(k)\times s(\bullet)$ is used to achieve the robust performance and the tracking capability. The weighted window function $T(k)\times s(\bullet)$ is shown in Figure 5.1(b). Inserting (5.12) and (5.13) into (5.3), we obtain

$$\begin{aligned}\mathbf{w}(k) = &\left(\gamma\, \mathbf{\Psi}_{\mathrm{MPE}}(k-1) + T(k) \times s\left(d(k) - \mathbf{w}^T(k)\mathbf{u}(k)\right)\mathbf{u}(k)\mathbf{u}^T(k)\right)^{-1} \\ &\times\left(\gamma\, \mathbf{\Phi}(k-1) + T(k) \times s\left(d(k) - \mathbf{w}^T(k)\mathbf{u}(k)\right)d(k)\mathbf{u}(k)\right)\end{aligned} \tag{5.18}$$

For a sudden system disturbance, $T(k)$ would also be very large for large filter length L, in which case (5.18) can be approximated as

$$\begin{aligned}\mathbf{w}(k) = &\left(T(k) \times s\left(d(k) - \mathbf{w}^T(k)\mathbf{u}(k)\right)\mathbf{u}(k)\mathbf{u}^T(k)\right)^{-1} \\ &\times\left(T(k) \times s\left(d(k) - \mathbf{w}^T(k)\mathbf{u}(k)\right)d(k)\mathbf{u}(k)\right)\end{aligned} \tag{5.19}$$

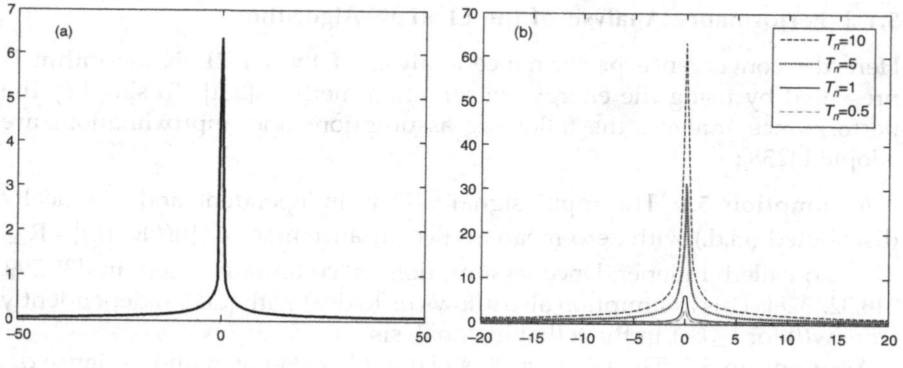

FIGURE 5.1
The functions $s(x)$ and $T_n \times s(x)$ with respect to x, where $p = 1$. (a) The function $s(x)$. (b) The function $T_n \times s(x)$ with $T_n = 10, 5, 1 \text{and} 0.5$ [238].

and therefore, the tracking capability of the algorithm would be retained. The steady-state analysis in Section 5.1.3 shows that the excess MSE of the ET-RLpN is independent of the adaptive gain factor $T(k)$.

Remark 5.2

The computational complexity of the ET-RLpN is compared with that of conventional algorithms in terms of the total number of additions, multiplications, comparisons, square roots, and median operation (MO). With the filter length L and estimation window N, the computational requirements for these adaptive filters with corresponding adaptive algorithms are summarized in Table 5.1. With only slightly more computations in the gain factor computation, the ET-RLpN behaves much better than the conventional RLS, RLM, and RLpN, especially for an abrupt change in system coefficients.

TABLE 5.1
Computational Complexity of Algorithms' Coefficients Update per Iteration

Algorithm	Multiplication	Addition	Square-root	Comparison	MO
RLS	$3L^2+4L+2$	$2L^2+2L$	0	0	0
RLM [294]	$3L^2+4L+6$	$2L^2+2L+1$	1	1	$O(N\log N)$
RLMP(p = 1) [295]	$3L^2+4L+4$	$2L^2+2L$	0	0	0
RRLS [296]	$3L^2+4L+5$	$2L^2+2L+2$	0	1	0
CAC-RLMP(p = 1) [222]	$2(3L^2+4L+4)+10$	$2(L^2+2L)+8$	0	1	0
ET-RLpN (p = 1)	$3L^2+4L+8$	$2L^2+2L+1$	**0**	**0**	**$O(N\log N)$**

5.1.3 Performance Analysis of the ET-RLpN Algorithm

Here the convergence performance analysis of the ET-RLpN algorithm is presented by using the energy conservation method [261]. To simplify the performance analysis, the following assumptions and approximations are adopted [238]:

Assumption 5.1: The input signal $\mathbf{u}(k)$ is independent and identically distributed (i.i.d.) with zero-mean and covariance matrix $E[\mathbf{u}(k)\mathbf{u}^T(k)] = \mathbf{R}_{\mathbf{uu}}$. This so-called independence assumption is commonly used in [35,299, 316,327,328]. This assumption also allows us to deal with $\{\mathbf{u}(k)\}$ independently from $\mathbf{w}(l)$ for $l \leq k-1$ in the following analysis.

Assumption 5.2: The noise $\{v(k)\}$ is i.i.d. with zero-mean and variance σ_v^2, and is independent of the input signal $\mathbf{u}(j)$ for all k, j.

Assumption 5.3: At steady state, $\mathbf{u}(k)$ is approximately independent of the *a priori* excess errors $e_a(k) = \tilde{\mathbf{w}}^T(k-1)\mathbf{u}(k)$, where $\tilde{\mathbf{w}}(k) = \mathbf{w}_o - \mathbf{w}(k)$ denotes the weight deviation vector. This is the so-called separation principle invoked in [261].

Assumption 5.4: The value of the forgetting factor β in (5.14) is close to one, so that $T(k)$ may be approximated by its mean value in the expressions.

5.1.3.1 Mean Stability

Applying the weight deviation vector, the estimation error $e(k)$ is rewritten as $e(k) = v^T(k-1)\mathbf{u}(k) - v(k)$.

Combining (5.13) and (5.14), we can rewrite the weight update equation as

$$\mathbf{w}(k) = \mathbf{w}(k-1) + \mu(k)\mathbf{\Omega}_{\mathrm{MPE}}(k-1)\,\mathrm{e}(k)\mathbf{u}(k) \tag{5.20}$$

where

$$\mu(k) = \frac{1}{\gamma\left|e(\mathrm{k})\right|^{2-p}/T(k) + \mathbf{u}^T(k)\mathbf{\Omega}_{\mathrm{MPE}}(k)\mathbf{u}(k)} \tag{5.21}$$

Using (5.20), the weight deviation vector recursion for the ET-RLpN can be written as

$$\tilde{\mathbf{w}}(k) = \tilde{\mathbf{w}}(k-1) + \mu(k)\mathbf{\Omega}_{\mathrm{MPE}}(k-1)\,\mathrm{e}(k)\mathbf{u}(k) \tag{5.22}$$

The mean convergence behavior of the coefficient vector can be determined by taking the expectation of both sides of (5.22)

$$E[\tilde{\mathbf{w}}(k)] = E\left[\tilde{\mathbf{w}}(k-1)\right] - E\left[\mu(k)\mathbf{\Omega}_{\mathrm{MPE}}(k-1)\mathbf{u}(k)\mathbf{u}^T(k)\tilde{\mathbf{w}}(k-1)\right] \tag{5.23}$$

where $E\left[\mu(k)\mathbf{\Omega}_{\mathrm{MPE}}(k-1)\mathbf{u}(k)\tilde{\mathbf{w}}(k)\right] \approx 0$, since $e_a(k) \approx \mathbf{u}^T(k)\tilde{\mathbf{w}}(k-1)$ during the transient state. Hence, the weight vector in the ET-RLpN can converge in the mean if and only if

$$0 < \lambda_{\max}\left(E\left[\mu(k)\boldsymbol{\Omega}_{\text{MPE}}(k-1)\mathbf{u}(k)\mathbf{u}^T(k)\right]\right) < 2 \tag{5.24}$$

where $\lambda_{\max}(\bullet)$ denotes the largest eigenvalue of a matrix.

Using the fact that $\lambda_{\max}(AB) < Tr(AB)$ in (5.24) and substituting (5.21) into (5.24), we obtain

$$\begin{aligned} 0 &< \lambda_{\max}\left(E\left[\frac{\boldsymbol{\Omega}_{\text{MPE}}(k-1)\mathbf{u}(k)\mathbf{u}^T(k)}{\gamma|e(\mathrm{k})|^{2-p}/T(k)+\mathbf{u}^T(k)\boldsymbol{\Omega}_{\text{MPE}}(k)\mathbf{u}(k)}\right]\right) \\ &< E\left[\frac{Tr\left(\mathbf{u}^T(\mathrm{k})\boldsymbol{\Omega}_{\text{MPE}}(k-1)\mathbf{u}(k)\right)}{\gamma|e(\mathrm{k})|^{2-p}/T(k)+\mathbf{u}^T(k)\boldsymbol{\Omega}_{\text{MPE}}(k)\mathbf{u}(k)}\right] \end{aligned} \tag{5.25}$$

Hence, the mean error weight vector of the ET-RLpN algorithm is convergent if the input signal is persistently exciting [261].

5.1.3.2 Steady State Analysis

In this subsection, the derivation of the excess mean square error (EMSE) of the ET-RLpN at steady state will be performed. The steady-state EMSE is defined as

$$\eta = \lim_{k\to\infty} E\left[e_a^2(k)\right] = \lim_{k\to\infty} E\left[\|\tilde{\mathbf{w}}(k-1)\|_R^2\right] \tag{5.26}$$

where (5.26) is from Assumption 5.1. By premultiplying both sides of (5.22) by $\mathbf{u}^T(k)$, we get

$$e_p(k) = e_a(k) - \|\mathbf{u}(k)\|^2 \mu(k)\boldsymbol{\Omega}_{\text{MPE}}(k-1)\mathrm{e}(k) \tag{5.27}$$

where $e_p(k) = \tilde{\mathbf{w}}^T(k)\mathbf{u}(k)$ is the *a posteriori* excess error.

Combining (5.22) and (5.27), we have

$$\tilde{\mathbf{w}}(k) + \mu(k)\boldsymbol{\Omega}_{\text{MPE}}(k-1)\frac{\mathrm{e}_a(k)\mathbf{u}(k)}{\|\mathbf{u}(k)\|^2_{\mu(k)\boldsymbol{\Omega}_{\text{MPE}}(k-1)}} = \tilde{\mathbf{w}}(k-1) + \mu(k)\boldsymbol{\Omega}_{\text{MPE}}(k-1)\frac{\mathrm{e}_p(k)\mathbf{u}(k)}{\|\mathbf{u}(k)\|^2_{\mu(k)\boldsymbol{\Omega}_{\text{MPE}}(k-1)}} \tag{5.28}$$

Taking the weighted Euclidean norms of both sides of (5.28), the fundamental weighted energy conservation relationship can be shown to be

$$\|\tilde{\mathbf{w}}(k)\|^2_{\mu^{-1}(k)\boldsymbol{\Omega}^{-1}_{\text{MPE}}(k-1)} + \frac{\mathrm{e}_a^2(k)}{\|\mathbf{u}(k)\|^2_{\mu(k)\boldsymbol{\Omega}_{\text{MPE}}(k-1)}} = \|\tilde{\mathbf{w}}(k-1)\|^2_{\mu^{-1}(k)\boldsymbol{\Omega}^{-1}_{\text{MPE}}(k-1)} + \frac{\mathrm{e}_p^2(k)}{\|\mathbf{u}(k)\|^2_{\mu(k)\boldsymbol{\Omega}_{\text{MPE}}(k-1)}} \tag{5.29}$$

Taking expectation on both sides of (5.29), we get

$$E\left[\|\tilde{\mathbf{w}}(k)\|^2_{\mu^{-1}(k)\Omega^{-1}_{\mathrm{MPE}}(k-1)}\right]+E\left[\frac{\mathrm{e}_a^2(k)}{\|\mathbf{u}(k)\|^2_{\mu(k)\Omega_{\mathrm{MPE}}(k-1)}}\right]$$
$$=E\left[\|\tilde{\mathbf{w}}(k-1)\|^2_{\mu^{-1}(k)\Omega^{-1}_{\mathrm{MPE}}(k-1)}\right]+E\left[\frac{\mathrm{e}_p^2(k)}{\|\mathbf{u}(k)\|^2_{\mu(k)\Omega_{\mathrm{MPE}}(k-1)}}\right] \tag{5.30}$$

By assuming (5.30) is operated at steady state when $k\to\infty$ and $E\left[\|\tilde{\mathbf{w}}(k)\|^2_{\mu^{-1}(k)\Omega^{-1}_{\mathrm{MPE}}(k-1)}\right]=E\left[\|\tilde{\mathbf{w}}(k-1)\|^2_{\mu^{-1}(k)\Omega^{-1}_{\mathrm{MPE}}(k-1)}\right]$, the above relationship can be got as follows

$$E\left[\frac{\mathrm{e}_a^2(k)}{\|\mathbf{u}(k)\|^2_{\mu(k)\Omega_{\mathrm{MPE}}(k-1)}}\right]=E\left[\frac{\mathrm{e}_p^2(k)}{\|\mathbf{u}(k)\|^2_{\mu(k)\Omega_{\mathrm{MPE}}(k-1)}}\right] \tag{5.31}$$

Substituting $e_p(k)=\tilde{\mathbf{w}}^T(k)\mathbf{u}(k)$ in (5.27) into (5.31), we obtain

$$E\left[\frac{\mathrm{e}_a^2(k)}{\|\mathbf{u}(k)\|^2_{\mu(k)\Omega_{\mathrm{MPE}}(k-1)}}\right]=E\left[\frac{\mathrm{e}_p^2(k)}{\|\mathbf{u}(k)\|^2_{\mu(k)\Omega_{\mathrm{MPE}}(k-1)}}\right]-2E\left[e_a(k)e(k)\right]+E\left[\|\mathbf{u}(k)\|^2_{\mu^{-1}(k)\Omega^{-1}_{\mathrm{MPE}}(k-1)}e(k)\right] \tag{5.32}$$

Thus, we obtain that at steady state

$$2E\left[e_a(\infty)e(\infty)\right]=E\left[\|\mathbf{u}(k)\|^2_{\mu(\infty)\Omega_{\mathrm{MPE}}(\infty)}e^2(\infty)\right] \tag{5.33}$$

i. **The term of the left side of (5.33)**

Considering Assumption 5.2 and $e(k)=e_a(k)+v(k)$, the term of the left side of (5.33) is

$$E\left[e_a(\infty)e(\infty)\right]=E\left[e_a^2(\infty)\right] \tag{5.34}$$

ii. **The term of the right side of (5.33)**

$$E\left[\|\mathbf{u}(k)\|^2_{\mu(\infty)\Omega_{\mathrm{MPE}}(\infty)}e^2(\infty)\right]=E\left[\|\mathbf{u}(k)\|^2_{\mu(\infty)\Omega_{\mathrm{MPE}}(\infty)}\left(e_a(\infty)+v(\infty)\right)^2\right]$$
$$=E\left[\|\mathbf{u}(k)\|^2_{\mu(\infty)\Omega_{\mathrm{MPE}}(\infty)}e_a^2(\infty)\right]+2E\left[e_a(\infty)+v(\infty)\|\mathbf{u}(k)\|^2_{\mu(\infty)\Omega_{\mathrm{MPE}}(\infty)}\right] \tag{5.35}$$
$$+E\left[\|\mathbf{u}(k)\|^2_{\mu(\infty)\Omega_{\mathrm{MPE}}(\infty)}v^2(\infty)\right]$$
$$\approx E\left[\|\mathbf{u}(k)\|^2_{\mu(\infty)\Omega_{\mathrm{MPE}}(\infty)}\right]E\left[e_a^2(\infty)\right]++E\left[\|\mathbf{u}(k)\|^2_{\mu(\infty)\Omega_{\mathrm{MPE}}(\infty)}\right]\sigma_v^2$$

where (5.35) is from Assumption 5.3 and the following approximations

$$E\left[\|\mathbf{u}(k)\|^2_{\mu(\infty)\Omega_{\text{MPE}}(\infty)} e_a^2(\infty)\right] \approx E\left[\|\mathbf{u}(k)\|^2_{\mu(\infty)\Omega_{\text{MPE}}(\infty)}\right] E\left[e_a^2(\infty)\right]$$

$$= E\left[\|\mathbf{u}(k)\|^2_{\mu(\infty)\Omega_{\text{MPE}}(\infty)} v^2(\infty)\right] \approx E\left[\|\mathbf{u}(k)\|^2_{\mu(\infty)\Omega_{\text{MPE}}(\infty)}\right] E\left[v^2(\infty)\right]$$

$$+E\left[e_a(\infty)v(\infty)\|\mathbf{u}(k)\|^2_{\mu(\infty)\Omega_{\text{MPE}}(\infty)}\right] \approx E\left[e_a(\infty)\|\mathbf{u}(k)\|^2_{\mu(\infty)\Omega_{\text{MPE}}(\infty)}\right] E\left[v(\infty)\right]$$

Inserting (5.34) and (5.35) into (5.33), we obtain

$$\eta = \frac{\kappa \sigma_v^2}{2-\kappa} \tag{5.36}$$

where

$$\kappa = E\left[\|\mathbf{u}(k)\|^2_{\mu(\infty)\Omega_{\text{MPE}}(\infty)}\right]$$

$$= E\left[\frac{T(k)Tr\left(\mathbf{u}(k)\mathbf{u}^T(k)\boldsymbol{\Omega}_{\text{MPE}}(\infty)\right)}{\gamma |e(k)|^{2-p} + T(k)Tr\left(\mathbf{u}(k)\mathbf{u}^T(k)\boldsymbol{\Omega}_{\text{MPE}}(\infty)\right)}\right] \tag{5.37}$$

Using the time average method, the approximation $\mathbf{E}[\boldsymbol{\Omega}_{\text{MPE}}(\infty)]$ is evaluated in Appendix **H** as

$$\mathbf{E}[\boldsymbol{\Omega}_{\text{MPE}}(\infty)] = \frac{1-\gamma}{T(\infty)\times E\left[|e(\infty)|^{p-2}\right]}\mathbf{R}^{-1} \tag{5.38}$$

If we now replace $\boldsymbol{\Omega}_{\text{MPE}}(\infty)$, $T(k)$ and $\mathbf{u}(k)\mathbf{u}^T(k)$ by their assumed mean values in (5.37), we obtain

$$\kappa = E\left[\frac{T(k)Tr\left(\mathbf{u}(k)\mathbf{u}^T(k)\boldsymbol{\Omega}_{\text{MPE}}((1-\gamma)L\infty)\right)}{E\left[|e(\infty)|^{2-p}\right] + \gamma |e(k)|^{2-p} + (1-\gamma)L}\right] \tag{5.39}$$

Remark 5.3

Note that $|e_a(k)| \ll |v(k)|$ at steady state for $0 \ll \lambda < 1$. Then, we have from (5.36) that

$$\eta = \frac{(1-\gamma)\mathrm{L}\times \mathrm{E}\left[1\Big/\left(E\left[|v(k)|^{2-p}\right]\gamma|v(k)|^{2-p} + (1-\gamma)L\right)\right]\sigma_v^2}{2-(1-\gamma)\mathrm{L}\times \mathrm{E}\left[1\Big/\left(E\left[|v(k)|^{2-p}\right]\gamma|v(k)|^{2-p} + (1-\gamma)L\right)\right]} \tag{5.40}$$

As one can see, the excess MSE (5.40) is independent of the adaptive gain factor $T(k)$.

5.1.3.3 Verification of MSE

In this section, experimental verification is provided for the EMSE formulas given by (5.40) in a system identification application for the ET-RLpN algorithm. The unknown system to be estimated in all experiments was an order L finite impulse response filter, and the weight vector of the unknown system was randomly generated and normalized to unit power. The initial weight vector of the adaptive filter was an all zero vector. We used correlated Gaussian input signal $\mathbf{u}(k)$, which was simulated by a first-order auto-regressive (AR) process. The theoretical steady-state MSE was calculated using the formula:

$$\text{MSE} = \sigma_V^2 + \text{EMSE}$$

where EMSE was given in (5.37) and was calculated by ensemble averaging. The experimental and theoretical MSE values are given in Table 5.2. As one can see, the experimental results agree quite well with the theoretical results and, in effect, the steady-state performance of the ET-RLpN algorithm can be accurately predicted.

5.1.4 Simulation Results

Here some simulation results are given to show the performance of the ET-RLpN algorithm in terms of robustness and tracking capability for a system identification model under different noise cases. The weight vector of the unknown system is randomly generated and normalized to unit energy.

5.1.4.1 Performance Comparison under Contaminated Gaussian Noise

The impulsive noise was modeled as $\eta(k) = b(k)v_b(k)$, where $b(k)$ is an i.i.d. Bernoulli random sequence, and $v_b(k)$ is an i.i.d. zero-mean Gaussian sequence. The comparisons of the ET-RLpN with the RLS, RLM, RLpN, and CAC-RLpN are shown in the case of a system suddenly changing environment. The unknown system coefficients were abruptly multiplied by –1 at

TABLE 5.2

MSE for the ET-RLpN Algorithm in dB with $L = 34$, $p = 1.2$, $\alpha = 0.8$, and $\lambda = 0.99$

	ET-RLpN	
Noise power	**Eq. (5.37)**	**Experimental**
$10^{-1.5}$	–14.64	–14.63
$10^{-2.0}$	–19.64	–19.62
$10^{-2.5}$	–24.64	–24.60
$10^{-3.0}$	–29.65	–29.61
$10^{-3.5}$	–34.65	–34.62

iteration 2500. The input was colored Gaussian process with a single pole at 0.8 and the order of the unknown system is 60. The measurement noise added to the desired signal was a zero-mean white Gaussian noise signal to achieve different SNRs in Figure 5.2(a), Figure 5.2(b), and Figure 5.3, respectively. One can see from Figure 5.3 that the ET-RLpN algorithm handles sudden system changes successfully and at the same time maintains its robustness with respect to impulsive noise. The ET-RLpN has better tracking performance compared with the RLpN and CAC-RLpN algorithms. In Figure 5.3, the impulse response of the unknown system is modeled as a first-order Markov model. As one can see from Figure 5.3, the ET-RLpN algorithm outperforms the CAC-RLpN algorithm in terms of steady-state MSD and tracking capability.

5.1.4.2 Performance Comparison under Alpha-Stable Noise

Instead of the conventional SNR, the following fractional-order SNR (FSNR) is used as the noise distortion measure [329], which is defined as:

$$\text{FSNR} = 10\log_{10}\left(\frac{E\left[|d(k)|^p\right]}{E\left[|v(k)|^p\right]}\right)$$

where $0 < p < \alpha$. The measurement noise added to the desired signal in Figure 5.4 was an alpha-stable process with $\alpha = 1.4$ and FSNR = 28 dB. The value of p was set to 1.1. Comparisons of the ET-RLpN with the RLS, RLM, RLpN, and CAC-RLpN in alpha-stable noise case are illustrated in Figure 5.4. As one can see, the ET-RLpN algorithm has better tracking performance than the RLS, RLM, and RLpN algorithms and yields the same steady-state MSD as the RLpN.

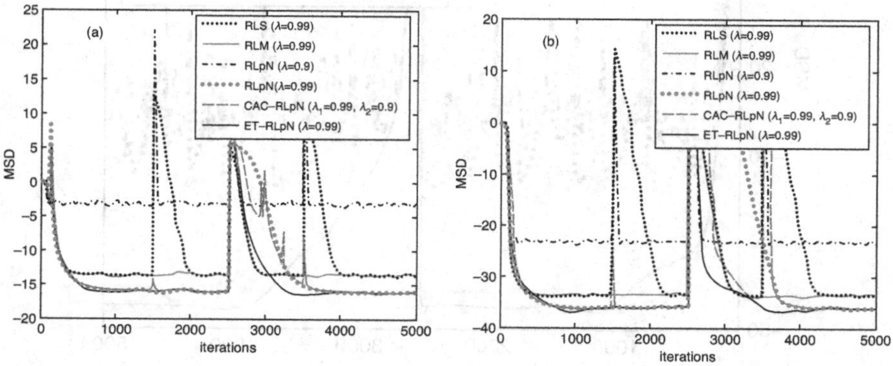

FIGURE 5.2
Performance comparisons of the ET-RLpN with the RLS, RLM, RLpN, and CAC-RLpN with CG noise in a stationary environment. (a) SNR = 20dB. (b) SNR = 40dB [238].

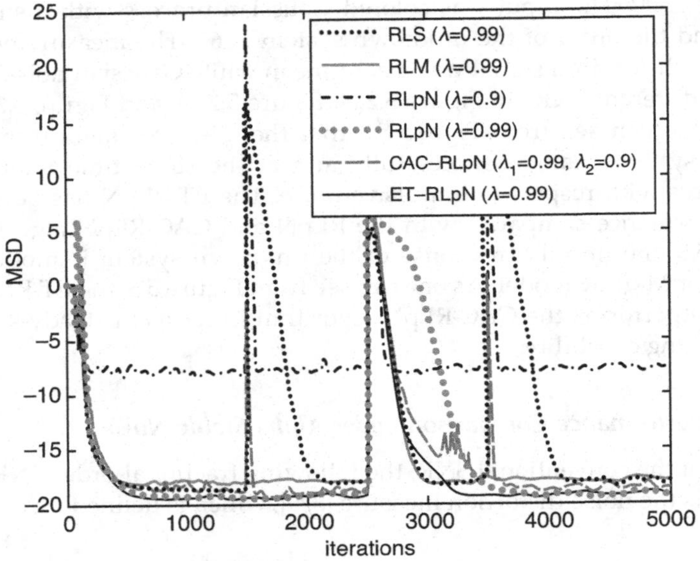

FIGURE 5.3
Performance comparisons of the ET-RLpN with the RLS, RLM, RLpN, and CAC-RLpN with CG noise in nonstationary environments with SNR = 25dB [238].

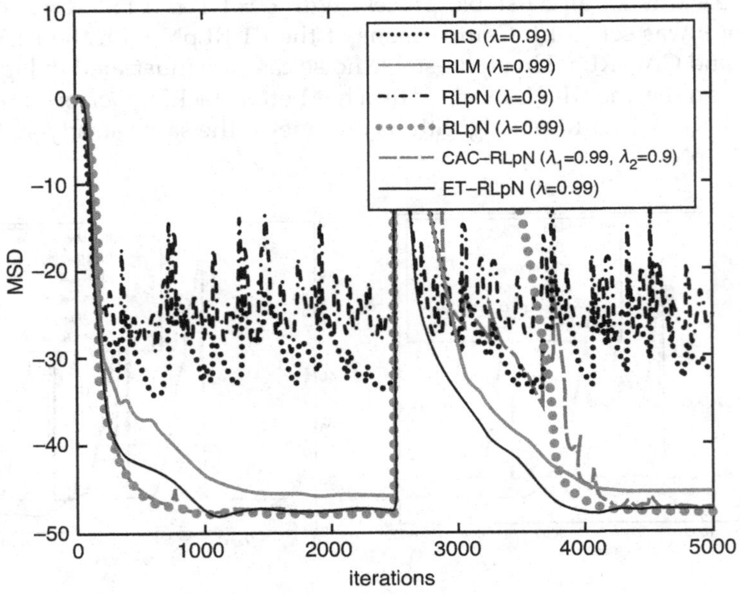

FIGURE 5.4
Performance comparisons in $\alpha - S$ noise. [238].

5.2 Sparsity-Aware Recursive Adaptive Filtering Algorithms under MMPE

Some sparsity-aware LMP algorithms with sparsity constraints are presented in Section 4.5. However, those methods still have flaws in convergence speed and steady-state performance due to adopting the gradient-based method to search the optimal solution. To address this problem, this section will present a sparse constrained recursive least mean p-power (RLMP) algorithm to identify the sparse parameters under non-Gaussian noise environments.

5.2.1 RLMP Algorithm with Sparsity Constraints

5.2.1.1 Derivation of the Algorithm

Here by making a simple modification over the MPE and by regularizing it with a sparse penalty term $\mathbf{S}(k)$, a cost function with sparsity penalty term can be defined as

$$J_{\mathrm{SRLMP}}(k) = \frac{1}{p}\sum_{i=0}^{k}\gamma^{k-i}\left|e(i)\right|^{p} + \rho\mathbf{S}(k) \tag{5.41}$$

where γ is forgetting factor close to but less than one, $\rho>0$ is called a sparse control factor, and $\mathbf{S}(k)$ represents a sparsity constraint function which can be selected as l_1-norm, weighted l_1-norm, and CIM mentioned above. One can get the optimal parameter $\mathbf{w}_o$ by solving the following optimization problem

$$\mathbf{w}_o = \arg\max_{\mathbf{w}} \mathbf{J}_{\mathrm{SRMP}}(k) \tag{5.42}$$

Consequently, we differentiate (5.40) with respect to $\mathbf{w}(k)$ and set the derivatives to zero yields

$$\sum_{i=0}^{k}\gamma^{k-i} f_{\mathrm{SRLMP}} e(i)\mathbf{u}(i) = \rho\mathbf{S}'(k) \tag{5.43}$$

where $f_{\mathrm{SRLMP}}(i) = \left|e(i)\right|^{p-2}$, and $\mathbf{S}'(k)$ denotes the derivative of $\mathbf{S}(k)$. After some manipulations in (5.43), we obtain

$$\sum_{i=0}^{k}\gamma^{k-i} f_{\mathrm{SRLMP}}\mathbf{u}(i)\mathbf{u}^{T}(i)\mathbf{w}(k) = \sum_{i=0}^{k}\gamma^{k-i} f_{\mathrm{SRLMP}}\mathbf{u}(i)d(i) - \rho\mathbf{S}'(k) \tag{5.44}$$

From (5.44), the weighted-auto-correlation matrix $\Psi(k)$ and weighted-cross-correlation vector $\Phi(k)$ are obtained as

$$\Psi(k)=\sum_{i=0}^{k}\gamma^{k-i}f_{\text{SRLMP}}\mathbf{u}(i)\mathbf{u}^{T}(i)$$
$$\Phi(k)=\sum_{i=0}^{k}\gamma^{k-i}f_{\text{SRLMP}}\mathbf{u}(i)d(i) \tag{5.45}$$

Using the matrix defined in (5.45), we have the matrix form of (5.44) as

$$\Psi(k)\mathbf{w}(k)=\Phi(k)-\rho\mathbf{S}'(k) \tag{5.46}$$

Then the estimate of optimal vector at each time instant k can be obtained by the following relation

$$\mathbf{w}(k)=\Psi^{-1}(k)\tilde{\Phi}^{s} \tag{5.47}$$

where $\tilde{\Phi}^{s}(k)=\Phi(k)-\rho\mathbf{S}'(k)$ denotes an extended weighted-cross-correlation vector with sparse constraint. To solve (5.44) and to avoid the large computational load of computing the inverse of the auto-correlation matrix $\Psi(k)$ for each time instant k, we express the $\Psi(k)$ in (5.45) recursively as

$$\Psi(k)=\gamma\sum_{i=0}^{k-1}\gamma^{k-i-1}f_{\text{SRLMP}}\mathbf{u}(i)\mathbf{u}^{T}(i)+f_{\text{SRLMP}}\mathbf{u}(k)\mathbf{u}^{T}(k) \tag{5.48}$$

As the iterations increasing ($k\to\infty$), the adaptive filtering algorithm will gradually approach to steady state, i.e., the $(k-1)^{\text{th}}$ weight coefficient $\mathbf{w}(k-1)$ closes to $\mathbf{w}(k)$. Under this condition, and we can obtain the recursive form of $\Psi(k)$ from (5.44) as

$$\Psi(k)=\gamma\Psi(k-1)+f_{\text{SRLMP}}\mathbf{u}(k)\mathbf{u}^{T}(k) \tag{5.49}$$

According to the similar method, one can obtain the recursive form of $\Phi(k)$ as

$$\Phi(k)=\gamma\Phi(k-1)+f_{\text{SRLMP}}\mathbf{u}(k)d(k) \tag{5.50}$$

Here we assume that the sign of the weight values does not change significantly in a single time step, i.e. $S'(k)\to S'(k-1)$, then we have

$$\begin{aligned}\tilde{\Phi}^{s}(k)&=\Phi(k)-\rho\mathbf{S}'(k)\\&=\gamma\Phi(k-1)+f_{\text{SRLMP}}(k)\mathbf{u}(k)d(k)-\rho\mathbf{S}'(k)\\&=\gamma\Phi(k-1)-\gamma\rho\mathbf{S}'(k-1)\\&\qquad+f_{\text{SRLMP}}(k)\mathbf{u}(k)d(k)-\rho\mathbf{S}'(k)+\gamma\rho\mathbf{S}'(k-1)\\&=\gamma\Phi^{s}(k-1)+f_{\text{SRLMP}}(k)\mathbf{u}(k)d(k)-\rho(1-\gamma)\mathbf{S}'(k-1)\end{aligned} \tag{5.51}$$

Now, we further define the following matrices

$$\mathbf{A}=\gamma\,\mathbf{\Psi}(k-1),\mathbf{B}=\mathbf{u}(k),\mathbf{C}=f_{\mathrm{SRLMP}}(k)\mathbf{I},\mathbf{D}=\mathbf{u}^{T}(k) \tag{5.52}$$

where $\mathbf{I}$ is an identity matrix, and using the matrix inversion lemma, we obtain the inverse of $\mathbf{\Psi}(k)$ as

$$\begin{aligned}\mathbf{\Psi}^{-1}(k)&=\gamma^{-1}\mathbf{\Psi}^{-1}(k-1)\frac{\gamma^{-1}\mathbf{\Psi}^{-1}(k-1)\mathbf{u}(k)\mathbf{u}^{T}(k)\gamma^{-1}\mathbf{\Psi}^{-1}(k-1)}{f_{\mathrm{SRLMP}}^{-1}(k)+\gamma^{-1}\mathbf{u}^{T}(k)\mathbf{\Psi}^{-1}(k-1)\mathbf{u}(k)}\\&=\gamma^{-1}\mathbf{\Psi}^{-1}(k-1)\frac{\gamma^{-1}f_{\mathrm{SRLMP}}(k)\mathbf{\Psi}^{-1}(k-1)\mathbf{u}(k)\mathbf{u}^{T}(k)\mathbf{\Psi}^{-1}(k-1)}{\gamma+f_{\mathrm{SRLMP}}(k)\mathbf{u}^{T}(k)\mathbf{\Psi}^{-1}(k-1)\mathbf{u}(k)}\end{aligned} \tag{5.53}$$

For a simple description of (5.53), we introduce extended gain vectors $\mathbf{\Omega}(k)$ and $\mathbf{K}(k)$ as

$$\mathbf{\Omega}(k)=\mathbf{\Psi}^{-1}(k) \tag{5.54}$$

$$\mathbf{K}(k)=\frac{f_{\mathrm{SRLMP}}(k)\mathbf{\Omega}(k-1)\mathbf{u}(k)}{\gamma+f_{\mathrm{SRLMP}}(k)\mathbf{u}^{T}(k)\mathbf{\Omega}(k-1)\mathbf{u}(k)} \tag{5.55}$$

Then (5.53) can be expressed as

$$\mathbf{\Omega}(k)=\lambda^{-1}\mathbf{\Omega}(k-1)-\mathbf{K}(k)\mathbf{u}^{T}(k)\mathbf{\Omega}(k-1) \tag{5.56}$$

Multiplying relation (5.55) by λ^{-1}, one can obtain another equivalent form of (5.55) as

$$\mathbf{K}(k)\left[1+\gamma^{-1}f_{\mathrm{SRLMP}}(k)\mathbf{u}^{T}(k)\mathbf{\Omega}(k-1)\mathbf{u}(k)\right]=\gamma^{-1}f_{\mathrm{SRLMP}}(k)\mathbf{\Omega}(k-1)\mathbf{u}(k) \tag{5.57}$$

According to (5.57) and using the result $\mathbf{\Omega}(k)$ in (5.56), the gain vector $\mathbf{K}(k)$ can be expressed in the following form

$$\begin{aligned}\mathbf{K}(k)&=f_{\mathrm{SRLMP}}(k)\left[\gamma^{-1}\mathbf{\Omega}(k-1)-\gamma^{-1}\mathbf{K}(k)\mathbf{u}^{T}(k)\mathbf{\Omega}(k-1)\right]\mathbf{u}(k)\\&=f_{\mathrm{SRLMP}}(k)\mathbf{\Omega}(k)\mathbf{u}(k)\end{aligned} \tag{5.58}$$

Now by inserting the (5.51) into (5.47), using (5.56) and (5.58), and after some simplifications, we obtain

$$\begin{aligned}\mathbf{w}(k)&=\mathbf{K}(k)d(k)\\&\quad+\left[\gamma^{-1}[\mathbf{I}-\mathbf{K}](k)\mathbf{u}^{T}(k)\mathbf{\Omega}(k-1)\right]\left[\gamma\tilde{\mathbf{\Phi}}^{s}(k-1)-\rho(1-\gamma)\mathbf{s}'(k-1)\right]\\&=\mathbf{w}(k-1)+\mathbf{K}(k)\left(d(k)-\mathbf{u}^{T}(k)\mathbf{w}(k-1)\right)\\&\quad-\gamma^{-1}\rho(1-\gamma)\left[\mathbf{I}-\mathbf{K}(k)\mathbf{u}^{T}(k)\right]\mathbf{\Omega}(k-1)\mathbf{S}'(k-1)\\&=\mathbf{w}(k-1)+\mathbf{K}(k)e_{a}(k)-\rho\gamma^{-1}(1-\gamma)\left[\mathbf{I}-\mathbf{K}(k)\mathbf{u}^{T}(k)\right]\mathbf{\Omega}(k-1)\mathbf{S}'(k-1)\end{aligned} \tag{5.59}$$

TABLE 5.3

Sparsity Constrained RLMP

Initialization: $\mathbf{w}(0) = \mathbf{0}, \quad \Omega(0) = \varepsilon^{-1}\mathbf{I}$
For $k = 1,2,\cdots$ Do
1. $y(k) = \mathbf{u}^T(k)\mathbf{w}(k-1)$
2. $e_a(k) = d(k) - y(k)$
3. $f_{\mathrm{SRLMP}}(i) = \lvert e(i)\rvert^{p-2}$
4. $\mathbf{K}(k) = \dfrac{f_{\mathrm{SRLMP}}(k)\boldsymbol{\Omega}(k-1)\mathbf{u}(k)}{\gamma + f_{\mathrm{SRLMP}}(k)\mathbf{u}^T(k)\boldsymbol{\Omega}(k-1)\mathbf{u}(k)}$
5. $\mathbf{w}(k) = \mathbf{w}(k-1) + \mathbf{K}(k)e_a(k) - \rho\gamma^{-1}(1-\gamma)\left[\mathbf{I} - \mathbf{K}(k)\mathbf{u}^T(k)\right]\Omega(k-1)\mathbf{S}'(k-1)$
6. $\Omega(k) = \gamma^{-1}\left[\mathbf{I} - \mathbf{K}(k)\mathbf{u}^T(k)\right]\Omega(k-1)$
End

where $e_a(k) = d(k) - \mathbf{u}^T(k)\mathbf{w}(k-1)$ is the *a priori* error signal. Then we can summarize the sparse constraint RLMP (SRLMP) algorithm in Table 5.3.

5.2.1.2 Special Cases of the SRLMP Algorithm

The SPC is a key component for designing SAF algorithms. The l_0-norm is an optimal SPC. However, the optimization of the l_0-norm is a NP-hard problem. For this reason, various approximations of l_0-norm are usually utilized as the SPC. The l_1-norm is a popular one of such approximations. One drawback of the l_1-norm is the non-smoothness at zero point. At present, the correntropy induced metric (CIM) has been introduced as a nice approximation of the l_0-norm for compressed sensing. Here we will mainly employ the l_1-norm and CIM as the SPC to develop sparse RLMP algorithms.

i. $\mathbf{S}(k) = \lVert \mathbf{w}(k)\rVert_1$. In this case, the gradient of $\mathbf{S}(k)$ is $\mathbf{S}(k) = \mathrm{sign}(\mathbf{w}(k))$, and we call the algorithm in Table 5.3 as ZARLMP.

ii. $\mathbf{S}(k) = \mathrm{CIM}^2(\mathbf{w}(k), 0) = \dfrac{1}{M\sigma_2\sqrt{2\pi}}\displaystyle\sum_{i=1}^{M}\left(1 - \exp\left(-\dfrac{w_i^2}{2\sigma_2^2}\right)\right)$. It has been shown that if $\lvert w_i\rvert > \varsigma$, $\forall w_i \neq 0$, then as $\sigma_2 \to 0$, one can get a solution arbitrarily close to that of the l_0-norm, where ς is a small positive number. The gradient in vector form is

$$\mathbf{S}'(k) = \frac{1}{M\sigma_2^3\sqrt{2\pi}}\mathbf{w}(k)\cdot{*}\exp\left(-\frac{\mathbf{w}(k)\cdot{*}\mathbf{w}(k)}{2\sigma_2^2}\right) \tag{5.60}$$

In this case, the algorithm is called CIMRLMP.

5.2.2 Simulation Results

In this section, Monte-Carlo (MC) simulations are done to confirm the performance of the sparse RLMP algorithms for sparse system parameters estimation under non-Gaussian and non-zero-mean noise environments. In particular, the SRLMP algorithm is compared with the LMP, RLMP, and sparsity-aware LMP algorithms (including ZALMP and CIMLMP). The impulse response of the sparse system model with $M = 30$ is illustrated in Figure 5.5.

First, the comparison results among all the algorithms in terms of the MSD under impulsive noise with alpha-stable distribution are shown. The convergence curves are shown in Figure 5.6. One can observe that the sparse type algorithms achieve faster convergence rate and better steady-state performance than the non-sparse algorithms (LMP and RLMP). In particular, the ZARLMP and CIMRLMP achieve better performance than other algorithms, and this confirms the advantage of the recursive method. To our delight, one can see that the CIMRLMP outperforms ZARLMP, and this result also reflects the superiority of the CIM approaching to l_0-norm.

Second, the performance of the sparse RLMP algorithm is further evaluated under noise with a mixture of different distributions. The noise model is $v(k) = (1 - a(k))A(k) + a(k)B(k)$, where $a(k)$ is an independent and identically distributed binary process with an occurrence probability, $A(k)$ is a noise process with smaller variance, and $B(k)$ is another noise process with substantially much larger variance to represent impulsive disturbances. In this case, $B(k)$ is

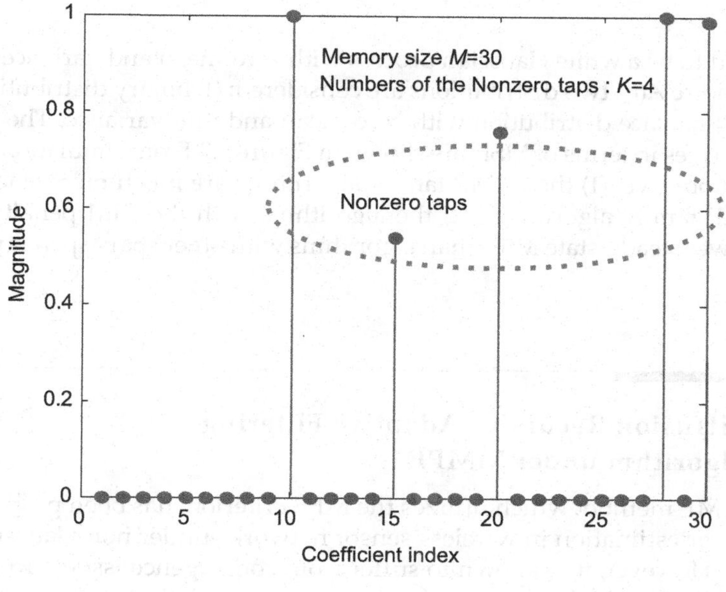

FIGURE 5.5
Impulse response of the sparse system.

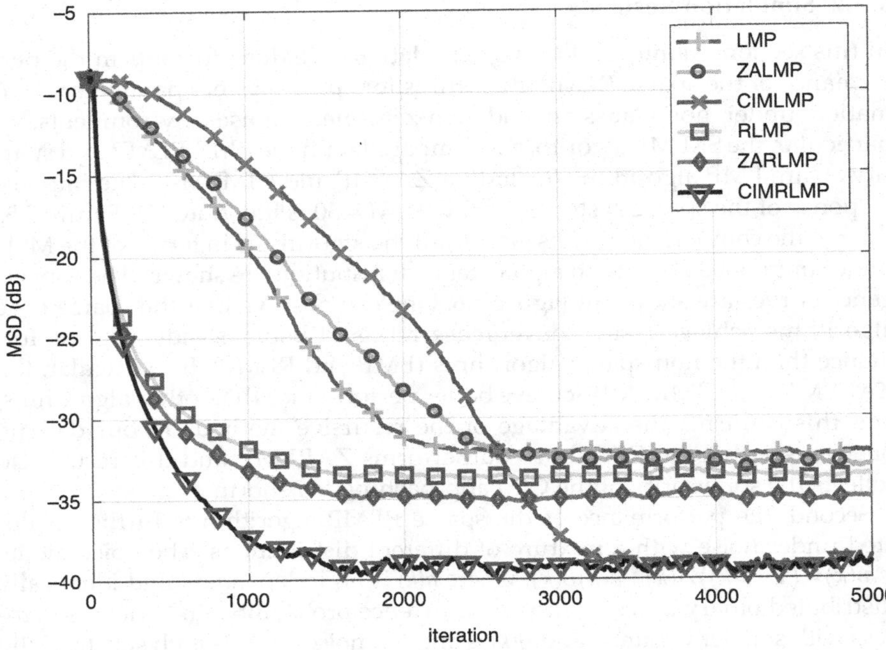

FIGURE 5.6
Tracking and steady-state behaviors.

assumed to be a white Gaussian process with zero-mean and variance 15, and for the noise $A(k)$, two distributions are considered: (1) binary distribution over {–1,1}; (2) Laplace distribution with zero-mean and unit variance. The convergence curves in terms of MSD are shown in Figure 5.7. From simulation results, one can observe: (1) the RLMP family algorithms are much more stable than the LMP family algorithms; (2) the algorithms with the CIM penalty terms have lower steady-state MSD than algorithms with other sparse penalty terms.

5.3 Diffusion Recursive Adaptive Filtering Algorithm under MMPE

The DLMP method, which utilizes the MPE criterion, has been proposed for parameter estimation in wireless sensor networks under non-Gaussian conditions. However, it is known to suffer from convergence issues. To address this problem and improve convergence performance, this section introduces a robust diffusion RLS-like algorithm based on the MPE, known as the

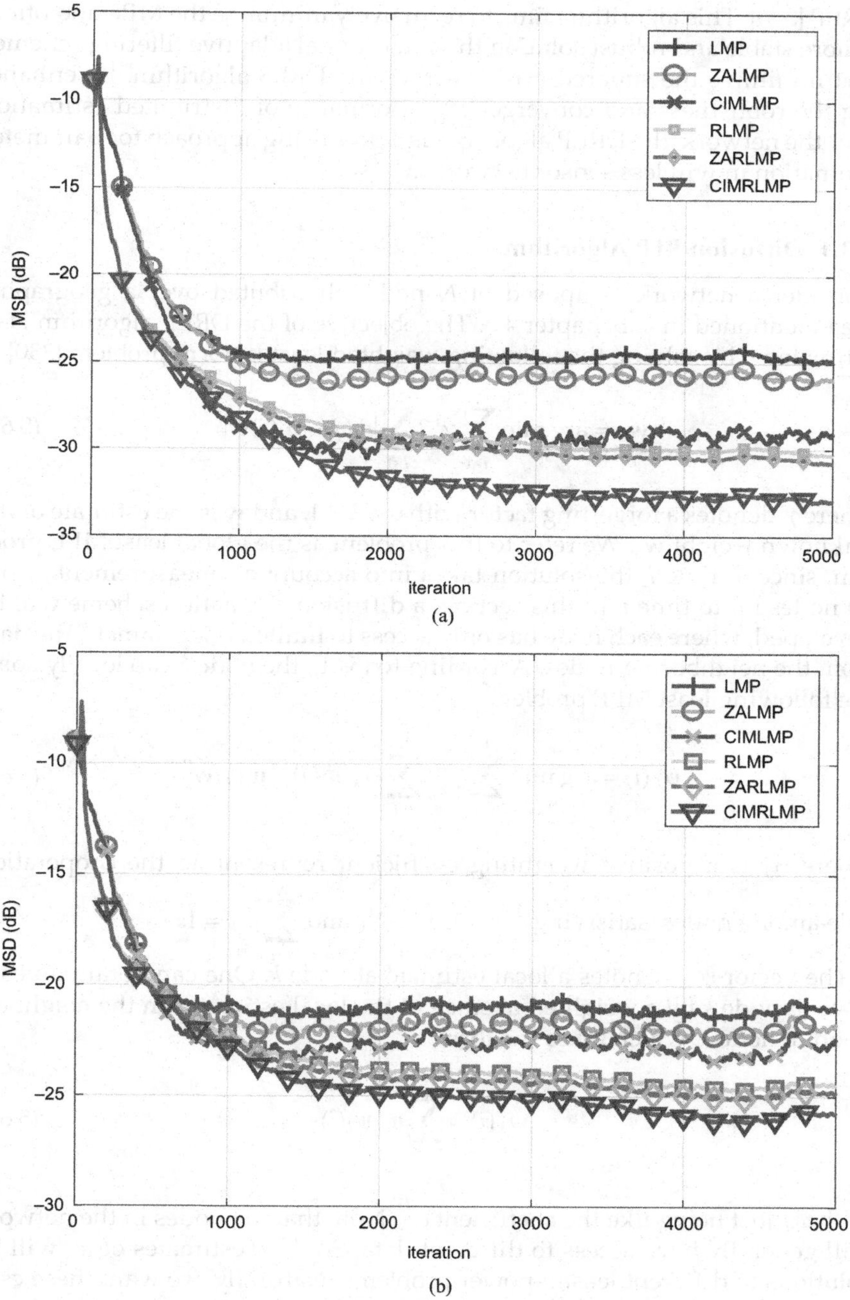

FIGURE 5.7
Convergence curves of all algorithms with different inner noise distributions: (a) Binary ($p = 2$); (b) Laplace ($p = 2.5$)

DRLP [330]. This algorithm aims to recursively minimize the MPE and offers a more stable and robust solution than traditional adaptive filtering schemes that minimize the squared error, such as the DRLS algorithm. By enhancing the robustness and convergence performance of distributed estimation over the network, the DRLP algorithm is a promising approach for parameter estimation in wireless sensor networks.

5.3.1 Diffusion RLP Algorithm

Consider a network composed of N nodes distributed over a geographic area mentioned in subchapter 4.5. The objective of the DRLP algorithm is to estimate $\mathbf{w}_o$ by solving the following weighted least l_p-norm problem [330]

$$\mathbf{w}_k = \arg\min_{\mathbf{w}} \sum_{j=1}^{k} \gamma^{i-j} \sum_{l=1}^{N} \left| d_l(j) - \mathbf{u}_l(j)\mathbf{w} \right|^p \tag{5.61}$$

where γ denotes a forgetting factor with $0 < \lambda \le 1$, and $\mathbf{w}$ is the estimate of the unknown weight $\mathbf{w}_o$. We refer to this problem as the global least MPE problem, since at time i, the solution takes into account all measurements from all nodes up to time i. In this section, a diffusion estimation scheme will be developed, where each node has only access to limited data, namely, the data from the neighboring nodes. According to (5.61), the node k can locally solve the following least MPE problem

$$\psi_k(i) = \arg\min_{\mathbf{w}} \sum_{j=1}^{k} \gamma^{i-j} \sum_{l=1}^{N} c_{l,k} \left| d_l(j) - \mathbf{u}_l(j)\mathbf{w} \right|^p \tag{5.62}$$

where $c_{l,k}$ is a positive weighting coefficient representing the cooperation rule among nodes, satisfying $c_{l,k} = 0.$ if $l \notin N_k$ and $\sum_{l=1}^{N} c_{l,k} = 1$.

The vector $\psi_{k,i}$ denotes a local estimate at node k. One can obtain an estimate at node k by a weighted average of the local estimates in the neighborhood of node k, which takes the form

$$\mathbf{w}_k(i) = \sum_{l \in N_k} a_{l,k} \psi_l(i) \tag{5.63}$$

with $a_{l,k}$ just being like the coefficient $c_{l,k}$. Note that the nodes in the network will generally have access to different data. So their estimates of w_o will be solutions to different least p-power problems. Naturally, we want these estimates to be close to the global least MPE solution of (5.57).

The equation (5.62) will be the least squares problem with non-regularized when $p = 2$. According to MPE-based aware (including RLP and LMP) algorithms, we know that the term only $|e_l(i)|^{p-2}$ is added compared with the MSE-based aware (including RLS and LMS) algorithms when $p \neq 2$. This term can be as a step size for MPE-based aware algorithms. When these algorithms have not converged, the $|e_l(i)|$ is large. This makes p-norm aware algorithms have fast convergence speed. When these algorithms have converged, the error $|e_l(i)|$ is small. This makes the misadjustment small. This character makes the MPE-based aware algorithms very suitable for addressing the signal-processing cases under impulsive noise.

For dealing with distributed estimation over network under impulsive noise environments, the DRLP algorithm can be used to collectively estimate $\mathbf{w}_o$ from individual measurements in two steps (including an incremental update and a spatial update stage) as follows:

i. At time i, the nodes communicate their measurements $d_k(i)$ and regressors $\mathbf{u}_k(i)$ with their neighbors and use these data to update their local estimates using RLP iterations (named incremental update). The resulting pre-estimates are named $\psi_k(i)$ as in (5.62).
ii. The nodes communicate their local pre-estimates with their neighbors and perform a weighted average to obtain the estimate $\mathbf{w}_k(i)$ (named spatial update).

The above update stage is similar to the DRLS [100]. The DRLP algorithm is shown schematically in Table 5.4, and the detailed derivation of this algorithm is omitted, which is similar to the DRLS algorithm.

TABLE 5.4

Diffusion Recursive Least p-Power Error Algorithm

Step1: Initialization: $w_{k,-1} = 0$ and $P_{k,-1} = \theta^{-1} I_M$, where $P_{k,-1}$ is an $M \times M$ matrix and $\theta > 0$, and I_M denotes an identity matrix.

Step2: Incremental update for every node k, repeat

$\psi_k(i) \leftarrow \mathbf{w}_k(i-1)$

$P_k(i) \leftarrow \lambda^{-1} P_k(i-1)$

For all $l \in N_k$

$$\mathrm{K}_k(i) = \frac{P_k(i)\mathbf{u}_l(i)}{\gamma + f(e_l(i)) c_{l,k} \mathbf{u}_l^T(i) P_k(i) \mathbf{u}_l(i)}$$

$$f(e_l(i)) = |e_l(i)|^{P-2} = |d_l(i) - \mathbf{u}_l(i)\psi_k(i)|^{p-2}$$

$$\psi_k(i) \leftarrow \psi_k(i) + f(e_l(i)) c_{l,k} \mathrm{K}_k(i) d_l(i) - \mathbf{u}_l(i)\psi_k(i)$$

$$P_k(i) \leftarrow P_k(i) + f(e_l(i)) c_{l,k} \mathrm{K}_k(i) \mathbf{u}_l(i) P_k(i)$$

end

Step3: Spatial update

$$\mathbf{w}_k(i) = \sum_{l \in N_k} a_{l,k} \psi_l(i)$$

Remark 5.4

The DRLP differs from the conventional DRLS in that the nonlinearity of error will be included in the incremental update stage, and it means that the complexity of the DRLP is more than one extra nonlinearity computation compared with DRLS. In addition, one can easily know that the DRLP is a generalized version of the DRLS, since it will reduce to the latter when $p = 2$.

5.3.2 Simulation Results

Here some simulation results are shown to demonstrate the performance of the DRLP algorithm. The topology of the network with 20 nodes is generated as a realization of the random geometric graph model as shown in Figure 5.8. The input regressors are zero-mean Gaussian, independent in time and space with size $M = 5$. In the following, we perform the simulation in distributed network parameter estimation scenarios under the alpha–stable noise environments. The performance of the DRLP is mainly compared with that of the DRLS, DLMP, and DMCC.

First, the convergence performance of the DRLP is investigated compared with that of other algorithms. Figure 5.9 gives the convergence curves in terms of MSD. We set the $p = 1$ for DLMP and DRLP algorithms. One can observe that DLMP, DMCC, and DRLP work well when large outliers occur, while DRLS fluctuates dramatically due to the sensitivity to the impulsive noises. As one can see from this result, the DRLP algorithm has excellent performance in convergence rate and accuracy compared with other methods and exhibits a significant improvement in robust performance under the impulsive noise environments. A similar performance in the steady-state behavior of the DRLP algorithms at each node k is obtained as shown in Figure 5.10. As expected, the DRLP performs better than all other algorithms included in this comparison.

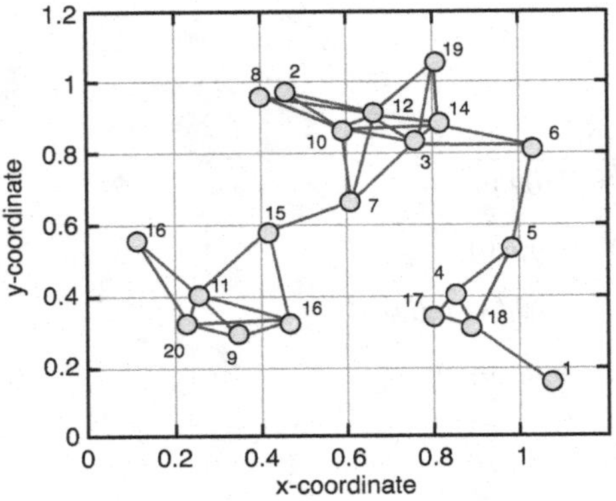

FIGURE 5.8
Network topology [330].

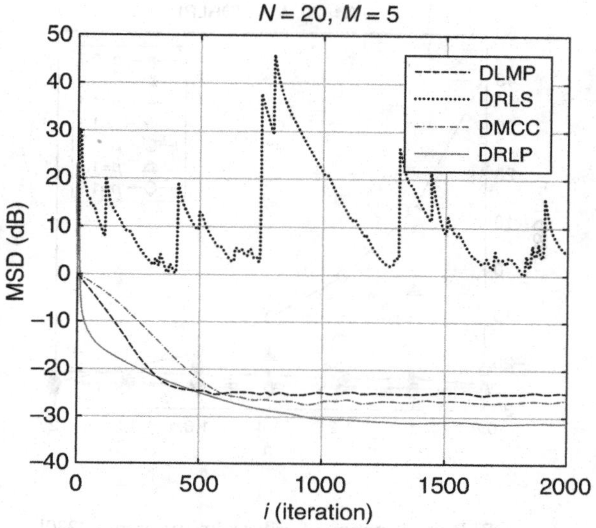

FIGURE 5.9
MSD curves of different algorithms [330].

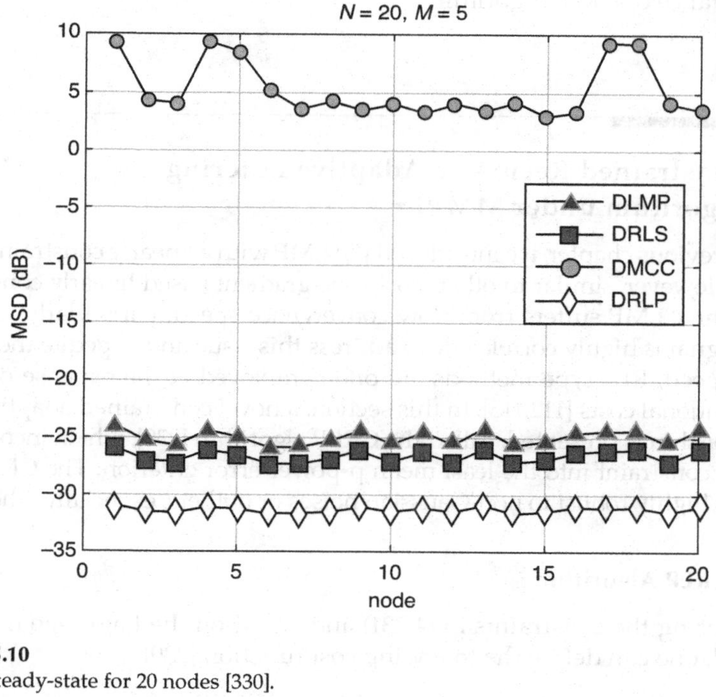

FIGURE 5.10
MSD at steady-state for 20 nodes [330].

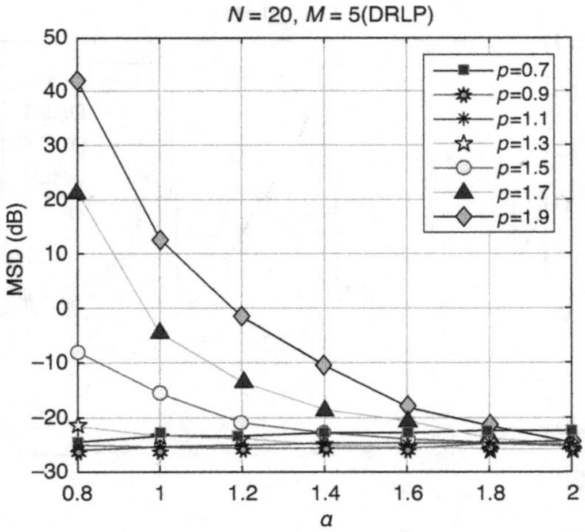

FIGURE 5.11
Steady-state MSD of the DRLP as a function of order p for different α [330].

Second, the joint effect of the p and noise power are tested in terms of different α on the performance of DRLP. From Figure 5.11, one can see that for $p = 1.9$ (which is close to 2), the larger the noise (the smaller α), the larger the MSD. This means that the impulsive noise has a great influence on the performance of DRLS algorithm.

5.4 Constrained Recursive Adaptive Filtering Algorithm under MMPE

In the previous chapter, we introduced the LMP with a linearly constrained algorithm. However, similar to other stochastic-gradient-based linearly constrained AFAs, the CLMP suffers from slow convergence speed, particularly when the input signal is highly correlated. To address this issue and expedite the convergence speed, RLS-type methods are often employed at the expense of higher computational costs [112,115]. In this section, a novel constrained adaptive filtering algorithm is presented, called the CRLP algorithm [239], which incorporates a linear constraint into the least mean p-power error criterion. The CRLP algorithm is highly robust to non-Gaussian noises or outliers, especially when $p < 2$.

5.4.1 CRLP Algorithm

Considering the constraints in (4.231) and based on the Lagrange multiplier method, one can define the following cost function [239]

$$J_k(i) = \sum_{i=1}^{k} |e(i)|^p + \sum_{i=1}^{K} \lambda_i \left(f_i - \mathbf{w}^T(k) c_i \right) \tag{5.64}$$

where f_i is the i-th element of vector f, c_i is the i-th column of matrix c and λ_i is a constant introduced by Lagrange multiplier. To derive the update rules of $\mathbf{w}(k)$, we first take the partial derivative of $J(k)$with respect to $\mathbf{w}(k)$:

$$\begin{aligned}
\frac{\partial J(\mathrm{k})}{\partial \mathbf{w}(\mathrm{k})} &= p \sum_{i=1}^{k} |e(i)|^{p-2} \left(d(i) - \mathbf{w}^T(k)\mathbf{u}(i) \right) \mathbf{u}^T(i) + \sum_{i=1}^{K} \lambda_i c_i^T \\
&= p \left(\sum_{i=1}^{n} |e(i)|^{p-2} d(i)\mathbf{u}^T(i) - \sum_{i=1}^{n} |e(i)|^{p-2} \mathbf{w}^T(k)\mathbf{u}(i)\mathbf{u}^T(i) \right) + p\lambda'^T \mathbf{C}^T
\end{aligned} \tag{5.65}$$

where $\lambda = [\lambda_1 \lambda_2, \cdots, \lambda_k]^T$ is a $k \times 1$ vector and $\lambda' = \lambda / p$. Then the following formulas can be defined

$$\mathbf{R}_{\mathbf{uu}}(k) = \sum_{i}^{k} |e(i)|^{p-2} \mathbf{u}(i)\mathbf{u}^T(i) \tag{5.66}$$

$$\mathbf{p}_{\mathbf{u}d}(k) = \sum_{i=1}^{k} |e(i)|^{p-2} d(i)\mathbf{u}^T(i) \tag{5.67}$$

It is worth noting that when $p = 2$, $\mathbf{R}_{\mathbf{uu}}(k)$ is the input signal weighted autocorrelation matrix and $\mathbf{p}_{\mathbf{u}d}(k)$ is the weighted input-desired signal cross-correlation vector. Substituting (5.66) and (5.67) into (5.65) yields

$$\frac{\partial J(k)}{\partial \mathbf{w}(k)} = p \left(\mathbf{p}_{\mathbf{u}d}^T(k) - \mathbf{w}^T(k)\mathbf{R}_{\mathbf{uu}}(k) + \boldsymbol{\lambda}'^T \mathbf{C}^T \right) \tag{5.68}$$

Setting the partial derivative in (5.68) to zero, we have

$$\mathbf{w}(k) = \mathbf{R}_{\mathbf{uu}}^{-1}(k) \left(\mathbf{p}_{\mathbf{u}d}(k) + \mathbf{C}\boldsymbol{\lambda}' \right) \tag{5.69}$$

To satisfy the linear constraints, combining (5.69) and (4.231), we obtain

$$\mathbf{C}^T \mathbf{R}_{\mathbf{uu}}^{-1}(k) \left(\mathbf{p}_{\mathbf{u}d}(k) + \mathbf{C}\boldsymbol{\lambda}' \right) = \mathbf{f} \tag{5.70}$$

Accordingly, one can obtain

$$\boldsymbol{\lambda}' = \left(\mathbf{C}^T \mathbf{R}_{\mathbf{uu}}^{-1}(k)\, \mathbf{C} \right)^{-1} \left(\mathbf{f} - \mathbf{C}^T \mathbf{R}_{\mathbf{uu}}^{-1}(k)\, \mathbf{p}_{\mathbf{u}d}(k) \right) \tag{5.71}$$

Finally, substituting (5.71) into (5.69) yields

$$\mathbf{w}(k)=\mathbf{R}_{\mathbf{uu}}^{-1}(k)\mathbf{p}_{\mathbf{u}d}(k)+\mathbf{R}_{\mathbf{uu}}^{-1}(k)\,\mathbf{C}\left(\mathbf{C}^T\mathbf{R}_{\mathbf{uu}}^{-1}(k)\mathbf{C}\right)^{-1} \times\left(\mathbf{f}-\mathbf{C}^T\mathbf{R}_{\mathbf{uu}}^{-1}(k)\mathbf{p}_{\mathbf{u}d}(k)\right) \tag{5.72}$$

Note that there are two terms in (5.72). The first term is the optimal solution without constraints

$$\mathbf{w}_{\mathrm{unc}}(k)=\mathbf{R}_{\mathbf{uu}}^{-1}(k)\mathbf{p}_{\mathbf{u}d}(k) \tag{5.73}$$

and the second term is correction term applied to make sure that the constraints are satisfied. To avoid computing $\mathbf{R}_{\mathbf{uu}}^{-1}(k)$ and $\mathbf{p}_{\mathbf{u}d}(k)$ directly, the iterative equations in [251] are adopted to update $\mathbf{R}_{\mathbf{uu}}^{-1}(k)$ and $\mathbf{p}_{\mathbf{u}d}(k)$, which are

$$\mathbf{R}_{\mathbf{uu}}^{-1}(k)=\left(\mathbf{I}-\frac{|e(k)|^{p-2}\,\mathbf{R}_{\mathbf{uu}}^{-1}(k-1)\mathbf{u}(k)\mathbf{u}^T(k)}{1+|e(k)|^{p-2}\,\mathbf{u}^T(k)\mathbf{R}_{\mathbf{uu}}^{-1}(k-1)\mathbf{u}(k)}\right)\mathbf{R}_{\mathbf{uu}}^{-1}(k-1) \tag{5.74}$$

$$\mathbf{p}_{\mathbf{u}d}(k)=\mathbf{p}_{\mathbf{u}d}(k-1)+|e(k)|^{p-2}\,d(k)\mathbf{u}(k) \tag{5.75}$$

Let

$$g(k)=\frac{|e(k)|^{p-2}\,\mathbf{R}_{\mathbf{uu}}^{-1}(k-1)\mathbf{u}(k)}{1+|e(k)|^{p-2}\,\mathbf{u}^T(k)\mathbf{R}_{\mathbf{uu}}^{-1}(k-1)\mathbf{u}(k)} \tag{5.76}$$

Then (5.74) can be simplified as

$$\mathbf{R}_{\mathbf{uu}}^{-1}(k)=\left(\mathbf{I}-\mathbf{g}(k)\mathbf{u}^T(k)\right)\mathbf{R}_{\mathbf{uu}}^{-1}(k-1) \tag{5.78}$$

By using (5.74) and after some simple manipulations, one can rewrite (5.76) as

$$g(k)=|e(k)|^{p-2}\,\mathbf{R}_{\mathbf{uu}}^{-1}(k)\mathbf{u}(k) \tag{5.79}$$

Substituting (5.73), (5.78) and (5.79) into (5.72), one can obtain

$$\mathbf{w}(k)=\mathbf{w}_{\mathrm{unc}}(k)+\mathbf{R}_{\mathbf{uu}}^{-1}(k)\mathbf{C}\left(\mathbf{C}^T\mathbf{R}_{\mathbf{uu}}^{-1}(k)\mathbf{C}\right)^{-1}\left(\mathbf{f}-\mathbf{C}^T\mathbf{w}_{\mathrm{unc}}(k)\right) \tag{5.80}$$

where $\mathbf{w}_{\mathrm{unc}}(k)=\mathbf{w}_{\mathrm{unc}}(k-1)+\mathbf{g}(k)e_{\mathrm{unc}}(k)$, $e_{\mathrm{unc}}(k)=d(k)-\mathbf{w}_{\mathrm{unc}}(k-1)\mathbf{u}(k)$.

When $k=0$, all the variables are set to zero except forward prediction error energy being set to E_0. Let $\mathbf{I}_L$ be an $\mathrm{L}\times\mathrm{L}$ identity matrix. Then we have $\mathbf{R}_{\mathbf{uu}}(k)=E_0\mathbf{I}_L$, $\mathbf{h}(0)=\mathbf{R}_{\mathbf{uu}}^{-1}(0)\mathbf{C}(\mathbf{C}^T\mathbf{R}_{\mathbf{uu}}^{-1}(0)\mathbf{C})^{-1}\mathbf{f}$.

The above algorithm is referred to as the CRLP algorithm as summarized in Table 5.5, which is equal to the CRLS algorithm when $p=2$.

TABLE 5.5

CRLP Algorithm

Parameters: C and f

Initialization: $\mathbf{R}_{\mathbf{uu}}(k) = \mathrm{E}_0\mathbf{I}_\mathrm{L}\,\mathbf{h}_{\mathrm{unc}}(0) = 0\ \mathbf{h}(0) = \mathbf{R}_{\mathbf{uu}}^{-1}(0)\mathbf{C}(\mathbf{C}^{\mathrm{T}}\mathbf{R}_{\mathbf{uu}}^{-1}(0)\mathbf{C})^{-1}\mathbf{f}$

Update: $e_{\mathrm{unc}}(k) = d(k) - \mathbf{w}_{\mathrm{unc}}(k-1)\mathbf{u}(k)$

$e(k) = d(k) - \mathbf{w}^{\mathrm{T}}(k) * \mathbf{u}(k)$

$$g(k) = \frac{|e(k)|^{p-2}\,\mathbf{R}_{\mathbf{uu}}^{-1}(k-1)\mathbf{u}(k)}{1 + |e(k)|^{p-2}\,\mathbf{u}^{T}(k)\mathbf{R}_{\mathbf{uu}}^{-1}(k-1)\mathbf{u}(k)}$$

$$\mathbf{R}_{\mathbf{uu}}^{-1}(k) = \left(\mathbf{I} - \mathbf{g}(\mathrm{k})\mathbf{u}^{T}(k)\right)\mathbf{R}_{\mathbf{uu}}^{-1}(k-1)$$

$$\mathbf{w}_{\mathrm{unc}}(k) = \mathbf{w}_{\mathrm{unc}}(k-1) + \mathbf{g}(k)e_{\mathrm{unc}}(k)$$

$$\mathbf{w}(k) = \mathbf{w}_{\mathrm{unc}}(k) + \mathbf{R}_{\mathbf{uu}}^{-1}(k)\mathbf{C}\left(\mathbf{C}^{T}\mathbf{R}_{\mathbf{uu}}^{-1}(k)\mathbf{C}\right)^{-1}\left(\mathbf{f} - \mathbf{C}^{T}\mathbf{w}_{\mathrm{unc}}(k)\right)$$

5.4.2 Simulation Results

In this part, the performance of the CRLP algorithm is compared with that of CLMS [109], CRLS [111], and CLMP [225] under mix-Gaussian noise, alpha-stable noise, Cauchy noise, and Laplace noise. The mean square deviation (MSD) is used as the performance measure. The input signal is zero-mean multivariate Gaussian distribution. The convergence performance of CLMS, CLMP, CRLS, and CRLP are studied. Simulation results are demonstrated in Figure 5.12.

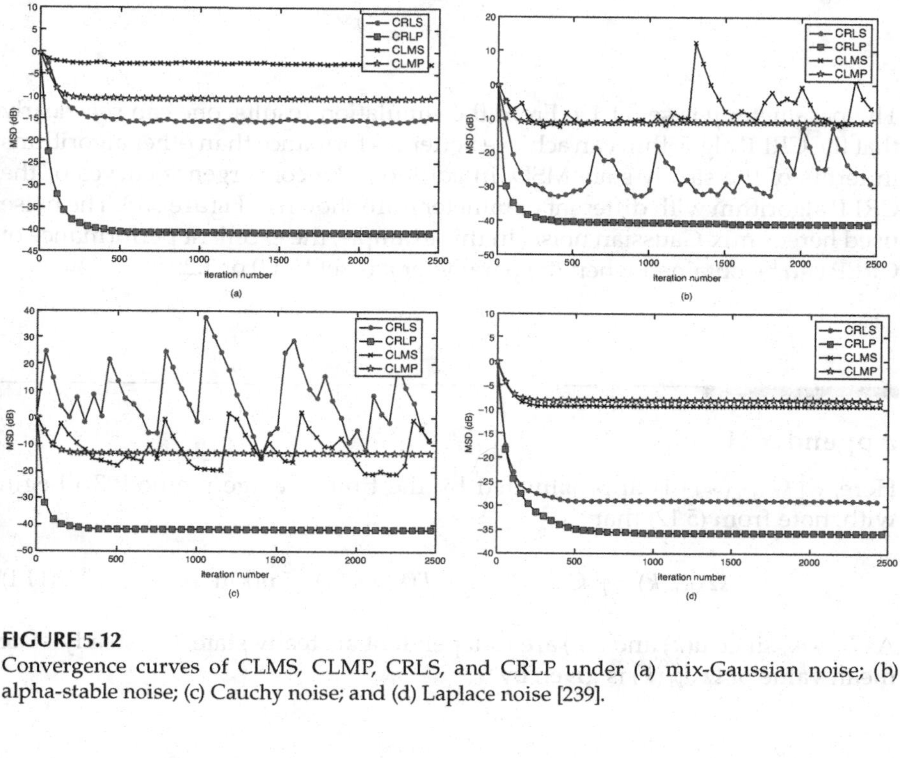

FIGURE 5.12
Convergence curves of CLMS, CLMP, CRLS, and CRLP under (a) mix-Gaussian noise; (b) alpha-stable noise; (c) Cauchy noise; and (d) Laplace noise [239].

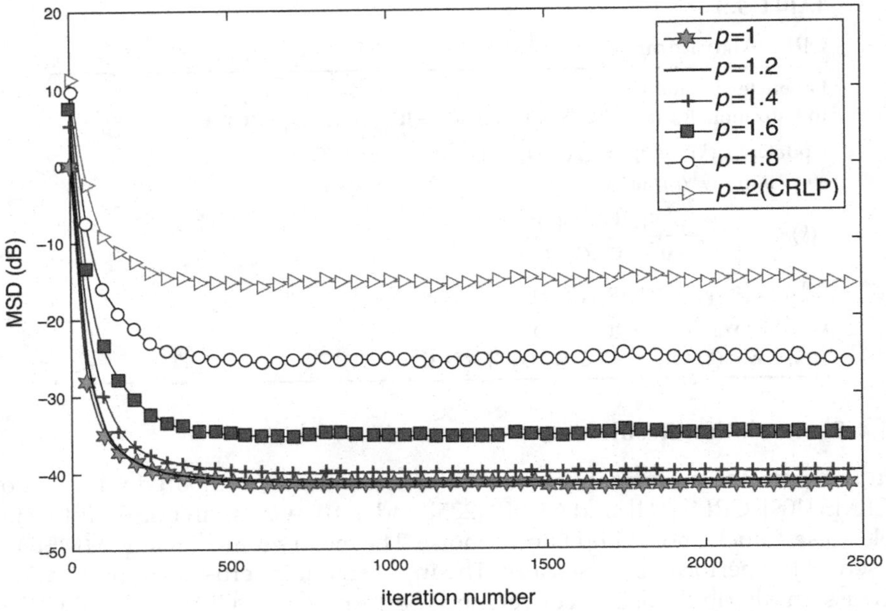

FIGURE 5.13
Convergence curves of CRLP for different p [239].

The parameter p is set at 1.4. From the simulation results, one can see clearly that the CRLP algorithm can achieve better performance than other algorithms in terms of the steady-state MSD. In addition, the convergence curves of the CRLP algorithm with different parameter p are shown in Figure 5.13. The noise used here is mix-Gaussian noise. In this example, the excellent performance of CRLP can be obtained when the parameter p is set to 1.0 or 1.2.

Appendix H

Here, $E\left[\boldsymbol{\Omega}_{\text{MPE}}(\infty)\right]$ is approximated by the time average method. To begin with, note from (5.12) that

$$\boldsymbol{\Omega}_{\text{MPE}}^{-1}(k) = \gamma\, \boldsymbol{\Omega}_{\text{MPE}}^{-1}(k-1) + T(k) \times \left|e(k)\right|^{p-2} \mathbf{u}(k)\mathbf{u}^{T}(k) \tag{H.1}$$

As $k \to \infty$, since $\mathbf{u}(k)$ and $e(k)$ are independent at steady state, the steady-state mean value of $\boldsymbol{\Omega}_{\text{MPE}}^{-1}(k)$ is given by

$$E\left[\mathbf{\Omega}_{\mathrm{MPE}}^{-1}(\infty)\right]=\gamma E\left[\mathbf{\Omega}_{\mathrm{MPE}}^{-1}(\infty)\right]+T(\infty)\times E\left[\left|e(\infty)\right|^{p-2}\right]E\left[\mathbf{u}(k)\mathbf{u}^{T}(k)\right] \quad \text{(H.2)}$$

Then we have

$$E\left[\mathbf{\Omega}_{\mathrm{MPE}}^{-1}(\infty)\right]=\frac{T(\infty)\times E\left[\left|e(\infty)\right|^{p-2}\right]}{1-\gamma}\mathbf{R} \quad \text{(H.3)}$$

and

$$E\left[\mathbf{\Omega}_{\mathrm{MPE}}(\infty)\right]=\frac{1-\lambda}{T(\infty)\times E\left[\left|e(\infty)\right|^{p-2}\right]}\mathbf{R}^{-1} \quad \text{(H.4)}$$

6

Nonlinear Filtering Algorithms under MMPE

Kernel adaptive filters (KAFs) and neural networks (NNs) are two popular nonlinear filters that have been extensively studied and applied in engineering. The classical KAFs, such as kernel recursive least squares (KRLS) and kernel least mean square (KLMS), utilize kernel methods to map the RLS and LMS into the feature space. However, like their linear counterparts, KRLS and KLMS are sensitive to non-Gaussian noises (e.g., outliers). To address this issue, researchers have developed various robust KAFs based on the MMPE criterion. This chapter will introduce several such algorithms, including kernel least mean p-power (KLMP) [240,241], kernel recursive least mean p-power (KRLP) [241], projected KLMP [244], and random Fourier feature extended KRLP [243].

Neural Networks with Random Weights (NNRW) are feedforward neural networks that utilize a non-iterative learning mechanism. Two NNRW models, Extreme Learning Machine (ELM) and Broad Learning System (BLS), have gained popularity. However, the original ELM and BLS models rely on minimizing the MSE, which cannot effectively address the non-Gaussian noise issue, as discussed in previous chapters. To improve robustness, several NNRW-based nonlinear filters have been developed, including LMP-ELM [249], RLMP-ELM [250], sparse RLMP-ELM [251], and MMPE-based BLS models [254,255]. This chapter will delve deeper into these robust ELM and BLS models.

6.1 Kernel Adaptive Filtering Algorithms under MMPE Criterion

This subchapter introduces two KLMS-type KAFs under MMPE, namely the KLMP and projected KLMP, which enhance the robustness of the KLMS in the presence of non-Gaussian noise. Furthermore, the recursion idea and kernel method are employed to develop robust recursive KAFs under MMPE, and we introduce the KRLP and Random Fourier Features Extended KRLP algorithms.

DOI: 10.1201/9781003176114-6

6.1.1 Kernel Least Mean p-Power Algorithm

For an input $\mathbf{u}(k)$ in the original space, the kernel-induced mapping $\varphi(\cdot)$ is employed to transform it into kernel space $\mathbf{F}$ as $\varphi(\mathbf{u}(k))$. Then, the following cost function is defined

$$J_{MPE}(k) = |e(k)|^p = |d(k) - \omega(k-1)\varphi(k)|^p \tag{6.1}$$

where $e(k)$ is the prediction error at iteration k and $\omega(k-1)$ denotes the estimate of the weight vector in $\mathbf{F}$ at iteration (k–1). Applying stochastic gradient descent algorithm on the new sample data $\{\varphi(k), d(k)\}$ yields

$$\begin{aligned} \omega(k) &= \omega(k-1) - \mu \frac{\partial J_{MPE}(k)}{\partial \omega(k-1)} \\ &= \omega(k-1) + \mu p\, |e(k)|^{p-2}\, e(k)\varphi(k) \\ &= \omega(k-1) + \eta\, |e(k)|^{p-2}\, e(k)\varphi(k) \end{aligned} \tag{6.2}$$

where η denotes the step size. Thus the output data $f_k(\cdot) = \omega(k)^T \varphi(\cdot)$ can be viewed as the composition of $\omega(k)$ and $\varphi(\cdot)$. However, the dimension of $\varphi(\cdot)$ is high and it is only implicitly known, leading to that an alternative way of carrying out the computation is required. To address this problem, the repeated application of the weight-update equation through iterations yields

$$\begin{aligned} \omega(k) &= \omega(k-1) + \eta |e(k)|^{p-2} e(k)\varphi(k) \\ &= \left[\omega(k-2) + \eta |e(k-1)|^{p-2} e(k-1)\varphi(k-1)\right] \\ &\quad + \eta |e(k)|^{p-2} e(k)\varphi(k) \\ &= \omega(k-2) + \eta[|e(k-1)|^{p-2} e(k-1)\varphi(k-1) \\ &\quad + |e(k)|^{p-2} e(k)\varphi(k)] \\ &\cdots \\ &= \omega(0) + \eta \sum_{j=1}^{k} |e(j)|^{p-2} e(j)\varphi(j) \end{aligned} \tag{6.3}$$

If the initial weight vector is set as $\omega(0) = 0$, we have

$$\omega(k) = \eta \sum_{j=1}^{k} |e(j)|^{p-2} e(j)\varphi(j) \tag{6.4}$$

Then the output of the system to a new input $\mathbf{u}(k)$ can be expressed by the inner products between transformed inputs

$$\langle \omega(k), \varphi(k) \rangle = \left[\eta \sum_{j=1}^{k} |e(j)|^{p-2} e(j)\varphi(j) \right] \varphi(k) = \eta \sum_{j=1}^{k} |e(j)|^{p-2} e(j) \langle \varphi(j), \varphi(k) \rangle \quad (6.5)$$

By the kernel trick, the filter output can be efficiently computed by

$$\langle \omega(k), \varphi(k) \rangle = \eta \sum_{j=1}^{k} |e(j)|^{p-2} e(j) \kappa \left(\varphi(j), \varphi(k) \right) \quad (6.6)$$

If f_k denotes the estimate of the input-output nonlinear mapping at time k, we have the following sequential learning rule in the original space:

$$f_{k-1}(\cdot) = \eta \sum_{j=1}^{k-1} |e(j)|^{p-2} e(j) \kappa(\mathbf{u}(j), \cdot) \quad (6.7)$$

$$f_{k-1}(\mathbf{u}(k)) = \eta \sum_{j=1}^{k-1} |e(j)|^{p-2} e(j) \kappa(\mathbf{u}(j), \mathbf{u}(k)) \quad (6.8)$$

$$e(k) = d(k) - f_{k-1}(\mathbf{u}(k)) \quad (6.9)$$

$$f_k(\cdot) = f_{k-1}(\cdot) + \eta |e(k)|^{p-2} e(k) \kappa(\mathbf{u}(k), \cdot) \quad (6.10)$$

We call this algorithm the KLMP, which can be viewed as the LMP in RKHS and can be considered as a generalization version of KLMS (which will reduce to KLMS when $p=2$). Based on the description above, the KLMP is summarized in Table 6.1.

TABLE 6.1

Kernel Least Mean p-power Algorithm

Initialization
$\mathbf{a}(1) = \eta d(1), f_1 = \mathbf{a}(1)k(\mathbf{u}(1), \cdot), \quad \sigma = \sigma_0, \eta = \eta_0$
Computation
while $\{u(k), d(k)\}$available do
$f_{k-1}(\mathbf{u}(k)) = \eta \sum_{j=1}^{k} |e(j)|^{p-2} e(j) \kappa\left(\mathbf{u}(j), \phi(k)\right)$
$e(k) = d(k) - f_{k-1}(\mathbf{u}(k))$
Compute the coefficient
$\mathbf{a}(k) = \eta |e(k)|^{p-2} e(k)$
End while

Remark 6.1

The KLMP is similar to the KLMS. In fact, they have almost the same computational complexity. The key difference between the two algorithms is that the KLMP uses the MPE as the cost function, and its coefficients contain a term $H(k)=|e(k)|^{p-2}$ at each iteration which is insensitive to outliers when $0 < p < 2$.

Remark 6.2

The KLMP can be regarded as a generalized version of the KAF because it can reduce to the KLMS and KLMAT when p is set at 2.0 and 3.0, respectively. Furthermore, one can obtain the kernel least absolute deviation (KLAD) and kernel least mean fourth (KLMF) algorithms by selecting $p = 1.0$, and $p = 4.0$, which are the kernelized versions of the LAD and LMF.

6.1.2 Projected KLMP Algorithm

The traditional KAF has a major drawback in that the functional representation of kernel-based algorithms increases linearly with the number of processed data, leading to higher memory and computational complexity. To address this issue, several sparsified kernel adaptive filters (SKAFs) have been proposed, utilizing various sparsification methods such as approximate linear dependency [116], coherence criterion (CC) [331], surprise criterion [119], vector quantization (VQ) method [125], fixed memory budget model [127], and sliding window method [117]. However, most of these SKAFs rely on the MSE criterion under Gaussian noise assumption, which may result in significant performance degradation when the assumption deviates from the underlying truth. In this subchapter, we present a novel SKAF, called the projected kernel least mean p-power algorithm (PKLMP), which is based on the MPE criterion and vector projection (VP) method.

6.1.2.1 PKLMP Algorithm

A simple online sparsification method, known as the VP method, has been proposed in [323–333] from the perspective of a feature space. VP is capable of projecting the transformed input data to its most relevant center (MRC) in a dictionary, similar to VQ. The VP method also utilizes the discarded redundant data to update the coefficients of the MRC in the dictionary. Moreover, VP can leverage the hidden information from the input data to refine the corresponding coefficients for improved filtering accuracy. In this subchapter, we introduce the PKLMP algorithm by applying the VP method to KLMP, which effectively reduces the growth of the structure [244].

Let

$$\begin{cases} j^* = \arg\max \cos(\varphi(k), D_j(k-1)) \\ \Theta = \cos(\varphi(k), D_{j^*}(k-1)) \end{cases} \tag{6.11}$$

where the cosine-relation function $\cos(\mathbf{x}, \mathbf{y}) = \frac{\langle \mathbf{x}, \mathbf{y} \rangle_F}{\|\mathbf{x}\|_F \|\mathbf{y}\|_F}$. Using the idea of VP, PKLMP can be obtained by projecting $\phi(k)$ on the weight-update equation $\omega(k) = \omega(k-1) + \eta |e(k)|^{p-2} e(k)\varphi(k)$, namely

$$\begin{cases} \omega(0) = 0 \\ e(k) = d(k) - \omega^T(k-1)\varphi(k) \\ \omega(\mathrm{k}) = \omega(k-1) + \eta |e(k)|^{p-2} e(k) P(\varphi(k)) \end{cases} \tag{6.12}$$

where $P(.)$ is the vector projection operator, and $P(\varphi(k)) = a(k)\varphi(D_{j^*}(k-1))$. Here, $a(k)$ is an approximation factor in VP and can be estimated as

$$a(k) = \begin{cases} \dfrac{\kappa(\mathbf{u}(k), D_{j^*}(k-1))}{\|\varphi(D_{j^*}(k-1))\|_F^2} & \text{if } \Theta \geq \tau_c \\ 1 & \text{otherwise} \end{cases} \tag{6.13}$$

where $D_{j^*}(k-1)$ is the jth entry of the dictionary $D(k-1)$ with M members at iteration $k-1$, and τ_c is a pre-selected coherence threshold, which provides a tradeoff between the filtering accuracy and computational complexity. The learning rule for PKLMP in the original input space can then be represented as [244]

$$\begin{cases} \omega(0) = 0 \\ e(k) = d(k) - f_{k-1}(\mathbf{u}(k)) \\ f_k = h_{-j^*}^k + h_{j^*}^k \\ h_{-j^*}^k = \sum\limits_{l=1, l\neq j^*}^{M} \omega_l(k-1)\kappa(D_l(k-1), \cdot) \\ h_{j^*}^k = \boldsymbol{\omega}_{j^*}(k-1) + \eta\, |\, e(k)\, |^{p-2}\, a(k) e(k) \kappa(D_l(k-1), \cdot) \end{cases} \tag{6.14}$$

where $h_{-j^*}^k$ contains all entries of $\omega(k)$ except for the j^*th, which is contained in $h_{j^*}^k$. The PKLMP algorithm is summarized in Table 6.2 [244].

TABLE 6.2

Projected Kernel Least Mean p-Power (PKLMP) Algorithm

Initialization: step size $\eta > 0$, Mercer kernel $\kappa(\cdot,\cdot)$, coherence threshold $0 < \tau_c < 1$, dictionary $\mathbf{D}(1) = \{\mathbf{u}(1)\}$, $\omega(1) = [\eta \mid d(1) \mid^{p-2} d(1)]^T$
Computation: while $\{\mathbf{u}(k), d(k)\}$ $(k \geq 2)$ available **do**
1. the output: $f(\mathbf{u}(k)) = \sum_{l=1}^{M} \omega_l (k-1)\kappa(D_l(k-1), \mathbf{u}(k))$
2. the estimation error: $e(k) = d(k) - f_{k-1}(\mathbf{u}(k))$
3. the cosine-relation between $\varphi(k)$ and $\varphi(D(k-1))$: $\Theta = \max_{1\leq j\leq M} \frac{\langle \varphi(k), \varphi(D_j(k-1))\rangle}{\| \varphi(k) \|_F \| \varphi(D_j(k-1)) \|_F}$
4. **if** $\Theta \geq \tau_c \geq \varepsilon_c$ keep the dictionary unchanged: $D(k) = D(k\text{-}1)$, $M \Leftarrow M$ update the coefficient of the MRC: $\omega_{j^*}(k) = \omega_{j^*}(k-1) + \eta \mid e(k) \mid^{p-2} a(k)e(k)$ with $a(k) = \frac{\kappa(\mathbf{u}(k), D_{j^*}(k-1))}{\| \varphi(D_{j^*}(k-1)) \|_F^2}$
5. **otherwise** change the dictionary: $D(k) = \{D(k\text{-}1), \mathbf{u}(k)\}$, $M \Leftarrow M+1$ update the kernel weight vector (KWV): $\omega(k) = [\omega^T(k-1), \eta \mid e(k) \mid^{p-2} e(k)]^T$
end while

The main steps of the algorithm are (3), (4), and (5). Based on the cosine-relation between $\phi(k)$ and $\phi(\boldsymbol{D}(k-1))$, PKLMP decides whether to change the dictionary or not. In step (4), PKLMP keeps the dictionary unchanged and uses $\eta|e(k)|^{p-2} a(k)e(k)$ to update the jth entry of KWV at instant $k-1$. In step (5), the dictionary absorbs $\boldsymbol{u}(k)$ as its new element, and KWV uses $\eta|e(k)|^{p-2} e(k)$ to expand itself.

Remark 6.3

> From the PKLMP algorithm given in Table 6.2, one can see that PKLMP only absorbs the samples that satisfy the significant criterion (i.e.,$\Theta < \varepsilon_c$). With the evolution of adaption, PKLMP achieves a SKAF with $|\mathbf{D}(k)|_c = M \ll k$. Although the vector projection is operated in **F**, as shown in (6.14), PKLMP can be effectively executed in **U**.

When $p = 2$, PKLMP becomes the PKLMS algorithm, and there are close relationships between the PKLMS algorithm and several existing ones, such as QKLMS and the modified QKLMS (MQKLMS) [334]. When a Gaussian kernel (with kernel size h) is applied to PKLMS, one can get

$$\begin{cases} a(k) = \kappa(\mathbf{u}(k), D_{j^*}(k-1)) \\ f_k = h^k_{-gj^*} + h^k_{Gj^*} \\ h^k_{Gj^*} = (\omega_{j^*}(k-1) + \eta \mid e(k) \mid^{p-2} e(k)\kappa(\mathbf{u}(k), D_{j^*}(k-1)) \times \kappa(D_j(k-1), \cdot)) \end{cases} \tag{6.15}$$

where $h^k_{-Gj^*}$ and $h^k_{Gj^*}$ are similar to $h^k_{-j^*}$ and $h^k_{j^*}$, respectively, except that a Gaussian kernel is used in the former. In this case, PKLMS has the same update formula as MQKLMS.

Remark 6.4

As one can see, PKLMP only considers the feature space compression and assumes that the corresponding outputs of the projected data are equal to those of MRC in the dictionary. In the impulsive noise cases, however, the outputs in a neighborhood may have large variations. Such disturbance in desired outputs may dramatically deteriorate the filtering accuracy. Hence, it is necessary to smooth the outputs to improve the performance, and thus a PKLMP with smoothed desired output (PKLMP-SD) has been developed in [244]. In addition, the centers of a dictionary are fixed in most sparsification methods. However, variable centers (VC) may lead to a better filtering performance for KAFs. When $\Theta < \varepsilon_c$, the dictionary remains unchanged, and $\mathbf{u}(k)$ is just replaced by $D_{j^*}(k-1)$. A simple method has been proposed to vary the centers of a dictionary, and a PKLMP with VC was developed in [244], called PKLMP-SD-VC.

6.1.2.2 Convergence Analyses

In this section, a stability analysis is provided, and then a mean-square convergence analysis for PKLMP is conducted with different values of p, i.e., converging in terms of MSE. We also analyze the mean convergence behavior of PKLMP and derive its convergence conditions.

Let f^o denote an unknown nonlinear model that needs to be estimated. According to the universal approximation property [335], there is an optimal vector $\omega^o \in F$ such that $f_o = \langle \omega^o, \varphi(k) \rangle_F$. Thus, we represent the desired signal $d(k)$ as

$$d(k) = f_o(\mathbf{u}(k)) + v(k) = \langle \omega^o, \varphi(k) \rangle_F + v(k) \tag{6.16}$$

where $v(k)$ is the disturbance noise.

6.1.2.2.1 Stability Analysis

According to (6.16), the estimation error $e(k)$ can be expressed as

$$e(k) = d(k) - \omega^T(k-1)\varphi(k) = e_a(k) + v(k) \tag{6.17}$$

where $e_a(k) = \tilde{\omega}^T(k-1)\varphi(k)$ denotes the *a priori* error, and $\tilde{\omega}(k-1) = \omega^o - \omega(k-1)$ is the weight error. Subtracting ω^o from both sides of the last line of (6.12), we have

$$\begin{aligned}&\underbrace{\boldsymbol{\omega}^o-\boldsymbol{\omega}(k)}_{\tilde{\boldsymbol{\omega}}(k)}=\underbrace{\boldsymbol{\omega}^o-\boldsymbol{\omega}(k-1)}_{\tilde{\boldsymbol{\omega}}(k-1)}-\eta\underbrace{|e(k)|^{p-2}e(k)}_{F(e(k))}\underbrace{P(\boldsymbol{\varphi}(k))}_{a(k)\boldsymbol{\varphi}_p(k)}\\&\Rightarrow\tilde{\boldsymbol{\omega}}(k)=\tilde{\boldsymbol{\omega}}(k-1)-\eta F(e(k))a(k)\boldsymbol{\varphi}_p(k)\end{aligned}\tag{6.18}$$

Squaring both sides of the last line of (6.18), and then taking the expectations, we have

$$\begin{aligned}&E\left[\|\tilde{\boldsymbol{\omega}}(k)\|_F^2\right]-E\left[\|\tilde{\boldsymbol{\omega}}(k-1)\|_F^2\right]\\&=E\left[(\eta a(k))F(e(k))^2\kappa_p(k)\right]\\&\quad-2E\left[\eta a(k)F(e(k))\tilde{\boldsymbol{\omega}}^T(k-1)\boldsymbol{\varphi}_p(k)\right]\end{aligned}\tag{6.19}$$

where $\kappa_p(k)=\left\langle\boldsymbol{\varphi}_p(k),\boldsymbol{\varphi}_p(k)\right\rangle_F$. To guarantee a converging solution, $E\left[\|\tilde{\boldsymbol{\omega}}(k)\|_F^2\right]-E\left[\|\tilde{\boldsymbol{\omega}}(k-1)\|_F^2\right]\le 0$ should be satisfied, which means that the value of η in PKLMP at $\forall k$ should satisfy

$$\begin{cases}E\left[\|\tilde{\boldsymbol{\omega}}(k)\|_F^2\right]-E\left[\|\tilde{\boldsymbol{\omega}}(k-1)\|_F^2\right]\le 0\\E\left[\eta a(k)F(e(k))\tilde{\boldsymbol{\omega}}^T(k-1)\boldsymbol{\varphi}_p(k)\right]>0\\0<\eta\le\dfrac{2E\left[a(k)F(e(k))\tilde{\boldsymbol{\omega}}^T(k-1)\boldsymbol{\varphi}_p(k)\right]}{=E\left[(a(k)F(e(k)))^2\kappa_p(k)\right]}\end{cases}\tag{6.20}$$

Remark 6.5

The last inequality in (6.20) presents a sufficient condition for mean-square convergence of PKLMP. However, in practice, it is difficult to know the upper bound exactly. Hence, it merely indicates that η needs to be small enough to make the algorithm converge and cannot be directly applied as a strict convergence condition for KAFs.

6.1.2.2.2 Convergence in Terms of Steady-State MSE

We define the EMSE $\xi=\lim_{k\to\infty}E[e_a(k)^2]$ and use it to study the steady-state MSE of PKLMP. Before proceeding, the following assumptions are listed, which are used in the rest of this part.

Assumption 6.1

The noise signal $v(k)$ follows multiple independent and identical distributions with zero-mean and the finite variance σ_v^2, and is independent of the input sequence $\{\mathbf{u}(k)\}$.

Assumption 6.2

The *a priori* error $e_a(k)$ with zero-mean is independent of $v(k)$

Assumptions 6.1 and 6.2 are commonly used in steady-state performance analysis for adaptive filters, e.g., in [125], and [272]. Taking the limits of both sides of (6.19), we get

$$\begin{aligned}&\lim_{k\to\infty} E\left[\|\tilde{\omega}(k)\|_F^2\right] - \lim_{k\to\infty} E\left[\|\tilde{\omega}(k-1)\|_F^2\right] \\ &= \lim_{k\to\infty} E\left[(\eta a(k)F(e(k)))^2 \kappa_p(k)\right] \\ &\quad - 2\lim_{k\to\infty} E\left[(\eta a(k)F(e(k))\tilde{\omega}^T(k-1)\varphi_p(k)\right]\end{aligned} \tag{6.21}$$

When the algorithm reaches the steady-state, we have $\lim_{k\to\infty} E\left[\|\tilde{\omega}(k)\|_F^2\right] = \lim_{k\to\infty} E\left[\|\tilde{\omega}(k-1)\|_F^2\right]$, and

$$\eta \lim_{k\to\infty} E\left[(a(k)F(e(k)))^2 \kappa_p(k)\right] = 2\lim_{k\to\infty} E\left[\eta a(k)F(e(k))\tilde{\omega}^T(k-1)\varphi_p(k)\right] \tag{6.22}$$

We further assume that $(a(k))^2\kappa_p(k)$ and $F(e(k))^2$ are asymptotically uncorrelated (the rationality of this assumption was discussed in [272]). Since $a(k)$ is only an approximation factor in VP, we can assume that $a(k)$ is asymptotically uncorrelated with $F(e(k))\tilde{\omega}^T(k-1)\varphi_p(k)$. In particular, if we set $a(n)=1$, this assumption becomes a truth. Then (6.22) becomes

$$\begin{aligned}&\frac{\eta}{2}\lim_{k\to\infty} E\left[(a(k))^2\kappa_p(k)\right]\lim_{k\to\infty} E\left[(F(e(k)))^2\right] = \lim_{k\to\infty} E[a(k)] \\ &\lim_{k\to\infty} E\left[a(k)F(e(k))\tilde{\omega}^T(k-1)\phi_p(k)\right]\end{aligned} \tag{6.23}$$

It is well known that adaptive filters based on LMP achieve different performances for different values of p. Hence, we derive the steady-state performance of PKLMP using a Taylor expansion method, which can decouple the correlation between $e_a(k)$ and $v(k)$ [212].

Since $F(e)=|e|^{p-2}$ and $e=e_a+v$, taking the Taylor expansion of $F(e)$ with respect to e_a around v, we get

$$\begin{cases} F(e) = F(v) + F'(v)e_a + \dfrac{1}{2}F''(v)e_a^2 + O(e_a^2) \\ F'(v) = (p-1)|v|^{p-2} \\ F''(v) = (p-1)(p-2)|v|^{p-2}\,v \end{cases} \tag{6.24}$$

where $F'(v)$ and $F''(v)$ denote the first and second derivatives, and $O(e_a^2)$ contains the third and higher-order terms. Hence, based on (6.24), we have

$$\begin{aligned} F(e(k))^2 &= F(v(k))^2 + 2F(v(k))F'(v(k))e_a(k) \\ &\quad + F(v(k))F''(v(k))e_a^2(k) + F'(v(k))^2 e_a^2(k) + O(e_a^2(k)) \\ &= |v(k)|^{2p-2} + 2(p-1)|v(k)|^{2p-4} e_a(k) + (p-1)(2p-3)e_a^2(k) + O(e_a^2(k)) \end{aligned} \tag{6.25}$$

and, based on Assumption 5.4, we have

$$\begin{aligned} &\lim_{k\to\infty} E\left[F(e(k))^2\right] \\ &= E\left[|v(k)|^{2p-2}\right] + E\left[O(e_a^2(k))\right] + (p-1)(2p-3)E\left[|v(k)|^{2p-4}\right]\lim_{k\to\infty} E\left[e_a(k)\right] \\ &= \xi_v^{2p-2} + (p-1)(2p-3)\xi_v^{2p-4}\xi \end{aligned} \tag{6.26}$$

where $\xi_v^p = E\left[|v|^p\right]$ and $E\left[O(e_a^2(k))\right]$ is ignored, and

$$\begin{aligned} &\lim_{k\to\infty} E\left[F(e(k))\tilde{\omega}^T(k-1)\varphi_p(k)\right] \\ &= (p-1)\xi_v^{p-2}\left(\xi + \lim_{k\to\infty} E\left[e_a(k)g_p(k)\right]\right) \end{aligned} \tag{6.27}$$

where $g_p(k) = \tilde{\omega}^T(k-1)(\varphi_p(k) + \varphi(k))$. Combining (6.23), (6.26), and (6.27), and after some calculations, we obtain

$$\begin{cases} \xi = \dfrac{\eta a_\infty^\kappa \xi_v^{p-2} - 2a_\infty \lim_{k\to\infty} E\left[e_a(k)g_p(k)\right]}{2a_\infty(p-1)\xi_v^{p-2} - \eta a_\infty^\kappa (p-1)(2p-3)\xi_v^{2p-4}} \\ a_\infty = \lim_{k\to\infty} E\left[a(k)\right] \\ a_\infty^\kappa = \lim_{k\to\infty} E\left[a^2(k)\kappa_p(k)\right] \end{cases} \tag{6.28}$$

When a Gaussian kernel is used in PKLMP, we get $\kappa_p(k) = 1$, and thus $a_\infty^\kappa \approx (a_\infty)^2$. In the steady-state, a_∞ can be approximated by ε_c. Hence, (6.28) can be rewritten as

$$\xi = \frac{\eta\varepsilon_c\xi_v^{p-2} - 2\lim_{k\to\infty} E\left[e_a(k)g_p(k)\right]}{2(p-1)\xi_v^{p-2} - \eta\varepsilon_c(p-1)(2p-3)\xi_v^{2p-4}} \tag{6.29}$$

One can know that $\kappa\left(D_{j^*}(k-1), \mathbf{u}(k)\right) \geq \varepsilon_c$ in the steady-state, and we also have

$$\begin{aligned}
&\left|E\left[\tilde{\omega}^T(k-1)\varphi(k)g_p(k)\right]\right| \\
&\leq E\left[\left\|\tilde{\omega}^T(k-1)\right\|_F^2\right]\underbrace{\left\|\varphi_p(k)-\varphi(k)\right\|_F}_{\sqrt{2-2\kappa(\varphi_p(k),\varphi(k))}} \\
&\overset{(a)}{\leq} E\left[\left\|\omega^o\right\|_F^2\right]\sqrt{2-2\varepsilon_c}
\end{aligned} \tag{6.30}$$

where the inequality (*a*) is obtained based on the fact that $\left\|\omega^T(k)\right\|_F^2 \leq \left\|\omega^T(k-1)\right\|_F^2 \leq \cdots \leq \left\|\omega^o(k)\right\|_F^2$. Therefore, injecting (6.30) into (6.29), we get

$$\begin{cases}
Low(p,\varepsilon_c) \leq \xi \leq U_p(p,\varepsilon_c) \\
Low(p,\varepsilon_c) = \max\left(\dfrac{\eta\varepsilon_c\xi_v^{2p-v} + 2C_{\varepsilon_c}^{\varpi^o}}{C_{\varepsilon_c}^{p\xi_v}}, 0\right) \\
U_p(p,\varepsilon_c) = \dfrac{\eta\varepsilon_c\xi_v^{2p-v} + 2C_{\varepsilon_c}^{\varpi^o}}{C_{\varepsilon_c}^{p\xi_v}} \\
C_{\varepsilon_c}^{\varpi^o} = E\left[\left\|\omega^o\right\|_F^2\right]\sqrt{2-2\varepsilon_c} \\
C_{\varepsilon_c}^{p\xi_v} = 2(p-1)\xi_v^{p-2} - \eta\varepsilon_c(p-1)(2p-3)\xi_v^{2p-4}
\end{cases} \tag{6.31}$$

where $Low(p,\varepsilon_c)$ and $Up(p,\varepsilon_c)$ denote the lower and upper bounds of ξ, respectively.

Remark 6.6

The EMSE ξ is derived in (6.29); however, it is hard to estimate the value of $\varphi_p(k)$ and then evaluate the EMSE exactly. Therefore, we provide the bounds of ξ in (6.31) to characterize the EMSE. In addition, for any given $p \geq 1$, it is not difficult to find ξ's bounds numerically from (6.31). And, from this formulation, we can obtain the following important observations:

(i) When $p=2$, (6.31) reduces to

$$\begin{cases} Low(2,\varepsilon_c) \le \xi \le U_p(p,\varepsilon_c) \\ Low(2,\varepsilon_c) = \max\left(\dfrac{\eta\varepsilon_c\sigma_v^2 + 2C_{\varepsilon_c}^{\varpi^o}}{C_{\varepsilon_c}^{p\xi_v}}, 0\right) \\ U_p(p,\varepsilon_c) = \dfrac{\eta\varepsilon_c\sigma_v^2 + 2C_{\varepsilon_c}^{\varpi^o}}{2-\eta\varepsilon_c} \end{cases} \tag{6.32}$$

From (6.32) one can see that, when $\varepsilon_c = 1$, the EMSE of PKLMP with $p=2$ is $\xi = \dfrac{\eta\sigma_v^2}{2-\eta}$, which is actually the EMSE for KLMS; (ii) When $1 \le p < 2$, we consider the fractional lower-order moment (FLOM) of $|e(k)|$ and the noise $v(k)$ is modeled by the alpha-stable distribution. The alpha-stable noise has infinite variance, with finite pth-order moments for $p < \alpha \in (0,2]$, where α denotes the characteristic factor that measures the tail heaviness of the distribution. Hence, based on the FLOM property of alpha-stable distribution, (6.31) becomes

$$\begin{cases} Low(p_<,\varepsilon_c) \le \xi \le U_p(p_<,\varepsilon_c) \\ Low(p_<,\varepsilon_c) = \max\left(\dfrac{\eta\varepsilon_c C_{\alpha 1}^{\gamma p_<} + 2C_{\varepsilon_c}^{\varpi^o}}{C_{\varepsilon_c}^{p\alpha}}, 0\right) \\ U_p(p_<,\varepsilon_c) = \dfrac{\eta\varepsilon_c C_{\alpha 1}^{\gamma p_<} + 2C_{\varepsilon_c}^{\varpi^o}}{C_{\varepsilon_c}^{p\alpha}} \\ C_{\varepsilon_c}^{p\alpha} = 2(p-1)(C_{\alpha 2}^{\gamma} - \eta\varepsilon_c(2p-3)C_{\alpha 3}^{\gamma p_<}) \\ C_{\alpha 1}^{\gamma p_<} = C(2p-2,\alpha)\gamma^{\frac{2p-2}{\alpha}} \\ C_{\alpha 2}^{\gamma p_<} = C(2p-2,\alpha)\gamma^{\frac{p-2}{\alpha}} \\ C_{\alpha 3}^{\gamma p_<} = C(2p-2,\alpha)\gamma^{\frac{2p-4}{\alpha}} \end{cases} \tag{6.33}$$

where "$p<$" means that p takes values over [1, 2), and

$$C(p,\alpha) = \frac{2^{p+1}\Gamma\left(\frac{p+1}{2}\right)\Gamma\left(-\frac{p}{\alpha}\right)}{\alpha\sqrt{\pi}\,\Gamma\left(-\frac{p}{2}\right)} \tag{6.34}$$

with $\Gamma(\cdot)$ denoting the Gamma function.

6.1.2.2.3 Mean Convergence Analysis

In this section, with some approximations, we analyze the mean convergence behavior of the PKLMP algorithm. Taking the expectation operations on both sides of the last row in (6.12), we obtain

$$\begin{aligned} &E\left[\boldsymbol{\omega}(k)\right] \\ &= E\left[\boldsymbol{\omega}(k-1)\right] + \eta \\ &\times E\left[\left|e(k)\right|^{p-2}(d(k) - \boldsymbol{\omega}^T(k-1)\boldsymbol{\varphi}(k))a(k)\boldsymbol{\varphi}_p(k)\right] \end{aligned} \tag{6.35}$$

Assume that $|e(k)|^{p-2}$ is uncorrelated with $(d(k) - \omega^T(k-1)\varphi(k))a(k)\varphi_p(k)$. This is a strong assumption, but it facilitates the analysis [211]. Together with the assumption of independence between the input sequence and weight vector, (6.35) can be represented as

$$E[\boldsymbol{\omega}(k)] = E[\boldsymbol{\omega}(k-1)] + \eta E[\,|\,e(k)\,|^{p-2}] \times a(k)(\mathbf{r}_{dp} - \mathbf{R}_{kp}E[\boldsymbol{\varphi}(k-1)]) \tag{6.36}$$

where $\mathbf{r}_{dp=}$ $E[d(k)\phi_p(k)]$, and $\mathbf{R}_{kp=}$ $E[\phi_p(k)\phi(k)^T]$ (assuming $\mathbf{R}_{kp}$ is positive definite).

Since $e(k) = e_a(k) + v(k)$, when the variance of $v(k)$ is larger than that of $e_a(k)$, we can assume that $e(k)$ is dominated by $v(k)$ in the transient state. Therefore, to make (6.36) numerically assessable, we can replace $|e(k)|^{p-2}$ by $|v(k)|^{p-2}$ in (6.36). Subtracting $E[\boldsymbol{\omega}^o]$ on both sides of (6.36), we obtain the following stochastic difference formula

$$\begin{aligned} \varsigma(k) &= (\mathbf{I} - \eta E[\,|\,v(k)\,|^{p-2}]a(k)\mathbf{R}_{kp}\varsigma(k-1) \\ &+ \eta E[\,|\,v(k)\,|^{p-2}\, a(k)(\mathbf{r}_{dp} - \mathbf{R}_{kp}E[\omega^o]) \end{aligned} \tag{6.37}$$

where $\varsigma(k) = E\left[\omega^o - \omega(k)\right]$, and $\mathbf{I}$ is an identity matrix. To guarantee the convergence of (6.37), the following condition needs to be satisfied $\left(1 - \eta E\left[\left|v(k)\right|^{p-2}\right]a(k)\lambda_{\max}\right) < 1$, i.e.,

$$0 < \eta < \frac{2}{E\left[\left|v(k)\right|^{p-2}\right] \times a(k)\lambda_{\max}} \tag{6.38}$$

where $\lambda_{\max}$ means the maximum eigenvalue of $\mathbf{R}_{kp}$. A more conservative condition can be obtained as

$$0 < \eta < \frac{2}{E\left[\left|v(k)\right|^{p-2}\right] \times a(k)\mathbf{R}_{kp}} \tag{6.39}$$

where $\mathrm{Tr}(\mathbf{R}_{kp})$ denotes the trace of $\mathbf{R}_{kp}$. Moreover, when no sparsification method is applied and a Gaussian kernel is used, (6.39) reduces to

$$0<\eta<\frac{2}{E\left[|v(k)|^{p-2}\right]} \tag{6.40}$$

Remark 6.7

For the following two types of noise $v(k)$, (6.40) can be further elaborated.
(i) The noise $v(k)$ has finite variance, i.e., $\sigma_{v^2}<\infty$. In this case, we can approximate $E\left[|v(k)|^{p-2}\right]$ as $E\left[|v(k)|^{2}\right]^{\frac{p-2}{2}}$. Hence, (6.40) becomes

$$0<\eta<\frac{2}{E\left[|v(k)|^{2}\right]^{\frac{p-2}{2}}} \tag{6.41}$$

which is a convergence condition for KLMP with $p\geq 2$.

(ii) The noise $v(k)$ has infinite variance, and more specifically we consider the α-SN. In this case, $E\left[|v(k)|^{p-2}\right]=C(p-2,\alpha)\gamma^{\frac{p-2}{\alpha}}$, where $C(p-2,\alpha)$ is defined in (6.34). Thus, (6.41) becomes

$$0<\eta<\frac{2}{C(p-2,\alpha)\gamma^{\frac{p-2}{2}}} \tag{6.42}$$

which is a convergence condition for KLMP with $p\in[1,2)$ under the alpha-stable noise environments.

6.1.2.3 Simulation Results

This section presents simulation results to showcase the effectiveness of PKLMP methods. The experiments include estimation of an artificial static function, prediction of short-term Chua's chaotic time series, and prediction of real-world Internet traffic and Sunspot number. Unless specified otherwise, a Gaussian kernel with a kernel size 1 is used in the experiments, and the testing MSE is used as the metric to evaluate the filtering performance. In [244], subclasses of PKLMP schemes were defined based on the value of the approximation factor $a(k)$, which is a crucial factor in PKLMP. When $a(k)=1$, the PKLMP scheme is referred to as hard PKLMP (HP-KLMP), while the original PKLMP with the exact $a(k)$ is called soft PKLMP (SP-KLMP). Similarly, one can obtain SP-KLMS and HP-KLMS for KLMS, and SP-KLMP-SD and HP-KLMP-SD for KLMP-SD.

6.1.2.3.1 Static Function Estimation

The following artificial function is considered

$$d(k) = \frac{\exp\left(\frac{(u(k)+1)^2}{-2}\right) + \exp\left(\frac{(u(k)-1)^2}{-2}\right)}{5} + v(k)$$

where the input samples $\{u(k)\} \in \mathbf{R}$ are generated following a white Gaussian process with zero-mean and variance σ_w^2, and the noise $\{v(k)\}$ is independent of $\{u(k)\}$.

6.1.2.3.1.1 Lower and Upper Bounds The accuracy of the lower and upper bounds (6.32) for SP-KLMP with $p=2$, i.e., SP-KLMS, is verified. For comparison, the corresponding lower and upper bounds for QKLMS in (6.38) are considered in this case. In this trial, $\{v(k)\}$ is a white Gaussian noise with zero-mean and variance $\sigma_v^2 = 0.04$. The step size is $\eta = 0.6$ for both algorithms. The coherence threshold ε_c is estimated via $\varepsilon_c = \exp(-0.5\varepsilon_U^2)$, where ε_U is the distance threshold of QKLMS. According to (6.32), the $E\left[\|\omega_o\|_F^2\right]$ needs to be estimated. In general, it is hard to get the optimal solution. Fortunately, the above model can be expressed as

$$\begin{cases} d(k) = f^o + v(k) \\ f^o = 0.29\left(\exp\left(-0.5(\cdot - 1)^2\right) + \exp\left(-0.5(\cdot + 1)^2\right)\right) \end{cases}$$

Hence $E\left[\|\omega_o\|_F^2\right] = \|f^o\|^2 = 0.0908$.

The EMSE for these two algorithms is given in Figure 6.1. One can have the following observations: (i) The EMSE indeed lies between the derived lower and upper bounds for SP-KLMS; (ii) The lower bound of SP-KLMS has very similar behavior to that of QKLMS; (iii) When ε_U increases from 0.35 to 10 (namely, ε_U decreases from 0.9406 to 0), SP-KLMS has smaller upper bound than QKLMS, which means the upper bound $U_p(\varepsilon_c)$ in (6.32) is more accurate; (iv) When $\varepsilon_U \approx 0$ (namely, $\varepsilon_c \approx 1$), the EMSE of SP-KLMS is close to that of KLMS, which can be theoretically estimated as $\xi = \frac{\eta\sigma_v^2}{2-\eta} = 0.0171$; (vi) When $\varepsilon_c \in (0.00073, 0.9406)$, SP-KLMS can realize smaller EMSE than KLMS. Furthermore, when $\varepsilon_c \in (0.1103, 0.9406)$, SP-KLMS outperforms QKLMS in terms of EMSE.

6.1.2.3.1.2 Filtering Performance The filtering performance for SPKLMP, HP-KLMP, SP-KLMP-SD, and HP-KLMP-SD is further tested under the impulsive noise. For comparisons, the NC and CC sparsification methods are applied to KLMP, and are denoted as NC-KLMP and CC-KLMP, respectively.

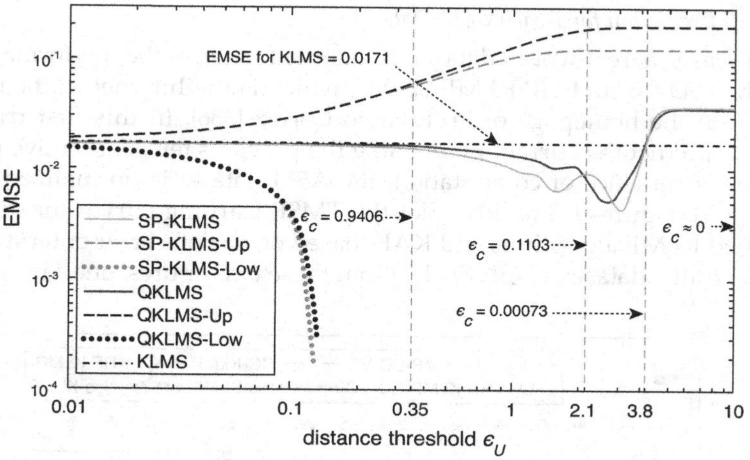

FIGURE 6.1
EMSEs for SP-KLMS and QKLMS versus the distance threshold $\varepsilon_U \in [0.01, 10]$ in SFE [244].

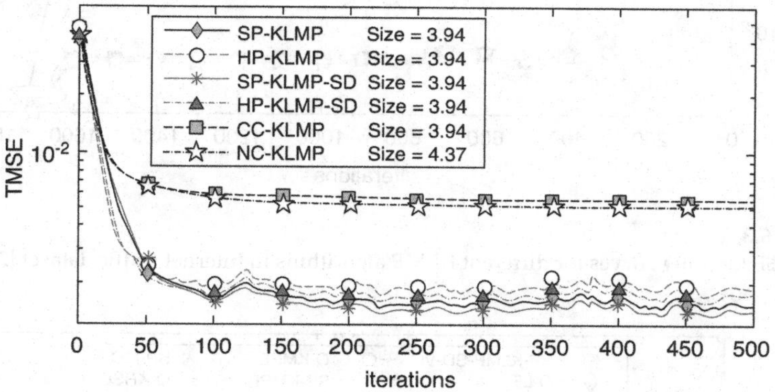

FIGURE 6.2
The TMSE learning curves for the six KLMPs in SFE [244].

Figure 6.2 plots the TMSE learning curves for these six KLMP algorithms and delivers the following messages: (i) Under similar initial convergence rates, the KLMP algorithms using VP outperform NC-KLMP and CC-KLMP in terms of filtering accuracy and the dictionary size; (ii) SP-KLMP-SD and HP-KLMP-SD achieve better filtering accuracy than SP-KLMP and HP-KLMP, respectively. This indicates that the smooth-desired output method does improve the filtering performance for KAF; (iii) Compared to the two HP-KLMPs, the two SP-KLMPs achieve smaller TMSE. Hence, we can conclude that the VP method can efficiently improve the filtering performance for KLMP.

6.1.2.3.2 Prediction for Real-World Data

In this case, a real-world dataset is adopted to test the performance of SP-KLMP-SD-VC and HP-KLMP-SD-VC, which is the Internet traffic named A5M from the homepage of Prof. Paulo Cortez [336]. In this first trial, the task is to predict the current point using the previous ten consecutive points. For the convenience of computation, the A5M dataset is normalized to the range [0, 1]. Figures 6.3 and 6.4 plot the TMSE learning curves for different sparsified KLMP algorithms and KAFs based on various error criteria for the Internet traffic dataset, respectively. From these two figures, one can see that:

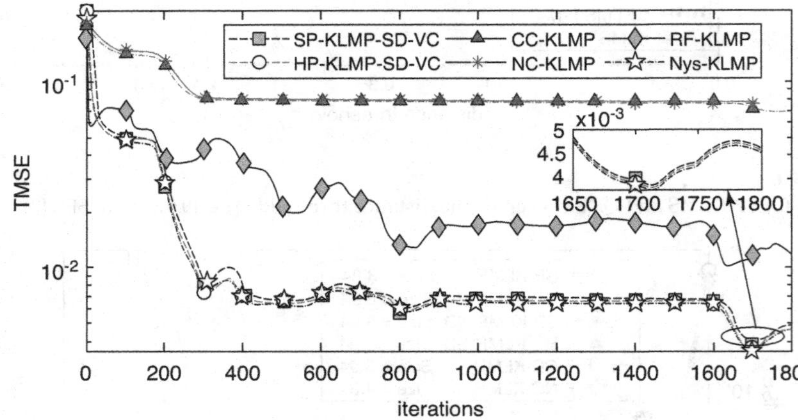

FIGURE 6.3
The TMSE learning curves for different KLMP algorithms in Internet traffic dataset [244].

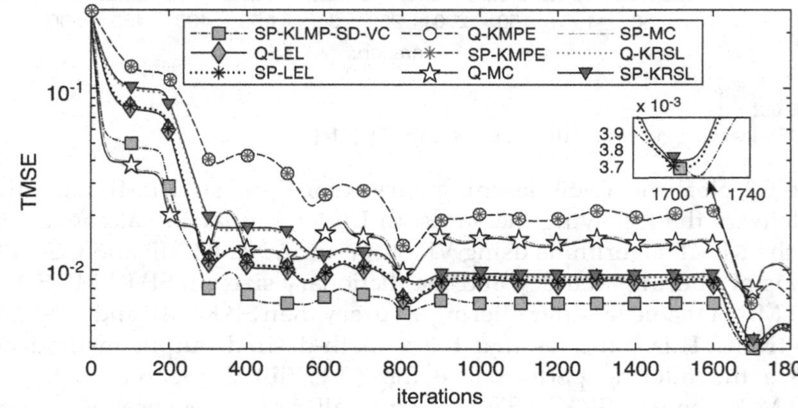

FIGURE 6.4
The TMSE learning curves of SP and quantized KAFs based on various error criteria in Internet traffic dataset [244].

(1) HP-KLMP-SD-VC and SP-KLMP-SD-VC can realize almost the same filtering performance as that of Nys-KLMP. These three algorithms outperform CC-KLMP, NC-KLMP, and RF-KLMP; (2) SP-KLMP-SD-VC achieves the best filtering accuracy among all KAFs.

6.1.3 Kernel Recursive Least Mean *p*-Power Algorithm

6.1.3.1 KRLP Algorithm

To derive the RLP algorithm in RKHS, an exponentially weighted mechanism is employed here to put more emphasis on recent data and to deemphasize data from the remote past, and the weight vector $\boldsymbol{\omega}(k)$ at each iteration needs to be solved recursively by the weighted cost function defined as [241]

$$J(\boldsymbol{\omega}(k)) = \min_{\Omega} \sum_{j=1}^{k} \gamma^{k-j} \left| d(j) - \boldsymbol{\omega}(k)^T \boldsymbol{\varphi}(j) \right|^p + \frac{1}{2} \gamma^k \beta \left\| \boldsymbol{\omega}(k) \right\|^2 \tag{6.43}$$

where γ denotes the forgetting factor in the interval [0 1], and $\mathbf{w}$ is the regularization factor. The second term is a norm penalizing term, whose role is to guarantee the existence of the inverse of the data auto-correlation matrix, especially during the initial update stages. Furthermore, the regularization term is weighted by γ, which deemphasizes regularization as time progresses. To obtain the optimal solution, the gradient decent method is employed as

$$\begin{aligned}
\frac{\partial J(\boldsymbol{\omega}(k))}{\partial \boldsymbol{\omega}(k)} &= -p \sum_{j=1}^{k} \gamma^{k-j} \left| d(j) - \boldsymbol{\varphi}^T(j)\boldsymbol{\omega}(k) \right|^{p-2} (d(j) - \boldsymbol{\varphi}^T(j)\boldsymbol{\omega}(k))\boldsymbol{\omega}(j) + \gamma^k \beta \boldsymbol{\omega}(k) \\
&\quad - p \sum_{j=1}^{k} \gamma^{k-j} \left| d(j) - \boldsymbol{\varphi}^T(j)\boldsymbol{\omega}(k) \right|^{p-2} d(j)\boldsymbol{\varphi}^T(j) \\
&\quad + p \sum_{j=1}^{k} \gamma^{k-j} \left| d(j) - \boldsymbol{\varphi}^T(j)\boldsymbol{\omega}(k) \right|^{p-2} \boldsymbol{\varphi}^T(j)\boldsymbol{\omega}(j)\boldsymbol{\omega}(k) + \gamma^k \beta \boldsymbol{\omega}(k) \\
&\quad - p \sum_{j=1}^{k} \gamma^{k-j} \left| d(j) - \boldsymbol{\varphi}^T(j)\boldsymbol{\omega}(k) \right|^{p-2} d(j)\boldsymbol{\varphi}^T(j) \\
&\quad + p \left(\sum_{j=1}^{k} \gamma^{i-j} \left| d(j) - \boldsymbol{\varphi}^T(j)\boldsymbol{\omega}(k) \right|^{p-2} \boldsymbol{\varphi}^T(j)\boldsymbol{\omega}(j) + \gamma^k \beta \right) \boldsymbol{\omega}(k)
\end{aligned} \tag{6.44}$$

Setting the above gradient to zero, we obtain the solution

$$\omega(k)=\left(\sum_{j=1}^{k}\gamma^{k-j}\left|d(j)-\varphi^{T}(j)\omega(k)\right|^{p-2}\varphi^{T}(j)\varphi(j)+\gamma^{k}\beta\right)^{-1}$$

$$\sum_{j=1}^{k}\gamma^{k-j}\left|d(j)-\varphi^{T}(j)\omega(k)\right|^{p-2}d(j)\varphi^{T}(j) \tag{6.45}$$

To describe it concisely, we introduce

$$\mathbf{d}(k)=\left[d(1),d(2),\ldots,d(k)\right] \tag{6.46}$$

$$\Phi(k)=\left[\varphi(1),\varphi(1),\ldots\varphi(k)\right] \tag{6.47}$$

$$\mathbf{B}(k)=diag\left[\gamma^{k-1}\left|d(1)-\omega(k)^{T}\varphi(1)\right|^{p-2},\gamma^{k-2}\left|d(2)-\omega(k)^{T}\varphi(2)\right|^{p-2},\ldots,\left|d(i)-\omega(k)^{T}\varphi(2)\right|^{p-2}\right] \tag{6.48}$$

Then we got the matrix form of the equation (6.45) at time k as

$$\omega(k)=\left(\Phi(k)\mathbf{B}(k)\Phi^{T}(k)+\gamma^{k}\beta\mathbf{I}\right)^{-1}\Phi(k)\mathbf{B}(k)\mathbf{d}(k) \tag{6.49}$$

Now, by using the matrix inversion lemma with the identifications $\beta\gamma^{k}\mathbf{I}\rightarrow A,\Phi(k)\rightarrow B,\mathbf{B}(k)\rightarrow C,\Phi(k)^{T}\rightarrow D$, we have

$$\left(\Phi(k)\mathbf{B}(k)\Phi^{T}(k)+\gamma^{k}\beta I\right)^{-1}\Phi(k)\mathbf{B}(k)=\Phi(k)\left(\Phi(k)\Phi^{T}(k)+\gamma^{k}\beta\mathbf{B}(k)^{-1}\right)^{-1} \tag{6.50}$$

Substituting the above result into (6.49) yields

$$\omega(k)=\Phi(k)(\Phi(\mathrm{k})\Phi^{T}(k)+\gamma^{k}\beta\mathbf{B}(k)^{-1})^{-1}\mathbf{d}(k) \tag{6.51}$$

Now the weight vector can be expressed explicitly as a linear combination of the input data in **F** as

$$\omega(k)=\Phi(k)\mathbf{a}(k) \tag{6.52}$$

where $\mathbf{a}(k)$ denotes the computable expansion coefficients vector of the weight by kernel trick, and it can be defined by

$$\mathbf{a}(k)=\Phi(k)\Phi^{T}(k)+\gamma^{k}\beta\mathbf{B}(k)^{-1}\mathbf{d}(k) \tag{6.53}$$

If we further denote

$$\mathbf{Q}(k) = \left(\Phi(k)\Phi^T(k) + \gamma^k \beta \mathbf{B}(k)^{-1}\right)^{-1} \tag{6.54}$$

where $\Phi(k) = \{\Phi(k-1), \varphi(k)\}$, then, we have

$$\mathbf{Q}(k) = \begin{bmatrix} \Phi(k-1)\varphi^T(k-1) + \gamma^k \beta \mathbf{B}(k-1)^{-1} & \Phi^T(k-1)\varphi(k) \\ \varphi^T(k)\Phi(k-1) & \varphi^T(k)\varphi(k) + \gamma^k \beta(|d(k) - \varphi^T(k)\Omega|^{p-2})^{-1} \end{bmatrix}$$

$$= \begin{bmatrix} \Phi(k-1)\Phi^T(k-1) + \gamma^k \beta \mathbf{B}(k-1)^{-1} & \Phi^T(k-1)\varphi(k) \\ \varphi^T(k)\Phi(k-1) & \varphi^T(k)\varphi(k) + \gamma^k \beta\theta(k) \end{bmatrix} \tag{6.55}$$

where $\theta(k) = \left(\left|d(k) - \varphi^T(k)\Omega(k)\right|^{p-2}\right)^{-1} = \left(\left|e(k)\right|^{p-2}\right)^{-1}$. One can easily observe that

$$\mathbf{Q}(k)^{-1} = \begin{bmatrix} \mathbf{Q}(k-1)^{-1} & \mathbf{h}(k) \\ \mathbf{h}(k)^T & \gamma^k \beta\theta(k) + \varphi^T(k)\varphi(k) \end{bmatrix} \tag{6.56}$$

where $\mathbf{h}(k) = \Phi^T(k-1)\varphi(k)$. By using the block matrix inversion formula, we obtain the update equation for the inverse of the growing matrix in (6.56) as

$$\mathbf{Q}(k) = r^{-1}(k)\begin{bmatrix} \mathbf{Q}(k-1)r(k) + Z(k)Z^T(k) & -z(k) \\ -z^T(k) & 1 \end{bmatrix} \tag{6.57}$$

where $z(k) = \mathbf{Q}(k-1)\mathbf{h}(k)$, and $r(k) = \gamma^k \beta\theta(k) + \varphi^T(k)\varphi(k) - z^T(k)\mathbf{h}(k)$.

Combining (6.53) and (6.57), we get

$$\mathbf{a}(k) = \mathbf{Q}(k)\mathbf{d}(k) = \begin{bmatrix} \mathbf{Q}(k-1) + Z(k)Z^T(k)r^{-1}(k) & -z(k)r^{-1}(k) \\ -z^T(k)r^{-1}(k) & r^{-1}(k) \end{bmatrix}\begin{bmatrix} \mathbf{d}(k-1) \\ d(k) \end{bmatrix}$$

$$= \begin{bmatrix} \mathbf{a}(k-1) - z(k)r^{-1}(k)e(k) \\ r^{-1}(k)e(k) \end{bmatrix} \tag{6.58}$$

Then the KRLP algorithm has been derived, as summarized in Table 6.3 [241].

TABLE 6.3

Kernel Recursive Least Mean p-power Algorithm

Initialization
$\mathbf{Q}(1) = (\lambda\chi + \kappa(\mathbf{u}(1), \mathbf{u}(1)))^{-1}, \mathbf{a}(1) = \mathbf{Q}(1)d(1)$
Computation
Iterate for $i > 1$
$\mathbf{h}(k) = [\kappa(\mathbf{u}(k), \mathbf{u}(1)), \ldots, \kappa(\mathbf{u}(k), \mathbf{u}(k-1))]^T$
$e(k) = d(k) - \mathbf{h}(k)^T \mathbf{a}(k-1)$
$z(k) - \mathbf{Q}(k-1)\mathbf{h}(k)$
$\theta(k) = (\lvert e(k) \rvert^{p-2})^{-1}$
$r(k) = \beta\gamma^k \theta(k) + \kappa(\mathbf{u}(k), \mathbf{u}(k)) - \mathbf{z}(k)^T \mathbf{h}(k)$
$\mathbf{Q}(k) = r(k)^{-1} \begin{bmatrix} \mathbf{Q}(k-1)r(k) + \mathbf{z}(k)\mathbf{z}(k)^T & -\mathbf{z}(k) \\ -\mathbf{z}(k)^T & 1 \end{bmatrix}$
$\mathbf{a}(k) = \begin{bmatrix} \mathbf{a}(k-1) - \mathbf{z}(k)r(k)^{-1} & e(k) \\ r(k)^{-1} & e(k) \end{bmatrix}$

Remark 6.8

The difference between the KRLS and KRLP is that the $\mathbf{B}(k)$ included in the matrix form of the weight in $\mathbf{F}$ contains the p-2 order moment of absolute of the error sequence.

Remark 6.9

The $r(k)$ in KRLP plays a key role, which guarantees the robustness of the algorithm against outliers. The error may be very large when outliers appear, and the value of $\theta(k)$ with $p<2$ will also be very large. This will lead to a very small value of $r^{-1}(k)$, which can reduce the adverse effects of the outliers on the update of the coefficients.

Remark 6.10

The KRLP will reduce to the KRLS algorithm when $p=2$. In addition, the computational complexity of the KRLP is almost the same as that of KRLS. The additional computational burden is the $H(e(k))$ in $r^{-1}(k)$ at each iteration.

Remark 6.11

Like many other KAFs, the KLMP and KRLP algorithms face the challenge of linearly growing structures with each sample, resulting in increased computational complexity and memory requirements. However, an efficient online quantization method has been developed in recent years to address this issue and obtain truly online algorithms. This method involves partitioning the input space into smaller regions and using a smaller body of data to represent the entire input data. The QKLMS [125] and QKRLS [120] algorithms have successfully applied this method. Inspired by this idea, quantized versions of the KLMP and KRLP can be derived, similar to the QKLMS and QKRLS algorithms. The updating process remains the same as the KLMP and KRLP, but the quantized data are used to update the coefficients.

6.1.3.2 Simulation Results

In this section, some simulation results on noisy CTSP are shown to demonstrate the performance of the KLMP and KRLP algorithms. Let the noise-free chaotic time series be $\{x(t), t=1,2,3, \ldots\}$. Then, we construct the input vector at time t as

$$In_{noisefree}(t) = \left[x(t), x(t-\tau), \ldots, x(t-(m-1)\tau)\right]^T \tag{6.59}$$

where m and τ are the embedded dimension and delay, respectively. Consider the prediction with ς step for the chaotic time series

$$x(t+\varsigma) = g(In_{noisefree}(t)) \tag{6.60}$$

where g denotes an unknown nonlinear prediction function to be estimated.

The chaotic time series is usually corrupted by some noises. We denote the noise at time t as $v(t)$. Then the observed value is

$$x_{noise}(t) = x(t) + v(t) \tag{6.61}$$

Now, the noisy input vector at time t is

$$In_{noise}(t) = \left[x_{noise}(t), x_{noise}(t-\tau), \cdots, x_{noise}(t-(m-1)\tau)\right]^T \tag{6.62}$$

The relationship between the input and output data will be

$$x_{noise}(t+\varsigma) = g(In_{noise}(t)) \tag{6.63}$$

In the present work, our primary focus is on alpha-stable noise. The SNR is a widely used metric for measuring noise levels, defined as the ratio of signal power to noise power. However, alpha-stable noise does not possess finite second-order statistics, rendering SNR an invalid distortion measure. To address this issue, a signal distortion measure based on fractional order statistics is used. Specifically, we employ the ratio of pth-order moments of the noise and signal where $0 < p < \alpha$, rather than their powers, as an alternative measure.

$$FSNR = 10\log 10\left(\frac{E\left[|d_k|^p\right]}{E\left[|v_k|^p\right]}\right) \approx \frac{\|d\|_{(p)}^p}{\|v_k\|_{(p)}^p} \tag{6.64}$$

where d represents the clean signal, v_k is the noise and $\| \, | v_k | \, \|_{(p)}^p$ is the l_p-norm, which is called fractional order SNR (FSNR). In the following simulation, we mainly use the FSNR as a measure for the noise level.The performance of the KLMP and KRLP are compared with some existing adaptive filtering algorithms (including KLMS, KRLS, KMCC, KRMCC, LMS, RLS, LMP, and RLP), and simulation studies are performed on the predictions of Mackey-Glass (MG) time series and Lorenz time series.

6.1.3.2.1 Mackey-Glass Time Series

MG chaotic time series is generated by the following delay differential equation

$$\frac{dx(t)}{dt} = bx(t) + \frac{ax(t-c)}{1+x(t-c)^{10}} \tag{6.65}$$

The $x(t)$ is the value of time series at time t. The parameters are set as $a = 2.4$, $b = 10$, and $c = 30$. We construct the training set and the testing set using the description of the data setting.

First, the convergence performance of the KAF under MMPE is presented in terms of testing MSE compared to other adaptive filters mentioned above. The noise level is set at FSNR=15, and the convergence curves are displayed in Figure 6.5. The results indicate that: (1) traditional linear adaptive filtering algorithms (LMS, LMP, RLS, and RLP) exhibit poor tracking performance for nonlinear prediction, while KAF algorithms achieve relatively notable steady-state MSE; (2) KLMP, KMCC, KRMCC, and KRLP demonstrate satisfactory performance, indicating that KAF based on MCC and FLOM are robust to impulsive noise, with KRLP achieving the lowest steady-state testing MSE. However, KLMS and KRLS exhibit poor performance under the MSE criterion due to their sensitivity to impulsive noise; (3) compared to stochastic gradient-based algorithms (i.e. KLMS, KMCC, and KLMP), recursive algorithms (i.e. KRLS, KRMCC, and KRLP) converge faster and achieve lower testing MSE. All testing MSE values at the final stage are summarized in Table 6.4, presented in the form of "average standard deviation". In addition,

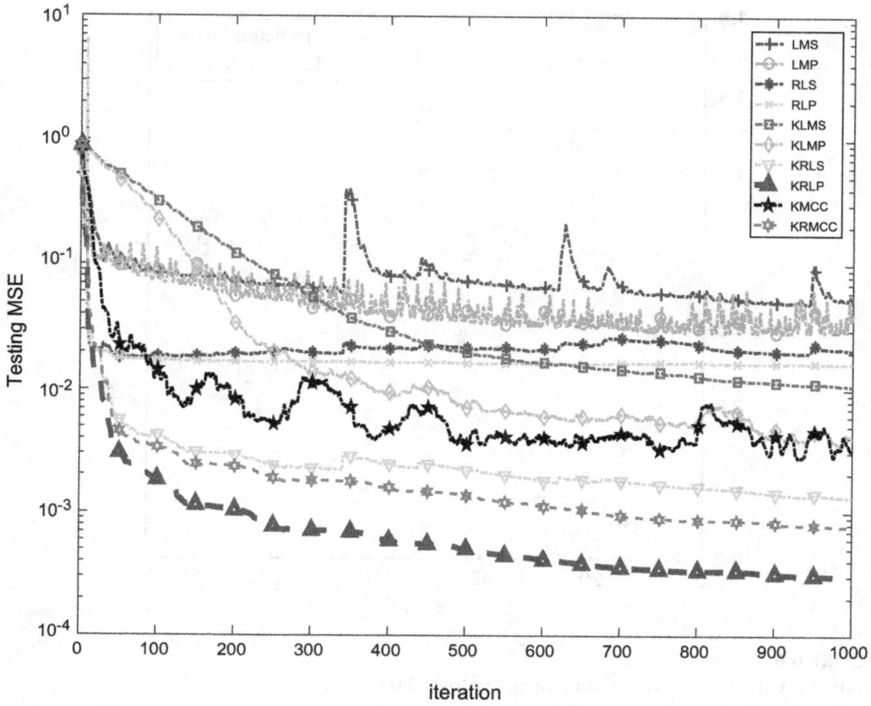

FIGURE 6.5
Convergence curves for Mackey-Glass CSTP [241].

TABLE 6.4
Testing MSEs at the Final Stage

Algorithm	Testing MSE
LMS	0.050158±0.0016757
LMP	0.035478±0.0074507
RLS	0.019416±0.00038433
RLP	0.016249±7.8646e-005
KLMS	0.011118±0.00023764
KLMP	0.0041952±0.00036143
KRLS	0.0014498±4.9794e-005
KRLP	0.0002998±4.6487e-006
KMCC	0.0034119±0.00053599
KRMCC	0.0007696±1.5224e-005

Figure 6.6 displays the predicted value with KRLP and the target value for the test stage, demonstrating that the predicted value coincides well with the actual value.

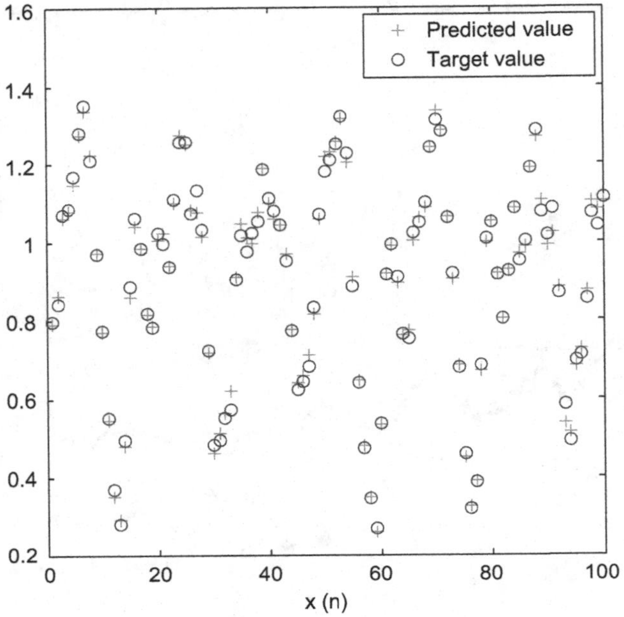

FIGURE 6.6
Predicted values of the KRLP and Target values [241].

Second, we demonstrate the impact of FSNR on the performance of KAFs under MMPE. We set the noise with different level FSNRs to present the convergence curves of KLMP and KRLP in Figure 6.7(a) and (b), respectively. Despite the occurrence of large outliers, both algorithms remain stable, with smaller FSNR resulting in stronger impulsive noises. These results indicate that KAFs under MMPE and FLOM exhibit remarkable robustness to outliers, with KRLP's recursive learning model demonstrating exceptional tracking ability. Consequently, we observe that KRLP outperforms the KLMP algorithm under the same FSNR.

Third, the study compares the performance of various KAFs such as KLMS, KLMP, KMCC, KRLS, KRMCC, and KRLP, along with their quantized versions, namely QKLMS, QKLMP, QKMCC, QKRLS, QKRMCC, and QKRLP. In addition, the research presents the performance comparison of KRLP with different algorithms and its network size growth curve in Figure 6.8. The results indicate that all four quantization version algorithms experience a certain loss in filtering accuracy when compared to the original algorithms, with QKRLP demonstrating the best performance. Furthermore, Figure 6.8(b) shows that by sacrificing a small amount of steady-state performance, the network size of QKRLP significantly decreases, resulting in a reduction in computational complexity during the training stage.

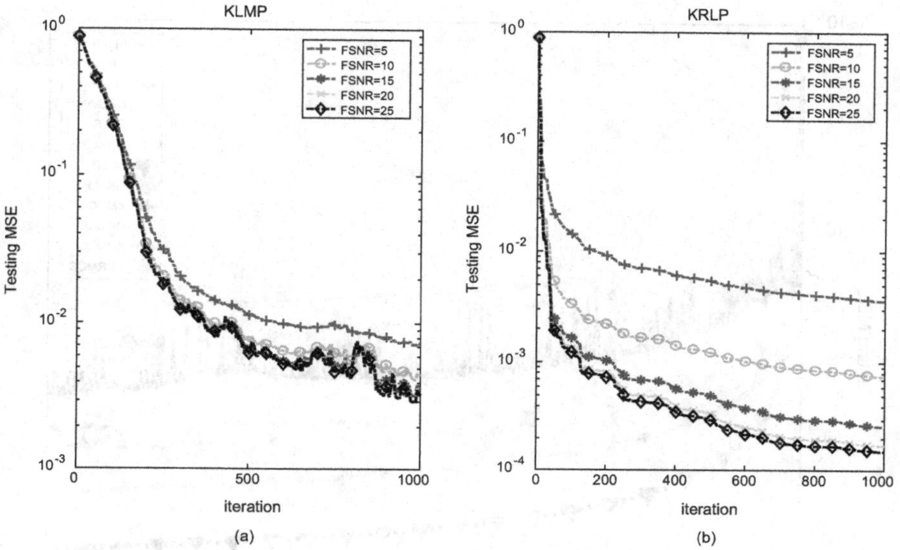

FIGURE 6.7
Convergence curves of the KLMP and KRLP with different values of FSNR [241].

6.1.3.2.2 Lorenz Time Series

The Lorenz system is noted for its unique butterfly shape with chaotic flow. The Lorenz time series is highly nonlinear, deterministic, and three-dimensional. Lorenz system is governed by the following set of differential equations

$$\begin{cases} \dfrac{dx}{dy} = -ax + yz \\ \dfrac{dy}{dt} = b(\mathrm{z} - \mathrm{y}) \\ \dfrac{dy}{dt} = -xy + cy - z \end{cases} \tag{6.66}$$

where the parameters are set as $a = 8/3$, $b = 10$, and $c = 28$.

In the case of predicting Lorenz chaotic time series, Figure 6.9 displays the convergence curves, revealing the following observations: (1) The KLMS and KLMP algorithms exhibit significant fluctuations in their convergence curves, with KLMP showing a smoother curve than KLMS. Meanwhile, the KMCC algorithm converges at a slower rate than the other algorithms. (2) The KRLS, KRMCC, and KRLP algorithms exhibit smoother convergence curves

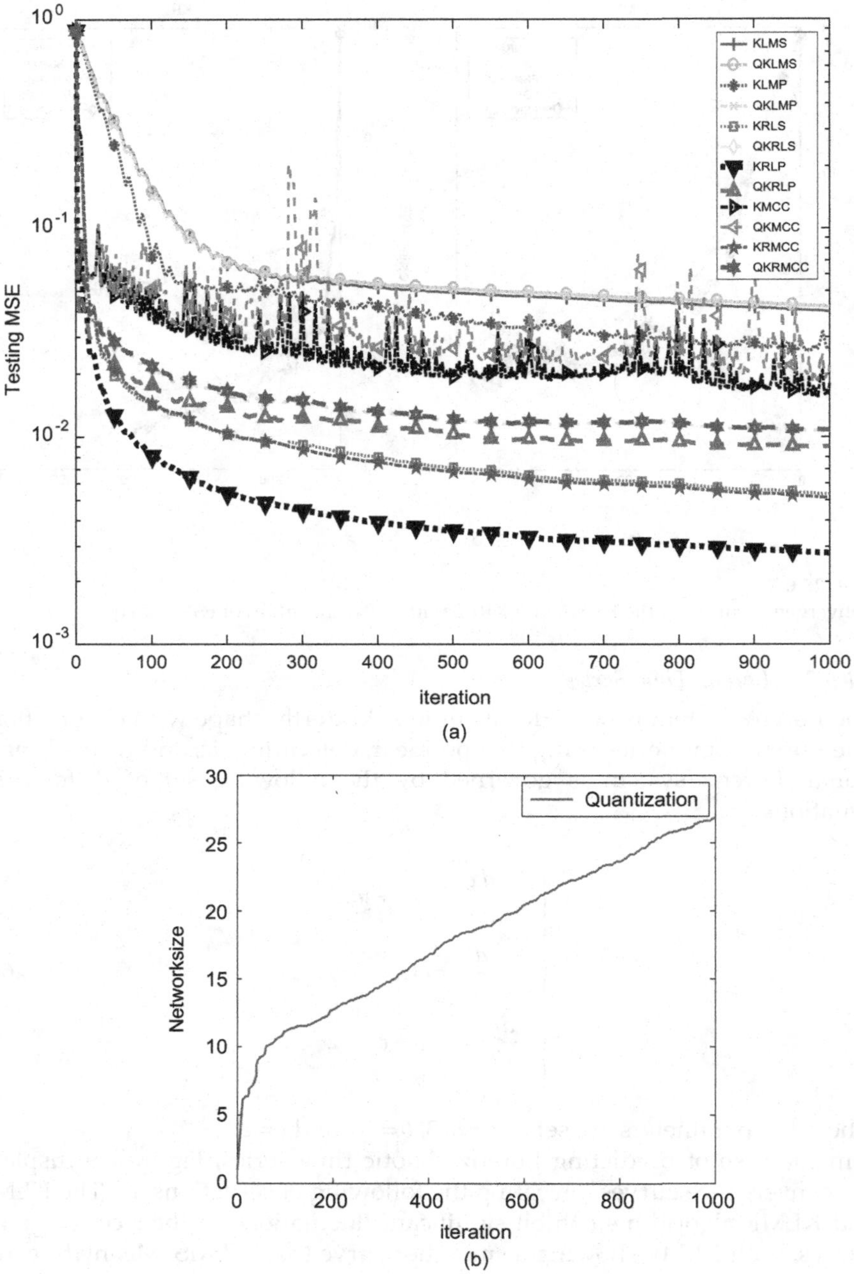

FIGURE 6.8
Performance comparison of KRLP with different algorithms and its network size growth curve: (a) Convergence curves; (b) network size growth curve [241].

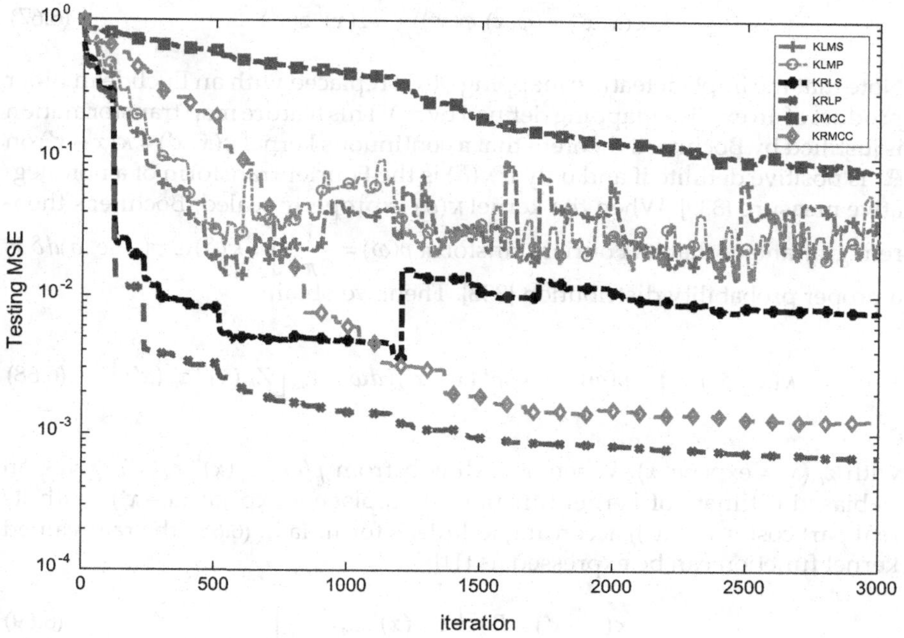

FIGURE 6.9
Convergence curves for Lorenz CSTP [241].

than KLMP and KLMS, with KRLP demonstrating a smaller steady-state testing MSE than KRLS and KRMCC.

6.1.4 Random Fourier Features Extended KRLP Algorithm

In this section, we introduce the RFF-EW-KRLP algorithm, which is an extension of the KRLP algorithm using random Fourier features (RFF). The RFF-EW-KRLP algorithm is designed to handle impulsive noises in the state-space model. It offers significant improvements in convergence rate, steady-state EMSE, and tracking ability while reducing computational complexity by approximating the kernel function. Moreover, the RFF-KRLS-type algorithm [337] can be considered as a special case of the RFF-EX-KRLP algorithm.

6.1.4.1 Approximation of Gaussian Kernel Functions with Random Fourier Features

The input data can be explicitly transformed into a low-dimensional Euclidean inner product space using a randomized feature map. Therefore, the kernel evaluation can be approximated by [338]:

$$\kappa(x,x') = \langle \varphi(x), \varphi(x') \rangle_H \approx z(x)^T z(x') \tag{6.67}$$

Note that the implicit feature mapping $\varphi(\cdot)$ is replaced with an Euclidean inner product of an explicit mapping defined by $z(\cdot)$. This feature map transformation is justified by Bochner's theorem that a continuous kernel $\kappa(x,x') = \kappa(x-x')$ on $\mathbb{R}^L$ is positive definite if and only if $\kappa(\delta)$ is the Fourier transform of a non-negative measure [339]. When the kernel $\kappa(\delta)$ is properly scaled, Bochner's theorem guarantees that its Fourier transform $p(\omega) = \frac{1}{2\pi^L}\int_{\mathbb{R}^L} \kappa(\delta)\exp(-jw^T\delta)d\delta$ is a proper probability distribution [338]. Then, we obtain

$$\kappa(x-x') = \int_{\mathbb{R}^L} p(\omega)\exp\left(jw^T(x-x')\right)d\omega = E_\omega\left[Z_\omega(x)^H z_\omega(x')\right] \tag{6.68}$$

with $z_\omega(x) = \exp(jw^T x)$. When ω is drawn from $p(\omega)$, $z_\omega(x)^H z_\omega(x')$, gives an unbiased estimate of kernel function. By replacing $\exp(jw^T(x-x'))$ with its real part $\cos(w^T(x-x'))$ according to Euler's formula in (6.68), the real-valued kernel function can be expressed as [11]

$$\kappa(x-x') = E_{\omega,b}\left[z_{\omega,b}(x)^T z_{\omega,b}(x')\right] \tag{6.69}$$

with $z_{\omega,b}(x) = \sqrt{2}\cos(\omega^T x + b)$, where ω is drawn from a real-valued probability density function (PDF) $p(\omega)$ and b is drawn from a uniform distribution on [0, 2π]. Subsequently, we only focus on approximating the translation invariant Gaussian kernel function using random feature mappings, i.e.,

$$\kappa(x,x') = \exp(-|x-x'|^2/2\sigma^2) \approx \frac{1}{D}\sum_{i=1}^{D} z_{\omega_i,b_i}(x) z_{\omega_i,b_i}(x') \tag{6.70}$$

where b_i can be obtained as same as (6.69), and ω_i is drawn from $p(\omega) = \left(\sigma\sqrt{2\pi}\right)^{-D}\exp\left(-\|\omega\|^2/2\sigma^2\right)$. Note that the PDF $p(\omega)$ is actually a multivariate Gaussian distribution $\omega \sim N(0_D, \sigma^2 \mathbf{I}_D)$[340]. Assume that the feature mapping $\mathbb{R}^L \to \mathbb{R}^D$, *i.e.*, $z_\Omega : x_i \to [z_1(x_i), \ldots, z_D(x_i)]^T$, is defined as

$$z_\Omega(x_i) = \sqrt{\frac{2}{D}}\left[\cos(\omega_1^T x_i + b_1), \ldots, \cos(\omega_D^T x_i + b_D)\right]^T \tag{6.71}$$

In general, a nonlinear function f can be written as a kernel expansion in terms of training data as

$$f = \sum_{i=1}^{n} \omega_i \kappa(\cdot, u_i) = \omega_n \kappa_n(\cdot)$$

By using (6.70) and (6.71), the kernelized input vector above can be approximately expressed by

$$\kappa_n(\cdot) = \begin{pmatrix} \kappa(\cdot, x_1) \\ \vdots \\ \kappa(\cdot, x_n) \end{pmatrix} \approx \begin{pmatrix} z_\Omega(x_i)^T z_\Omega(\cdot) \\ \vdots \\ z_\Omega(x_n)^T z_\Omega(\cdot) \end{pmatrix} \tag{6.72}$$

Since the kernelized input can be approximated in arbitrary precision with the sufficient order D [340,341], the nonlinear function f can be reformulated as

$$f = \omega_n^T \kappa_n(\cdot) \approx \begin{pmatrix} \omega_1 \\ \vdots \\ \omega_n \end{pmatrix} \begin{pmatrix} z_\Omega(x_i)^T z_\Omega(\cdot) \\ \vdots \\ z_\Omega(x_i)^T z_\Omega(\cdot) \end{pmatrix} = z_\Omega(\cdot)^T \theta_n^k \tag{6.73}$$

with the modified $D \times 1$ dimensional weight vector $\theta_n^k \triangleq [z_\Omega(\omega_1), \ldots, z_\Omega(\omega_n)]\omega_n$. Using (6.73), the framework of the representative KAFs can be deduced with sufficiently large D, which means that we can get rid of the curse of dimensionality without using an online dictionary built by existing sparsification criteria.

6.1.4.2 Fourier Features Extended KRLP Algorithm

6.1.4.2.1 Problem Formula

Based on (6.73), one can obtain the transformed data pairs $\{(z_\Omega(x_i), y_i)\}_{i=1}^n$ with the explicit map $z_\Omega(\cdot) \in \mathbb{R}^D$. The input-output data are assumed to be related via the linear measurement [337]

$$y_i(\infty) = z_\Omega(x_i)^T \theta_i^* + v_i \tag{6.74}$$

where the weight coefficient vector θ_i^* is updated according to the state-space model

$$\theta_{i+1}^* = F_i \theta_i^* + q_i^* \tag{6.75}$$

Observe that the form of (6.74) is the linear model, whereas it is the approximation of the nonlinear model actually because of the nonlinear transformed inputs. Here, measurement noise v_i is assumed to be heavy-tailed impulsive noise modeled by the symmetric alpha-stable (SαS) distribution. The process disturbance q_i^* is assumed to follow the zero-mean Gaussian distribution with covariance matrix Q_i. The time-variant transition matrix F_i is assumed to be known and shifts the state of weight vector θ_i^* from time instant i to time instant i+1. This is with the goal to estimate the initial state of weight vector θ_0^* and the state model disturbance signals $\{q_i^*\}_{i=0}^{n}$.

6.1.4.2.2 Random Fourier Features EX-KRLP Algorithm

To effectively suppress the adverse impact of impulsive noise z_i on stability, we consider the cost function of the regularized least p-power as follows:

$$\min_{\{\theta_0^*,q_i^*\}_{i=0}^{n}} \left\{ (\theta_0^*)^T U_0^{-1}\theta_0^* + \sum_{i=0}^{n} r_i^{-1} \left| d_i - z_\Omega(\boldsymbol{u}_i)^T \theta_i^* \right|^p + \sum_{i=0}^{n} (\boldsymbol{q}_i^*)^T T_i^{-1} \boldsymbol{q}_i^* \right\} \tag{6.76}$$

where U_0 is the positive-definite regularization matrix, r_i and T_i are the positive weighting factors and the positive-definite weighting matrices, respectively. Then, (6.76) can be rewritten as [24]

$$\min_{\{\theta_0^*,q_i^*\}_{i=0}^{n}} \left\{ (\theta_0^*)^T U_0^{-1}\theta_0^* + \sum_{i=0}^{n} \wedge_i^{-1} \left| d_i - z_\Omega(\boldsymbol{u}_i)^T \theta_i^* \right|^2 + \sum_{i=0}^{n} (\boldsymbol{q}_i^*)^T \boldsymbol{T}_i^{-1} \boldsymbol{q}_i^* \right\} \tag{6.77}$$

with $\wedge_i = r_i \left| d_i - z_\Omega(\boldsymbol{u}_i)^T \theta_i^* \right|^{2-p}$. Let us introduce the following column vectors:

$$\Theta^* = col\{\theta_0^*, \boldsymbol{q}_i^*, \ldots \boldsymbol{q}_n^*\} \tag{6.78}$$

$$\boldsymbol{d} = col\{d_0, d_1, \ldots, d_n\} \tag{6.79}$$

as well as the matrices:

$$\wedge^{-1} = diag\{\wedge_0, \ldots, \wedge_n\} \tag{6.80}$$

$$\Pi^{-1} = diag\{\boldsymbol{U}_0, \boldsymbol{T}_1, \ldots, \boldsymbol{T}_n\} \tag{6.81}$$

By combining (6.78), (6.79), (6.80), and (6.81), (6.77) can be reformulated as the regularized weighting least-squares cost function

$$\min_{\Theta^*} \left\{ \left\| \boldsymbol{d} - \boldsymbol{H}\Theta^* \right\|_{\wedge}^2 + \Theta^{*T} \Pi \Theta^* \right\} \tag{6.82}$$

$$
\boldsymbol{H}=\begin{pmatrix}
z_\Omega(\boldsymbol{u}_0)^T & & & & \\
z_\Omega(\boldsymbol{u}_1)^T\phi(1,0) & z_\Omega(\boldsymbol{u}_1)^T & & & \\
z_\Omega(\boldsymbol{u}_2)^T\phi(2,0) & z_\Omega(\boldsymbol{u}_2)^T\phi(2,1) & z_\Omega(\boldsymbol{u}_i)^T & & \\
\vdots & \vdots & \vdots & \ddots & \\
z_\Omega(\boldsymbol{u}_n)^T\phi(n,0) & z_\Omega(\boldsymbol{u}_n)^T\phi(n,1) & z_\Omega(\boldsymbol{u}_n)^T\phi(n,2) & \ldots z_\Omega(\boldsymbol{u}_n)^T & 0
\end{pmatrix} \tag{6.83}
$$

where the weighted Euclidean norm is defined as $\|\boldsymbol{u}\|_W^2=\boldsymbol{u}^T\boldsymbol{W}\boldsymbol{u}$ for any column vector $\boldsymbol{u}$, and the block lower triangular matrix $\boldsymbol{H}$ is given by (6.83) with the matrices $\phi(i,y)$ being defined as [25]

$$
\phi(i,y)=\begin{cases}\boldsymbol{F}_{i-1}\boldsymbol{F}_{i-2}\cdots\boldsymbol{F}_j(i>j)\\ \mathbf{I}_D(i=j)\end{cases} \tag{6.84}
$$

Note that (6.83) is actually not the general regularized weighting least-squares cost function due to $\wedge_i$ that depends on θ_i^*. The solution of (6.82) is given by $\Theta_{|n}=col\{\theta_{0|n},\boldsymbol{q}_{0|n},\ldots,\boldsymbol{q}_{n|n}\}$, where $\theta_{i|i-1}$ denotes the estimate of θ_i^* based on the observation set $\{d_j\}_{j=0}^{i-1}$ up to time $i-1$ in $\boldsymbol{d}$. Since determining $\Theta_{|n}$ is equivalent to $\{\theta_{0|n},\boldsymbol{q}_{i|n}\}_{i=0}^{n}$, we start to derive the random Fourier features EX-KRLP (RFFEX-KRLP) algorithm by recursively computing $\theta_{i|i-1}$ from Kalman filter theory [261]. By exploiting the recursive estimation formula with innovations [261], we have

$$
\theta_{i+1|i}=\theta_{i+1|i-1}+E\{e_i\theta_{i+1}^*\}e_i/r_{e,i} \tag{6.85}
$$

where the innovation used in Kalman Filter theory is given by

$$
e_i=d_i-z_\Omega(\boldsymbol{u}_i)^T\theta_{i|i-1} \tag{6.86}
$$

with $r_{e,i}=E\{e_i^2\}$. By combining (6.75), the first term on the right-hand side of (6.85) can be rewritten as

$$
\theta_{i+1|i-1}=\boldsymbol{F}_i\theta_{i+1|i-1}+\boldsymbol{q}_i=\boldsymbol{F}_i\theta_{i|i-1} \tag{6.87}
$$

due to $q_i \perp d_j$ with $j \leq i-1$. Substituting (6.87) into (6.85), we obtain the recursion for computing the innovations of (6.86):

$$\theta_{i+1|i} = F_i\theta_{i+1|i-1} + k_{p,i}e_i \tag{6.88}$$

with the initial conditions $\theta_{0|-1} = 0_D, e_0 = d_0$, and the gain vector $k_{p,i} = E\left\{e_i\theta^*_{i+1}\right\} r_{e,1}^{-1}$. To further evaluate $r_{e,1}$, we define the state-estimation error:

$$\theta_{i|i-1} = \theta^*_i + \theta_{i|i-1} \tag{6.89}$$

Correspondingly, its covariance matrix can be defined as

$$P_{i|i-1} = E\left\{\tilde{\theta}_{i|i-1}\tilde{\theta}^T_{i|i-1}\right\} \tag{6.90}$$

Using the measure model (6.74) and the state-estimation error $\tilde{\theta}_{i|i-1}$ in (6.89), the innovation (6.86) can be reformulated as

$$e_i = d_i - z_\Omega(u_i)^T \theta_{i|i-1} = z_\Omega(u_i)^T \tilde{\theta}_{i|i-1} + z_i \tag{6.91}$$

Substituting (6.75), (6.89) into (6.91) and using the uncorrelated relationship between e_i and $\theta_{i|i-1}$ [261], the gain vector is given by

$$\begin{aligned} k_{p,i} &= E\left\{e_i(\mathrm{F}_i\,\theta^*_i) + \mathrm{q}^*_i\right\}/r_{e,i} \\ &= F_iE\left\{e_i\theta^*_i\right\}/r_{e,i} \\ &= F_iE\left\{e_i(\tilde{\theta}_{i|i-1} + \theta_{i|i-1})\right\}/r_{e,i} \\ &= F_iE\left\{(z_\Omega(u_i)^T\tilde{\theta}_{i|i-1} + z_i)\tilde{\theta}_{i|i-1}\right\}/r_{e,i} \\ &= F_iP_{i|i-1}z_\Omega(u_i)/r_{e,i} \end{aligned} \tag{6.92}$$

Since $\tilde{\theta}_{i|i-1}$ is uncorrelated with z_i, post-multiplying (6.91) by its transpose and taking the expectation, we get

$$r_{e,i} = \wedge_i + z_\Omega(u_i)^T P_{i|i-1}z_\Omega(u_i) \tag{6.93}$$

with $\wedge_i = E\left\{z_i^2\right\}$. In the following, we focus on the evaluation of matrix $P_{i|i-1}$. Post-multiplying (6.91) by its transpose and taking the expectation, we obtain

$$X_{i+1} = F_iX_iF_i^T + Q_i \tag{6.94}$$

with the covariance matrix $\boldsymbol{X}_i = E\left\{\boldsymbol{\theta}_i^*(\boldsymbol{\theta}_i^*)^T\right\}$ for $\boldsymbol{\theta}_i^*$, where the process disturbance $\boldsymbol{q}_i^*$ is supposed to be independent of $\boldsymbol{\theta}_i^*$. Likewise, the innovation e_i is independent of $\tilde{\boldsymbol{\theta}}_{i|i+1}$. Post-multiplying (6.89) by its transpose and taking the expectation, we further get

$$\hat{\boldsymbol{X}}_{i+1} = \boldsymbol{F}_i\hat{\boldsymbol{X}}_i\boldsymbol{F}_i^T + r_{e,i}\boldsymbol{k}_{p,i}\boldsymbol{k}_{p,i}^T \tag{6.95}$$

with the covariance matrix $\hat{\boldsymbol{X}}_i = E\left\{\boldsymbol{\theta}_{i|i-1}\boldsymbol{\theta}_{i|i-1}^T\right\}$ for $\boldsymbol{\theta}_{i|i-1}$, and the initial condition $\hat{\boldsymbol{X}}_i = 0_{D\times D}$. According to (6.90), we post-multiply $\boldsymbol{\theta}_i^* = \tilde{\boldsymbol{\theta}}_{i|i-1} + \boldsymbol{\theta}_{i|i-1}$ by its transpose and take the expectation, yielding

$$\boldsymbol{X}_i = \boldsymbol{P}_{i|i-1} + \hat{\boldsymbol{X}}_i \tag{6.96}$$

where the orthogonal relation $\tilde{\boldsymbol{\theta}}_{i|i-1} \perp \boldsymbol{\theta}_{i|i-1}$ is invoked. Substituting (6.94) and (6.95) into (6.96) is based on the observations up to time $i+1$, the recursion of the covariance matrix of state-estimation error can be described as

$$\begin{aligned}\boldsymbol{P}_{i+1|i} &= \boldsymbol{X}_{i+1} - \hat{\boldsymbol{X}}_{i+1} = \boldsymbol{F}_i\boldsymbol{X}_i\boldsymbol{F}_i^T + \boldsymbol{Q}_i - \boldsymbol{F}_i\hat{\boldsymbol{X}}_i\boldsymbol{F}_i^T - r_{e,i}\boldsymbol{k}_{p,i}\boldsymbol{k}_{p,i}^T \\ &= \boldsymbol{F}_i\boldsymbol{P}_{i|i-1}\boldsymbol{F}_i^T + \boldsymbol{Q}_i - r_{e,i}\boldsymbol{k}_{p,i}\boldsymbol{k}_{p,i}^T\end{aligned} \tag{6.97}$$

with the initial condition of $\boldsymbol{P}_{0|-1} = 0_{D\times D}$.

We shall now turn to compute the set $\left\{q_{i|n}\right\}_{i=0}^{n}$ based on $\boldsymbol{\theta}_{i+1|i}$ in (6.86). For (6.83), using (6.86) and the definition of $\Theta_{|n}$, we have the following compact recursive equation

$$\begin{aligned}\Theta_{|i} &= \Theta_{|i-1} + E\left\{e_i\Theta^*\right\}e_i/r_{e,i} \\ &= \Theta_{|i-1} + E\left\{\Theta^*\tilde{\theta}_{i|i-1}^T\right\}z_\Omega(\boldsymbol{u}_i)e_i/r_{e,}\end{aligned} \tag{6.98}$$

where $\Theta_{|i} = col\left\{\boldsymbol{\theta}_{0|i}, q_{0|i}, q_{1|i}, \ldots, q_{i-1|i},\ 0, \ldots, 0\right\}$ with $\boldsymbol{q}_{j|i} = 0$ for $j \geq i$. Then we introduce the auxiliary variable $\rho_{i|n} = \sum_{j=1}^{n} \phi_p(j,i)^T z_\Omega(\boldsymbol{u}_i)e_i/r_{e,i}$ with

$$\phi_p(i,j) = \begin{cases}\boldsymbol{F}_{p,i-1}\boldsymbol{F}_{p,i-2}\ldots\boldsymbol{F}_{p,j} & i > j \\ \boldsymbol{I}_D & i = j\end{cases} \tag{6.99}$$

Last, the recursions of $\boldsymbol{q}_{j|n}$ are given by [6.89]

$$\boldsymbol{\theta}_{o|n} = \boldsymbol{U}_0\rho_{0|n} \tag{6.100}$$

$$\boldsymbol{q}_{j|n} = \boldsymbol{Q}_j\rho_{j+1|n} \tag{6.101}$$

$$\rho_{j|n} = F_{p,j}^T \rho_{j+1|n} + z_\Omega(u_i) e_i / r_{e,i} \tag{6.102}$$

With $F_{p,i} = F_i - k_{p,i} z_\Omega(u_i)$.

The RFF-EX-KRLP algorithm is summarized in Table 6.5.

Remark 6.12

Note that the presented recursive expressions of the RFF-EX-KRLP algorithm are evidently different from the existing KRLS-type expressions. The reason is that the derived algorithm for the state-space model (6.75) is a more general fixed-order form which has good tracking ability.

Remark 6.13

If $\theta_{0|i}$ is denoted by θ_i and setting$P_{i+1|1} = \lambda^i P_i$, the RFF-EXKRLP algorithm will reduce to the standard random Fourier features exponentially weighted KRLP (RFF-EW-KRLP) algorithm. Under the above special choices, the RFF-EW-KRLP algorithm can be also derived from the cost function (6.76) by multiplying λ^n on it. The procedure of the RFF-EW-KRLP algorithm is listed in Table 6.6.

TABLE 6.5

Random Fourier Features EX-KRLP Algorithm

1: Initialization:
Select the order p, the parameters of kernel function; b_i and ω_i are drawn from a uniform $[0,2\pi]$ and a multivariate Gaussian distribution $\omega \sim N(0_D, \xi^2 I_D)$, respectively. Initialize the weight vector $\theta_{0|-1} = 0$, $P_{0|-1} = U_0$, and $\rho_{n+1|n} = 0$.

2: Input: $\{(x_i, y_i)\}, i = 1,2,\ldots,n$

3: for $i = 1,2,\ldots,n$ do

$$r_{e,i} = r_i \mid e_i \mid^{2-p} + z_\Omega(\mathbf{u}_i)^T \mathbf{P}_{i|i-1} z_\Omega(\mathbf{u}_i)$$

$$k_{p,i} = \mathbf{F}_i \mathbf{P}_{i|i-1} z_\Omega(\mathbf{u}_i) / r_{e,i}$$

$$e_i = d_i - \theta_{i|i-1} z_\Omega(\mathbf{u}_i);$$

$$\theta_{i+1|i} = \mathbf{F}_i \theta_{i|i-1} + k_{p,i} e_i$$

$$\mathbf{P}_{i+1|i} = \mathbf{F}_i \mathbf{P}_{i|i-1} \mathbf{F}_i^T + Q_i - r_{e,i} k_{p,i} k_{p,i}^T$$

4: end for

5: for $i = n, n-1,\ldots,1,0$ do

$$\rho_{j|n} = F_{p,j}^T \rho_{j+1|n} + z_\Omega(u_i) e_i / r_{e,i}$$

6: end for

7: Obtain $\theta_{0|n} = F_{p,i}^T \rho_{0|n} = Q_i \rho_{i+1|n}$ for $0 \le i \le n$.

TABLE 6.6

Random Fourier Features EW-KRLP Algorithm

1: Initialization:
Select the order p, the parameters of kernel function; b_i and ω_i are drawn from uniform $[0,2\pi]$ and a multivariate Gaussian distribution $\omega \sim N(0_D, \xi^2 I_D)$, respectively. Initialize the weight vector $\theta_0 = 0$ and $P_0 = U_0$.

2: Input: $\{(u_i, d_i)\}, i = 1,2,\ldots,n$

3: for $i = 1,2,\ldots,n$ **do**

$$k_{p,i} = \frac{\lambda^{-1}|e_i|^{p-2} P_{i-1} z_\Omega(u_i)}{1+\lambda^{-1}|e_i|^{p-2} z_\Omega(u_i)^T P_{i-1} z_\Omega(u_i)};$$

$$e_i = y_i - \theta_{i=1}^T z_\Omega(u_i);$$

$$\theta_i = \theta_{i-1} + k_i e_i;$$

$$P_i = \lambda^{-1}(P_{i-1} - k_{p,i}\, z_\Omega(u_i)^T P_{i-1});$$

4: end for

TABLE 6.7

RFF-EX-KRLP Algorithm and Its Variants

Parameter settings	$p = 2$	$0 < p < 2$
$F_i \neq I_D$ and $q_i^* \neq 0_D$	RFF-EX-KRLS	RFF-EX-KRLP
$F_i = I_D$ and $q_i^* = 0_D$	RFF-EW-KRLS ($0 < \lambda < 1$)	RFF-EW-KRLP ($0 < \lambda < 1$)
	RFF-KRLS ($\lambda = 1$)	RFF-KRLP ($\lambda = 1$)

Moreover, Table 6.7 illustrates that the other RFF-KRLS-type algorithms can be regarded as the particular cases of the RFF-EW-KRLP algorithm with special choices for parameters setting. Additionally, the computational complexity of the RFF-EX-KRLP algorithm is less than those of the conventional KAF algorithms with kernel approximation instead of the time-consuming exponential operation, which results in a little performance degradation as shown in simulations.

6.1.4.3 Simulation Results

In this section, two simulation examples are provided to show the better performance of the EW-KRLP algorithm compared to the KLMS, KLMP, and RFF-EX-KRLS algorithms in the nonstationary environment contaminated by the impulsive noise.

6.1.4.3.1 Nonstationary Nonlinear System Identification

Here we consider an ideal nonstationary nonlinear system consisting of the time-varying optimum weight vector. Figure 6.10(a) shows that the

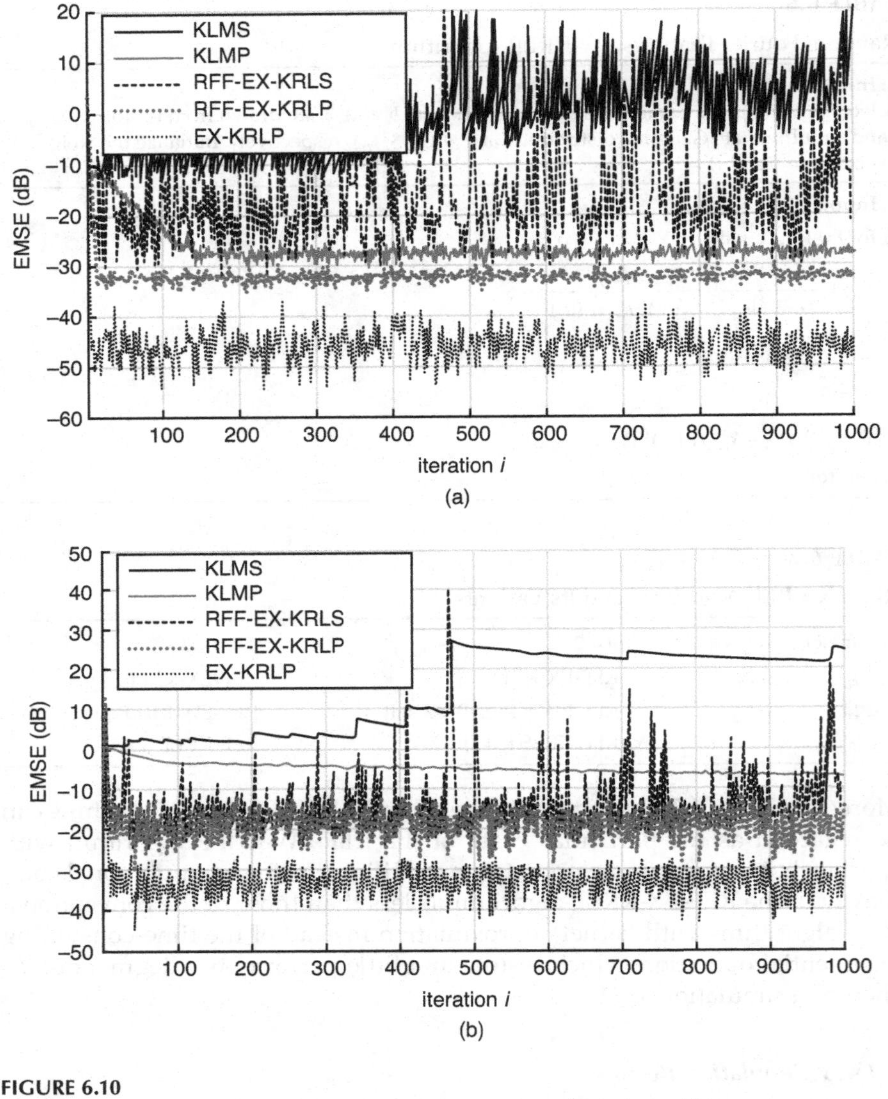

FIGURE 6.10
Simulation results for nonstationary nonlinear system identification: (a) Learning curves of different algorithms in terms of EMSE; (b) Learning curves of different algorithms in terms of MSD [337].

RFF-EX-KRLP algorithm is able to achieve more smooth and lower learning curve of the EMSE in comparison to the KLMS and RFF-EX-KRLS algorithms. Figure 6.10(b) shows that the RFF-EW-KRLP algorithm performs the robust convergence performance of MSD by dynamically scaling down the weight vector coefficients affected by the impulsive noise, which demonstrates its robustness, faster convergence rate, and lower steady-state MSD. In contrast, the KLMS-type and KRLS-type algorithms developed from the second statistic of error under the Gaussian assumption suffer from the sever divergence of weight coefficients and performance degradation. Meanwhile, it should be pointed out that the performance of the EX-KRLP algorithm is much better than that of RFF-EX-KRLP due to the kernel approximation with insufficient order D.

6.1.4.3.2 Lorenz Chaotic Time-Series Prediction

Here the Lorenz chaotic time series prediction is further used as the performance testing case.It is shown in Figure 6.11(a) that the RFF-EX-KRLP has faster convergence rate and lower steady-state EMSE than KLMP in the prediction of Lorenz chaotic time-series corrupted by the impulsive noise. However, the KLMS and RFF-EXKRLS algorithms failed to provide accuracy predictions. One can see from Figure 6.12(b) that the lengths of online dictionaries for KLMS and KLMP are overlapping and almost identical to the order D for RFF-EX-KRLS and RFF-EX-KRLP, that is to say, the computation overhead is much less for the latter two without exponential operation. Consequently, the RFF-EW-KRLP algorithm is able to significantly improve its tracking ability and robust convergence performance, simultaneously.

6.2 ELM Models under MMPE

ELM is a fast learning model that is designed to train a single-layer feedforward network with hidden neuron weights that are randomly initialized and fixed. This approach is markedly different from other training algorithms, such as the back-propagation algorithm and its improved algorithms [170,171], which require the tuning of hidden neuron weights. ELM offers several advantages over full parameter determination algorithms, including fast learning speed, universal approximation capability [172,173], and a unified learning paradigm for regression and classification [174]. However, ELM may perform poorly in situations where data are non-Gaussian, as the MSE criterion used to construct its cost function only takes into account the second-order statistics and is based on the Gaussian assumption. To address this issue, several robust ELM models based on the LMP error criterion have been developed, including LMP-ELM [249], RLMP-ELM [250], and sparse RLMPELM [251], which will be discussed in this subchapter.

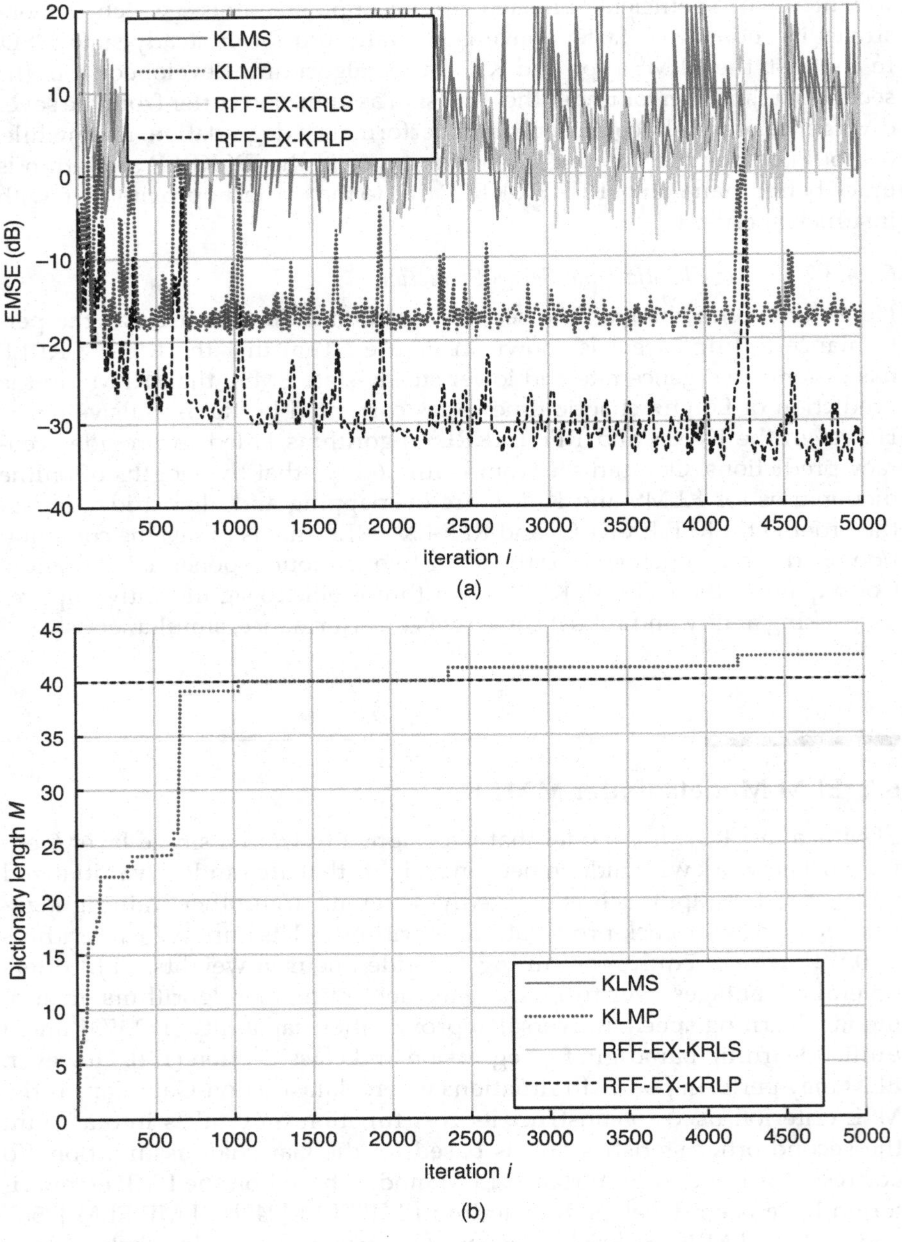

FIGURE 6.11
Simulation results of Lorenz chaotic time-Series prediction: (a) Learning curves of different algorithms in terms of EMSE; (b) Evolution of the length of online dictionary [337].

6.2.1 Extreme Learning Machine

The ELM is a type of single hidden layer feedforward neural network learning algorithm. One of its distinguishing features is that the parameters of hidden layer nodes can be given randomly or artificially and do not require adjustment. This unique approach allows for the learning process to focus solely on calculating the output weight, resulting in a unique optimal solution. ELM is highly efficient, with fast computing speeds and excellent generalization abilities. Its practicality, simplicity, and effectiveness make it a popular choice for classification, regression, clustering, and feature learning. For a visual representation of the ELM structure, please refer to Figure 6.12.

With a quadratic cost function, ELM usually requires no iterative tuning and the global optimum can thus be solved in a batch mode. Given N distinct training samples $\{x_k, t_k\}_{k=1}^{N}$, with $\mathbf{x}_k = [x_{k1}, x_{k2}, \ldots, x_{kM}]^T \in \mathbb{R}^M$ being the input vector and $t_k \in \mathbb{R}$ being target response, the output of a standard SLFN with L hidden nodes will be

$$y_k = \sum_{j=1}^{L} \beta_i f(\mathbf{w}_j \cdot \mathbf{x}_k + b_j) \tag{6.103}$$

where $f()$ is an activation function, $\mathbf{w}_j = [w_{j1}, w_{j2}, \ldots, w_{jd}] \in \mathbb{R}^d$ and $b_j \in \mathbb{R}$ $(j = 1, 2, \ldots, L)$ are the learning parameters of the ith hidden node,

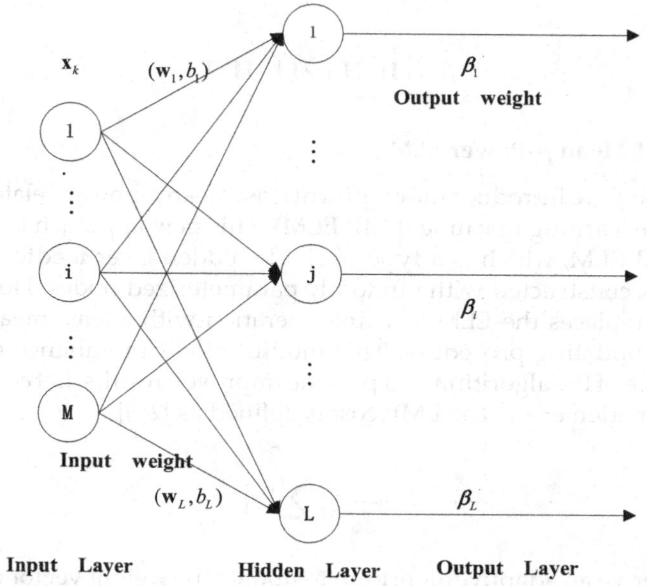

FIGURE 6.12
Basic structure of Extreme Learning Machine.

$\mathbf{w}_j \cdot \mathbf{x}_i$ denotes the inner product of $\mathbf{w}_j$ and $\mathbf{x}_i$, and $\beta_j \in \mathbb{R}$ represents the weight parameter connecting the jth hidden node to the output node. The above equation can be written in a vector form as

$$\mathbf{Y} = \mathbf{H}\boldsymbol{\beta} \tag{6.104}$$

where $\mathbf{Y} = (y_1, y_2, \ldots, y_N)^T$, $\boldsymbol{\beta} = (\beta_1, \ldots, \beta_N)^T$ and

$$\mathbf{H} = \begin{pmatrix} f(\mathbf{w}_1 \cdot \mathbf{x}_1 + b_1), & \cdots & f(\mathbf{w}_L \cdot \mathbf{x}_1 + b_L) \\ \vdots & \ddots & \vdots \\ f(\mathbf{w}_1 \cdot \mathbf{x}_N + b_1), & \cdots & f(\mathbf{w}_L \cdot \mathbf{x}_N + b_L) \end{pmatrix} \tag{6.105}$$

represents the output matrix of the hidden layer. In general, the output weight vector $\boldsymbol{\beta}$ can be solved by minimizing the regularized MSE (or LS) loss

$$J(\boldsymbol{\beta}) = \sum_{k=1}^{N} e_k^2 + \lambda \|\boldsymbol{\beta}\|_2^2 = \|\mathbf{H}\boldsymbol{\beta} - \mathbf{T}\|_2^2 + \lambda \|\boldsymbol{\beta}\|_2^2 \tag{6.106}$$

where $e_k = t_k - y_k$ is the error between the kth target response and the kth actual output, $\lambda \geq 0$ stands for the regularization parameter to prevent overfitting, and $\mathbf{T} = (t_1, \ldots, t_N)^T$ is the target response vector. With a pseudo-inversion operation, one can easily obtain a unique solution under the loss (6.106), that is

$$\boldsymbol{\beta} = \left[\mathbf{H}^T\mathbf{H} + \lambda \mathbf{I}\right]^{-1} \mathbf{H}^T\mathbf{T} \tag{6.107}$$

6.2.2 Least Mean *p*-Power ELM

In this section, we introduce the empirical least mean p-power-related sequential extreme learning machine (LMP-ELM). This new approach is built upon the original ELM, which is a type of single hidden layer feedforward neural network constructed with randomly parameterized nodes. However, the LMP-ELM replaces the ELM learning operation with a least mean p-power sequential updating procedure. This modification can enhance the overall performance of the algorithm and provide improved results. Let $e_k = tk - f(\mathbf{x}_k)$ be the estimation error. The LMP cost is defined as [249]

$$J_{LMP} = \frac{1}{N} \sum_{k=1}^{N} |e_k|^p \tag{6.108}$$

One can derive an adaptive algorithm to update the weight vector of the ELM under MPE. A simple stochastic gradient-based algorithm is

$$\boldsymbol{\beta}_{k+1} = \boldsymbol{\beta}_k - \eta \frac{\partial J}{\partial \boldsymbol{\beta}_k} \tag{6.109}$$

where the parameter η is the step size. When the instantaneous value of the cost in (6.109) is used to replace its ensemble averaged value, we have

$$\boldsymbol{\beta}_{k+1} = \boldsymbol{\beta}_k - \eta \frac{\partial |e_k|^p}{\partial \boldsymbol{\beta}_k} \tag{6.110}$$

The following expression can be obtained

$$\boldsymbol{\beta}_{k+1} = \boldsymbol{\beta}_k - \eta p |e_k|^{p-2} e_k \frac{\partial |e_k|}{\partial \boldsymbol{\beta}_k} \tag{6.111}$$

Using $\frac{\partial e_k}{\partial \boldsymbol{\beta}_k} = -G_k$, equation (6.111) can be further expressed as

$$\boldsymbol{\beta}_{k+1} = \boldsymbol{\beta}_k - \eta p |e_k|^{p-2} e_k G_k \tag{6.112}$$

Remark 6.14

It is evident that the weight vector is updated sequentially with each kth input, using a carefully chosen step size parameter. In fact, a method for online sequential learning of the batch ELM has been devised in [174]. This approach, known as OS-ELM, employs the MSE cost function to train the weight vector while applying the LMP cost function. Additionally, instead of using recursive least square learning algorithms, a stochastic gradient algorithm is utilized to obtain the optimal weight in OS-ELM [249].

Remark 6.15

When $p=2$ is chosen, the LMP algorithm reduces to the stochastic gradient LMS algorithm. A stochastic gradient LMS based ELM-like method is introduced where an auto-encoder that uses nonorthogonal random hidden parameters plays a role of the ELM-SLFN. We may consider the LMPELM algorithm as the stochastic gradient LMS algorithm with a time-varying step size $\eta_k = \frac{\eta p |e_k|^{p-2}}{2}$. Compared with [342], the LMP-ELM has a faster convergence speed and a smaller misadjustment due to the time-varying step size. When the LMP-ELM algorithm does not converge to its optimum, the error $|e_k|$ is large. This makes step size large and the LMP-ELM algorithm has fast convergence speed. When the LMP-ELM algorithm converges to its optimum, the error $|e_k|$ is small. This makes step size and misadjustment small.

Remark 6.16

Similar to OS-ELM, the output weight learning phase of the LMP-ELM can commence in a chunk-by-chunk learning mode as follows:

$$\beta_{k+B_k} = B_k + \eta p \sum_{l=0}^{B_k-1} |e_{k+l}|^{p-2} e_{k+l} G_{k+l} \tag{6.113}$$

Also, the error terms can be expressed by

$$e_{k+l} = t_{k+l} - f(\mathbf{x}_{k+l}), k = 0,1,\ldots,B_{k-1} \tag{6.114}$$

where $f(\mathbf{x}_{k+l}) = \beta_k^T G_{k+l}$, and B_k is the block length. The chunk length, i.e., the number B_k of the inputs in the kth chunk, does not need to be the same as B_{k-1}.

6.2.3 Recursive Least Mean *p*-Power ELM

This subchapter introduces the recursive least mean p-power (RLMP) related sequential ELM (RLMP-ELM) algorithm. The RLMP-ELM is based on the primitive ELM model, which constructs a randomly parameterized SLFN. However, the ELM learning operation is replaced by a recursive least mean p-power sequential updating procedure in RLMP-ELM. This algorithm maintains the computational simplicity of the ELM, but introduces an LMP error criterion that aims to minimize the p powers of the error, providing a mechanism to update the output weights sequentially. The RLMP-ELM algorithm has the same computational complexity as ELM and OS-ELM under the same architecture.

6.2.3.1 *RLMP-ELM Algorithm*

In the following parts, we will present the detailed process of the RLMP-ELM algorithm which updates the weight vector of the ELM under the LMP error criterion. For a linear system, the RLMP algorithm is an extension of the RLS algorithm with cost function (6.108). The optimal solution β_N for minimizing J_{LMP} can be obtained by differentiating (6.108) with respect to β_N and setting the derivatives to zero. The derivatives can be obtained as [250]

$$\frac{\partial J_{LMP}}{\partial \beta_N} = \frac{1}{N} \sum_{k=1}^{N} p |e_k|^{p-2} e_k \frac{\partial e_k}{\partial \beta_N} \tag{6.115}$$

Substituting $e_k = t_k - \beta_N^T G_k$ into (6.115) yields

$$\frac{\partial J_{LMP}}{\partial \boldsymbol{\beta}_N} = \frac{1}{N}\sum_{k=1}^{N} p|e_k|^{p-2}\left(t_k - \boldsymbol{\beta}_N^T \boldsymbol{G}_k\right)\boldsymbol{G}_k \tag{6.116}$$

Setting $\frac{\partial J_{LMP}}{\partial \boldsymbol{\beta}_N} = 0$, (6.116) can be further written as

$$\sum_{k=1}^{N} |e_k|^{p-2} \boldsymbol{G}_k \boldsymbol{G}_k^T \boldsymbol{\beta}_N = \sum_{k=1}^{N} |e_k|^{p-2} t_k \boldsymbol{G}_k \tag{6.117}$$

We define

$$\mathbf{R}_N = \sum_{k=1}^{N} |e_k|^{p-2} \boldsymbol{G}_k \boldsymbol{G}_k^T \tag{6.118}$$

$$\boldsymbol{P}_N = \sum_{k=1}^{N} |e_k|^{p-2} t_k \boldsymbol{G}_k \tag{6.119}$$

Here, we set $\boldsymbol{G}_N = [\boldsymbol{G}_1, \ldots, \boldsymbol{G}_N]$ and call $\boldsymbol{R}_N$ and $\boldsymbol{P}_N$ the p-power correlation matrix of $\boldsymbol{G}_N$ and p-power cross-correlation vector between $\boldsymbol{G}_N$ and $\boldsymbol{T}$, respectively.

According to (6.117), the following relationship holds

$$\boldsymbol{R}_N \boldsymbol{\beta}_N = \boldsymbol{P}_N \tag{6.120}$$

Then we get the optimal solution

$$\boldsymbol{\beta}_N = \boldsymbol{R}_N^{-1} \boldsymbol{P}_N \tag{6.121}$$

Equations (6.118) and (6.119) can be further written as

$$\begin{aligned} \boldsymbol{R}_N &= \sum_{k=1}^{N-1} |e_k|^{p-2} \boldsymbol{G}_k \boldsymbol{G}_k^T + |e_k|^{p-2} \boldsymbol{G}_N \boldsymbol{G}_N^T \\ &= \boldsymbol{R}_{N-1} + |e_N|^{p-2} \boldsymbol{G}_N \boldsymbol{G}_N^T \end{aligned} \tag{6.122}$$

$$\begin{aligned} \boldsymbol{P}_N &= \sum_{k=1}^{N-1} |e_k|^{p-2} t_k \boldsymbol{G}_k + |e_N|^{p-2} t_N \boldsymbol{G}_N \\ &= \boldsymbol{P}_{N-1} + |e_N|^{p-2} t_N \boldsymbol{G}_N \end{aligned} \tag{6.123}$$

Substituting (6.123) into (6.121), we get

$$\boldsymbol{\beta}_N = \mathbf{R}_N^{-1}(\boldsymbol{P}_{N-1} + |e_N|^{p-2} t_N \boldsymbol{G}_N) \tag{6.124}$$

According to (6.121), one can obtain

$$P_{N-1} = \mathbf{R}_{N-1}\boldsymbol{\beta}_{N-1} \tag{6.125}$$

From (6.122), we have

$$\mathbf{R}_{N-1} = \mathbf{R}_N - |e_N|^{p-2} G_N G_N^T \tag{6.126}$$

Substituting (6.126) into (6.125), and taking the result of P_{N-1} into (6.124) yields

$$\begin{aligned}\boldsymbol{\beta}_N &= \boldsymbol{R}_N^{-1}\left[(\boldsymbol{R}_N - |e_N|^{p-2} G_N G_N^T)\boldsymbol{\beta}_{N-1} + |e_N|^{p-2} t_N G_N\right] \\ &= \boldsymbol{R}_N^{-1}\left(\boldsymbol{R}_N\boldsymbol{\beta}_{N-1} - |e_N|^{p-2} G_N G_N^T \boldsymbol{\beta}_{N-1} + |e_N|^{p-2} t_N G_N\right) \\ &= \boldsymbol{\beta}_{N-1} + |e_N|^{p-2} \boldsymbol{R}_N^{-1} G_N (t_N - G_N^T \boldsymbol{\beta}_{N-1})\end{aligned} \tag{6.127}$$

The equation for updating $\boldsymbol{\beta}_N$ can be obtained as

$$\boldsymbol{\beta}_N = \boldsymbol{\beta}_{N-1} + |e_N|^{p-2} \mathbf{R}_N^{-1} G_N (t_N - \boldsymbol{\beta}_{N-1}^T G_N) \tag{6.128}$$

Applying the matrix inversion lemma and according to (6.122), we can obtain a recursive form to compute the inverse of $\mathbf{R}_N$

$$\mathbf{R}_N^{-1} = \left(\mathbf{I} - \frac{|e_N|^{p-2} \mathbf{R}_{N-1}^{-1} G_N}{1 + |e_N|^{p-2} G_N^T \mathbf{R}_{N-1}^{-1} G_N} G_N^T\right)\mathbf{R}_{N-1}^{-1} \tag{6.129}$$

To design an ELM, we first need to select the types of nodes and the corresponding activation functions and the hidden node number $\tilde{N}$. Suppose the data $\aleph = \left\{(\mathbf{x}_k, t_k) \middle| \mathbf{x}_k \in \mathbf{R}^n, t_k \in \mathbf{R}, k = 1, \ldots N_0\right\}$ arrive already and the new data follow sequentially. The initial value $\boldsymbol{\beta}_0$ is set to zero. In addition, we set

$$\mathbf{H}_0 = \begin{bmatrix} G_1^T \\ \vdots \\ G_{N_0}^T \end{bmatrix} = \begin{bmatrix} g(\mathbf{x}_1; \mathbf{c}_1, a_1) & \cdots & g(\mathbf{x}_1; \mathbf{c}_{\tilde{N}}, a_{\tilde{N}}) \\ & \vdots & \\ g(\mathbf{x}_{N_0}; \mathbf{c}_1, a_1) & \cdots & g(\mathbf{x}_{N_0}; \mathbf{c}_{\tilde{N}}, a_{\tilde{N}}) \end{bmatrix}_{N_0 \times \tilde{N}} \tag{6.130}$$

$$\boldsymbol{T}_0 = [t_1, t_2 \ldots t_{N0}]^T \tag{6.131}$$

$$\boldsymbol{E}_0 = \begin{bmatrix} |t_1|^{\frac{p}{2}-1} \ldots 0 \\ \vdots \\ 0 \ldots |t_{N_0}|^{\frac{p}{2}-1} \end{bmatrix}_{N_0 \times N_0} \tag{6.132}$$

Setting $\mathbf{M}_0 = \mathbf{E}_0\mathbf{H}_0$, we get

$$\mathbf{R}_0 = \mathbf{M}_0^T\mathbf{M}_0 \tag{6.133}$$

$$P_0 = \mathbf{H}_0^T T_0 \tag{6.134}$$

$$R_0^{-1} = (\mathbf{M}_0^T\mathbf{M}_0)^{-1} \tag{6.135}$$

$$\beta_0 = (\mathbf{M}_0^T\mathbf{M}_0)^{-1}\mathbf{H}_0^T T_0 \tag{6.136}$$

For the new arriving data ($\mathbf{x}_1$, t_1), one can compute the hidden layer output $\boldsymbol{H}_1$ and update the parameters of the output layer according to (6.128) and (6.129):

$$H_1 = \left[g(\mathbf{x}_1;\mathbf{c}_1;a_1), g(\mathbf{x}_1;\mathbf{c}_2;a_2),\ldots, g(\mathbf{x}_1;\mathbf{c}_{\tilde{N}};a_{\tilde{N}})\right] \tag{6.137}$$

$$\mathbf{R}_1^{-1} = \left(\mathbf{I} - \frac{|e_1|^{p-2}\,\mathbf{R}_0^{-1}\boldsymbol{H}_1^T}{1+|e_1|^{p-2}\,\boldsymbol{H}_1\mathbf{R}_0^{-1}\boldsymbol{H}_1^T}\boldsymbol{H}_1\right)\mathbf{R}_0^{-1} \tag{6.138}$$

$$\boldsymbol{\beta}_1 = \boldsymbol{\beta}_0 + |e_1|^{p-2}\,\mathbf{R}_1^{-1}\boldsymbol{H}_1^T(t_1 - \boldsymbol{H}_1\boldsymbol{\beta}_0) \tag{6.139}$$

where $e_1 = t_1 - \boldsymbol{H}_1\boldsymbol{\beta}_0$. If there arrive any other new data, the equations (6.138)–(6.139) can be repeated.

Similar to OS-ELM, the RLMP-ELM consists of two phases, namely initialization phase and sequential learning phase. In the initialization, the value of $\boldsymbol{\beta}_0$ can be estimated based on a small chunk of samples. The RLMP-ELM algorithm is summarized in Table 6.8.

6.2.3.2 Simulation Results

In this section, the performance of the RLMP-ELM learning algorithm is compared with that of ELM and OS-ELM on some benchmark datasets in regression and time series prediction tasks. To validate the effectiveness of the RLMP-ELM algorithm with different values of p, four types of training data were utilized. In addition, training data with different noise distributions were used to demonstrate that selecting an appropriate p value based on the distribution features can lead to better performance. To effectively show the superior performance of the RLMP-ELM algorithm, both Gaussian and non-Gaussian noises were considered. Gaussian noises were added to the noise-free training set or real data to generate the Gaussian training set, while symmetry alpha-stable (SαS) noise, the sum of independent SαS and Gaussian random noise, and uniform noise were used to create the SαS training set, SαSG training set, and uniform training set, respectively. A 'SinC' example is presented to show the learning behavior of the RLMP-ELM algorithm. Here, the 'SinC' function is given by

TABLE 6.8
RLMP-ELM Algorithm

Step 1 **Initialization Phase:** Initialize the learning using a small chunk of initial training data $\aleph_0 = \{(\mathbf{x}_k, \mathbf{t}_k)\}_{k=1}^{N_0}, N_0 \geq \tilde{N}$ a. Assign random input weights $\mathbf{c}_i$ and bias a_i (for additive hidden nodes) or center $\mathbf{c}_i$ and impact factor a_i (for RBF hidden nodes), $i = 1,\ldots,\tilde{N}$. b. Calculate the initial hidden layer output matrix β_0 according to (6.131) and T_0 based on equation (6.132). c. estimate the initial output weight β_0 according to (6.137). Set the training step k=1.
Step 2 **Sequential Learning Phase:** a. Obtain the current training data ($\mathbf{x}_k$, t_k). b. Calculate the partial hidden layer output matrix $H_k = \left[g(\mathbf{x}_k;\mathbf{c}_1,a_1), g(\mathbf{x}_k;\mathbf{c}_2,a_2),\ldots,g(\mathbf{x}_k;\mathbf{c}_{\tilde{N}},a_{\tilde{N}})\right]$ c. Calculate the error term $e_k = t_k - H_k\beta_{k-1}$. d. Calculate the output weight β_k $\mathbf{R}_k^{-1} = \mathbf{R}_{k-1}^{-1} - \dfrac{\lvert e_k\rvert^{p-2}\,\mathbf{R}_{k-1}^{-1}H_k^T}{1+\lvert e_k\rvert^{p-2}\,H_k\mathbf{R}_{k-1}^{-1}H_k^T}H_k\mathbf{R}_{k-1}^{-1}$ $\beta_k = \beta_{k-1} + \lvert e_k\rvert^{p-2}\,\mathbf{R}_k^{-1}H_k^T e_k$ e. If there is any new training data, set k=k+1 and go to step 2. Otherwise, the algorithm is ended.

$$y(x) = \begin{cases} \sin(x)/x & x \neq 0 \\ 0 & x = 0 \end{cases}$$

where the input x is uniformly randomly distributed on the interval (–10, 10). This function has also been used in [257] to generate the synthetic data.

For each type of dataset, the model selection procedure is first performed to determine the optimal architecture of the SLFN, that is, the number of hidden nodes. Then, the performance of RLMP-ELM is illustrated by comparing with ELM and OS-ELM.

6.2.3.2.1 Model Selection

Here the estimation of the optimal architecture of the network is called model selection. For RLMP-ELM, ELM, and OS-ELM algorithms, the optimal number of hidden units needs to be determined. The model selection procedure for a Gaussian training set is outlined as follows. For the RLMP-ELM algorithm, the optimal p value is selected from the range of 1.1–2, with an interval of 0.1. For the ELM or OS-ELM algorithms, the training process is conducted with varying numbers of hidden nodes, chosen from the range of 2–50, with an interval of 2. The Monte Carlo method is utilized, and over 200 trials are performed for each number of hidden nodes. In each trial, random zero-mean Gaussian noises with a variance of 0.16 are generated and added to all training samples to create the Gaussian training set. After each trial, the testing set without any noise is used to evaluate the algorithm's performance. The average performance is presented in Figure 6.13(a). The Root Mean Square Error (RMSE) of the testing set is used as the criterion to assess the algorithm's performance.

The model selection procedures for SαS and SαSG training sets are the same as those used for the Gaussian training set. The performances of all algorithms based on these two training sets are shown in Figure 6.13(b) and (c), respectively. In both figures, the lowest validation error is achieved when the number of hidden nodes of the ELM algorithm and the other algorithms is within the range of [171,175]. Furthermore, the RMSE curves of RLMP-ELM algorithms with $p=1.2$ and $p=1.6$ are smoother compared to ELM, OS-ELM, and RLMP-ELM algorithm with $p=2$. This implies that RLMP-ELM algorithms with $p=1.2$ and $p=1.6$ are not sensitive to the network size, even when the training samples are disturbed by the symmetry alpha-stable random noise. The same model selection procedures were applied to both the uniform training set and the training set with large uniform noise distributed in the range of [−0.4, 0.4]. The performance curves of all algorithms are presented in Figure 6.13(d), where the five curves are almost identical. It is evident from the figure that the algorithms with a higher number of hidden nodes achieved the lowest validation error. In addition, the RMSE curves of all algorithms were smooth, indicating that these algorithms were not sensitive to the network size when the training samples were disturbed by uniform random noise.

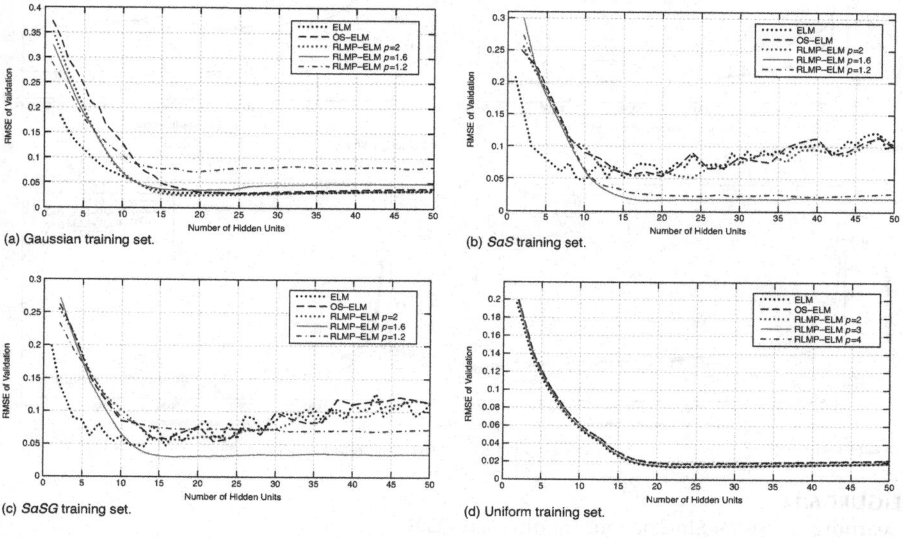

(a) Gaussian training set.

(b) *SaS* training set.

(c) *SaSG* training set.

(d) Uniform training set.

FIGURE 6.13
Model selection for SinC on four training sets [250].

6.2.3.2.2 Performance Evaluation

In this section, the performance of RLMP-ELM algorithms with different p values is discussed. According to the analysis above, 20 is selected as the optimal number of hidden nodes for RLMP-ELM, ELM and OS-ELM algorithms. The convergence curves in terms of the validation RMSE are illustrated in Figure 6.14 for different training sets. As one can observe from Figure 6.14(a), all algorithms are robust to the Gaussian noises. Apart from the RLMP-ELM algorithm with $p=1.2$, other algorithms almost have the same convergence rate and the stable testing performance. The convergence rate and stable performance of the RLMP-ELM algorithm with $p=1.2$ are not as well as those of other algorithms. Figure 6.14(b) illustrates that both RLMP-ELM algorithms with $p=1.2$ and $p=1.6$ are robust to the symmetry alpha-stable random noises so that the curves remain stable. Furthermore, these two curves are below other two curves so that the stable testing RMSEs of the corresponding algorithms are lower. It is clear from Figure 6.14(c) that both RLMP-ELM algorithms with $p=1.2$ and $p=1.6$ are robust to the symmetry alpha-stable random noise and Gaussian noise, but the stable performance of the algorithm with $p=1.2$ is worse than that of the algorithm with $p=1.6$ because of the Gaussian random noises. The convergence rate of the RLMP-ELM algorithm with $p=1.6$ is the fastest among all algorithms. In Figure 6.14(d), it can

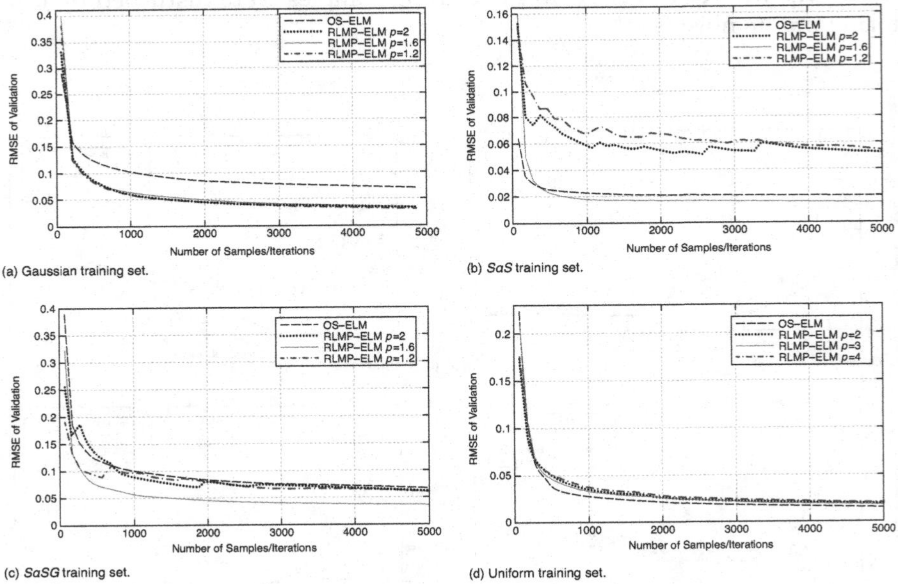

(a) Gaussian training set. (b) *SaS* training set. (c) *SaSG* training set. (d) Uniform training set.

FIGURE 6.14
Learning curves for SinC on four training sets [250].

be seen that all algorithms are robust to the uniform noises and the curves remain steady. Compared with other algorithms, the RLMP-ELM algorithm with p=4 has the best convergence performance. Other algorithms almost have the same convergence rate and stable testing performance. Here, the p value is usually bounded above by a certain positive number.
The results show that the RLMP-ELM can obtain better performance in non-Gaussian situations than ELM and OS-ELM algorithms. Furthermore, the RLMP-ELM has several interesting and significant features:

i. For Gaussian distributed data, the RLMP-ELM with $p=2$ can achieve better generalization performance and more accurate results.
ii. For non-Gaussian heavy-tailed data, the RLMP-ELM with $0<p<2$ can obtain better generalization performance and more accurate results.
iii. For non-Gaussian light-tailed data, the RLMP-ELM with $2<p$ can get better generalization performance and more accurate results.

How to determine the exact value of p, with which the RLMP-ELM achieves the best performance is however still not clear.

6.2.4 Sparse RLMP-ELM

To determine the optimal number of hidden neurons, the ELM model is trained dynamically, with the number of hidden neurons being adjusted during the training process. This is achieved through the use of techniques such as incremental ELM (I-ELM) [343–345] and bidirectional ELM (B-ELM) [346], which ensure that only appropriate neurons are added to the network, resulting in a more compact model. In addition, a sparse Bayesian approach [180] is presented, which automatically tunes most of the output weights to zero with an assumed prior distribution, resulting in a more compact ELM model. Another approach is the sparse ELM (S-ELM) [347], which involves solving a quadratic programming problem analytically, resulting in a significant reduction in storage space and testing time. Both of these sparse models are designed specifically for classification problems. Inspired by the adaptive filters with sparse constraint term and sparse ELM algorithms mentioned above, this subchapter mainly presents an online sparse ELM algorithm (namely, sparse recursive least mean p-power ELM (SRLMP-ELM)) by incorporating a sparsity penalty term into the least mean p-power error criterion as the cost function, while the more initial hidden neurons are selected first and the parameters of hidden layer are randomly generated as in the conventional ELM. In SRLMP-ELM, a cost function, i.e., the sparse LMP(SLMP) error criterion, provides a mechanism to update the output weights sequentially and automatically tune some parameters of the output weights to zeros. The SLMP error criterion aims to minimize the combination of the mean p-power of the errors and a sparsity penalty constraint of the output weights. For real industrial system requirements, the proposed online learning algorithm is able to provide higher accuracy, more compact model, and better generalization ability than ELM and online sequential ELM, whereas the non-Gaussian noises impact the processes, especially impulsive noises.

6.2.4.1 Sparse Recursive Least Mean p-power ELM

In this section, the SRLMP-ELM algorithm is derived to update the weight vector of the ELM under the SLMP error criterion. According to the description of ELM in preliminary, the output of an ELM can be seen as a general linear system $\beta^T g = t$. For this general linear system, the SRLMP algorithm is the extension of the recursive least square (RLS) algorithm with the MPE cost function. The cost function of SRLMP algorithm is defined as regularizing LMP error criterion by a sparse penalty term as follows [251]:

$$J_{SRLMP} = \frac{1}{N}\sum_{k=1}^{N} \gamma^{N-k} |e_k|^p + \rho S_N \tag{6.140}$$

where e_k is the error in kth sample time and $e_k = t_k - \beta_N^T g_k$. S_N denotes a sparsity penalty constraint (SPC) and l_1-norm is selected as the SPC here

$$S_N = \|\beta_N\|_1 \tag{6.141}$$

Substituting (6.141) into (6.140) yields

$$J_{SRLMP} = \frac{1}{N}\sum_{k=1}^{N} \lambda^{N-k} |e_k|^p + \rho \|\beta_N\|_1 \tag{6.142}$$

In theory, it has been proved by some results of convex function in [201] that every minimum of LMP error criterion is a global minimum while $p \geq 1$. Thus the performance function J_{SLMP} has a global minimum while S_N is a convex function. Since l_1-norm is a convex function, J_{SLMP} has a global minimum. The optimal solution β_N for minimizing J_{SLMP} can be obtained by differentiating (6.143) with respect to β_N and setting the derivatives to zero. The derivatives are

$$\begin{aligned}\frac{\partial J_{SRLMP}}{\partial \boldsymbol{\beta}_N} &= \frac{1}{N}\sum_{k=1}^{N}\gamma^{N-k}\frac{\partial |e_k|^p}{\partial \boldsymbol{\beta}_N} + \rho\frac{\partial \|\boldsymbol{\beta}_N\|_1}{\partial \boldsymbol{\beta}_N} \\ &= \frac{1}{N}\sum_{k=1}^{N}\gamma^{N-k}\frac{\partial |e_k|^p}{\partial e_k}\cdot\frac{\partial e_k}{\partial \boldsymbol{\beta}_N} + \rho\frac{\partial \|\boldsymbol{\beta}_N\|_1}{\partial \boldsymbol{\beta}_N}\end{aligned} \tag{6.143}$$

Then we have

$$\frac{\partial J_{SRLMP}}{\partial \boldsymbol{\beta}_N} = \frac{1}{N}\sum_{k=1}^{N}\gamma^{N-k}p|e_k|^{p-2}e_k\frac{\partial e_k}{\partial \boldsymbol{\beta}_N} + \rho\frac{\partial \|\boldsymbol{\beta}_N\|_1}{\partial \boldsymbol{\beta}_N} \tag{6.144}$$

Substituting $e_k = t_k - \boldsymbol{\beta}_N^T g_k$ into (6.144) yields

$$\frac{\partial J_{SRLMP}}{\partial \boldsymbol{\beta}_N} = \frac{1}{N}\sum_{k=1}^{N}\gamma^{N-k}p|e_k|^{p-2}(t_k - \boldsymbol{\beta}_N^T g_k)g_k + \rho\frac{\partial \|\boldsymbol{\beta}_N\|_1}{\partial \boldsymbol{\beta}_N} \tag{6.145}$$

Setting $\frac{\partial J_{SLMP}}{\partial \boldsymbol{\beta}_N} = 0$, one can obtain

$$\frac{1}{N}\sum_{k=1}^{N}\gamma^{N-k}|e_k|^{p-2}g_k g_k^T\boldsymbol{\beta}_N = \frac{1}{N}\sum_{k=1}^{N}\gamma^{N-k}|e_k|^{p-2}t_k g_k + \rho\frac{\partial \|\boldsymbol{\beta}_N\|_1}{\partial \boldsymbol{\beta}_N} \tag{6.146}$$

Let

$$\boldsymbol{\Psi}_N = \sum_{k=1}^{N}\gamma^{N-k}|e_k|^{p-2}g_k g_k^T \tag{6.147}$$

and

$$\boldsymbol{\Phi}_N = \sum_{k=1}^{N}\gamma^{N-k}|e_k|^{p-2}t_k g_k^T \tag{6.148}$$

Here, we set $G_N = [g_{1,\ldots,}g_n]$, then Ψ_N and Φ_N can be rewritten as

$$\boldsymbol{\Psi}_N = G_N\begin{bmatrix}\gamma^{N-1}|e_1|^{p-2} & \cdots & 0 \\ & \vdots & \\ 0 & \cdots & |e_N|^{p-2}\end{bmatrix}G_N^T \tag{6.149}$$

and

$$\Phi_N = G_N \begin{bmatrix} \gamma^{N-1}|e_1|^{p-2} & \cdots & 0 \\ & \vdots & \\ 0 & \cdots & |e_N|^{p-2} \end{bmatrix} T \tag{6.150}$$

Here Ψ_N and Φ_N are called as the sparse p-Power correlation matrix of G_N and the sparse p-Power cross-correlation vector of G_N and T, respectively. They serve a similar purpose as the conventional correlation matrix of G_N and the cross-correlation vector of G_N and T. Furthermore, we set

$$\Upsilon_N = \Phi_N + \rho \frac{\partial \|\boldsymbol{\beta}_N\|_1}{\partial \boldsymbol{\beta}_N} \tag{6.151}$$

Considering (6.146)–(6.151), the following relationship can be obtained

$$\Psi_N \boldsymbol{\beta}_N = \Upsilon_N \tag{6.152}$$

The optimal solution $\boldsymbol{\beta}_N$ is

$$\boldsymbol{\beta}_N = \Psi_N^{-1} \Upsilon_N \tag{6.153}$$

The (6.147), (6.148), and (6.151) can be further written as

$$\begin{aligned} \Psi_N &= \gamma \sum_{k=1}^{N-1} \gamma^{N-1-k} |e_k|^{p-2} g_k g_k^T + |e_N|^{p-2} g_N g_N^T = \gamma \, \psi_{N-1} + |e_N|^{p-2} g_N g_N^T \\ \Phi_N &= \gamma \sum_{k=1}^{N-1} \gamma^{N-1-k} |e_k|^{p-2} t_k g_k^T + |e_N|^{p-2} t_N g_N^T = \gamma \, \Phi_{N-1} + |e_N|^{p-2} t_N g_N^T \end{aligned} \tag{6.154}$$

$$\begin{aligned} \Upsilon_N &= \gamma \, \Phi_{N-1} + |e_N|^{p-2} t_N g_N + \rho \frac{\partial \|\boldsymbol{\beta}_N\|_1}{\partial \boldsymbol{\beta}_N} \\ &= \gamma \, \Phi_{N-1} + \gamma\rho \frac{\|\boldsymbol{\beta}_{N-1}\|_1}{\partial \boldsymbol{\beta}_N} + |e_N|^{p-2} t_N g_N + \rho \frac{\partial \|\boldsymbol{\beta}_N\|_1}{\partial \boldsymbol{\beta}_N} - \gamma\rho \frac{\|\boldsymbol{\beta}_{N-1}\|_1}{\partial \boldsymbol{\beta}_N} \\ &= \gamma \, \Upsilon_{N-1} + |e_N|^{p-2} t_N g_N + \rho \frac{\partial \|\boldsymbol{\beta}_N\|_1}{\partial \boldsymbol{\beta}_N} - \lambda\rho \frac{\|\boldsymbol{\beta}_{N-1}\|_1}{\partial \boldsymbol{\beta}_N} \end{aligned} \tag{6.155}$$

Here we assume that $\partial\|\boldsymbol{\beta}_N\|_1/\partial\boldsymbol{\beta}_N$ does not change significantly in a single time step, i.e., $\partial\|\boldsymbol{\beta}_N\|_1/\partial\boldsymbol{\beta}_N$ approaches to $\partial\|\boldsymbol{\beta}_N\|_1/\partial\boldsymbol{\beta}_N$. Hence, (6.155) can be approximated by

$$\Upsilon_N = \gamma\Upsilon_{N-1} + |e_N|^{p-2} t_N g_N + \rho(1-\gamma)\frac{\partial\|\boldsymbol{\beta}_{N-1}\|}{\partial\boldsymbol{\beta}_N} \tag{6.156}$$

Substituting (6.156) into (6.153), we get

$$\boldsymbol{\beta}_N = \boldsymbol{\Psi}_N^{-1}\left[\gamma\Upsilon_{N-1} + |e_N|^{p-2} t_N g_N + \rho(1-\gamma)\frac{\partial\|\boldsymbol{\beta}_{N-1}\|_1}{\partial\boldsymbol{\beta}_N}\right] \tag{6.157}$$

Considering (6.154) and applying the matrix inversion lemma, then we get

$$\boldsymbol{\Psi}_N^{-1} = \lambda^{-1}\boldsymbol{\Psi}_{N-1}^{-1}\left(I - \frac{|e_N|^{p-2} g_N g_N^T \boldsymbol{\Psi}_{N-1}^{-1}}{\lambda + |e_N|^{p-2} g_N^T \boldsymbol{\Psi}_{N-1}^{-1} g_N}\right) \tag{6.158}$$

To get a compact expression of (6.158), we define Ω_N and $\mathbf{K}_N$ as

$$\Omega_N = \boldsymbol{\Psi}_N^{-1}$$

$$\mathbf{K}_N = \frac{|e_N|^{p-2}\Omega_{N-1} g_N}{\gamma + |e_N|^{p-2} g_N^T \Omega_{N-1} g_N} \tag{6.159}$$

Then we obtain

$$\Omega_N = \lambda^{-1}(\mathrm{I} - \mathbf{K}_N g_N^T)\Omega_{N-1} \tag{6.160}$$

where Ω_N and $\mathbf{K}_N$ are the extended Kalman gain vectors similar to those in RLS. Thus, (6.157) can be rewritten as

$$\begin{aligned}
\boldsymbol{\beta}_N &= \gamma^{-1}(\mathrm{I} - \mathbf{K}_N g_N^T)\Omega_{N-1}\left[\gamma\Upsilon_{N-1}\right] + |e_N|^{p-2} t_N g_N + \rho(1-\lambda)\frac{\partial\|\boldsymbol{\beta}_{N-1}\|_1}{\partial\boldsymbol{\beta}_N} \\
&= (\mathrm{I} - \mathbf{K}_N g_N^T)\left[\Omega_{N-1}\Upsilon_{N-1} + \gamma^{-1}\Omega_{N-1}|e_N|^{p-2} t_N g_N + \rho\gamma^{-1}(1-\lambda)\Omega_{N-1}\frac{\partial\|\boldsymbol{\beta}_{N-1}\|_1}{\partial\boldsymbol{\beta}_N}\right] \\
&= \Omega_{N-1}\Upsilon_{N-1} - \mathbf{K}_N\Omega_{N-1}\Upsilon_{N-1} + \gamma^{-1}|e_N|^{p-2}(I - K_N g_N^T)\Omega_{N-1} t_N g_N \\
&\quad + \rho\gamma^{-1}(1-\gamma)(I - K_N g_N^T)\Omega_{N-1}\frac{\partial\|\boldsymbol{\beta}_{N-1}\|_1}{\partial\boldsymbol{\beta}_N} \\
&= \boldsymbol{\beta}_{N-1} + \mathbf{K}_N(\mathrm{t}_N - g_N^T\boldsymbol{\beta}_{N-1}) + \rho\gamma^{-1}(1-\gamma)(I - \mathbf{K}_N g_N^T)\Omega_{N-1}\frac{\partial\|\boldsymbol{\beta}_{N-1}\|_1}{\partial\boldsymbol{\beta}_N}
\end{aligned} \tag{6.161}$$

The equation for updating β_N can be expressed by

$$\beta_N = \beta_{N-1} + e_N \mathbf{K}_N + \rho\gamma^{-1}(1-\gamma)(\mathbf{I} - \mathbf{K}_N g_N^T)\Omega_{N-1} \frac{\partial \|\beta_{N-1}\|_1}{\partial \beta_N} \tag{6.162}$$

Furthermore, the derivative $\partial\|\beta_N\|_1 / \partial\beta_N$ is sign(β_N). The function for updating β_N can be obtained

$$\beta_N = \beta_{N-1} + e_N \mathbf{K}_N + \rho\gamma^{-1}(1-\gamma)(\mathbf{I} - \mathbf{K}_N g_N^T)\Omega_{N-1}\text{sign}(\beta_{N-1}) \tag{6.163}$$

Considering again the description of ELM in preliminary, one can find that there is a standard SLFN and N arbitrary distinct samples $(\mathbf{x}_k, \boldsymbol{t}_k)$ in the algorithm. The SLFN with $\tilde{N}$ hidden nodes with activation function $g(\mathbf{x})$ and the hidden layer output matrix is $g_k = \left[g(\mathbf{x}_k; a_1, b_1), g(\mathbf{x}_k; a_2, b_2), \ldots, g(\mathbf{x}_k; a_{\tilde{N}}, b_{\tilde{N}})\right]^T$. The SRLMP-ELM algorithm is summarized in Table 6.9.

Remark 6.17

The computation complexity of the SRLMP-ELM algorithm is discussed. For the $\tilde{N}$ hidden units and N-length training sequence, the total training complexity of the SRLMP-ELM is $O(N\tilde{N}^2)$. The same computation complexity can thus be observed comparing that of $O(N\tilde{N}^2)$ in the primitive ELM matrix inversion and of $O(N\tilde{N}^2)$ in the OS-ELM [175,348]. But since the data are processed sequentially in SRLMP-ELM and OS-ELM, they cost more time than the ELM algorithm. However, the more compact model can be obtained by SRLMP-ELM through the sparse penalty constraint. Thus, the running time can be reduced in the testing phase which will be illustrated by the following simulation results.

TABLE 6.9

SRLMP-ELM Algorithm

1. Assign random input weights a_i and bias b_i (for additive hidden nodes) or center a_i and impact factor b_i (for RBF hidden nodes), $i = 1, \ldots, \tilde{N}$. Initialize $\beta_0 = 0, \Omega_0 = \mathbf{I}_{\tilde{N}\times\tilde{N}}, \lambda, \rho$ and p. Set the training step $k = 1$.

2. Obtain the current training data $(\mathbf{x}_k, t_k)$
3. Calculate the hidden layer output matrix $g_k = \left[g(\mathbf{x}_k; c_1, a_1), g(\mathbf{x}_k; c_2, a_2), \ldots, g(\mathbf{x}_k; c_{\tilde{N}}, a_{\tilde{N}})\right]^T$
4. Calculate the error term $e_k = t_k - \beta_{k-1}^T g_k$
5. Calculate the gain vector $\mathbf{K}_k = \dfrac{|e_k|^{p-2}\,\Omega_{k-1} g_k}{\lambda + |e_k|^{p-2}\, g_k^T \Omega_{k-1} g_k}$
6. Calculate the output weight β_k
7. $\beta_k = \beta_{k-1} + e_k \mathbf{K}_k + \rho\gamma^{-1}(1-\gamma)(\mathrm{I} - \mathbf{K}_k g_k^T)\Omega_{k-1}\text{sign}(\beta_{k-1})$
8. Update Ω_k
9. $\Omega_k = \gamma^{-1}(\mathbf{I} - \mathbf{K}_k g_k^T)\Omega_{k-1}$
10. If there is any new training sample, set $k = k + 1$ and go to 2. Otherwise, the algorithm is terminated.

6.2.4.2 Simulation Results

In this section, the performance of the SRLMP-ELM learning algorithm is compared with that of ELM, OS-ELM, and RLMP-ELM on a few regression problems. To validate the effectiveness of the SRLMP-ELM with different p and ρ values, the training samples with different noise distributions are employed to demonstrate that optimal performance can be achieved by selecting p and ρ based on the characteristics of the noise distribution. The SαS distribution, a well-known non-Gaussian distribution, is particularly useful for modeling impulsive noise with heavy-tailed distributions. To effectively illustrate the good performance of the SRLMP-ELM algorithm, Gaussian and non-Gaussian datasets are considered in the study. For Gaussian dataset, Gaussian noises are added to the noise-free training set or real data to generate training samples, called as Gaussian training set. Some non-Gaussian datasets, such as symmetry alpha-stable ($S\alpha S$) noise, sum of independent $S\alpha S$ and Gaussian random noise ($S\alpha SG$), and uniform noises are used to create training samples. They are called as $S\alpha S$ training set, $S\alpha SG$ training set, and uniform training set, respectively. The details of the validation process are shown in the following sections.

6.2.4.2.1 SinC Dataset

To illustrate the compact size of the SRLMP-ELM network model, we make model selection procedure first for each type of dataset to determine the optimal architecture, which is the number of the hidden nodes. Then the performance of SRLMP-ELM is illustrated by comparing it with ELM, OS-ELM, and RLMP-ELM algorithms.

6.2.4.2.1.1 Model Selection For ELM, OS-ELM, and RLMP-ELM algorithms, the optimal number of hidden units needs to be determined. What's more, the initial network size of SRLMP-ELM should be determined by the model selection. In order to illustrate the good performance of the SPLMP-ELM algorithm, the number of hidden units of OS-ELM, RLMP-ELM and the initial number of hidden nodes are selected as same as the one of the ELM algorithms. Thus the model selection procedure is focused on the performance of the ELM algorithm with different hidden nodes while the training datasets are Gaussian and non-Gaussian separately. The result of the model selection is shown in Figure 6.15.

For the ELM algorithm with Gaussian training dataset, random zero-mean Gaussian noises with variance 0.16 are created and added to all training samples to generate the Gaussian training set in each trial. The average performance is shown with a green curve in Figure 6.15. The RMSE of the testing set is used as the criterion of the ELM's performance. For the other three training datasets, the model selection procedures are the same as that of the Gaussian

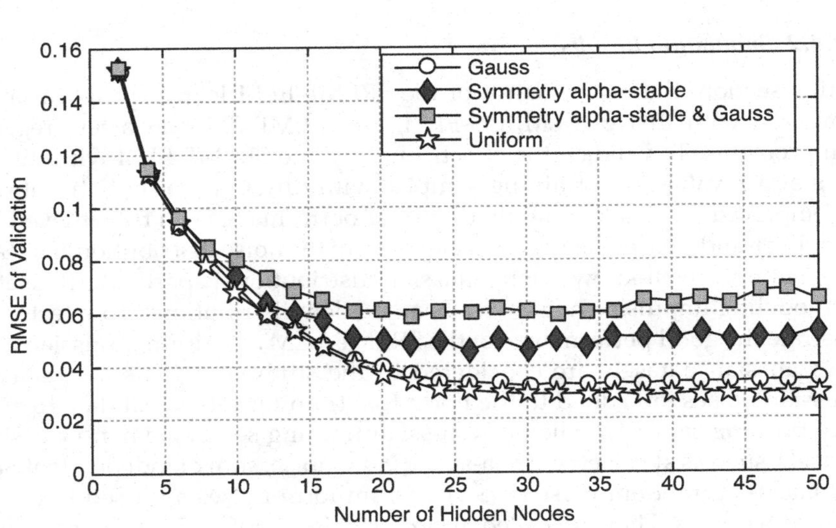

FIGURE 6.15
ELM model selection for SinC based on four types of training sets [251].

training set. The performances of ELM with these three different training datasets are illustrated in blue, red, and yellow curves in Figure 6.15, separately. As observed from the figure, the lowest validation errors are achieved when the number of hidden nodes of ELM is above 24 for the Gaussian and uniform training sets. One can also see that RMSE curves for these two training datasets are smooth. It implies that ELM is not sensitive to the network size while the outputs of training data are stained by Gaussian and uniform noises. For $S\alpha S$ and $S\alpha SG$ training sets, the curves are not smooth and ELM is slightly sensitive to the network size for the outputs with $S\alpha S$ and $S\alpha SG$ noises. But the lowest validation errors are achieved when the number of hidden nodes of ELM is in the range. According to the result of the model selection, 30 hidden units are chosen for ELM, OS-ELM, RLMP-ELM, and the initial hidden units of the SRLMP-ELM algorithm.

6.2.4.2.1.2 Performance Evaluation In this part, the performance of SRLMP-ELM algorithms with different values of p and ρ is discussed for SinC example. According to the analysis above, 30 is selected as the optimal number of hidden nodes for ELM, OS-ELM, and RLMP-ELM algorithms, in addition to the initial number of hidden units for SRLMP-ELM. The details of the comparison about SRLMP-ELM algorithm with different values p and ρ, ELM, OS-ELM and RLMP-ELM algorithms are summarized in Table 6.10.

TABLE 6.10

Performance Comparison for SinC Example [251]

Noise Type	Algorithms			Training RMSE	Training Dev	Training Time(s)	Validation RMSE	Validation Dev	Validation Time(s)	#nodes
Gauss	ELM			0.3996	0.0045	0.0310	0.0326	0.0045	0.0183	30
	OS-ELM			0.4031	0.0049	0.7198	0.0411	0.0065	0.0216	30
	SRLMP-ELM	$p=1.6$	$\rho=0$	0.4000	0.0038	0.7539	0.0279	0.0064	0.0198	30
			RLMP-ELM							
			$\rho=0.3$	0.3999	0.0041	0.7839	0.0307	0.0039	0.0146	24
			$\rho=1.2$	0.4007	0.0043	0.7620	0.0313	0.0069	0.0106	20
		$p=2.0$	$\rho=0$	0.3994	0.0045	0.7609	0.0307	0.0053	0.0187	30
			RLMP-ELM							
			$\rho=0.3$	0.3996	0.0037	0.7638	0.0292	0.0042	0.0145	25
			$\rho=1.2$	0.3998	0.0039	0.7740	0.0331	0.0058	0.0096	21
$S\alpha S$	ELM			0.6320	0.4170	0.0313	0.0503	0.0275	0.0153	30
	OS-ELM			0.7096	0.4403	0.7873	0.0527	0.0278	0.0146	30
	SRLMP-ELM	$p=1.6$	$\rho=0$	0.6750	0.4413	0.8134	0.0151	0.0027	0.0164	30
			RLMP-ELM							
			$\rho=0.3$	0.6016	0.4473	0.8006	0.0182	0.0046	0.0113	19
			$\rho=1.2$	0.7186	0.4607	0.8706	0.0203	0.0035	0.0102	17
		$p=2.0$	$\rho=0$	0.5984	0.4130	0.8000	0.0445	0.0220	0.0155	30
			RLMP-ELM							
			$\rho=0.3$	0.5620	0.3966	0.8813	0.0459	0.0125	0.0115	20
			$\rho=1.2$	0.6113	0.3927	0.8682	0.0468	0.0144	0.0096	17

(Continued)

TABLE 6.10 (*Continued*)

Performance Comparison for SinC Example [251]

Noise Type	Algorithms			Training RMSE	Training Dev	Training Time(s)	Validation RMSE	Validation Dev	Validation Time(s)	#nodes
$S\alpha SG$	ELM			0.7096	0.4403	0.8173	0.0752	0.0278	0.0156	30
	OS-ELM			0.7883	0.3587	0.7895	0.0274	0.0046	0.0166	30
	SRLMP-ELM	p=1.6	$\rho=0$	0.7648	0.3589	0.0302	0.0603	0.0237	0.0149	30
			RLMP-ELM							
			$\rho=0.3$	0.7020	0.3330	0.8453	0.0297	0.0048	0.0130	24
			$\rho=1.2$	0.7192	0.3168	0.8433	0.0319	0.0049	0.0117	22
		p=2.0	$\rho=0$	0.8317	0.4088	0.8828	0.0617	0.0234	0.0157	30
			RLMP-ELM							
			$\rho=0.3$	0.7674	0.3528	0.8730	0.0631	0.0249	0.0135	25
			$\rho=1.2$	0.7441	0.3248	0.8858	0.0644	0.0227	0.0118	22
Uniform	ELM			0.3452	0.0021	0.0315	0.0292	0.0043	0.0165	30
	OS-ELM			0.3456	0.0027	0.8153	0.0317	0.0068	0.0156	30
	SRLMP-ELM	p=2.0	$\rho=0$	0.3458	0.0021	0.8469	0.0289	0.0043	0.0142	30
			RLMP-ELM							
			$\rho=0.3$	0.3461	0.0023	0.8002	0.0295	0.0059	0.0122	26
			$\rho=1.2$	0.3467	0.0033	0.8163	0.0306	0.0062	0.0111	21
		p=4.0	$\rho=0$	0.3467	0.0021	0.7973	0.0261	0.0052	0.0153	30
			RLMP-ELM							
			$\rho=0.3$	0.3472	0.0021	0.7892	0.0275	0.0060	0.0123	22
			$\rho=1.2$	0.3481	0.0028	0.7955	0.0313	0.0069	0.0101	17

As observed from Table 6.10, the accuracies of SRLMP-ELM with different ρ and p, ELM and OS-ELM based on the Gaussian training dataset are similar to each other. One can conclude from the results in Table 6.10 that (1) All algorithms are robust to the Gaussian distribution data. There is an obvious difference that the training time consumed by ELM is much less than those cost by other algorithms. (2) For the SαS training dataset, the SRLMP-ELM algorithm with p=1.6 and ρ value in the range of [0.3,1.2] can obtain better accuracy and more compact model than other algorithms. (3) For $S\alpha SG$ training dataset, the performance of SRLMP-ELM algorithm with p=1.6 and ρ value in the range of [0.3,1.2] is better than other algorithms. (4) For uniform training dataset, the SRLMP-ELM algorithm with $p=4$, $\rho=0$, obtains the lowest testing RMSE of 0.261 because the uniform data are bounded. The best accuracy is only slightly better than those of other algorithms. However, the SRLMP-ELM algorithm with p=4 and ρ value in the range of [0.3,1.2] obtains a more compact model and less testing time than other algorithms. In all, one can observe from the simulation results of SinC case that the SRLMP-ELM algorithm with appropriate p and ρ can obtain better accuracy, more compact model and less testing time on non-Gaussian dataset than ELM and OS-ELM algorithms. The SRLMP-ELM with ρ in the range of [0.3,1.2] needs fewer hidden nodes and less testing time than RLMP-ELM with the same p value, while their learning accuracies are similar.

6.2.4.2.2 Time Series of Internet Traffic

To further illustrate the performance of SRLMP-ELM, the time series prediction is conducted with the internet traffic data. The related data can be downloaded from Paulo Cortez's home page, and the specific download address can refer to [251]. The goal is to predict the value of the current sample using the previous ten consecutive samples. All the datasets are normalized into [0,1]. The detailed performances of each algorithm for the training datasets with different noises are illustrated in Table 6.11. One can observe from the table that (1) for Gaussian noise case, the SRLMP-ELM with p=1.6, ρ=0.3 and p=2.0, ρ=0.3 can obtain the similar accuracy as ELM while with only 15 hidden nodes. (2) For $S\alpha S$ training dataset, the testing RMSEs of SRLMP-ELM with p=1.6 and ρ in the range of [0,1.2] are smaller than those of other algorithms. The testing accuracy of SRLMP-ELM with p=1.6 and ρ=0.3 is 0.0303, which is almost half of that of ELM while the number of hidden units is only 11 and the testing time is 0.0048 second. SRLMP-ELM with p=1.6 and ρ=1.2 can obtain the testing accuracy 0.0377 with 8 hidden nodes and 0.0035 second testing time. (3) For SαSG training dataset, the testing RMSEs of SRLMP-ELM with p=1.6 and ρ in the range of [0,1.2] are a little smaller than those of other algorithms for the case of Gaussian noise. The better accuracy and more compact model can be obtained by SRLMP-ELM with p=1.6, ρ=0.3 and p=1.6, ρ=1.2, which is similar to $S\alpha S$ training dataset. (4) For uniform

TABLE 6.11
Performance Comparison for Time Series of Internet Traffic [251].

Noise Type	Algorithms			Training			Validation			#nodes
				RMSE	Dev	Time(s)	RMSE	Dev	Time(s)	
Gauss	ELM			0.4063	0.0057	0.4805	0.0481	0.0335	0.0058	18
	OS-ELM			0.4004	0.0045	0.5031	0.0376	0.0064	0.0048	18
	SRLMP-ELM	p=1.6	$\rho = 0$	0.3991	0.0043	0.0249	0.0394	0.0064	0.0060	18
			RLMP-ELM							
			$\rho = 0.3$	0.4015	0.0047	0.4885	0.0402	0.0084	0.0042	15
			$\rho = 1.2$	0.4023	0.0045	0.5008	0.0630	0.0089	0.0029	10
		p=2.0	$\rho = 0$	0.4010	0.0049	0.4849	0.0349	0.0058	0.0063	18
			RLMP-ELM							
			$\rho = 0.3$	0.4006	0.0041	0.5182	0.0404	0.0065	0.0047	15
			$\rho = 1.2$	0.4035	0.0037	0.5008	0.0664	0.0078	0.0030	10
SαS	ELM			0.6565	0.4449	0.0182	0.0589	0.0318	0.0069	18
	OS-ELM			0.6748	0.4523	0.4750	0.0588	0.0321	0.0075	18
	SRLMP-ELM	p=1.6	$\rho = 0$	0.6861	0.4654	0.4883	0.0260	0.0105	0.0067	18
			RLMP-ELM							
			$\rho = 0.3$	0.5800	0.4119	0.5028	0.0303	0.0196	0.0048	11
			$\rho = 1.2$	0.6710	0.4541	0.4856	0.0377	0.0220	0.0035	8
		p=2.0	$\rho = 0$	0.6656	0.4324	0.4783	0.0462	0.0220	0.0065	18
			RLMP-ELM							
			$\rho = 0.3$	0.6952	0.4639	0.4581	0.0578	0.0209	0.0043	11
			$\rho = 1.2$	0.6696	0.4467	0.4863	0.0659	0.0251	0.0032	6

(Continued)

TABLE 6.11 (*Continued*)

Performance Comparison for Time Series of Internet Traffic [251].

Noise Type	Algorithms			Training			Validation			#nodes
				RMSE	Dev	Time(s)	RMSE	Dev	Time(s)	
SαSG	ELM			0.7836	0.3596	0.0162	0.0669	0.0281	0.0053	18
	OS-ELM			0.7978	0.3690	0.4941	0.0695	0.0362	0.0060	18
	SRLMP-ELM	p=1.6	$\rho=0$	0.8245	0.3936	0.4952	0.0382	0.0069	0.0065	18
			RLMP-ELM							
			$\rho=0.3$	0.8422	0.4149	0.4850	0.0438	0.0087	0.0044	16
			$\rho=1.2$	0.8162	0.4018	0.4933	0.0640	0.0115	0.0039	11
		p=2.0	$\rho=0$	0.8348	0.3765	0.4941	0.0498	0.0233	0.0061	18
			RLMP-ELM							
			$\rho=0.3$	0.8569	0.4028	0.4847	0.0531	0.0204	0.0048	16
			$\rho=1.2$	0.7458	0.3377	0.4984	0.0696	0.0234	0.0036	11
Uniform	ELM			0.3462	0.0025	0.0170	0.0384	0.0096	0.0059	18
	OS-ELM			0.3460	0.0026	0.4950	0.0390	0.0036	0.0056	18
	SRLMP-ELM	p=2.0	$\rho=0$	0.3466	0.0025	0.4947	0.0369	0.0071	0.0057	18
			RLMP-ELM							
			$\rho=0.3$	0.3469	0.0024	0.4879	0.0388	0.0072	0.0041	16
			$\rho=1.2$	0.3500	0.0028	0.5862	0.0438	0.0084	0.0033	10
		p=4.0	$\rho=0$	0.3474	0.0023	0.5655	0.0336	0.0064	0.0062	18
			RLMP-ELM							
			$\rho=0.3$	0.3475	0.0027	0.5140	0.0361	0.0074	0.0044	16
			$\rho=1.2$	0.3411	0.0030	0.5218	0.0416	0.0092	0.0029	10

training dataset, the testing accuracy of SRLMP-ELM with $p=4.0$ and ρ in the range [0,1.2] is slightly better than that of other algorithms. The lowest testing RMSE is obtained by SRLMP-ELM with $p=4.0$, $\rho=0$. The similar accuracy and more compact model are obtained by SRLMP-ELM.

6.3 BLS Models under MMPE

The BLS, like the ELM, has random weights and biases from the input layer (or feature layer) to the middle layer, but obtains the weights and biases from the middle layer to the output layer by solving the pseudo-inverse. However, there are some key differences between the two models. Firstly, the BLS has a connection between the feature layer (or input layer) and the output layer, while the ELM does not. Second, the input layer of the BLS is a direct input feature, whereas the input layer of the ELM is the data themselves. The standard BLS uses the MMSE as the default optimization criterion for training the network output weights. However, this can lead to degraded performance in complicated noise environments, particularly when data are contaminated by outliers. To address this issue, alternative optimization criteria, such as l_1-norm with different regularization terms and MCC, have been proposed to train the output weights of the BLS. This chapter will focus on two robust BLS models, including BLS under MMPE and BLS under mixture MPE [254,255], which can perform well in both Gaussian and non-Gaussian noise environments.

6.3.1 Broad Learning System

Here the original BLS model is first briefly reviewed. The basic architecture of BLS can be seen from Figure 6.16. Herein, $\mathbf{X}=\left[\mathbf{x}_1^T,\mathbf{x}_2^T,\ldots,\mathbf{x}_N^T\right]^T \in \mathbb{R}^{N\times d}$ is the input pattern matrix, $\mathbf{Y}=\left[\mathbf{y}_1^T,\mathbf{y}_2^T,\ldots,\mathbf{y}_N^T\right]^T \in \mathbb{R}^{N\times C}$ is the output matrix. Based on $\mathbf{X}$, we first obtain M groups of feature pattern matrices: $\mathbf{Z}_1,\mathbf{Z}_2,\ldots,\mathbf{Z}_M$. Each of them has the form of

$$\mathbf{Z}_i=\phi_i(\mathbf{X}\mathbf{W}_{ei}+\beta_{ei}), i=1,2,\ldots,M \tag{6.164}$$

where $\mathbf{W}_{ei}$ and β_{ei} are weights and biases, respectively. They can be generated randomly and then slightly tuned by a sparse auto-encoder. The ϕ_i is usually a linear transformation. Concatenating all feature pattern matrices together, we have

$$\mathbf{Z}^M=\left[\mathbf{Z}_1,\mathbf{Z}_2,\ldots,\mathbf{Z}_M\right] \tag{6.165}$$

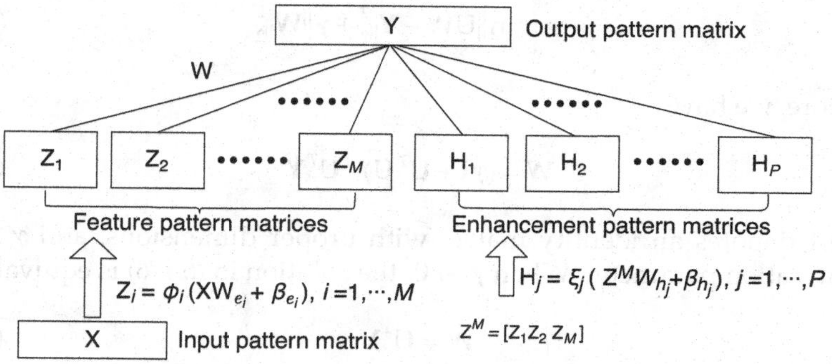

FIGURE 6.16
Basic architecture of broad learning system.

Further, based on $\mathbf{Z}^M$, we obtain P groups enhancement pattern matrices: $\mathbf{H}_1, \mathbf{H}_2, \ldots, \mathbf{H}_P$. Every enhancement pattern matrix is obtained by

$$\mathbf{H}_j = \xi_j(\mathbf{Z}^M \mathbf{W}_{hj} + \beta_{hj}), i = 1, 2 \ldots, P \tag{6.166}$$

where $\mathbf{W}_{hj}$ and β_{hj} are also randomly generated weights and biases, respectively; ξ_j is an activation function with various selections, such as the commonly used sigmoidal function and tangent function can be considered. These enhancement pattern matrices can be cascaded into one, that is

$$\mathbf{H}^P = [\mathbf{H}_1, \mathbf{H}_2, \ldots, \mathbf{H}_P] \tag{6.167}$$

By concatenating $\mathbf{Z}^M$ and $\mathbf{H}^P$, a new input pattern matrix $\mathbf{U}$ is obtained in the following way

$$\mathbf{U} = [\mathbf{Z}^M, \mathbf{H}^P] \in \mathbb{R}^{N \times L} \tag{6.168}$$

where L is the number of columns of $\mathbf{U}$ with its value depending on the total number of hidden layer nodes of the system. Finally, the estimated outputs in the matrix form can be expressed by

$$\hat{\mathbf{Y}} = \mathbf{U}\mathbf{W} \tag{6.169}$$

where $\mathbf{W} \in \mathbb{R}^{L \times c}$ is a matrix to store output weights. Since all the feature pattern matrices and enhancement pattern matrices are fixed during training, the learning task is now reduced to that of estimating the optimal $\mathbf{W}$. In the standard BLS, the following l_2-norm-based optimization model is applied to determine $\mathbf{W}$

$$\arg\min_{\mathbf{W}} \|\mathbf{UW}-\mathbf{Y}\|_2^2 + \gamma\|\mathbf{W}\|_2^2 \tag{6.170}$$

Therefore, we have

$$\mathbf{W} = (\gamma\mathbf{I} + \mathbf{U}^T\mathbf{U})^{-1}\mathbf{U}^T\mathbf{Y} \tag{6.171}$$

where $\mathbf{I}$ denotes an identity matrix with proper dimensions, and γ is the regularization parameter. When $\gamma \to 0$, the solution in (6.176) is equivalent to

$$\mathbf{W} = \mathbf{U}^{+}\mathbf{Y} \tag{6.172}$$

where $\mathbf{U}^{+} = \lim_{\gamma\to 0}(\gamma\mathbf{I} + \mathbf{U}^T\mathbf{U})^{-1}\mathbf{U}^T$ denotes the pseudo-inverse of $\mathbf{U}$. For regression, $C=1$, and the $\mathbf{W}$ is a vector with dimension $L\times 1$; for classification, $C>1$, and the $\mathbf{W}$ is a matrix with dimension $L\times C$.

The original BLS model employs the least square with the regularization method to obtain the optimal $\mathbf{W}$, which enhances the training process but compromises the robustness of the model. This is due to the use of MSE as the cost function, which usually fails to effectively mitigate the bad impact of non-Gaussian noise interferences or outliers. To overcome this limitation, the MPE and mixture MPE criteria can be used as alternative costs to develop more robust BLS models.

6.3.2 BLS under MMPE Criterion

6.3.2.1 Derivation of Least p-norm Based BLS

This section will introduce the BLS under MPE, called least p-norm-based BLS(LP-BLS), and the derivation of LP-BLS is summarized as:

First, we consider C = 1 case, i.e., the general regression problem. When C = 1, the output is a vector with dimension $N \times 1$ as [254]

$$\mathbf{y} = \left[y_1, y_2, \ldots, y_N\right]^T \in \mathbb{R}^{N\times 1} \tag{6.173}$$

Meanwhile, the weight w to be solved is a vector with $L \times 1$ dimension. Here, we define the estimated error vector as

$$e = \left[e_1, \ldots, e_n \ldots e_N\right] \tag{6.174}$$

where $e_n = y_n - \boldsymbol{u}_n^T w$ denotes the estimate error of the nth sample. $\boldsymbol{u}_n \in \mathbb{R}^{1\times L}$ is the nth row vector in the transformed input matrix $\mathbf{U} \in \mathbb{R}^{N\times L}$, $y_n \in \mathbb{R}$ represents the nth element of the output vector $\mathbf{y}$. According to the definition of the MPE criterion, and combining the regularization term $\frac{\lambda}{2}\|w\|_2^2$ yields the following optimal problem

$$\arg\min_{w} = \left(\sum_{n=1}^{N} |e_n|^p + \frac{\lambda}{2} \|w\|_2^2 \right) \tag{6.175}$$

Further, we can define the cost function as

$$J(w) = \sum_{n=1}^{N} |e_n|^p + \frac{\lambda}{2} \|w\|_2^2 \tag{6.176}$$

To solve the problem in (6.175) by using the gradient descent method, we compute the gradient of the (6.176) as

$$\begin{aligned} \frac{\partial J(w)}{\partial w} &= -\sum_{n=1}^{N} p|e_n|^{p-1} sign(e_n) u_n + \lambda w \\ &= \sum_{n=1}^{N} p|e_n|^{p-2} (u_n w - y_n) u_n + \lambda w \\ &= \mathbf{U} \Lambda_w^- (\mathbf{U} w - \mathbf{y}) + \lambda w \end{aligned} \tag{6.177}$$

where

$$\Lambda_w^- = pdiag\left[|e_1|^{p-1}, |e_2|^{p-1}, \ldots |e_N|^{p-1} \right] \tag{6.178}$$

Then we have

$$w = (\mathbf{U}^T \Lambda_w^+ \mathbf{U} + \rho \mathbf{I})^{-1} \mathbf{U}^T \Lambda_w^+ \mathbf{y} \tag{6.179}$$

where

$$\begin{aligned} \rho &= \frac{\lambda}{p} \\ \Lambda_w^+ &= \frac{\Lambda_w^-}{p} \end{aligned} \tag{6.180}$$

A weighted diagonal matrix Λ_w^+ is introduced into (6.179) compared with (6.171) When p=2, The solutions in (6.179) will degenerate into a classical regularization least-squares solution due to $\Lambda_w^+ = \mathbf{I}$. As we all know, using regularized least squares for solving the output weight is the optimal choice under Gaussian noise, but it is difficult to suppress the negative effects of non-Gaussian noises or outliers. In this case, the robustness of the system can be enhanced by change the p value. For example, when some training samples are disturbed by outliers, the corresponding estimation error is also very large (the amplitude is usually larger than or even much larger than 1). Then, setting $1<p<2$, the amplitude of the diagonal elements corresponding to the abnormal sample can be limited below 1, which can partly weaken the negative impact of the abnormal data. In the following, considering the case of $C>1$ and C as an integer, and it corresponds to the general classification problem. Then the system output $\mathbf{Y}$ is a matrix with $N \times C$ dimension, and the corresponding weight $\mathbf{W}$ to be solved is a matrix with $L \times C$ dimension. According to the principle of matrix segmentation, the $\mathbf{Y}$ and $\mathbf{W}$ are represented as the following block matrix form:

$$\mathbf{Y} = \left[\mathbf{y}_1^T, \mathbf{y}_2^T, \ldots, \mathbf{y}_C^T\right]^T \in \mathbb{R}^{N\times C} \tag{6.181}$$

$$\mathbf{W} = \left[w_1, w_2, \ldots w_C\right]^T \in \mathbb{R}^{L\times C} \tag{6.182}$$

where $\mathbf{y}_1^T, \mathbf{y}_2^T, \ldots, \mathbf{y}_C^T$ are column vectors with $N\times 1$ dimension, $w_1, w_2, \ldots w_C$ are column vectors with $L\times 1$ dimension. Using the results in (6.179), we have

$$w_c = (\mathbf{U}^T \Lambda_{w_c}^{+} \mathbf{U} + \rho \mathbf{I})^{-1} \mathbf{U}^T \Lambda_{w_c}^{+} y_c, \quad c = 1, 2, \ldots, C \tag{6.183}$$

Inserting (188) into (6.182), we can get the output weight **W.** One can easily find that both (6.179) and (6.183) are an implicit equation related to w or w_c when $p \neq 2$, and thus we need to select an effective strategy to find their true solutions. Fixed-point iteration [349] is a method of solving the solution of the implicit equations in an iterative way, and here we use the fixed-point iterative strategy to solve (6.179) and (6.183). Taking (6.179) as an example. Let

$$f(w) = (\mathbf{U}^T \Lambda_{w}^{+} \mathbf{U} + \rho \mathbf{I})^{-1} \mathbf{U}^T \Lambda_{w}^{+} \mathbf{y} \tag{6.184}$$

Then, the fixed-point iterative equation can be expressed as

$$w(t) = f(w(t-1)) \tag{6.185}$$

where $w(t)$ is the solution of the output weight at iteration time t. In general, an initialized weight vector $w(0) \in \mathbb{R}^{L\times 1}$ should be given randomly, and setting a maximum upper limit value of iterations such that one can achieve the goal of the starting and terminating the iteration process.

6.3.2.2 Simulation Results

6.3.2.2.1 Regression Problem

We first also use the common task of fitting the SinC function curve to verify the regression performance of the LP-BLS. In this case, 500 training samples and testing samples are generated by the SinC model, and the outputs of the training samples are added to different noise interferences, including (1) Gaussian noise with zero mean and variance 0.01; (2) uniform noise distributed on [−0.4, 0.4]; (3) impulsive noise. The performance of the LP-BLS is compared with that of the original BLS, and testing RMSE (TRMSE) is used as the evaluation criterion. For the construction of hidden layer nodes, only one set of enhancement transformations is used, i.e. $m=1$. Meanwhile, the grid search method used in [350,351] is employed to determine the number of feature mapping groups(k),the number of feature nodes corresponding to each group of feature mapping (q), and the number of enhancement nodes corresponding to each group of enhancement transformation(r). The search ranges for these three parameters are 1: 1:10, 1: 1:10, and 1: 2:100, respectively. In addition, the regularization parameters for all the algorithms are fixed to 2^{-30}. Table 6.12 shows the TRMSE and related parameters settings for BLS and LP-BLS in different noise interference environments. According to the results shown in Table 6.12, one knows that the LP-BLS can achieve the best TRMSE when the value of the p is set at 2, 2.5, and 1.5 for Gaussian, uniform, and impulsive noise environments, respectively. But for BLS, its performance will degrade significantly under impulsive noise interference cases. These results suggest that, in comparison with BLS, LP-BLS can be a better choice when the statistical properties of noise are unknown.

6.3.2.2.2 EEG Classification

The data set IVa from BCI competition III was utilized for binary EEG classification application. This data set comprises EEG signals from 5 healthy subjects who performed motor-imagery tasks involving their right hand and foot. The training trials for subjects aa, al, av, aw, and ay were 168, 224, 84, 56, and 28, respectively, while their corresponding test trials were 112, 56, 196, 224, and 252, respectively. Similar to [250], the time segments from 0.5 to 2.5 seconds after visual cues that triggered motor-imagery activities were selected as trials. The common spatial pattern (CSP) [352] was then applied to extract features from the filtered EEG signals. As recommended in [250],

TABLE 6.12

TRMSE and Node Parameters Setting of BLS and LP-BLS under Different Noises

Method	Gaussian		Uniform		Impulsive	
	TRMSE	(k,q,r)	TRMSE	(k,q,r)	TRMSE	(k,q,r)
BLS	0.0135	(4,2,41)	0.0261	(2,3,39)	0.1885	(1,7,87)
LP-BLS(p=1.5)	0.0137	(7,1,35)	0.0318	(2,2,83)	0.0071	(7,1,87)
LP-BLS(p=2.0)	0.0135	(4,2,41)	0.0261	(2,3,39)	0.1885	(1,7,87)
LP-BLS(p=2.5)	0.0142	(3,2,99)	0.0236	(2,3,39)	0.4760	(2,1,1)
LP-BLS(p=3.0)	0.0174	(5,2,41)	0.0351	(4,2,79)	0.2810	(4,2,39)

TABLE 6.13
Best Classification Results of Different Algorithms on Dataset IVa of BCI Competition III

Method	Subjects Number					Mean Value (%)
	aa	al	av	aw	ay	
LDA	66.07	96.43	47.45	71.88	49.60	66.29
KNN	64.29	98.21	50.00	58.93	53.97	65.08
SVM	66.96	98.21	43.88	62.95	50.40	64.48
ELM	68.66	97.86	50.10	62.54	49.52	65.74
BLS	66.70	96.96	50.15	67.46	50.91	66.44
LP-BLS	67.14	98.93	50.36	72.95	51.31	68.14

three pairs of filters were used to construct the feature vectors, resulting in each feature vector having six dimensions. The obtained feature vectors, along with their corresponding labels from the training set, were used to train the classifier.

In this case, the performance of the LP-BLS is compared with that of the original BLS, Linear Discriminant Analysis (LDA) [353], *k*-Nearest Neighbor (KNN) [354], Support Vector Machine (SVM) [355], ELM for the task of the classification. To obtain the best classification accuracy of the corresponding algorithm, the grid search method is used to determine their parameter configurations. Table 6.13 presents the best classification accuracy obtained by the different algorithms on the public dataset. The classification accuracy corresponding to each subject in the table was obtained by averaging based on ten Monte Carlo runs. As shown in Table 6.13, LP-BLS achieved the highest classification accuracy on all three subjects except aa and ay. Although BLS shows stronger classification ability than LDA, KNN, SVM, and ELM, the classification accuracy is still lower than LP-BLS.

6.3.3 BLS under Mixture MMPE Criterion

To enhance the performance of the standard BLS in non-Gaussian noise environments, the mixed-norm-based BLS (MN-BLS) is introduced in this section.

6.3.3.1 Model Formulation

The mixed-norm-based BLS model can be formulated as [255]

$$\arg\min_{\mathbf{W}} \lambda\|\mathbf{Y}-\mathbf{UW}\|_1 + (1-\lambda)\|\mathbf{Y}-\mathbf{UW}\|_2^2 + \gamma_1\|\mathbf{W}\|_1 + \gamma_2\|\mathbf{W}\|_2^2 \tag{6.186}$$

where $\gamma_2 > 0$ is the regularization parameter for alleviating the issue of overfitting, $\gamma_1 > 0$ is also the regularization parameter but it is used for encouraging the weight values to be sparse, and $\lambda \in [0, 1]$ is the mixed parameter which is key to make the BLS keep the good performance in various noise environments. For example, when we set $\lambda = 1$, (6.186) is actually an l_1-norm-based optimization model with double regularization terms. As we know, it is robust to outliers. When $\lambda = 0$, (6.191) degenerates to the traditional l_2-norm optimization model with double regularization terms and is expected to get a satisfactory classification performance in Gaussian noise or noise-free environments.

6.3.3.2 Model Optimization

Since the l_1-norm is introduced, the objective function in (6.186) cannot be solved analytically. To address this computational problem, the ALM method is adopted here. First, we write (6.186) as

$$\begin{aligned} &\arg\min_{\mathbf{W}} \lambda \|\mathbf{E}\|_1 + (1+\lambda)\|\mathbf{F}\|_2^2 + \gamma_1 \|\mathbf{Q}\|_1 + \gamma_2 \|\mathbf{W}\|_2^2 \\ &s.t. \qquad \mathbf{E} = \mathbf{Y} - \mathbf{U}\mathbf{W}, \mathbf{E} = \mathbf{F}, \mathbf{Q} = \mathbf{W} \end{aligned} \tag{6.187}$$

Then, the augmented Lagrangian function of (6.187) can be expressed by

$$\begin{aligned} &\arg\min_{\mathbf{W}} \lambda \|\mathbf{E}\|_1 + (1+\lambda)\|\mathbf{F}\|_2^2 + \gamma_1 \|\mathbf{Q}\|_1 + \gamma_2 \|\mathbf{W}\|_2^2 \\ &+\mathbf{C}_1^T(\mathbf{Y} - \mathbf{U}\mathbf{W} - \mathbf{E}) + \mathbf{C}_2^T(\mathbf{E} - \mathbf{F}) + \mathbf{C}_3^T(\mathbf{W} - \mathbf{Q}) + \\ &\frac{\mu}{2}\left(\|\mathbf{Y} - \mathbf{U}\mathbf{W} - \mathbf{E}\|_2^2 + \|\mathbf{E} - \mathbf{F}\|_2^2 + \|\mathbf{W} - \mathbf{Q}\|_2^2\right) \end{aligned} \tag{6.188}$$

where $\mathbf{C}_1$, $\mathbf{C}_2$, and $\mathbf{C}_3$ are Lagrange multipliers; $\mu \geq 0$ is the penalty parameter. According to ALM, the unknown variables **W**, **E**, **F**, **Q**, $\mathbf{C}_1$, $\mathbf{C}_2$, and $\mathbf{C}_3$ can be updated in an alternatively iterative manner. Now, we give the details of how to optimize them with the ALM method.

i) Optimize **W**: Fix the other variables and remove the irrelevant terms. Then, the optimization of **W** is equivalent to

$$\begin{aligned} &\arg\min_{\mathbf{W}} \gamma_2 \|\mathbf{W}\|_2^2 + \mathbf{C}_1^T(\mathbf{Y} - \mathbf{U}\mathbf{W} - \mathbf{E}) + \mathbf{C}_3^T(\mathbf{W} - \mathbf{Q}) + \\ &\frac{\mu}{2}\left(\|\mathbf{Y} - \mathbf{U}\mathbf{W} - \mathbf{E}\|_2^2 + \|\mathbf{W} - \mathbf{Q}\|_2^2\right) \end{aligned} \tag{6.189}$$

After some simple calculations, the update equation for **W** can be obtained by

$$\mathbf{W} = \left[(2\gamma_2/\mu + 1)\mathbf{I} + \mathbf{U}^T\mathbf{U}\right]^{-1}\left(\mathbf{U}^T\mathbf{Y} + \mathbf{P}\right) \tag{6.190}$$

where $\mathbf{P} = \mathbf{Q} - \mathbf{U}^T\mathbf{E} + (\mathbf{U}^T\mathbf{C}_1 - \mathbf{C}_3)/\mu$.

ii) Optimize **E**: By fixing the other variables and removing the irrelevant terms, we establish the optimization problem for **E** as

$$\arg\min_{\mathbf{E}} \lambda\|\mathbf{E}\|_1 + \mathbf{C}_1^T(\mathbf{Y} - \mathbf{U}\mathbf{W} - \mathbf{E}) + \mathbf{C}_2^T(\mathbf{E} - \mathbf{F}) + \frac{\mu}{2}\left(\|\mathbf{Y} - \mathbf{U}\mathbf{W} - \mathbf{E}\|_2^2 + \|\mathbf{E} - \mathbf{F}\|_2^2\right) \tag{6.191}$$

Further, the update equation for **E** has the form of

$$\mathbf{E} = \frac{1}{2} S_{\lambda/\mu}\left(\mathbf{C}_1/\mu - \mathbf{C}_2/\mu + \mathbf{Y} - \mathbf{U}\mathbf{W} + \mathbf{E}\right) \tag{6.192}$$

where $S_{\lambda/\mu}$ is the soft thresholding operator. Specifically, if we use $\odot$ to denote the element-wise multiplication of two matrices, S can be described by

$$S_\kappa(\mathbf{A}) = \max\{|\mathbf{A}| - \kappa, 0\} \odot \operatorname{sign}(\mathbf{A}) \tag{6.193}$$

iii) Optimize **F**: When the variables **W**, **E**, **Q**, $\mathbf{C}_1$, $\mathbf{C}_2$, and $\mathbf{C}_3$ are fixed and the irrelevant terms are removed, the optimization of **F** can be achieved by solving

$$\arg\min_{\mathbf{F}} (1+\lambda)\|\mathbf{F}\|_2^2 + \gamma_1\|\mathbf{Q}\|_1 + \mathbf{C}_2^T(\mathbf{E} - \mathbf{F}) + \frac{\mu}{2}\left(\|\mathbf{E} - \mathbf{F}\|_2^2\right) \tag{6.194}$$

Then, the update equation for **F** is

$$\mathbf{F} = 1/(1 - 2\lambda + \mu)(\mathbf{C}_2 + \mu\mathbf{E}) \tag{6.195}$$

iv) Optimize **Q**: When the variables **W**, **E**, **F**, $\mathbf{C}_1$, $\mathbf{C}_2$, and $\mathbf{C}_3$ are fixed and the irrelevant terms are removed, the optimization of **Q** is equivalent to

$$\arg\min_{\mathbf{Q}} \gamma_1\|\mathbf{Q}\|_1 + \mathbf{C}_3^T(\mathbf{W} - \mathbf{Q}) + \frac{\mu}{2}\left(\|\mathbf{W} - \mathbf{Q}\|_2^2\right) \tag{6.196}$$

Hence, one can obtain

$$\mathbf{Q} = S_{\gamma 1/\mu}(\mathbf{W} + \mathbf{C}_3/\mu) \tag{6.197}$$

v) Update parameters $\mathbf{C}_1$, $\mathbf{C}_2$, *and* $\mathbf{C}_3$: At every iteration, $\mathbf{C}_1$, $\mathbf{C}_2$, and $\mathbf{C}_3$ can be updated by

$$\begin{aligned} \mathbf{C}_1 &= \mathbf{C}_1 + \mu(\mathbf{Y} - \mathbf{U}\mathbf{W} - \mathbf{E}) \\ \mathbf{C}_2 &= \mathbf{C}_2 + \mu(\mathbf{E} - \mathbf{F}) \\ \mathbf{C}_3 &= \mathbf{C}_3 + \mu(\mathbf{W} - \mathbf{Q}) \end{aligned} \tag{6.198}$$

So far, we have given the update equations for **W**, **E**, **F**, **Q**, $\mathbf{C}_1$, $\mathbf{C}_2$ and $\mathbf{C}_3$. These equations should be utilized in an alternatively iterative manner until the convergence condition is satisfied or the iteration reaches its maximum. Finally, the MN-BLS is summarized in Table 6.14, in which the N_f, N_w and N_e are the number of feature mapping groups, the number of feature nodes in each feature mapping group, and the number of enhancement nodes, respectively.

6.3.3.3 Simulation Results

Here the MN-BLS is applied to both binary and multiclass motor-imagery classifications to evaluate its performance. Other than the BLS, two commonly used classifiers including naive Bayes [356] and ELM [169] were also chosen as the benchmark for comparison. Similar to [184] and [188], the appropriate grid research method was adopted to determine the network parameters for BLS, MN-BLS, and ELM, while the regularization parameters in them are set to 2^{-30} empirically.

6.3.3.3.1 Binary EEG Classification

The data set IVa of BCI competition III is considered again in the binary EEG classification application. Here the subjects of aa, al, av, aw, and ay are denoted as S01, S02, S03, S04, and S05, respectively. Other simulation settings are the same as the Section 6.3.2. The classification results of different classifiers are shown in Table 6.15. It can be seen that, although ELM and BLS perform better on subjects S01, S03, and S04, their overall performances are worse than MN-BLS.

6.3.3.3.2 Multiclass EEG Classification

To further test the performance of MN-BLS on multiclass EEG classification cases, the data set IIa of BCI competition IV [357] is chosen as the benchmark

TABLE 6.14

Mixed-Norm-Based BLS

Input: Training set {**X**,**Y**}. **Output**: Output weight matrix **W**.

1. **Parameters setting**: Network parameters: N_f, N_w, N_e; mixed parameter: λ; regularization parameters: γ_1, γ_2; termination tolerance ε; maximum iteration number T.
2. **Initialization**: Set **E**, **Q**, **F**, **W**, **C1**, **C2** and **C3** to be zero matrices with appropriate dimensions;
3. **for** $t = 1; \ldots; T$ **do**
4. Update W according to (6.195);
5. Update E according to (6.197);
6. Update F according to (6.200);
7. Update Q according to (6.202);
8. Update C1, C2 and C3 according to (6.203);
9. Until $\|\mathbf{W}(t+1) - \mathbf{W}(t)\|_2^2 < \varepsilon$.
10. end for

TABLE 6.15

Test Classification Accuracies of Different Classifiers on the Data Set IVa of BCI Competition III

Subject	Classifier			
	Naive Bayes	ELM	BLS	MN-BLS
S01	65.18	68.66	66.70	67.50
S02	98.21	97.86	96.96	99.46
S03	46.94	50.10	50.15	50.00
S04	60.71	62.54	67.46	64.91
S05	52.38	49.56	50.91	53.61
Mean	64.69	65.75	66.44	67.10

data set. It comprises EEG signals from nine subjects who performed left hand, right hand, foot, and tongue motor-imageries. There are 22 electrodes for EEG recordings. For each subject, a training and a testing set are available, and both of them contain 72 trials for every class. Without loss of generality, we also chose the time segments from 0.5 to 2.5 seconds after visual cues as trials, and each trial was bandpass filtered in 8–30 Hz with a fifth-order Butterworth filter. To extract the features from the filtered signals effectively and reliably, the one-versus-one (OVO) approach [358] was combined with CSP. As one can see from the classification results of different classifiers in Table 6.16, the MN-BLS can achieve higher classification accuracy on most subjects, and it also has the highest average classification accuracy on all subjects.

TABLE 6.16

Test Classification Accuracies of Different Classifiers on the Data Set IIa of BCI Competition IV

Subject	Classifier			
	Naive bayes	ELM	BLS	MN-BLS
S01	68.75	72.05	74.24	**75.94**
S02	**46.53**	43.23	42.67	41.67
S03	70.14	74.06	**79.58**	73.54
S04	58.33	54.41	58.85	**59.72**
S05	27.43	35.97	39.38	**39.86**
S06	41.67	42.78	42.88	**43.02**
S07	60.07	**75.80**	70.73	72.12
S08	67.01	73.61	80.49	**81.98**
S09	74.65	75.94	75.31	**80.49**
Mean	57.18	60.87	62.68	**63.15**

7

Adaptive Filtering Algorithms under Mixture MMPE

Many AFAs that use a single norm as the cost function to update weight vectors may suffer from poor convergence performance, particularly the LMS and LMF algorithms, as their convergence properties are highly sensitive to the proximity of adaptive weights to optimal solutions [359]. To address this sensitivity issue and improve misadjustment performance, a convex combination of different norms can provide an efficient approach. The least mean mixed-norm (LMMN) algorithm [274, 360,361] is a typical example that combines the benefits of the LMS and LMF algorithms and has been successfully applied to system identification, echo cancellation, and power quality enhancement in distribution systems. To solve system identification under non-Gaussian noises, researchers have proposed robust mixed-norm (RMN) and normalized RMN algorithms in [275,362], which are modifications of the LMMN algorithm that replace the fourth-order error norm with the first-order one. In addition, several kernel or diffusion AFAs have been developed by researchers [268,363–365] to take advantage of the mixed norm. All of these algorithms can be viewed as special cases of the mixture MPE criterion (or general mixture-norm (GMN) criterion) based AFAs [268].

This chapter introduces the concept of the GMN criterion, which involves a convex linear combination of MPE with p and q norms, controlled by a scalar-mixing parameter. This approach is utilized to develop different mixed-norm AFAs [268]. Furthermore, the chapter presents linear, sparsity aware, diffusion, and kernel adaptive algorithms based on GMN [268], each with different p and q values. These algorithms are designed for sparse system identification, distributed estimation, and nonlinear time series prediction.

7.1 Adaptive Filtering Algorithm under Mixture MMPE

To obtain an AFA based on GMN, we consider the system identification problem aforementioned. We seek the optimal weight by minimizing the mixture MPE criterion in (3–24). The instantaneous estimate of the gradient of the mixture MMPE is

DOI: 10.1201/9781003176114-7

$$\begin{aligned}\frac{\partial J_{GMN}(\mathbf{w}(k))}{\partial \mathbf{w}(k)} &= -\omega \,|\, e(k) \,|^{p-1} \operatorname{sign}(e(k))\mathbf{u}(k) - (1-\omega) \,|\, e(k) \,|^{q-1} \operatorname{sign}(e(k))\mathbf{u}(k) \\ &= -(\omega \,|\, e(k) \,|^{p-1} + (1-\omega) \,|\, e(k) \,|^{q-1}) \operatorname{sign}(e(k))\mathbf{u}(k) \\ &= -H(k) \operatorname{sign}(e(k))\mathbf{u}(k)\end{aligned} \tag{7.1}$$

where $H(k)$ is a key factor to the algorithm based on GMN. By using the stochastic gradient descent method, one can obtain the following update equation

$$\begin{aligned}\mathbf{w}(k) &= \mathbf{w}(k-1) - \mu \frac{\partial J_{GMN}(\mathbf{w}(k))}{\partial \mathbf{w}(k)} \\ &= \mathbf{w}(k-1) + \mu H(k) \operatorname{sign}(e(k))\mathbf{u}(k)\end{aligned} \tag{7.2}$$

where μ is the step size. The adaptive algorithm in (7.2) is referred to as the GMN algorithm [268].

Remark 7.1

The GMN algorithm can be viewed as a generalized version of some existing adaptive filtering algorithms (e.g. LMS, LMF, LMP, LAD, LMMN, and RMN), and it will reduce to these algorithms by choosing certain parameters. Table 7.1 shows that different p and q result in different AFAs.

7.2 Special Cases of Linear Adaptive Filtering Algorithms under Mixture MMPE

In this subchapter, the LMMN and RMN algorithms as the special cases of the mixture MMPE (or GMN) based algorithms are presented. Furthermore, a mixed controlled $l_2 - l_p$ adaptive filtering algorithm is introduced, which can also be viewed as a generalized version of the LMMN and RMN algorithms.

TABLE 7.1

Different p and q Result in Different Adaptive Filtering Algorithms

	LMS	LMF	LAD	LMMN	RMN	GMN
p	2	4	1	2	2	$p>1$
q	2	4	1	4	1	$q>1$

7.2.1 LMMN Adaptive Filtering Algorithm

The LMMN algorithm is initially introduced as a convex, linear combination of the norms, which is regulated by a scalar-mixing parameter [360,361]. In addition, the convergence analysis is presented using the averaging analysis and the total stability theorem [359]. To obtain a misadjustment expression, a second-order asymptotic analysis is conducted through the method of ordinary differential equation. The optimal choice for the mixture parameter is revealed by examining the variation of the misadjustment. Notably, the derivation does not rely on the restrictive assumption that the adaptive weights are statistically independent of the input signal. The theoretical misadjustment expression is experimentally verified to be accurate.

7.2.1.1 LMMN Algorithm

Adaptive filtering applied to a system identification problem in the presence of additive measurement noise $v(k)$ is illustrated in Figure 4.1. The input signal $\mathbf{u}(k)$ is sent to the adaptive filter and the unknown system is assumed to be time-invariant and to have an FIR structure. The objective is to minimize the difference between the desired signal $d(k)$ and the output of the adaptive filter according to some optimization criteria. When the SNR at the output of the unknown system is low, $\tilde{v}(k)$ significantly disturbs the adaptive weights $\mathbf{w}(k) \in \mathbb{R}^N$, and leads to poor performance. The degradation in performance is closely linked to the statistical features of $\tilde{v}(k)$ [200].

The cost function minimized in the LMMN algorithm is a linear mixture of $J_2(k) \triangleq E\{e^2(k)\}$ and $J_4(k) \triangleq 1/4E\{e^4(k)\}$ as follows [359]:

$$J(k) = \frac{\delta}{2} J_2(k) + \frac{1-\delta}{4} J_4(k) \tag{7.3}$$

where $\delta \in [0,1]$ is the mixture parameter which can be shown to be convex in $\mathbf{w}(k)$. The gradient vector which defines the search direction is

$$\nabla J(k) \triangleq \frac{\delta J(k)}{\delta \mathbf{w}(k)} = -E\{e(k)\{\delta + (1-\delta)e^2(k)\}\mathbf{u}(k)\} \tag{7.4}$$

Now a stochastic gradient algorithm based on an instantaneous estimate of $\nabla J(k)$ can be derived, leading to the update equation of the LMMN algorithm is

$$\mathbf{w}(k+1) = \mathbf{w}(k) + \mu e(k)\{\delta + (1-\delta)e^2(k)\}\mathbf{u}(k) \tag{7.5}$$

where μ is the step size. Note that, if $\delta = 1$, (7.5) reduces to the LMS update and if $\delta = 0$, the LMF update is got, and with intermediate values of δ, the two algorithms are merged.

7.2.1.2 Local Exponential Stability of the LMMN Algorithm

The stability analysis is carried out by using the deterministic total stability theorem (TST) [366], the concept of slow time variation and averaging analysis [366–368]. To facilitate the following feature analysis of the LMMN algorithm, some notations and assumptions on the signals are introduced as [359]:

The *a priori* error is

$$e(k) = v(k) - \mathbf{u}^T(k)\tilde{\mathbf{w}}(k) \tag{7.6}$$

The mismatch of the unknown system and $\mathbf{w}_o$ produces the error

$$e_{\text{opt}}(k) = d(k) - \mathbf{u}^T(k)\mathbf{w}_o \tag{7.7}$$

Therefore, the effective measurement noise due to $e_{\text{opt}}(k)$ and $\tilde{v}(k)$ is given by

$$v(k) = e_{\text{opt}}(k) + \tilde{v}(k) \tag{7.8}$$

The following assumptions on the signals are made to facilitate convergence analysis.

Assumption 7.1: $\mathbf{u}(k)$, $d(k)$ and $v(k)$ are zero-mean and statistically stationary.

Assumption 7.2: $\mathbf{u}(k)$ and $d(k)$ are jointly stationary.

Assumption 7.3: $v(k)$ is symmetrically distributed, $\sigma_v^2 \triangleq E\{v^2(k)\}$, $\zeta_v^4 \triangleq E\{v^4(k)\}$ and $\xi_v^6 \triangleq E\{v^6(k)\}$ exist.

Assumption 7.4: $\mathbf{u}(k)$ and $v(k)$ are uniformly bounded:

$$\chi \triangleq \sup_k \|\mathbf{u}(k)\|_\infty < \infty \quad \varpi \triangleq \sup_k |v(k)| < \infty \tag{7.9}$$

Assumption 7.5: $\mathbf{u}(k)$ and $v(k)$ are statistically independent.

The update equation (7.8) falls into the category of recursive updates of the form

$$\mathbf{w}(k+1) = \mathbf{w}(k) + \mu F\{\mathbf{u}(k)\} g\{e(k)\} \tag{7.10}$$

where $F(\cdot): \mathbb{R}^N \rightarrow \mathbb{R}^N$ and $g(\cdot): \mathbb{R} \rightarrow \mathbb{R}$ are given by

$$F\{\mathbf{u}(k)\} = \mathbf{u}(k) \tag{7.11}$$

$$g\{e(k)\} = e(k)\{\delta + (1-\delta)e^2(k)\} \tag{7.12}$$

This general update law is analyzed inf a deterministic setting in [319,320]. The analysis in [368] emphasizes the importance of the SiGn-preservation property (SGPP) that $g\{e(k)\}$ must satisfy, which ensures that the adaptive algorithm will always descend on the error-performance surface (EPS).

From (7.12), $g\{e(k)\}$ is clearly sign preserving. Therefore, one can at least anticipate local exponential stability (LES) of the LMMN algorithm for sufficiently small μ.

The TST enables us to perform a local-stability analysis of nonlinear systems and has been stated in different forms in [367,368]. It argues that if the exponential stability (ES) of the linearized system can be established, the nonlinear system will be LES and small perturbations around the optimum, which might be caused by noise or other nonidealities, will not cause instability.

7.2.1.2.1 Linearization

The linearization system does not involve any noise-related terms [368], so we work with $\tilde{g}_{\mathbf{uw}}(k) \triangleq g\{e(k)\}|_{d=0}$. Let $\mathbf{w}_*$ be a stable stationary point on the EPS. The Jacobian matrix corresponding to $F\{\mathbf{u}(k)\}g\{e(k)\}$ at $\mathbf{w} = \mathbf{w}_*$ is defined by

$$B_{\mathbf{w}_*}(k) \triangleq F\{u(k)\} \left. \frac{\partial \tilde{g}_{\mathbf{uw}}^T(k)}{\partial \mathbf{w}(k)} \right|_{\mathbf{w}(k)=\mathbf{w}_*} \tag{7.13}$$

and the corresponding linearized system around $\mathbf{w}_*$ is given by

$$\mathbf{w}(k+1) = \{I - \mu B_{\mathbf{w}_*}(k)\}\mathbf{w}(k) \tag{7.14}$$

For convenience, we rewrite (7.14) as

$$\mathbf{w}(k+1) = \left[I - \mu c(k, \mathbf{u}, \mathbf{w}_*) F\{\mathbf{u}(k)\}\mathbf{u}^T(k)\right]\mathbf{w}(k) \tag{7.15}$$

In the LMMN algorithm, $F\{\mathbf{u}(k)\} = \mathbf{u}(k)$ and

$$\tilde{g}_{\mathbf{uw}}(k) = -\mathbf{u}^T(k)\mathbf{w}(k)[\delta + (1-\delta)\{\mathbf{u}^T(k)\mathbf{w}(k)\}^2] \tag{7.16}$$

By substituting (7.16) into (7.13), we have

$$B_{\mathbf{w}_*}(k) = c(k, \mathbf{u}, \mathbf{w}_*)\mathbf{u}(k)\mathbf{u}^T(k) \tag{7.17}$$

where

$$c(k, \mathbf{u}, \mathbf{w}_*) = [\delta + 3(1-\delta)\{\mathbf{u}^T(k)\mathbf{w}_*\}^2] \tag{7.18}$$

7.2.1.2.2 Slow Time Variation and Averaging

The slow-time-variation lemma [366] indicates that, in (7.18), if μ is small, the time variation of $\{I - \mu B_{\mathbf{w}_*}(k)\}$ is also slow. This enables us to work with the time-invariant system $B_{\mathbf{w}_*}(k)$ for each k. $B_{\mathbf{w}_*}(k)$ is a rank-1 matrix due to the outer product $\mathbf{u}(k)\mathbf{u}^T(k)$. However, the averaging theorem [366] states that, if the averaged system

$$\bar{\mathbf{w}}(k+1) = \{I - \mu \bar{B}_{\mathbf{w}*}(k)\}\bar{\mathbf{w}}(k) \tag{7.19}$$

where

$$\bar{B}_{\mathbf{w}*}(k) = \frac{1}{K}\sum_{s=0}^{K-1} B_{\mathbf{w}*}(k+s) \tag{7.20}$$

is exponentially stable, so too is (7.17).

Let us state the theorem in [368] in a slightly different form.

Theorem 7.1

Consider the general update (7.10) for a K-periodic [366] input signal $\mathbf{u}(k)$ and let $\mathbf{w}_*$ be a stable stationary point of (7.10). Assume that $F(\cdot)$ and $g(\cdot)$ are memoryless, sign preserving and $g(\cdot)$ is differentiable at $\mathbf{w} = \mathbf{w}_*$. If there exists $\beta > \alpha > 0$ and $p > 0$ such that

$$\beta > \mathrm{Re}\left[\lambda_i\left\{\sum_{s=0}^{K-1} c(k+s,\mathbf{u},\mathbf{w}_*)\mathbf{u}(k+s)\mathbf{u}^T(k+s)\right\}\right] > \alpha > 0 \tag{7.21}$$

then there is a $\mu \in (0,\mu^*)$ such that the linearized system of (7.18) is exponentially stable and convergent to $\mathbf{w}_*$.

Clearly the LMMN algorithm has a stable stationary point at $\mathbf{w}_* = \mathbf{0}_N$. Taking into account the switching capability of the LMMN algorithm to the LMF algorithm for $\delta = 0$, we have two extreme cases.

Case 1 ($\delta \in (0,1]$): By substituting $\mathbf{w}_* = \mathbf{0}_N$ into (7.19), the ES condition of the linearized system in (7.21) becomes

$$\beta > \mathrm{Re}\left[\lambda_i\left\{\sum_{s=0}^{K-1} \mathbf{u}(k+s)\mathbf{u}^T(k+s)\right\}\right] > \alpha > 0 \tag{7.22}$$

which is the deterministic persistency-of-excitation (PE) condition for the LMS algorithm.

The following corollary from [368] directly applies for the other case ($\delta = 0$)

Corollary 7.1

Consider the update in (7.10) and the corresponding linearized system in (7.15) for a K-periodic [353] input signal $\mathbf{u}(k)$ and assume that $F(\cdot)$ and $g(\cdot)$ satisfy the requirements in Theorem 7.1. Suppose that $c(k,\mathbf{u},\mathbf{0}_N) = 0$

Define the connected region $D = \{\mathbf{w} : \|\mathbf{w}\| < r, r > 0\}$. If $c(k,\mathbf{u},\mathbf{w})$ is continuous and nondecreasing in D and if for some $\beta > \alpha > 0$ and $p > 0$, then there exists a $\mu \in (0,\mu^*)$ such that the linearized system (7.15) is exponentially stable in D.

Case 2 ($\delta = 0$, LMF algorithm): Substituting $\theta_* = \mathbf{0}_N$ $\delta = 0$, into (7.19), we have $c(k,\mathbf{u},\mathbf{0}_N) = 0$. From Corollary 7.1, the ES condition for this case is the same as (7.22), but the convergence is to a region around $\theta_* = \mathbf{0}_N$ rather than $\mathbf{w}_* = \mathbf{0}_N$ itself.

7.2.1.2.3 *Total Stability*

The LES of the LMMN algorithm is tied to the ES of the linearized system (7.15) and (7.19), through the total stability theorem [366]. For the LMMN algorithm, it is necessary to show that the regularity conditions in the TST are satisfied in the presence of noise. Let us state the TST in [366] for convenience.

Theorem 7.2 (total stability)

Consider the ordinary difference equation

$$\mathbf{w}(k+1) = A(k)\mathbf{w}(k) + H\{k,\mathbf{w}(k)\} + L\{k,\mathbf{w}(k)\} \tag{7.23}$$

where $A(k) \in \mathbb{R}^{N\times N}$ is the transition matrix of the linearized system, $\mathbf{w}(k) \in \mathbb{R}^N$. Let $D = \{\mathbf{w} : \|\mathbf{w}\| < r, r > 0\}$. Assume that $\forall \mathbf{w} \in D, A(k), H, \{k,\mathbf{w},(k)\}$ and $L\{k,\mathbf{w}(k)\}$ are bounded functions of time. If $\forall k \geq 0, \forall(\mathbf{w},\mathbf{w}_1,\mathbf{w}_2) \in D, \exists(\beta_1,\beta_2) > 0$

$$(B1)\, H(k,\mathbf{0}_N) = \mathbf{0}_N \tag{7.24}$$

$$(B2)\, \|H(k,\mathbf{w}_1) - H(k,\mathbf{w}_2)\| \leq \beta_1 \|\mathbf{w}_1 - \mathbf{w}_2\| \tag{7.25}$$

$$(B3)\, \|L(k,\mathbf{w})\| \leq \beta_2 r \tag{7.26}$$

$$(B4)\, \|L(k,\mathbf{w}_1) - L(k,\mathbf{w}_2)\| \leq \beta_2 \|\mathbf{w}_1 - \mathbf{w}_2\| \tag{7.27}$$

and the linearized system

$$\mathbf{w}(k+1) = A(k)\mathbf{w}(k) \tag{7.28}$$

is exponentially stable, then (7.23) is locally exponentially stable in D.

In (7.23), $H\{k,\mathbf{w}(k)\}$ and $L\{k,\mathbf{w}(k)\}$ represent the difference between the actual update and the linearized update (7.15) around $\mathbf{w}_* = \mathbf{0}_N$. Such terms are due to the nonidealities in the adaptation such as noise or mild nonlinearities. In the LMMN algorithm, from (7.10) to (7.12), and (7.19), we have

$$H\{k,\mathbf{w}(k)\}+L\{k,\mathbf{w}(k)\}=\mu[g\{e(k)\}\mathbf{u}(k)-c(k,\mathbf{u},\mathbf{0}_N)\mathbf{u}(k)\mathbf{u}^T(k)\mathbf{w}(k)] \tag{7.29}$$

By using (7.6) in (7.29), we can identify the terms of $H\{k,\mathbf{w}(k)\}$ and $L\{k,\mathbf{w}(k)\}$ as follows:

$$H\{k,\mathbf{w}(k)\}=-\mu(1-\delta)\{\mathbf{u}(k)^T\mathbf{w}(k)\}$$
$$\left[3v^2(k)-3v(k)\mathbf{w}(k)^T\mathbf{w}(k)+\{\mathbf{u}(k)^T\mathbf{w}(k)\}^2\right]\mathbf{u}(k) \tag{7.30}$$

$$L\{k,\mathbf{w}(k)\}=\mu\{\delta+(1-\delta)v^2(k)\}v(k)\mathbf{u}(k) \tag{7.31}$$

The regularity conditions (7.24)–(7.27) are shown to be valid in Appendix I, and therefore the LMMN algorithm has been shown to be LES.

7.2.1.3 Steady-State Behavior of LMMN

The steady-state performance of the LMMN algorithm is clearly a function of the mixture parameter δ. The steady-state performance is quantified in terms of the misadjustment, which is defined as

$$M=\frac{1}{\sigma_w^2}\lim_{k\to\infty}E\left[\{\mathbf{u}^T(k)\mathbf{w}(k)\}^2\right] \tag{7.32}$$

While deriving the misadjustment, it is assumed that the adaptive weights fluctuate around their optimal values in a small region. It is also helpful to assume that the step size is small. Under these conditions we have from (7.6),

$$e^3(k)\simeq v^3(k)-3v^2(k)\mathbf{u}^T(k)\mathbf{w}(k) \tag{7.33}$$

which simplifies the analysis [51]. Therefore, the update in (7.5) can be approximated as

$$\mathbf{w}(k+1)\simeq\left[I-\mu\{\delta+3(1-\delta)v^2(k)\}\mathbf{u}(k)\mathbf{u}^T(k)\right]\mathbf{w}(k)$$
$$+\mu\{\delta+(1-\delta)v^2(k)\}v(k)\mathbf{u}(k) \tag{7.34}$$

Let us define the weight-error-correlation matrix as $\Theta(k) \triangleq \mathbf{w}(k)\mathbf{w}^T(k)$. By using (7.34), a recursion for $\Theta(k)$ can be computed by

$$\begin{aligned}\Theta(k+1) = &\left[I-\mu\left\{\delta+3(1-\delta)v^2(k)\right\}\mathbf{u}(k)\mathbf{u}^T(k)\right]\Theta(k)\\ &\times\left[I-\mu\left\{\delta+3(1-\delta)v^2(k)\right\}\mathbf{u}(k)\mathbf{u}^T(k)\right]\\ &+\mu\left\{\delta+(1-\delta)v^2(k)\right\}v(k)\\ &\times\left[I-\mu\left\{\delta+3(1-\delta)v^2(k)\right\}\mathbf{u}(k)\mathbf{u}^T(k)\right]\theta(k)\mathbf{u}^T(k)\\ &+\mu\left\{\delta+(1-\delta)v^2(k)\right\}v(k)\mathbf{u}(k)\theta^T(k)\\ &\times\left[I-\mu\left\{\delta+3(1-\delta)v^2(k)\right\}\mathbf{u}(k)\mathbf{u}^T(k)\right]\\ &+\mu^2\left\{\delta+(1-\delta)v^2(k)\right\}^2 v^2(k)\mathbf{u}(k)\mathbf{u}^T(k)\end{aligned} \tag{7.35}$$

The ordinary differential equation (ODE) method in [369,370] can be used to associate the stochastic difference equation in (7.35) with a deterministic differential equation. The ODE method assumes that the step size is very small. During the conversion to a differential equation, the second and third terms on the right-hand side of (7.35) disappear due to Assumptions 7.3 and 7.5. We also ignore $\mu^2\mathbf{u}(k)\mathbf{u}^T(k)\Theta(k)\mathbf{u}(k)\mathbf{u}^T(k)$, which is implicit in the first time, since the element $\Theta(k)$ are very small and $\mu^2 \ll 1$. This approximation is more formally discussed in [370]. Therefore, we consider

$$\begin{aligned}\Theta(k+1)-\Theta(k) = -\mu[\{\delta+3(1-\delta)v^2(k)\}\\ \times\{\Theta(k)\mathbf{u}(k)\mathbf{u}^T(k)+\mathbf{u}(k)A^T(k)\Theta(k)\}\\ +\mu\{\delta+(1-\delta)v^2(k)\}^2 v^2(k)\mathbf{u}(k)\mathbf{u}^T(k)]\end{aligned} \tag{7.36}$$

Let us define $t_k \triangleq \mu k$ and make the ODE conversion [356] to obtain

$$\begin{aligned}\dot{\Theta}(t+\mu) = \{\delta+3(1-\delta)\sigma_v^2\}\{\Theta(t)R+R\Theta(t)\}\\ +\mu E\{[\delta+(1-\delta)v^2(k)]^2 v^2(k)\}R\end{aligned} \tag{7.37}$$

where $\Theta(t)$ is the continuous-time derivative of $\Theta(t)$ and $R \triangleq E\{\mathbf{u}(k)\mathbf{u}^T(k)\}$ is the input-signal auto-correlation matrix. We have

$$\lim_{t\to\infty}\Theta(t+\mu) = \mathbf{0}_{N\times N} \tag{7.38}$$

Therefore, as $t \to \infty$, (7.37) becomes

$$\{\delta+3(1-\delta)\sigma_v^2\}\{\Theta(\infty)R+R\Theta(\infty)\}$$
$$=\mu\mathrm{E}\{[\delta+(1-\delta)v^2(k)]^2v^2(k)\}R \tag{7.39}$$

whose solution is

$$\Theta(\infty)=\frac{\mu\delta^2\sigma_v^2+2\delta(1-\delta)\zeta_v^4+(1-\delta)^2\xi_u^6}{\delta+3(1-\delta)\sigma_v^2}I \tag{7.40}$$

The above expression indicates that the elements of $\Theta(k)$ are asymptotically uncorrelated. In this case, the definition of the misadjustment of (7.32) is simplified by $(I/\sigma_v^2)E\{u^2(k)\}\,\mathrm{tr}[\Theta(\infty)]$ and therefore, by using (7.40). The misadjustment of the LMMN algorithm is given by

$$M=\frac{\mu N}{2\sigma_v^2}\frac{\delta^2\sigma_v^2+2\delta(1-\delta)\zeta_v^4+(1-\delta)^2\xi_v^6}{\delta+3(1-\delta)\sigma_v^2}E\{u^2(k)\} \tag{7.41}$$

Note that (7.41) reduces to the misadjustment of the LMS algorithm for $\delta=1$ and to the misadjustment of the LMF algorithm for $\delta=0$.

By using (7.41), one can also obtain an expression for the asymptotic value of the LMMN cost function in (7.3), which is defined as

$$J_{\min}\triangleq\lim_{k\to\infty}\frac{\delta}{2}J_2(k)+\frac{1-\delta}{4}J_4(k) \tag{7.42}$$

For this purpose, we write

$$E\{e^2(k)\}=\sigma_v^2+E\{[\mathbf{u}^T(k)\mathbf{w}(k)]^2\} \tag{7.43}$$

and make the approximation

$$E\{e^4(k)\}=\zeta_v^4+6\sigma_v^2E\{[\mathbf{u}^T(k)\mathbf{w}(k)]^2\} \tag{7.44}$$

similar to (7.13) and after invoking Assumptions 7.3 and 7.5. By substituting (7.43) and (7.44) into (7.42) and using (7.32), we obtain

$$J_{\min}\approx\frac{\delta}{2}\sigma_v^2+\frac{1-\delta}{4}\zeta_v^4+\left\{\frac{\delta}{2}+\frac{3(1-\delta)}{2}\sigma_v^2\right\}\sigma_v^2M \tag{7.45}$$

where M is given in (7.41). The accuracy of (7.41) is verified in the following simulations.

7.2.1.4 Simulation Results

7.2.1.4.1 Testing Misadjustment Against δ

Here the unknown system impulse response is chosen as {0.1, 0.2, 0.3, 0.4, 0.5, 0.4, 0.3, 0.2, 0.1}. The step size of the LMMN algorithm is $\mu=1\times10^{-4}$. The mixing parameter is chosen as 10 equispaced points in [0, 1]. The input signal is zero-mean and uniformly distributed with unity power. The noise signal is also zero-mean and it is obtained by adding a Gaussian distributed noise of power $\sigma_{v1}^2 = 0.1$ and a uniform distributed noise of power $\sigma_{v2}^2 = 0.1$. The SNR is 9.29 dB. The corresponding theoretical values are obtained from (7.41). The variation of the misadjustment with respect to δ is shown in Figure 7.1, where it is clear that (7.41) is a reliable indicator of the misadjustment of the LMMN algorithm. Furthermore, the misadjustment curve has a well defined minimum for $\delta = 0.41$. Therefore, by choosing $\delta = 0.41$, it is expected that the LMMN algorithm will perform better than the LMS ($\delta = 0$) and LMF ($\delta = 1$) algorithms.

Some results supporting the above observation are reported in [274]. The noise signal is created as an additive mixture of two components with different statistical distributions. The experimental results indicate that there can be a value of $\delta \in [0,1]$ for which the misadjustment is a minimum depending on the statistical properties of the noise signal.

7.2.1.4.2 Performance Comparison

This case further considers the unknown system of the previous experiment to be identified by using the LMS, LMF, and LMMN algorithms. The input signal $\mathbf{u}(k)$ is zero-mean and uniformly distributed with unity power. The noise signal $v(k)$ is obtained by adding two noise signals, $v_1(k)$ and $v_2(k)$. The noise-signal distributions, SNR, and the adaptation parameters μ and δ are presented in Table 7.2. The experimental values of the misadjustment are

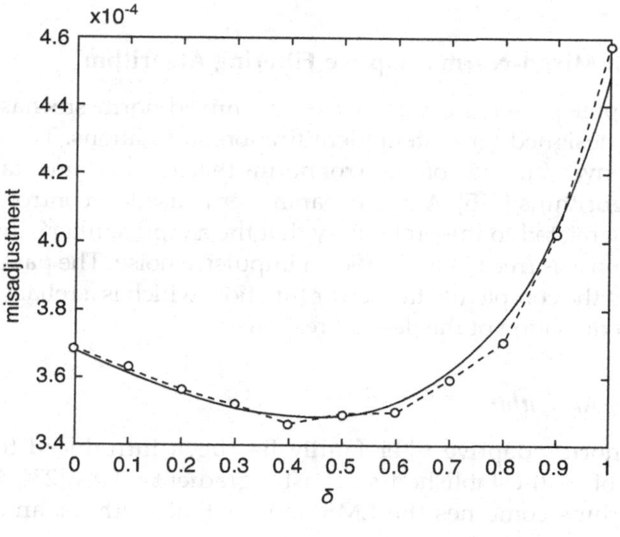

FIGURE 7.1
The variation of the misadjustment with respect to δ [359].

TABLE 7.2
Misadjustments of LMMN, LMS, and LMF Algorithms [359]

$w_1(k)$	$w_2(k)$	SNR (dB)	Algorithm	δ	μ	Misadjustment (dB)
Gaussian	Uniform	−1.30	LMS	1	2×10^{-2}	−9.89
$\sigma_{w1}^2=0.15$	$\sigma_{w2}^2=1.00$		LMF	0	8×10^{-3}	−8.29
			LMMN	0.41	8×10^{-3}	−10.82
Gaussian	Bernoulli	−1.58	LMS	1	2×10^{-2}	−9.94
$\sigma_{w1}^2=0.20$	$\sigma_{w2}^2=1.00$		LMF	0	8×10^{-3}	−10.95
			LMMN	0.83	8×10^{-3}	−13.60
Laplacian	Uniform	−1.06	LMS	1	2×10^{-2}	−10.07
$\sigma_{w1}^2=0.08$	$\sigma_{w2}^2=1.00$		LMF	0	8×10^{-3}	−9.61
			LMMN	0.50	8×10^{-3}	−11.62
Laplacian	Bernoulli	−1.58	LMS	1	2×10^{-2}	−9.73
$\sigma_{w1}^2=0.20$	$\sigma_{w2}^2=1.00$		LMF	0	5×10^{-3}	−12.57
			LMMN	0.40	5×10^{-3}	−14.23

computed from (7.35) by averaging over 100 Monte Carlo trials. The results in Table 7.2 demonstrate clearly the potential of the LMMN algorithm for practical problems when the measurement noise has mixed statistical features. The corresponding convergence behavior of the adaptive weights is illustrated in Figure 7.2, where the system mismatch, which is defined as the trace of the weight-error-correlation matrix (i.e. $\mathrm{tr}[\Theta(\infty)]$), is shown. The LMMN algorithm exhibits superior steady-state behavior than the LMS and LMF algorithms.

7.2.2 Robust Mixed-Norm Adaptive Filtering Algorithm

This subchapter presents a member of the mixed-norm stochastic gradient AFA family, designed for system identification applications. The algorithm is based on a convex function of the error norms that are fundamental to the LMS and LAD algorithms [275]. A scalar parameter is used to control the mixture and is closely related to the probability that the adaptive filter's instantaneous desired response is free from significant impulsive noise. The parameter is calculated using the complementary error function, which is a reliable estimate of the standard deviation of the desired response.

7.2.2.1 RMN Algorithm

The mixed-norm adaptive filter family has been introduced to merge the advantages of well-established stochastic gradient AFAs [274, 361]. One of these algorithms combines the LMS and LMF algorithms, and its benefits

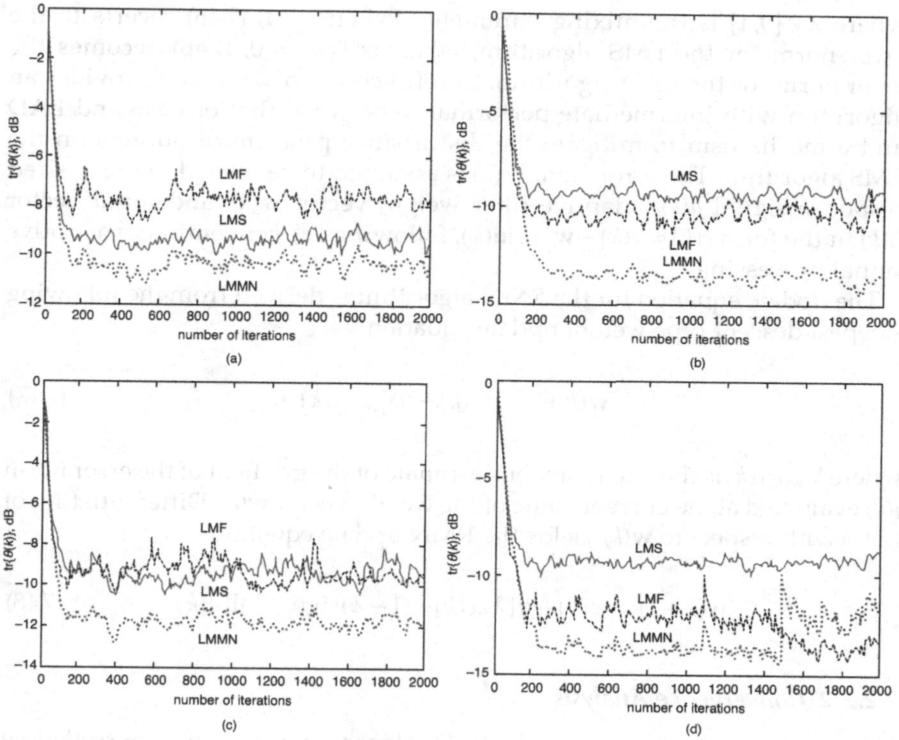

FIGURE 7.2
System mismatch against number of iterations for: (a) Gaussian uniform noise; (b) Gaussian-Bernoulli mixture noise; (c) Laplacian-uniform noise; (d) Laplacian-Bernoulli noise mixture [359].

have been confirmed in [359]. In this study, we present a new mixed-norm AFA that is specifically designed for system identification applications where the desired response is corrupted by impulsive or heavy-tailed distribution noise. The mean value of $v(k)$ is zero, and its variance is determined by $\text{var}\{v(k)\} = \sigma_v^2$ [275].

This estimate will serve as a guide for the RMN adaptation algorithm [275], which combines the conventional LMS and LAD algorithms, also known as the sign-error LMS or pilot LMS [371], with stochastic gradient algorithms. The rationale behind this approach is that, for identical initial convergence rates, the LMS algorithm generally yields a more precise final solution with less misadjustment when there is no impulsive noise. However, it is highly sensitive to outliers. On the other hand, the LAD algorithm is less accurate with higher misadjustment, but it is more robust to outliers when the adaptation gain is fixed [372]. The algorithm is based on minimizing a convex combination of error norms.

$$J(k) = \lambda E\{e^2(k)\} + (1-\lambda)E\{|e(k)|\} \tag{7.46}$$

where $\lambda \in [0,1]$ is the mixing parameter. When $\lambda = 1$, (7.46) reverts to the error norm for the LMS algorithm, whereas for $\lambda = 0$, (7.46) becomes the error norm for the LAD algorithm. Careful choice of λ thereby provides an algorithm with intermediate performance between that of LMS and LAD and a mechanism to mitigate the disturbance problem of outliers on the LMS algorithm. The error signal $e(k)$ is assumed to be related to the desired response signal $d(k)$, adaptive filter weight vector $\mathbf{w}(k)$, and input vector $\mathbf{u}(k)$ in the form $e(k) = d(k) - \mathbf{w}^T(k)\mathbf{u}(k)$, following the convention in adaptive signal processing.

The update equation for the RMN algorithm is derived from the following steepest descent type weight update equation

$$\mathbf{w}(k+1) = \mathbf{w}(k) - \mu \hat{\nabla}_{\mathbf{w}(k)} J(k) \tag{7.47}$$

where $\hat{\nabla}_{\mathbf{w}(k)} J(k)$ is the instantaneous estimate of the gradient of the error norm $J(k)$ evaluated at the current value of the weight vector $\mathbf{w}(k)$. Differentiating of (7.46) with respect to $\mathbf{w}(k)$ yields the RMN update equation

$$\mathbf{w}(k+1) = \mathbf{w}(k) + \mu\{\lambda 2e(k) + (1-\lambda)\,\mathrm{sign}[e(k)]\}\mathbf{u}(k) \tag{7.48}$$

7.2.2.2 Convergence Analysis

The convergence properties of the RMN algorithm are mainly controlled by the adaptation gain parameter μ and mixing parameter λ. An analysis based on [373], in which $E\{\mathrm{sign}[e(k)]\mathbf{u}(k)\} = \sqrt{(2/\pi)}[1/\sigma_e(k)]E\{e(k)\mathbf{u}(k)\}$, for small μ, where $\sigma_e(k)$ is the standard deviation of the error sequence, yields the following sufficient conditions on μ for convergence of the mean

$$0 < \mu < \frac{2}{\left(2\lambda + [1-\lambda]\sqrt{\dfrac{2}{\pi\sigma_v^2}}\right) N\sigma_u^2} \tag{7.49}$$

where σ_v^2 and σ_u^2 are, respectively, the measurement and input noise powers, and N is the length of the adaptive filter. The RMN algorithm, with fixed λ requires at each iteration only two more additions than the LMS algorithm and may be initialized with a null vector. To complete the definition of the RMN algorithm, it is necessary to relate the outlier trimmed sliding window standard deviation estimate to the mixing parameter λ. The time-varying nature of the standard deviation estimate implies that the mixing parameter is a function of the sample index k. The selection of $\lambda(k)$ is based upon the following probability, where $d(k)$ has a symmetric distribution, and d_0 is positive:

$$\begin{aligned}\lambda(k) &\equiv \text{Prob}\{d(k) > d_0 \cup d(k) < -d_0\} \\ &= 2(1 - \text{Prob}\{d(k) \le d_0\}) \\ &= 2\left[1 - \int_{-\infty}^{d_0} f_D(x)\,dx\right] \\ &= 2\left[\frac{1}{2} - \int_0^{d_0} f_D(x)\,dx\right]\end{aligned} \tag{7.50}$$

where $f_D(x)$ is the distribution of the desired response. Specializing this to the case that $d(k)$ has a zero-mean Gaussian distribution with standard deviation equal to the estimate as

$$\hat{\sigma}_d(k) = \sqrt{\frac{1}{N_w - 3}\underline{o}^T T \underline{o}} \tag{7.51}$$

where the N_w most recent values of the desired response are ordered in terms of their amplitudes, i.e., $\underline{d} = [d(k), d(k-1), \ldots, d(k - N_w + 1)]^t$, $\underline{o} = \text{ord}(\underline{d})$ in which the elements of $\underline{o}$ are algebraically ordered from smallest to largest, the minimum and maximum values of which are discounted by application of a diagonal trimming matrix, $T = \text{diag}[0, 1, 1, \ldots, 1, 1, 0]$ that nullifies the first and last elements of $\underline{o}$. Then we have

$$\begin{aligned}\lambda(k) &= 2\left[\frac{1}{2} - \int_0^{|d(k)|/\hat{\sigma}_d} \frac{1}{\sqrt{2\pi}} e^{-x^2/2}\,dx\right] \\ &= 2\text{erfc}\left[\frac{|d(k)|}{\hat{\sigma}_d}\right]\end{aligned} \tag{7.52}$$

where erfc(·) is the complementary error function. The probability that the instantaneous desired response contains significant impulsive noise is then approximated as $[1 - \lambda(k)]$. The underlying concept is that the LMS algorithm is progressively replaced by the LAD algorithm as the likelihood of an outlier increases. The value of this can be supported by considering the weight propagation equation for the RMN algorithm. With the definition of the weight error vector $\tilde{\mathbf{w}}(k)$, (7.49) becomes

$$\tilde{\mathbf{w}}(k+1) = [I - 2\mu\lambda(k)\mathbf{u}(k)\mathbf{u}(k)^t]\tilde{\mathbf{w}}(k) + \{2\mu\lambda(k)v(k) + \mu[1 - \lambda(k)]\text{sign}[e(k)]\}\mathbf{u}(k) \tag{7.53}$$

If the adaptive filter is assumed to be close to the optimal solution at time sample k, i.e., $\tilde{\mathbf{w}}(k) \approx 0$, the arrival of a large impulsive noise component, $v(k)$, implies that the second term will dominate the right-hand side of (7.53), so that

$$\tilde{\mathbf{w}}(k+1) \approx \{2\mu\lambda(k)v(k) + \mu[1 - \lambda(k)]\text{sign}[v(k)]\}\mathbf{u}(k) \tag{7.54}$$

For the LMS case, i.e., $\lambda(k)=1$, $E\{\tilde{\mathbf{w}}(k+1)^T\tilde{\mathbf{w}}(k+1)\}=4\mu^2N\sigma_v^2\sigma_u^2$, whereas when the LAD algorithm is used, $\lambda(k)=0$, $E\{\tilde{\mathbf{w}}(k+1)^t\tilde{\mathbf{w}}(k+1)\}=\mu^2N\sigma_u^2$, and the weight error vector norm is independent of the impulsive noise statistics. Although the form of (7.52) is not well suited to real-time implementation, it could easily be replaced by a look-up table, or the calculation of $\lambda(k)$ could be restricted to those sample instances for which the ratio $|d(k)|/\hat{\sigma}_d$ is large.

7.2.2.3 Simulation Results

Here a system identification simulation results are demonstrated to show the performance of the RMN algorithm. The desired response signal is formed by inputting white Gaussian distributed noise of unit power to a FIR filter, and to test the algorithms severely, independent Gaussian distributed noise of fixed variance is added to its output so that the effective SNR is 0 dB prior to the addition of the impulsive noise. The desired responses produced in this manner are shown in Figure 7.3. The length of the sliding window standard deviation estimator N_w is 10, so that the probability that more than one significant impulse lies within the window is very small. The adaptation gain for the LMS algorithm is set at 0.01, whereas for the RMN algorithm, it is 0.018 and 0.04 for the LAD algorithm. Such setting of the parameters can guarantee that in simulation the initial convergence rates of the three

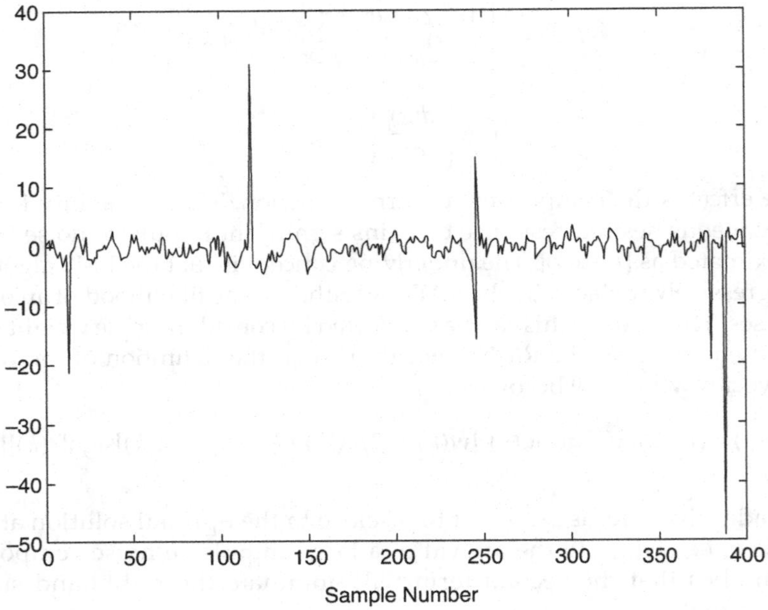

FIGURE 7.3
Desired responses signal contaminated with the impulsive noise [275].

algorithms were visually identical when no impulsive noise was presented. This is demonstrated in Figure 7.4, which shows the log normalized weight error vector norm averaged across ten independent trials. It is clear that the performance of the RMN algorithm is almost identical to that of LMS, both much improved on that of the LAD algorithm. Finally, in Figure 7.5, the log normalized weight error vector norm is shown for simulations with the existence of impulsive noises and constant for each trial. It is evident how the behavior of the conventional LMS algorithm deteriorates the performance, whereas the RMN algorithm is unaffected and remains much improved than LAD algorithm.

7.2.3 A Mixed l_2-l_p Adaptive Filtering Algorithm

AFAs designed through the minimization of equation l_2-l_4 have a disadvantage when the absolute value of the error is greater than one. This will make the algorithm go unstable unless either a small value of the step size or a large value of the controlling parameter is chosen such that this unwanted instability disappears. We know that the l_p-norm ($p\neq 2$), performs better than the MSE criterion when the noise statistics are non-Gaussian. l_p-norm-based minimization algorithms for signal parameter estimation or minimization

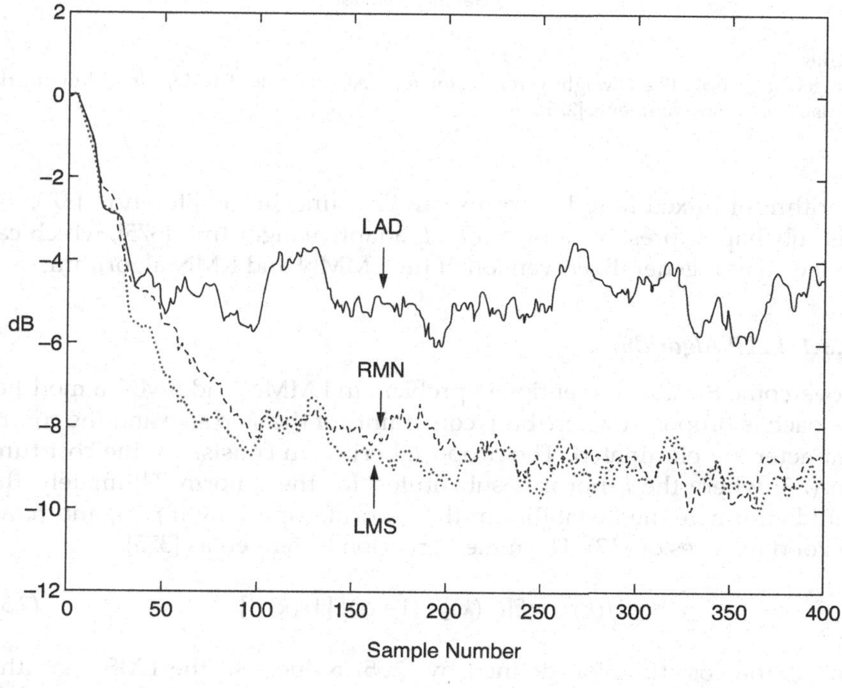

FIGURE 7.4
Averaged log normalized weight error vector for LAD, LMS, and RMN algorithms [275].

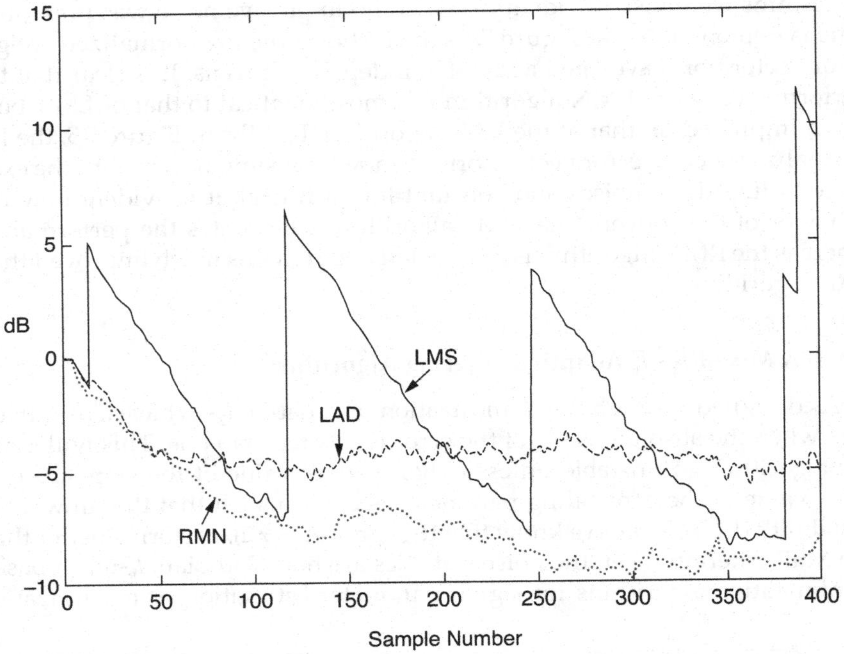

FIGURE 7.5
Averaged log normalized weight error vector for LAD, LMS, and RMN algorithms in the impulsive noise environments [275].

algorithm of mixed l_1 and l_2 norms can be found in the literature [373,374]. This subchapter presents a mixed $l_2 - l_p$ adaptive algorithm [375], which can be viewed as a generalized version of the LMMN and RMN algorithms.

7.2.3.1 L2LP Algorithm

To overcome the above-mentioned problem in LMMN and RMN, a modified approach is proposed where both constraints of the step size and the control parameter are eliminated. The proposed criterion consists of the cost function (7.3) where the l_p-norm is substituted for the l_4-norm. Ultimately, this should eliminate the instability in the l_4-norm, especially if $p<4$, and retain the good features of (7.3). The mixed criterion is defined as [375]:

$$J(k) = \alpha E[e^2(k)] + (1-\alpha)E[\,|e(k)|^p] \tag{7.55}$$

If $p = 2$, the cost function defined by (7.55) reduces to the LMS algorithm whatever the value of α in the range [0, 1] for which the unimodality of the cost function is preserved.

For $\alpha = 0$, the algorithm reduces to the l_p-norm adaptive algorithm, and if $p = 1$ the RMN algorithm is obtained.

For $p<2$, l_p gives less weight for larger error and this tends to reduce the influence of aberrant noise, while it gives relatively larger weight to smaller errors and this will improve the tracking capability of the algorithm.

Using the gradient descent method, the updating scheme of the L2LP algorithm based on (7.55) is given by:

$$\mathbf{w}(k+1) = \mathbf{w}(k) + \mu[\alpha e(k) + p(1-\alpha)\,|\,e(k)\,|^{(p-1)}\ \text{sign}(e(k))]\mathbf{u}(k) \qquad (7.56)$$

and sufficient condition for convergence in the mean of this algorithm can be shown to be given by:

$$0 < \mu < \frac{2}{\{\alpha + p(p-1)(1-\alpha)E[\,|\,v(k)\,|^{p-2}]\}\,\text{Tr}\{\mathbf{R}\}} \qquad (7.57)$$

where tr{$\mathbf{R}$} is the trace operation of the auto-correlation matrix $\mathbf{R}$.

In general, the step size is chosen small enough to ensure the convergence and produce less misadjustment error. The error surface of (7.55) is analyzed as:

i. Case $p = 2$: Let the input auto-correlation matrix be $\mathbf{R} = E[\mathbf{u}(k)\mathbf{u}^T(k)]$, and the cross-correlation vector that describes the cross-correlation between the received signal ($\mathbf{u}(k)$) and the desired data, $\mathbf{p} = E[\mathbf{u}(k)d(k)]$. The error function can be more conveniently expressed as follows:

$$J(k) = \sigma_u^2 - 2\mathbf{w}^T(k)\mathbf{p} + \mathbf{w}^T(k)\mathbf{R}\mathbf{w}(k) \qquad (7.58)$$

It is clear from (7.58) that the MSE is precisely a quadratic function of the components of the tap coefficients, and the shape associated with it is hyperparaboloid. The adaptive process continuously adjusts the tap coefficients, seeking the bottom of this hyperparaboloid.

$$\mathbf{w}_o = \mathbf{R}^{-1}\mathbf{p} \qquad (7.59)$$

ii. Case $p \neq 2$: It can be shown as well that the error-function for the (7.55) will have a global minimum since the latter one is a convex function. As in the (7.58), the adaptive process will continuously seek the bottom of the error function of the (7.55).

7.2.3.2 Convergence Analysis

Usual assumptions, which can be found in the literature [5, 261] and can also be justified in many practical instances, are used during the convergence analysis of the mixed controlled L2LP algorithm.

7.2.3.2.1 First Moment Behavior of the Weights

We start by evaluating the statistical expectation of both sides of (7.56) which looks after subtracting $\mathbf{w}_o$ of both sides to give [375]:

$$\tilde{\mathbf{w}}(k+1) = \tilde{\mathbf{w}}(k) + \mu(\alpha e(k) + (1-\alpha)\,\mathrm{sign}(e(k)))\mathbf{u}(k) \tag{7.60}$$

After substituting the error $e(k)$ and taking the expectation of both sides results, we obtain

$$E[\tilde{\mathbf{w}}(k+1)] = [\mathbf{I} - \alpha\mu\mathbf{R}]E[\tilde{\mathbf{w}}(k)] + \mu(1-\alpha)E[\mathrm{sign}(e(k))\mathbf{u}(k)] \tag{7.61}$$

It is to show that the misalignment vector will converge to the zero vector if the step-size, μ, is given by

$$0 < \mu < \frac{2}{\left[\alpha + (1-\alpha)\sqrt{\dfrac{2}{\pi J_{\min}}}\right]\lambda_{\max}}$$

where $\lambda_{\max}$ is the largest eigenvalue of the auto-correlation matrix $\mathbf{R}$, since in general $\mathrm{Tr}\{\mathbf{R}\} \gg \lambda_{\max}$ and $J_{\min}$ is the minimum MSE.

7.2.3.2.2 Second Moment Behavior of the Weights

From (7.60), the following expression can be got:

$$\begin{aligned}\tilde{\mathbf{w}}(k+1)\tilde{\mathbf{w}}^T(k+1) = \tilde{\mathbf{w}}(k)\tilde{\mathbf{w}}^T(k) &+ \mu[\alpha e(k) + (1-\alpha)\,\mathrm{sign}(e(k))] \\ &\times\left[\tilde{\mathbf{w}}(k)\mathbf{u}^T(k) + \mathbf{u}(k)\tilde{\mathbf{w}}^T(k)\right] + \mu^2 \\ &\times\left[\alpha^2 e^2(k) + 2\alpha(1-\alpha)|e(k)| + (1-\alpha)^2\right]\mathbf{u}(k)\mathbf{u}^T(k)\end{aligned} \tag{7.62}$$

Let $\mathbf{K}(k) = E[\tilde{\mathbf{w}}(k)\tilde{\mathbf{w}}^T(k)]$ define the second moment of the misalignment vector. Therefore the above equation, after taking the expectation of both sides of it, becomes as follows:

$$\begin{aligned}\mathbf{K}(k+1) = \mathbf{K}(k) &+ \mu\alpha\{E[\tilde{\mathbf{w}}(k)\mathbf{u}^T(k)e(k)] + E[\mathbf{u}(k)\tilde{\mathbf{w}}^T(k)e(k)]\} \\ &+ \mu(1-\alpha)\{E[\tilde{\mathbf{w}}(k)\mathbf{u}^T(k)\,\mathrm{sign}(e(k))] + E[\mathbf{u}(k)\tilde{\mathbf{w}}^T(k)\,\mathrm{sign}(e(k))]\} \\ &+ \mu^2\{\alpha^2 E[\mathbf{u}(k)\mathbf{u}^T(k)e^2(k)] + 2\alpha(1-\alpha)E[\mathbf{u}(k)\mathbf{u}^T(k)\,|\,e(k)\,|\,] + (1-\alpha)^2\mathbf{R}\}\end{aligned}$$

Two cases can be considered for the step size μ so that the weight vector converges in the mean square sense.

i. Case $i \neq j$: a sufficient condition for mean square convergence is given by

$$0 < \mu < \frac{1}{\left[\alpha + (1-\alpha)\sqrt{\dfrac{2}{\pi J_{\min}}}\right]\mathrm{Tr}\{\mathbf{R}\}} \tag{7.63}$$

ii. Case $i = j$: the convergence in the mean square sense can be given by

$$0 < \mu < \frac{2\left[\alpha + (1-\alpha)\sqrt{\frac{2}{\pi}}\frac{1}{\sigma_{e_n}}\right]}{\left[\alpha^2 - 2\alpha(1-\alpha)\sqrt{\frac{2}{\pi}}\frac{1}{\sigma_{e_n}}\right]\lambda_i} \tag{7.64}$$

Remark 7.2

Note that if $\alpha = 0$, the denominator of expression (7.64) will be zero and therefore will make μ take any value in the range of positive numbers, a contradiction with the ranges of values for the step sizes of LMS and LMF algorithms. Moreover, any value for α in [0, 1] will make the step size μ set by (7.64) less than zero, also this condition is discarded. This concludes that it is safer to use a more realistic range for the step size μ for convergence in the mean square described by the range of (7.63) which will guarantee stability regardless of the value of α, and therefore will be considered here.

7.2.3.3 Simulation Results

Here the performance analysis of the mixed controlled L2LP adaptive algorithm is investigated in an unknown system identification problem for different values of p and different values of the mixing parameter α. The simulations reported here are based on an FIR channel system identification

Figure 7.6 depicts the convergence behavior of this algorithm for different values of α and $p = 1$ in a white Gaussian noise, Laplacian noise, and uniform noise, respectively. As one can see from this figure, the best performance is obtained when $\alpha = 0.8$. This makes sense as the resultant algorithm is steering toward the LMS algorithm. Also, as one can see from these figures that the best performance, as far as the noise statistics are concerned, is obtained when the noise environment is Laplacian, then Gaussian, and finally uniform. This makes sense as the update algorithm is biased to the sign error LMS algorithm as α approaches zero. If one compares the performance of this algorithm when the noise statistics are Laplacian, one sees clearly that an enhancement in performance is obtained and about a 2dB improvement is achieved for all values of α.

The situation changes when $p = 4$ as reported in Figure 7.7 which depicts the convergence behavior of this algorithm for different values of α in a white Gaussian noise, Laplacian noise, and uniform noise, respectively. As one can see from this figure, the best performance is obtained when $\alpha = 0.2$ as the L2LP algorithm is mostly LMF in this case. More importantly, the best noise statistics for this scenario is when the noise is uniformly distributed. Similarly as above if one compares the performance of the L2LP algorithms under different values of α when the noise statistics are uniform, one sees

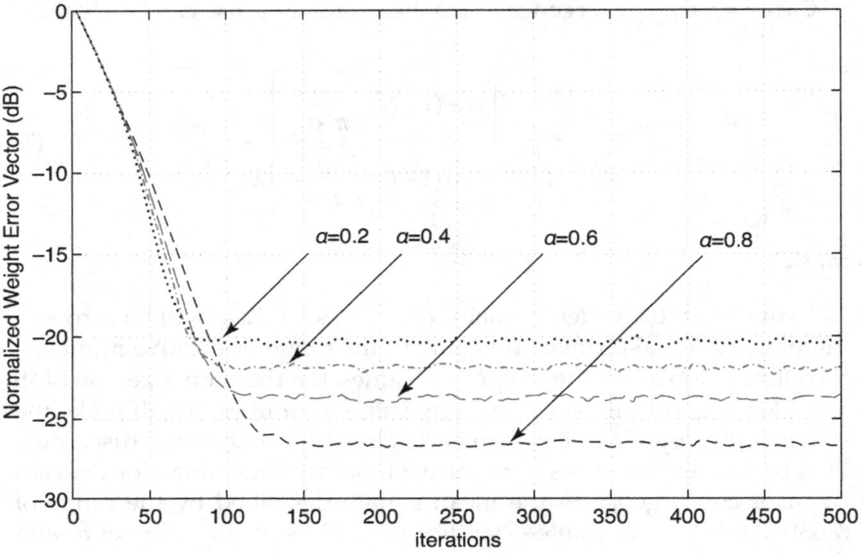

FIGURE 7.6
Effect of α on the convergence behavior of the L2LP algorithm in an AWGN noise environment scenario for $p = 1$ [375].

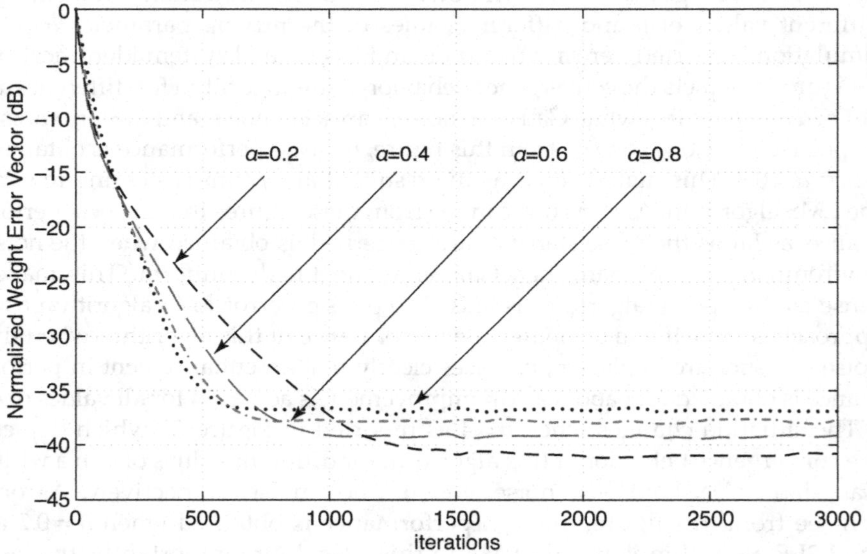

FIGURE 7.7
Effect of α on the convergence behavior of the L2LP algorithm in an AWGN noise environment scenario for $p = 4$ [375].

clearly that an enhancement in performance is obtained and about a 2dB improvement is achieved for all values of α. Also, one can notice that the worst performance is obtained when the noise is Laplacian distributed.

Next, to assess further the performance of the L2LP algorithm for the same steady-state value, two different cases are considered, for $p = 1$ and $p = 4$ when $\alpha = 0.8$. Figure 7.8a illustrates the learning behavior of the L2LP algorithm for $p = 1$. As one can see from this figure that the best performance is obtained with Laplacian noise while the worst performance is obtained with uniform noise environment. In the case of $p = 4$, as reported in Figure 7.8b which depicts the learning behavior of the L2LP algorithm in the different noise environments, it can be seen that the best performance is obtained with uniform noise environment. The Laplacian noise results in the worst performance when compared to Gaussian and uniform noise environments.

7.3 Sparsity-Aware Adaptive Filtering Algorithms under Mixture MMPE

The conventional L2LP channel estimation algorithm, which utilizes the performance of LMS and *lp* adaptive algorithms, can enhance the estimation performance in system identification scenarios. However, it is evident that this algorithm fails to leverage the inherent sparse structure of an unknown FIR system. To address this limitation, this subchapter introduces sparse adaptive system estimation algorithms that rely on a mixed controlled l_2 and *lp*-norm error criterion, and employ zero-attracting theory to exploit the sparse structure information of the system [276]. These algorithms implement the sparsity-aware characteristic through l_1-norm penalty, correntropy-induced metric penalty, and log-sum function constraint, which enable them

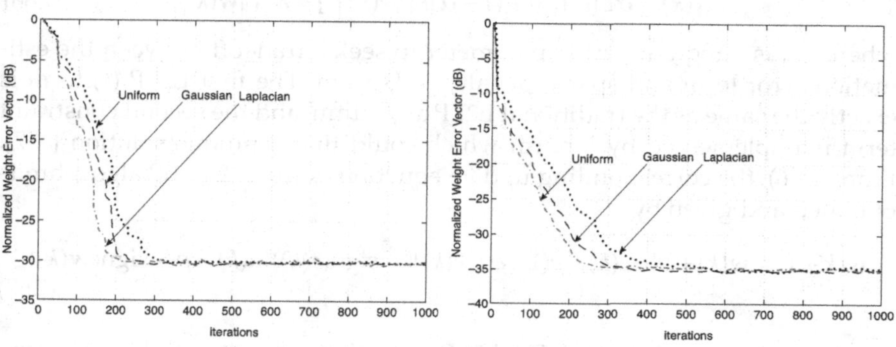

FIGURE 7.8
Learning curves of the L2LP algorithm in different noise environments: (a) the parameters are set as $p = 1$ and $\alpha = 0.8$; (b) The parameters are set as $p = 4$ and $\alpha = 0.8$ [375].

to exploit the in-nature sparseness of the channels. The sparsity-aware algorithms incorporate desired zero attractors in their iterations to expedite the convergence. The previously introduced sparsity-aware algorithms can be considered as a special case of the sparse L2LP adaptive algorithms.

7.3.1 Sparse Mixed L2LP Algorithm

In this part, three sparse adaptive L2LP algorithms are presented for sparse system estimation applications, which are realized by integrating an l_1-norm, a log sum function and a correntropy-induced metric (CIM) penalty into the cost function of the traditional L2LP algorithm, respectively. In order to achieve a clearer and more intuitive understood, the sparse adaptive L2LP estimation *algorithm* is summarized as [276,376]

$$\mathbf{w}(k+1) = \mathbf{w}(k) + \mu[\alpha e(k) + p(1-\alpha)\,|\,e(k)\,|^{p-1}\,\mathrm{sign}(e(k))]\mathbf{u}(k) + spc \tag{7.65}$$

where sparse penalties spc are obtained by using the differential of the l_1-norm, log sum function or CIM to exploit the in-nature sparse structure of system. From the update equation, the objective of the sparse adaptive L2LP algorithms is realized by incorporating different sparse penalties into the cost function of the L2LP algorithm to take advantage of sparse property for improving the estimate performance.

7.3.1.1 Zero-Attracting L2LP Algorithm

Similar to the previously presented sparse adaptive filtering algorithms, an l_1-norm is integrated into the cost function of the traditional L2LP algorithm to construct the zero-attracting L2LP (ZA-L2LP) algorithm to exploit the channel sparsity. The cost function of the ZA-L2LP algorithm is written as [276]

$$J_{ZA}(k) = \alpha E[e^2(k)] + (1-\alpha)E[\,|\,e(k)\,|^p\,] + \lambda_{ZA}\left\|\mathbf{w}(k)\right\|_1 \tag{7.66}$$

where λ_{ZA} is a regularization parameter to seek a tradeoff between the estimation error term and sparse penalty of l_1-norm. The first L2LP (k) term is exactly the same as the traditional L2LP algorithm, and the second constraint term is implemented by l_1-norm which could find a unique solution [377]. From (7.66), the corresponding update equation of the ZA-L2LP algorithm is obtained and given by

$$\mathbf{w}(k+1) = \mathbf{w}(k) + \mu[\alpha e(k) + p(1-\alpha)\,|\,e(k)\,|^{p-1}\,\mathrm{sign}(e(k))]\mathbf{u}(k) + \rho_{ZA}\,\mathrm{sign}(\mathbf{w}(k)) \tag{7.67}$$

where μ is the step-size of the ZA-L2LP algorithm, and $\rho_{RZA} = \mu\lambda_{RZA}$ is a positive parameter, which is a zero-attracting strength control parameter. Note that the difference between the ZA-L2LP algorithm and the traditional L2LP algorithm is that there is an additional sparse penalty term $\rho_{ZA}\,\mathrm{sign}(\mathbf{w}(k))$, which can speed up the convergence by attracting the zero and near-zero taps to zero when most of the unknown FIR channel coefficients are non-dominants. Hence, the additional sparse penalty term acts as a zero attractor.

7.3.1.2 Reweighting Zero-Attracting L2LP Algorithm

Since all the taps are enforced to zero by $\rho_{ZA}\,\mathrm{sign}(\mathbf{w}(k))$ uniformly, the ZA-L2LP algorithm cannot very well distinguish zero and non-zero taps. Therefore, the performance of ZA-L2LP algorithm may be reduced in less sparse channel impulse response. Based on the principle of compressed sensing (CS) [378] and the previously presented L2LP and ZA-L2LP algorithms, a log-sum function is used for substituting the l_1-norm in the ZA-L2LP algorithm to create a reweighted ZA-L2LP (RZA-L2LP) algorithm to further exploit the sparsity of the channels. Therefore, the RZA-L2LP algorithm is implemented by adding a log-sum function penalty into the basic L2LP algorithm to provide a reweighting factor. Then, the cost function of the RZA-L2LP algorithm is given by

$$J_{RZA}(k) = \alpha E[e^2(k)] + (1-\alpha)E[\,|\,e(k)\,|^p\,]\lambda_{RZA}\sum_{i=1}^{N}\log(1+\varepsilon_{RZA}\|\mathbf{w}(k)\|_1) \quad (7.68)$$

where λ_{RZA} and ε_{RZA} are positive parameters. λ_{RZA} is used to tradeoff the system sparsity and system estimation performance, and the reweighting factor ε_{RZA} plays a role to further exploit channel sparsity in comparison with the ZA-L2LP algorithm. We note that the RZA-L2LP can give smaller force on dominant taps, and the corresponding update equation is given by

$$\begin{aligned}
\mathbf{w}(k+1) &= \mathbf{w}(k) - \mu\frac{\partial L_{RZA}(k)}{\partial \mathbf{w}(k)} \\
&= \mathbf{w}(k) + \mu[\alpha e(k) + p(1-\alpha)\,|\,e(k)\,|^{(p-1)}\,\mathrm{sign}(e(k))]\mathbf{u}(k) \\
&\quad - \mu\lambda_{RZA}\frac{\partial\sum_{i=1}^{N}\log\left(1+\varepsilon_{RZA}\|w_i(k)\|_1\right)}{\partial \mathbf{w}(k)} \\
&= \mathbf{w}(k) + \mu[\alpha e(k) + p(1-\alpha)\,|\,e(k)\,|^{(p-1)}\,\mathrm{sign}(e(k))]\mathbf{u}(k) \\
&\quad - \rho_{RZA}\sum_{i=1}^{N}\frac{\mathrm{sign}\left(\|w_i(k)\|_1\right)}{1+\varepsilon_{RZA}\|w_i(k)\|_1} \\
&= \mathbf{w}(k) + \mu[\alpha e(k) + p(1-\alpha)\,|\,e(k)\,|^{(p-1)}\,\mathrm{sign}(e(k))]\mathbf{u}(k) - \rho_{RZA}\frac{\mathrm{sign}\left(\|\mathbf{w}(k)\|_1\right)}{1+\varepsilon_{RZA}\|\mathbf{w}(k)\|_1}
\end{aligned} \quad (7.69)$$

where $\rho_{RZA} = \mu\lambda_{RZA}\varepsilon_{RZA}$ is a zero-attracting control factor parameter, which depends on the step size μ, regularization parameter λ_{RZA} and reweighting factor ε_{RZA}. In the sparse penalty term of (7.69), one can see that the multi-path channel estimation will be close to zeros if the magnitudes of $\mathbf{w}(k)$ are smaller than $1/\varepsilon_{RZA}$. Therefore, the selection of ε_{RZA} is critical to improve the performance of the RZA-L2LP algorithm. In addition, similar to l_1-norm, $\rho_{RZA}\dfrac{\text{sign}\left(\|\mathbf{w}(k)\|_1\right)}{1+\varepsilon_{RZA}\|\mathbf{w}(k)\|_1}$ can achieve a better convergence performance than the ZA-L2LP algorithm by selecting the appropriate ρ_{RZA} and ε_{RZA}. That is to say, the RZA-L2LP algorithm can provide a superior performance in comparison with the ZA-L2LP algorithm when appropriate parameters are selected.

7.3.1.3 CIM L2LP Algorithm

Following the idea of zero-attracting technology, a CIM-penalized L2LP algorithm can be designed to continue exploit the sparsity of the unknown FIR channel. Based on ZA-technique and CIM theory, we construct a modified cost function by combining the cost function of the L2LP algorithm and CIM criterion, which is expressed as [276]

$$J_{CIM}(k) = \alpha E[e^2(k)] + (1-\alpha)E[|e(k)|^{(p)}] + \lambda_{CIM} CIM(\mathbf{w}(k), 0) \tag{7.70}$$

where λ_{CIM} is a regulation parameter to tradeoff the estimation misalignment and sparsity of $\mathbf{w}(k)$. One can easily find that the first term plays a robust role to estimate multi-path channel while the CIM sparse penalty plays a role in exploiting the sparsity of the unknown FIR channel.

An update equation of the CIM-based L2LP (CIM-L2LP) algorithm is derived as follows:

$$\begin{aligned} w_i(k+1) &= w_i(k) - \mu\frac{\partial L_{RZA}(k)}{\partial \mathbf{w}(k)} \\ &= w_i(k) + \mu[\alpha e(k) + p(1-\alpha)|e(k)|^{(p-1)}\,\text{sign}(e(k))]\mathbf{u}(k) \\ &\quad - \mu\lambda_{CIM}\frac{\partial CIM^2(\mathbf{w}(k),0)}{\partial \mathbf{w}(k)} \\ &= w_i(k) + \mu[\alpha e(k) + p(1-\alpha)|e(k)|^{(p-1)}\,\text{sign}(e(k))]\mathbf{u}(k) \\ &\quad - \rho_{CIM}\frac{1}{N\sigma^3\sqrt{2\pi}} w_i(k)\exp\left(-\frac{w_i(k)^2}{2\sigma^2}\right) \end{aligned} \tag{7.71}$$

where $\rho_{CIM} = \mu\lambda_{CIM}\varepsilon_{CIM}$ is a positive parameter. The matrix vector form of the update equation of the CIM-L2LP algorithm can be rewritten as

$$\begin{aligned}\mathbf{w}(k+1) &= \mathbf{w}(k) - \mu\frac{\partial L_{RZA}(k)}{\partial \mathbf{w}(k)} \\ &= \mathbf{w}(k) + \mu[\alpha e(k) + p(1-\alpha)\,|\,e(k)\,|^{(p-1)}\ \mathrm{sign}(e(k))]\mathbf{u}(k) \\ &\quad - \rho_{CIM}\frac{1}{N\sigma^3\sqrt{2\pi}}\mathbf{w}(k)\exp\left(-\frac{\mathbf{w}(k)^2}{2\sigma^2}\right)\end{aligned} \tag{7.72}$$

Here one can see that the CIM-L2LP algorithm can be applied in systems to estimate the channel performance by selecting a suitable kernel bandwidth σ to render the CIM approach close to the l_0-norm. Compared with the previous L2LP algorithm, the complexity of the CIM-L2LP algorithm is very low.

7.3.2 Convergence Analysis

The convergence analysis of the sparse L2LP algorithms is presented in this part. By using the method of approximation, similar assume analysis methods are used in [5,38] to present a simple analysis of the algorithms. Firstly, the mean and variance of the input signal $\mathbf{u}(k)$ are set to 0 and σ_u^2, respectively. Next, we assume that the noise signal $v(k)$ is zero-mean and identically distributed, and is independent of the input signal $\mathbf{u}(k)$.

7.3.2.1 First Moment Behavior of the Weight Error Vector

The convergence evaluation of the ZA-L2LP algorithm is presented first. Subtracting $\mathbf{w}_0(k)$ on both sides of the update equation of the ZA-L2LP algorithm in (7.67), we get

$$\tilde{\mathbf{w}}(k+1) = \tilde{\mathbf{w}}(k) + \mu[\alpha e(k) + p(1-\alpha)\,|\,e(k)\,|^{p-1}\ \mathrm{sign}(e(k))]\mathbf{u}(k) - \rho_{ZA}\,\mathrm{sign}(\mathbf{w}(k)) \tag{7.73}$$

Substituting $e(k)$ into equation (7.73) and taking the expectation operator on its both sides, we obtain

$$E[\tilde{\mathbf{w}}(k+1)] = (\mathbf{I} - \mu\mathbf{R})E[\tilde{\mathbf{w}}(k)] + \mu(1-\alpha)E[\mathrm{sign}(e(k))\mathbf{u}(k)] - \rho_{ZA}E[\mathrm{sign}(\mathbf{w}(k))] \tag{7.74}$$

where the $E[\mathbf{u}(k)\mathrm{sign}(e(k))]$ is calculated by using Price's theorem [379], which is given by

$$\begin{aligned}E[\mathrm{sign}(e(k))\mathbf{u}(k)] &= \sqrt{\frac{2}{\pi}}\frac{1}{\sigma_k}E[e(k)\mathbf{u}(k)] \\ &= \sqrt{\frac{2}{\pi}}\frac{1}{\sigma_k}E[v(k)\mathbf{u}(k) - \mathbf{u}(k)\mathbf{u}^T(k)\tilde{\mathbf{w}}(k)] \\ &= \sqrt{\frac{2}{\pi}}\frac{1}{\sigma_k}\mathbf{R}E[\tilde{\mathbf{w}}(k)]\end{aligned} \tag{7.75}$$

Combining (7.74) and (7.75), we obtain

$$E[\tilde{\mathbf{w}}(k)] = \left\{ I - \mu\left[\alpha + (1-\alpha)\sqrt{\frac{2}{\pi}}\frac{1}{\sigma_k}\right]\mathbf{R} \right\} E[\tilde{\mathbf{w}}(k)] - \mathbf{B} \tag{7.76}$$

where $\mathbf{B} = \rho_{ZA}E[\text{sign}(\mathbf{w}(k))]$, which is a bounded vector based on [380]. From (7.76), one can easily get μ when the misalignment vector converges, and μ is given by

$$0 < \mu < \frac{2}{\left[\alpha + (1-\alpha)\sqrt{\frac{2}{\pi J_{\min}}}\right]\text{Tr}\{\mathbf{R}\}} \tag{7.77}$$

where Tr{**R**} denotes a trace of the matrix **R**. As a result, if the convergence of the ZA-L2LP algorithm occurs, the simpler conditional in the mean is

$$0 < \mu < \frac{2}{\left[\alpha + (1-\alpha)\sqrt{\frac{2}{\pi J_{\min}}}\right]\lambda_{\max}} \tag{7.78}$$

where $J_{\min}$ is the minimum of MSE and $\lambda_{\max}$ is the largest eigenvalue of the auto-correlation matrix **R**. Thus, the root mean squared estimation error σ_k can be obtained

$$\sigma_k > \sqrt{\frac{2}{\pi}}\left[\frac{\mu(1-\alpha)\lambda_{\max}}{2-\mu\alpha\lambda_{\max}}\right] \tag{7.79}$$

It is straightforward from (7.79) that, a valid condition for the ZA-L2LP to converge is to meet the following equation

$$0 < \mu < \frac{2}{\alpha\lambda_{\max}}$$

Meanwhile, one can see that the convergence of the LMS is achieved for $\alpha = 1$.

7.3.2.2 Second Moment Behavior of the Weight Error Vector

Here the misalignment vector $\mathbf{K}(k) = E[\tilde{\mathbf{w}}(k)\tilde{\mathbf{w}}^T(k)]$ is also defined to analyze the second moment of the ZA-L2LP algorithm. From (7.73), we get the following expression:

$$\begin{aligned}\tilde{\mathbf{w}}(k+1)\tilde{\mathbf{w}}^T(k+1) &= \tilde{\mathbf{w}}(k)\tilde{\mathbf{w}}^T(k) + \mu[\alpha e(k) + (1-\alpha)\text{sign}(e(k))][\tilde{\mathbf{w}}(k)\mathbf{u}^T(k) + \mathbf{u}(k)\tilde{\mathbf{w}}^T(k)] \\ &+ \mu^2[\alpha^2 e^2(k) + 2\alpha(1-\alpha)\,|\,e(k)\,| + (1-\alpha)^2]\mathbf{u}(k)\mathbf{u}^T(k) \\ &- \rho_{ZA}\,\text{sign}(e(k))[\tilde{\mathbf{w}}(k) + \tilde{\mathbf{w}}^T(k)] \\ &- \mu\rho_{ZA}[\alpha e(k) + (1-\alpha)\text{sign}(e(k))]\text{sign}(\mathbf{w}(k))[\mathbf{u}(k) + \mathbf{u}^T(k)] + \rho_{ZA}^2\end{aligned} \tag{7.80}$$

Taking expectation on both sides of (7.80), one can obtain

$$\begin{aligned}\mathbf{K}(k+1) = \mathbf{K}(k) &+ \mu\alpha\left\{E\left[\tilde{\mathbf{w}}(k)\mathbf{u}^T(k)e(k)\right] + E\left[\mathbf{u}(k)\tilde{\mathbf{w}}^T(k)e(k)\right]\right\} \\ &+ \mu(1-\alpha)\left\{E\left[\tilde{\mathbf{w}}(k)\mathbf{u}^T(k)\text{sign}(e(k))\right] + E\left[\mathbf{u}(k)\tilde{\mathbf{w}}^T(k)\text{sign}(e(k))\right]\right\} \\ &+ \mu^2\left\{e^2 E\left[\mathbf{u}(k)\mathbf{u}^T(k)e^2(k)\right] + 2\alpha(1-\alpha)E\left[\mathbf{u}(k)\mathbf{u}^T(k)\,|\,e(k)\,|\right] + (1-\alpha)^2\mathbf{R}\right\} \\ &- \rho_{ZA}\left\{E\left[\tilde{\mathbf{w}}(k)\text{sign}(\mathbf{w}(k))\right] + E\left[\tilde{\mathbf{w}}^T(k)\text{sign}(\mathbf{w}(k))\right]\right\} \\ &- \mu\rho_{ZA}\left\{\alpha E\left[\tilde{\mathbf{w}}(k)e(k)\text{sign}(\mathbf{w}(k))\right] + (1-\alpha)E\left[\mathbf{u}^T(k)\text{sign}(e(k))\,\text{sign}(\mathbf{w}(k))\right]e(k)\right\} \\ &+ \rho_{ZA}^2\end{aligned} \tag{7.81}$$

Then, we will evaluate the mean in (7.81) under the assumptions of zero mean and independent Gaussian of $v(k)$ and $\mathbf{u}(k)$. Thus, we get

$$\begin{aligned}&E\left[\tilde{\mathbf{w}}(k)\mathbf{u}^T(k)e(k)\right] = -\mathbf{K}(k)\mathbf{R} \\ &E\left[\mathbf{u}(k)\tilde{\mathbf{w}}^T(k)e(k)\right] = -\mathbf{R}\mathbf{K}(k) \\ &E[\tilde{\mathbf{w}}(k)\mathbf{u}^T(k)\text{sign}(e(k))] = -\sqrt{\frac{2}{\pi}}\frac{1}{\sigma_k}\mathbf{K}(k)\mathbf{R} \\ &E\left[\mathbf{u}(k)\mathbf{u}^T(k)e^2(k)\right] = E\left[\mathbf{u}(k)\mathbf{u}^T(k)(v(k) - \tilde{\mathbf{w}}^T(k)\mathbf{u}(k))\right] = \mathbf{R}\{J_{\min} + tr(\mathbf{K}(k)\mathbf{R})\} \\ &E\left[\mathbf{u}(k)\mathbf{u}^T(k)\,|\,e(k)\,|\right] = E\left[\mathbf{u}(k)\mathbf{u}^T(k)e(k)\,\text{sign}(e(k))\right] \\ &\qquad = \sqrt{\frac{2}{\pi}}\frac{1}{\sigma_k}E\left[\mathbf{u}(k)\mathbf{u}^T(k)e^2(k)\right] \\ &\qquad = \sqrt{\frac{2}{\pi}}\frac{1}{\sigma_k}\mathbf{R}\{J_{\min} + tr(\mathbf{K}(k)\mathbf{R})\}\end{aligned} \tag{7.82}$$

Substituting (7.82) into (7.81), we have

$$\mathbf{K}_{n+1} = \mathbf{K}_n \left\{ I - \mu \left[\alpha + (1-\alpha)\sqrt{\frac{2}{\pi}}\frac{1}{\sigma_k} \right] \mathbf{R} \right\}$$
$$+\mu^2 \mathbf{R} \left\{ (1-\alpha)^2 + \left[\alpha^2 - 2\alpha(1-\alpha)\sqrt{\frac{2}{\pi}}\frac{1}{\sigma_k} \right] \times \left[J_{\min} + tr(\mathbf{K}(k)\mathbf{R}) \right] \right\} \quad (7.83)$$
$$-\mu \left[\alpha + (1-\alpha)\sqrt{\frac{2}{\pi}}\frac{1}{\sigma_k} \right] \mathbf{R}\mathbf{K}(k) + \mathbf{Q}(k)$$

where $J_{\min}$ denotes the minimum of MSE, and $\mathbf{Q}(k)$ is a simplified form given by

$$\mathbf{Q}(k) = -\rho_{ZA} \left\{ E\left[\tilde{\mathbf{w}}(k)\text{sign}(\mathbf{w}(k)) \right] + E\left[\tilde{\mathbf{w}}^T(k)\text{sign}(\mathbf{w}(k)) \right] \right\}$$
$$- \mu\rho_{ZA} \left\{ \alpha E\left[\mathbf{u}(k)e(k)\text{sign}(\mathbf{w}(k)) \right] + (1-\alpha)E\left[\mathbf{u}^T(k)\text{sign}(e(k))\text{sign}(\mathbf{w}(k)) \right] \right\} + \rho_{ZA}^2 \quad (7.84)$$

Now, we assume that $\mathbf{R}$ is positive-defined [261] with eigenvalues λ_i and hence, the auto-correlation matrix can be written as $\mathbf{R} = \mathbf{P}\Lambda\mathbf{P}^T$, where Λ is a diagonal matrix of eigenvalues λ_i, and $\mathbf{P}$ represents orthogonal matrix. One can obtain

$$\mathbf{G}(k) = \mathbf{P}\mathbf{K}(k)\mathbf{P}^T \quad (7.85)$$

Thus (7.33) can be derived as

$$\mathbf{G}(k+1) = \mathbf{G}(k) \left\{ I - \mu \left[\alpha + (1-\alpha)\sqrt{\frac{2}{\pi}}\frac{1}{\sigma_k} \right] \Lambda \right\}$$
$$+\mu^2 \Lambda \left\{ (1-\alpha)^2 + \left[\alpha^2 - 2\alpha(1-\alpha)\sqrt{\frac{2}{\pi}}\frac{1}{\sigma_k} \right] \times \left[J_{\min} + Tr(\Lambda\mathbf{G}(k)) \right] \right\} \quad (7.86)$$
$$-\mu \left[\alpha + (1-\alpha)\sqrt{\frac{2}{\pi}}\frac{1}{\sigma_k} \right] \Lambda\mathbf{G}(k) + \mathbf{P}^T\mathbf{Q}(k)\mathbf{P}$$

where $\mathbf{Q}(k)$ is bound, and hence, the value of step size μ will guarantee the convergence of the ZA-L2LP algorithm. Then, (7.86) can be converted to

$$g_{i,j}(k+1) = g_{i,j}(k) \left\{ \mathbf{I} - \mu \left[\alpha + (1-\alpha)\sqrt{\frac{2}{\pi}}\frac{1}{\sigma_k} \right] (\lambda_i + \lambda_j) \right\}$$
$$+ \mu^2 \lambda_i \left\{ (1-\alpha)^2 + \left[\alpha^2 - 2\alpha(1-\alpha)\sqrt{\frac{2}{\pi}}\frac{1}{\sigma_k} \right] \times \left[J_{\min} + \sum_{i=1}^{N} \lambda_i g_{i,j}(k) \right] \right\} \delta_{i,j} \quad (7.87)$$

where $\delta_{i,j} = 1$ when $i = j$, otherwise $\delta_{i,j} = 0$, and $g_{i,j}(k)$ is the (i, j)th element of the matrix $\mathbf{G}(k)$. When $i = j$, equation (7.87) can be rewritten as

$$g_{i,j}(k+1) = g_{i,j}(k)\left\{1-2\mu\left[\alpha+(1-\alpha)\sqrt{\frac{2}{\pi}}\frac{1}{\sigma_k}\right]\lambda_i + \mu^2\left[\alpha^2-2\alpha(1-\alpha)\sqrt{\frac{2}{\pi}}\frac{1}{\sigma_k}\right]\lambda_i^2\right\}$$
$$+\mu^2\lambda_i\left\{(1-\alpha)^2+\left[\alpha^2-2\alpha(1-\alpha)\sqrt{\frac{2}{\pi}}\frac{1}{\sigma_k}\right]\times\left[J_{\min}+\sum_{i=1,i\neq j}^{N}\lambda_i g_{i,j}(k)\right]\right\} \tag{7.88}$$

Accordingly, we can obtain the step size of the ZA-L2LP algorithm to guarantee convergence under the mean square occurs, which can be written as

$$0<\mu<\frac{2\left[\alpha+(1-\alpha)\sqrt{\frac{2}{\pi}}\frac{1}{\sigma_k}\right]}{\alpha^2-2\alpha(1-\alpha)\sqrt{\frac{2}{\pi}}\frac{1}{\sigma_k}\lambda_i} \tag{7.89}$$

Note that the step size μ does not take any value for convergence in mean square, such as $\eta = 0$, which is inconsistent with any value of η given in [0, 1]. Therefore, this condition is discarded. Next, we will consider the case of $i \neq j$, (7.86) reduces to

$$g_{i,j}(k+1) = g_{i,j}(k)\left\{1-2\mu\left[\alpha+(1-\alpha)\sqrt{\frac{2}{\pi}}\frac{1}{\sigma_k}\right]\lambda_i+\lambda_j\right\} \tag{7.90}$$

Therefore, the adaptive ZA-L2LP algorithm is stable if μ meets

$$0<\mu<\frac{2}{\left[\alpha+(1-\alpha)\sqrt{\frac{2}{\pi}}\frac{1}{\sigma_k}\right](\lambda_i+\lambda_j)} \tag{7.91}$$

At last, a sufficient condition to guarantee the convergence of the ZA-L2LP algorithm is

$$0<\mu<\frac{2}{\left[\alpha+(1-\alpha)\sqrt{\frac{2}{\pi}}J_{\min}\right]\mathrm{Tr}\{\mathbf{R}\}} \tag{7.92}$$

The result in (7.92) shows that for $\forall_\alpha \in [0, 1]$, the convergence of the ZA-L2LP algorithm is guaranteed. The RZA-L2LP and CIM-L2LP algorithms can also be analyzed in detail based on the convergence of the ZA-L2LP algorithm.

7.3.3 Simulation Results

Here some simulation results are presented to evaluate the performance of the sparse L2LP algorithms by averaging over 100 independent MC runs for getting each point. The length of sparse channel **w** is set to $N = 16$ and its dominant taps are randomly allocated in the length of sparse channel, which is subject to $E\left[\|w\|_2^2\right] = 1$. Moreover, the values of dominant taps follow a random distribution. The SNR is chosen to be 10 dB in all simulations.

7.3.3.1 Convergence and Steady-State Analysis

In this case, the sparse L2LP algorithms are compared with the LMS, LMF, LMS/F [367] and their corresponding sparse algorithms. Therefore, the convergence rate of the sparse L2LP algorithms is first evaluated in comparison with the LMS, L2LP, ZA-LMS, and RZA-LMS algorithms. The simulation parameters are set to achieve almost the same steady-state error. The convergence curves are shown in Figure 7.9. It shows that the ZA-L2LP, RZA-L2LP, and CIM-L2LP are faster than the LMS, L2LP, and sparse LMS algorithms when they have almost the same steady-state error. Moreover, the CIM-L2LP algorithm achieves the fastest convergence rate in all the mentioned algorithms.

The steady-state behaviors of the sparse L2LP algorithms in a multi-path wireless communication channel with different sparsity levels are further validated. The steady-state behaviors of the sparse L2LP algorithms of the ZA-L2LP, RZA-L2LP, and CIM-L2LP are provided in Figure 7.11 (a)-(c) for $K = 1$, $K = 2$ and $K = 4$, respectively. In Figure 7.11a, one can see that the sparse

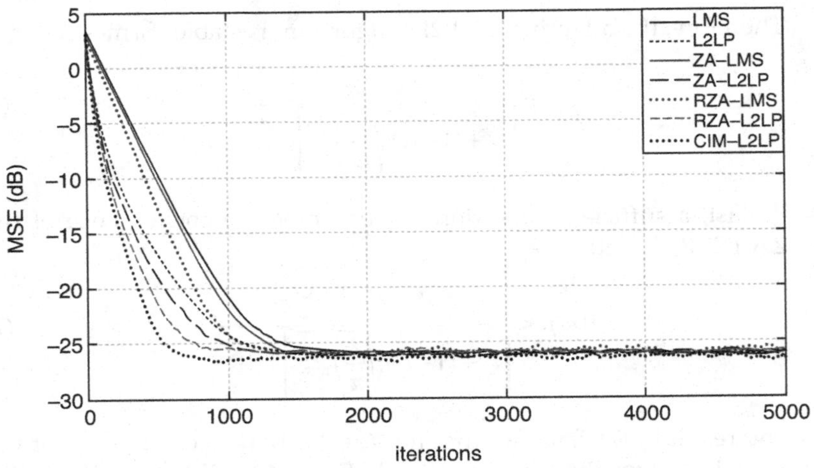

FIGURE 7.9
Convergence comparisons between the sparse L2LP and LMS algorithms [276].

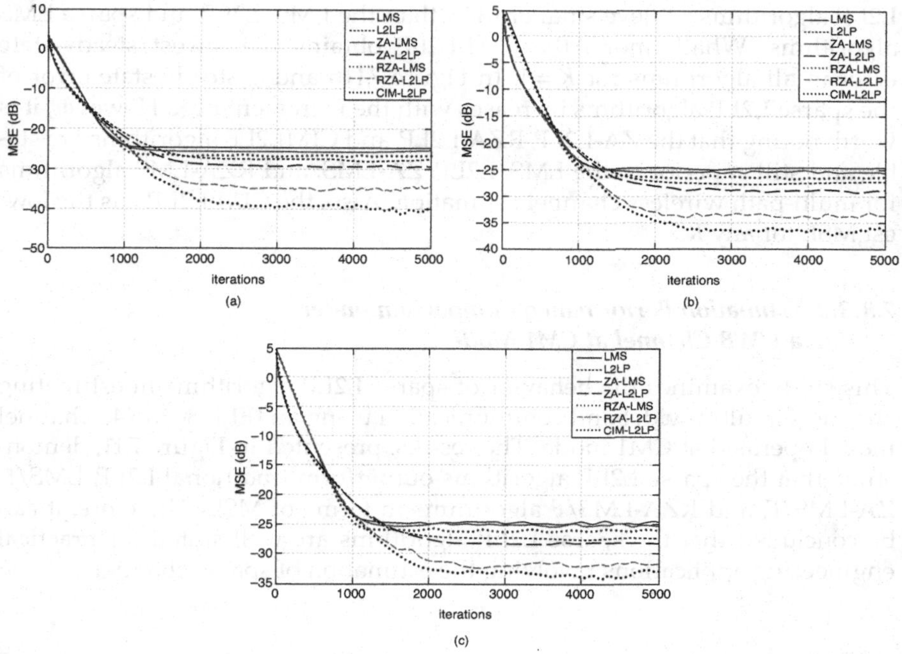

FIGURE 7.10
Steady-state behavior comparisons between the sparse L2LP and LMS algorithms for (a) $K=1$, (b) $K=2$, and (c) $K=4$ [276].

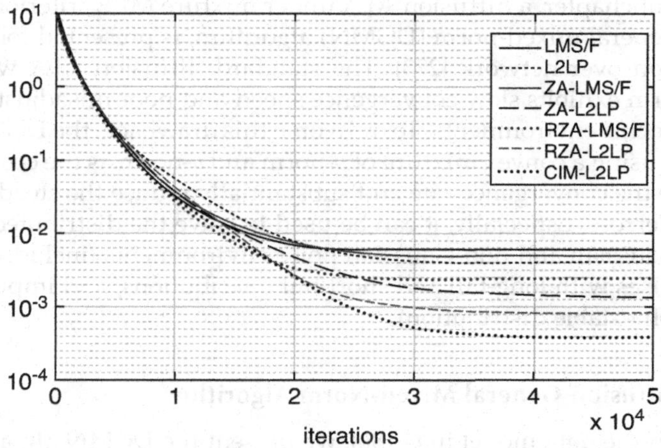

FIGURE 7.11
Estimation behavior of the sparse L2LP algorithms for estimating a UWB channel at CM1 mode [276].

L2LP algorithms achieve smaller MSE than the LMS, L2LP, and sparse LMS algorithms. What's more, the CIM-L2LP obtains the lowest steady-state error in all algorithms for $K = 1$. In Figure 7.11b and c, steady-state error of the sparse L2LP algorithms increases with the increment of K. However, it is worth noting that the ZA-L2LP, RZA-L2LP, and CIML2LP algorithms are stable and still superior to the LMS, L2LP, ZA-LMS, and RZA-LMS algorithms for multi-path wireless channel estimation. Also, the CIM-L2LP has the lowest MSE for any K.

7.3.3.2 Estimation Performance Comparison under a UWB Channel at CM1 Mode

This study examines the behavior of sparse L2LP algorithms in estimating channels in ultra-wideband communication using an IEEE 802.15.4a channel model operated at CM1 mode. The results, presented in Figure 7.11, demonstrate that the sparse L2LP algorithms outperform traditional L2LP, LMS/F, ZA-LMS/F, and RZA-LMS/F algorithms in terms of MSE. Therefore, it can be concluded that the sparse L2LP algorithms are well suited for practical engineering applications involving the estimation of sparse channels.

7.4 Diffusion Adaptive Filtering Algorithm under Mixture MMPE

In this subchapter, a diffusion AFA under mixture MPE criterion, called diffusion general mixed-norm (DGMN) algorithm, is presented for distributed estimation over network [273]. The standard diffusion AFA with a single error norm exhibits slow convergence speed and poor misadjustment under some specific environments. To overcome this drawback, the DGMN is developed by using a convex mixture of p-norm and q-norm as the cost function to improve the convergence rate and substantially reduce the steady-state coefficients' errors. Especially, it can be used to solve the distributed estimation under Gaussian and non-Gaussian noise environments (including measurement noises with long-tail and short-tail distributions, and impulsive noises with alpha-stable distribution).

7.4.1 Diffusion General Mixed-Norm Algorithm

Based on the data model in (4.186), we present the DGMN algorithm in this part. For each node k, the following local cost function based on GMN is defined [273]

$$J_k^{\text{local}}(\mathbf{w}_k) = \sum_{i \in N_k} \alpha_{l,k} J_{GMN}(e_i(i)) = \sum_{l \in N_k} \alpha_{l,k} J_{GMN}\left(d_l(i) - \mathbf{w}_k^T \mathbf{u}_l(i)\right) \tag{7.93}$$

where $\{\alpha_{l,k}\}$ represents some non-negative combination coefficients satisfying $\sum_{l \in N_k} \alpha_{l,k} = 1$, and $\alpha_{l,k} = 0$ if $l \notin N_k$, and

$$J_{GMN}(e_l(i)) = \omega \frac{1}{p}|e_l(i)|^p + (1-\omega)\frac{1}{q}|e_l(i)|^q = \omega \frac{1}{p}|d_l(i) - \mathbf{w}^T \mathbf{u}_l(i)|^p + (1-\omega)\frac{1}{q}|d_l(i) - \mathbf{w}^T \mathbf{u}_l(i)|^q$$

Then we take the derivative of (7.93) as

$$\nabla J^{\text{local}}(\mathbf{w}_k) = \sum_{l \in N_k} \alpha_{l,k} \frac{\partial J_{GMN}(e_l(i))}{\partial \mathbf{w}_k} = -\sum_{l \in N_k} \alpha_{l,k}\left[\omega |e_l(i)|^{p-1} + (1-\omega)|e_l(i)|^{q-1}\right]\text{sign}(e_l(i))\mathbf{u}_l(i) \tag{7.94}$$

By (7.94), the estimate of $w_\circ$ at node k can thus be obtained by the gradient descent method as

$$\mathbf{w}_k(i) = \mathbf{w}_k(i-1) - \mu_k \nabla J^{\text{local}}(\mathbf{w}_k) = \mathbf{w}_k(i-1) + \mu_k \sum_{l \in N_k} \alpha_{l,k}\left[\omega |e_l(i)|^{p-1} + (1-\omega)|e_l(i)|^{q-1}\right]\text{sign}(e_l(i))\mathbf{u}_l(i) \tag{7.95}$$

where $\mathbf{w}_k(i)$ is the estimate of $\mathbf{w}_\circ$ at time index i, and μ_k is the step size for node k which is assumed to be unchanged throughout iterations. The adapt-to-combine (ATC) scheme is an outstanding strategy for DAF algorithms, and it first updates the local estimates using the adaptive algorithm and then the estimates of the neighbors are fused together. Here we mainly focus on developing the ATC version of DGMN algorithms. For each node k, the intermediate estimates can be computed by

$$\psi_k(i-1) = \sum_{l \in N_k} \beta_{l,k} \mathbf{w}_l(i-1) \tag{7.96}$$

where $\psi_k(i-1)$ stands for an intermediate estimate offered by node k at instant time$_{i-1}$, and $\beta_{l,k}$ represents a weight with which a node should share its intermediate estimate $w_1(i-1)$ with the node k. Applying all the intermediate estimates, the nodes update their estimates by

$$\varphi_k(i)=\psi_k(i-1)+\mu_k\sum_{i\in N_k}\alpha_{l,k}\left(\omega\left|e_l(i)\right|^{p-1}+(1-\omega)\left|e_l(i)\right|^{q-1}\right)\text{sign}\left(e_l(i)\right)\mathbf{u}_l(i) \quad (7.97)$$

The update process in (7.97) is represented as incremental step. The combination coefficients $\{\alpha_{l,k}\}$ determine which nodes should enjoy together their measurements $\{d_l(i),\mathbf{u}_l(i)\}$ with node k.

After obtaining the intermediate estimates, the linear combination is then performed as

$$\mathbf{w}_k(i)=\sum_{l\in N_k}\delta_{l,k}\varphi_l(i) \quad (7.98)$$

The above formula (7.98) is a convex combination of intermediate estimates from incremental step (7.97) fed by spatially distinct data $\{d_l(i),u_l(i)\}$, and it is denoted as diffusion step. The coefficients $\{\delta_{l,k}\}$ determine which nodes should share their intermediate estimates $\varphi_l(i)$ with node k, and it plays the same role with the $\{\delta_{l,k}\}$.

Based on the analysis above, we can obtain the following general DGMN algorithm by combining (7.96)–(7.98)

$$\begin{cases}\psi_k(i-1)=\sum_{l\in N_k}\beta_{l,k}\mathbf{w}_l(i-1) & \text{diffusion } I\\ \varphi_k(i)=\psi_k(i-1) & \\ \quad+\mu_k\sum_{l\in N_k}\alpha_{l,k}\left(\omega\left|e_l(i)\right|^{p-1}+(1-\omega)\left|e_l(i)\right|^{q-1}\right)sign\left(e_l(i)\right)\mathbf{u}_l(i) & \text{incremented}\\ \mathbf{w}_k(i)=\sum_{l\in N_k}\delta_{l,k}\varphi_l(i) & \text{diffusion } II\end{cases} \quad (7.99)$$

The different methods for choosing the combination weights $\beta_{l,k}$, $\alpha_{l,k}$ $\delta_{l,k}$ can be found in [95], and are not discussed in detail here.

Here the combination coefficients $\beta_{l,k}$, $\alpha_{l,k}$, $\delta_{l,k}$ in (7.99) are considered as non-negative real values, and they correspond to the $\{l,k\}$ entries of matrices $\boldsymbol{P}_1$, $\boldsymbol{P}_2$, and $\boldsymbol{P}_3$, respectively. Further, they satisfy

$$\mathbf{1}^T\mathbf{P}_1=\mathbf{1}^T,\mathbf{1}^T\mathbf{P}_2=\mathbf{1}^T,\mathbf{1}^T\mathbf{P}_3=\mathbf{1}^T$$

where $\mathbf{1}$ denotes the $N\times 1$ vector with unit entries. Then, we give the concrete form of the diffusion GMN algorithm.

First, we choose $\mathbf{P}_1 = \mathbf{I}$, the algorithm (7.99) will reduce to the uncomplicated DGMN version as

$$\begin{cases} \varphi_k(i) = \mathbf{w}_k(i-1) + \mu_k \sum_{l \in N_k} \alpha_{l,k} \left(\omega |e_l(i)|^{p-1} + (1-\omega)|e_l(i)|^{q-1} \right) \text{sign}(e_l(i)) \mathbf{u}_l(i) \\ \text{where} \quad e_l(i) = d_l(i) - \mathbf{u}_k^T(i) \mathbf{w}_k(i-1) \\ \mathbf{w}_k(i) = \sum_{l \in N_k} \delta_{l,k} \varphi_l(i) \end{cases} \tag{7.100}$$

Furthermore, the algorithm (7.100) will reduce to the DGMN with no measurement exchange when $P_1 = \mathbf{I}$, $P_2 = \mathbf{I}$ and is given as

$$\begin{cases} \varphi_k(i) = \mathbf{w}_k(i-1) + \mu_k \left(\omega |e_l(i)|^{p-1} + (1-\omega)|e_l(i)|^{q-1} \right) \text{sign}(e_l(i)) \mathbf{u}_l(i) \\ \text{where} \quad e_k(i) = d_k(i) - \mathbf{u}_k^T(i) \mathbf{w}_k(i-1) \\ \mathbf{w}_k(i) = \sum_{l \in N_k} \delta_{l,k} \varphi_l(i) \end{cases} \tag{7.101}$$

The equations of (7.100) and (7.101) are similar to the ATC diffusion LMS (ATCDLMS), and the diffusion LMS (DLMS), respectively. Especially, the algorithms with no measurement exchange with lower computation complexity compared with (7.100) have been shown excellent performance, and most of the existing DAF algorithms select this form. To this end, we mainly focus on the DGMN algorithm represented by (7.101).

Remark 7.3

Clearly, the DGMN in (7.101) can be viewed as the DLMS with a variable step size. $\mu_k \left(\omega |e_k(i)|^{p-1} + (1-\omega)|e_k(i)|^{q-2} \right)$ In addition, no exchange of data is needed during the adaptation of the step size, which makes the communication cost relatively low.

Remark 7.4

The choice of mixing parameter is an open question. Here a method is introduced to obtain the parameter adaptively by the sigmoidal function of the error at node k. The detailed form is

$$\omega_k(i) = \frac{1}{1 + \exp\left(-e_k^2(i)\right)} \tag{7.102}$$

In this case, $\omega_k(i)$ is thresholded so that it belongs to the range of values [0, 1]. This method can gain two advantages, including (i) it can avoid the setting of the mixed-norm parameter by human; (ii) the impact of the various components can be adjusted automatically with the variation of the output error. The validity and efficiency of the method in (7.102) will be shown by the simulation results.

7.4.2 Convergence Performance Analysis

In this part, the convergence and the steady-state behavior of the DGMN algorithm are studied. For tractable analysis, the following assumptions are firstly given [273].

Assumption 7.6: All input regressors $\{\mathbf{u}_k(i)\}$ arise from Gaussian sources with zero-mean and spatially and temporally independent.

Assumption 7.7: The error nonlinearity $f(e_l(i))$ is independent of the input regressors $\mathbf{u}_k(i)$, error components $e_k(i)$ and measurement noise $v_k(i)$.

Assumption 7.8: At the steady state, the error signal $e_k(i)$ is approximately equal to the noise component $v_k(i)$.

Assumption 7.6 may not be realistic in general, it is usually applied in the analysis of adaptive filtering algorithms with error nonlinearity [302,331] and the family of diffusion LMS algorithms. Although Assumption 7.7 cannot really hold for the mixed-norm algorithm because the $f(e_l(i))$ is usually a function of the error, so independence cannot hold in general. However, the step-sizes in adaptive filtering [379,381,382] are often assumed to be independent of the input regressors and Assumption 7.7 is used in the mean square performance analysis for VSS diffusion LMS algorithm and the results match empirical results well. The $f(e_l(i))$ can be regarded as a VSS method, since this assumption is feasible for the analysis of the DGMN. Assumption 7.8 is valid because the step-size generally changes very slowly compared to the input regressors and the error signals.

7.4.2.1 *Signal Modeling*

To perform the analysis, the error quantities for node k are defined as follows:

$$\tilde{\mathbf{w}}_k(i) = \mathbf{w}_\circ - \mathbf{w}_k(i), \quad \tilde{\varphi}_k(i) = \mathbf{w}_\circ - \varphi_k(i) \tag{7.103}$$

and then we can obtain the global network and intermediate network weight error vectors as

$$\tilde{\mathbf{w}}(i) = \left[\tilde{w}_1(i),\ldots,\tilde{w}_N(i)\right]^T, \quad \tilde{\varphi}(i) = \left[\tilde{\varphi}_1(i),\ldots,\tilde{\varphi}_N(i)\right]^T \tag{7.104}$$

We also introduce the diagonal matrix $\Upsilon = \mathrm{diag}\{\mu_1, \mu_2, \ldots, \mu_N\}$, and let

$$S(i) = \mathrm{diag}\left(\omega\left|e_1(i)\right|^{p-2} + (1-\omega)\left|e_1(i)\right|^{q-2}\right), \ldots, \omega\left|e_N(i)\right|^{p-2} + (1-\omega)\left|e_N(i)\right|^{q-2} \tag{7.105}$$

We further define the extended weighting matrices as

$$P_3 = \mathbf{P}_3 \otimes I_L \tag{7.106}$$

where ⊗ denotes Kronecker product operations. Similar to other works, we define

$$\mathbf{D}(i) = \text{diag}\left\{\mathbf{u}_1^T(i)\mathbf{u}_1(i),\dots,\mathbf{u}^T{}_N(i)\mathbf{u}_N(i)\right\} \tag{7.107}$$

$$\mathbf{G}(i) = \text{col}\left\{\mathbf{u}^T{}_1(i)v_1(i),\dots,\mathbf{u}^T{}_N(i)v_N(i)\right\} \tag{7.108}$$

According to the definition above, we have

$$\tilde{\boldsymbol{\varphi}}(i) = \tilde{\mathbf{w}}(i-1) - \Upsilon S(i)\left[\mathbf{D}(i)\tilde{\mathbf{w}}(i-1) + \mathbf{G}(i)\right] \tag{7.109}$$

$$\tilde{\mathbf{w}}(i) = P_3^T \tilde{\boldsymbol{\varphi}}(i) \tag{7.110}$$

Or, equivalently

$$\begin{aligned}\tilde{\mathbf{w}}(i) &= P_3^T\tilde{\mathbf{w}}(i-1) - P_3^T\Upsilon S(i)\left[D(i)\tilde{\mathbf{w}}(i-1) + \mathbf{G}(i)\right] \\ &= P_3^T\left[I - \Upsilon S(i)\mathbf{D}(i)\right]\tilde{\mathbf{w}}(i-1) - P_3^T\Upsilon S(i)\mathbf{G}(i)\end{aligned} \tag{7.111}$$

To analyze the mean and mean square performance, we define

$$\begin{aligned}\bar{\mathbf{D}} = E\mathbf{D}(i) &= \text{diag}\left\{E\left[u_1^T(i)u_1(i),\dots,E\left[u_N^T(i)u_N(i)\right]\right]\right\} \\ &= \text{diag}\left\{\mathbf{R}_{u,1},\dots,\mathbf{R}_{u,N}\right\}\end{aligned} \tag{7.112}$$

$$\bar{\mathbf{G}} = E\mathbf{G}(i)\mathbf{G}^T(i) = \text{diag}\left\{\sigma^2{}_{v,1}\mathbf{R}_{u,1},\dots,\sigma^2{}_{v,N}\mathbf{R}_{i,N}\right\} \tag{7.113}$$

where $\mathbf{R}_{u,k} = E\left[u_k(i)u_k^T(i)\right]$ denotes a weighted auto-correlation matrix of the input signal, and $\sigma^2{}_{v,k} = E\left[v_k(i)v_k^T(i)\right]$ is a weighted variance of the noise.

7.4.2.2 Mean Performance

From Assumption 7.6, *D*(*i*) is independent of $\tilde{\mathbf{w}}(i-1)$, which depends on regression up to time $i-1$. Using Assumption 7.7, and taking expectation of (7.111) yields

$$\begin{aligned}E\left[\tilde{\mathbf{w}}(i)\right] &= E\left[P_3^T\left[I - \Upsilon S(i)\mathbf{D}(i)\right]\tilde{\mathbf{w}}(i-1) - P_3^T\Upsilon S(i)\mathbf{G}(i)\right] \\ &= P_3^T\left[I - \Upsilon E\left[S(i)\mathbf{D}(i)\right]\right]\tilde{\mathbf{w}}(i-1) - P_3^T\Upsilon E\left[S(i)\mathbf{G}(i)\right] \\ &= P_3^T\left[I - \Upsilon \mathbf{S}(i)\bar{\mathbf{D}}\right]E\left[\tilde{\mathbf{w}}(i-1)\right]\end{aligned} \tag{7.114}$$

where

$$\mathbf{S}(i)=E[S(i)]$$
$$=\mathrm{diag}\left\{E\left\{\omega|e_1(i)|^{p-2}+(1-\omega)|e_1(i)|^{q-2}\right\}\cdots E\left\{\omega|e_N(i)|^{p-2}+(1-\omega)|e_N(i)|^{q-2}\right\}\right\}$$
$$=\mathrm{diag}\{\mathbf{S}_1(i),\ldots,\mathbf{S}_N(i)\} \tag{7.115}$$

The above equation (7.114) describes the mean behavior of the DGMN algorithm. Employing the Lemma 1 in [92] and (7.114), we have asymptotic unbiasedness if, and only if the matrix $I-\Upsilon\mathbf{S}(i)\mathbf{D}$ is stable. Thus, we require $I-\mu_k\mathbf{S}_k(i)\mathbf{R}_{u,k}$ to be stable for all k, which is equivalent to $|1-\mu_k\mathbf{S}_k(i)\lambda_{\max}(\mathbf{R}_{u,k})|<1$ and we obtain

$$0<\mu_k<\frac{2}{\mathbf{S}_k(i)\lambda_{\max}(\mathbf{R}_{u,k})},\quad k=1,2,\cdots,N \tag{7.116}$$

Using inequality $|a+b|\le|a|+|b|$ and Holder inequality [43,44] $|\tilde{\mathbf{w}}(i)\mathbf{u}|\le\|\tilde{\mathbf{w}}(i)\|_1\|\mathbf{u}\|_\infty$, we have

$$|e_k(i)|^{p-2}\le\left\||\tilde{\mathbf{w}}_k(i)\mathbf{u}_k(i)|+|v_k(i)|\right\|\le\left\|\|\tilde{\mathbf{w}}_k(i)\|_1\|\mathbf{u}_k(i)\|_\infty+|v_k(i)|\right\|^{p-2} \tag{7.117}$$

$$|e_k(i)|^{q-2}\le\left\||\tilde{\mathbf{w}}_k(i)\mathbf{u}_k(i)|+|v_k(i)|\right\|\le\left\|\|\tilde{\mathbf{w}}_k(i)\|_1\|\mathbf{u}_k(i)\|_\infty+|v_k(i)|\right\|^{q-2} \tag{7.118}$$

By (7.117) and (7.118), (7.116) can be reformulated as

$$0<\mu_k<\frac{2}{E\left[\omega\left\{\left\|\|\tilde{\mathbf{w}}_k(i)\|_1\|\mathbf{u}_k(i)\|_\infty+|v_k(i)|\right\|^{p-2}\right\}+(1-\omega)\left\{\left\|\|\tilde{\mathbf{w}}_k(i)\|_1\|\mathbf{u}_k(i)\|_\infty+|v_k(i)|\right\|^{q-2}\right\}\right]\lambda_{\max}(\mathbf{R}_{u,k})} \tag{7.119}$$

As a result, the algorithm will be stable when the step size is within the bound of (7.119).

Notice that in the absence of mixed-norm of error, i.e. $p=q=2$, the range of step-size values that guarantee the convergence of the DGMN reduces to

$$0<\mu_k<\frac{2}{\lambda_{\max}(\mathbf{R}_{u,k})},\quad k=1,2,\ldots,N$$

and it coincides with that of the DLMS algorithm.

7.4.2.3 Mean Square Performance

This part further studies the mean-square convergence performance of the DGMN algorithm. To do so, we draw lessons from the energy conservation analysis. Then, taking the expectation of the squared weighted norm $\tilde{\mathbf{w}}(i)$ in (7.111), we obtain [273]

$$\begin{aligned} E\|\tilde{\mathbf{w}}(i)\|_{\Sigma}^{2} &= E\|\tilde{\mathbf{w}}(i-1)\|_{[I-\Upsilon S(i)D(i)]\Sigma(I-\Upsilon S(i)D(i))}^{2} \\ &+E\left[\mathbf{G}^{T}(i)S(i)\Upsilon P_3\Sigma P^{T}{}_{3}\Upsilon S(i)\mathbf{D}(i)\right] \end{aligned} \tag{7.120}$$

where Σ is any hermitian positive definite matrix. Under Assumption 7.6, the equation (7.120) can be rewritten as a variance relation in the following form

$$\begin{aligned} E\|\tilde{\mathbf{w}}(i)\|_{\Sigma}^{2} &= E\|\tilde{\mathbf{w}}(i-1)\|_{\Sigma}^{2}+Tr\left[\Sigma P_3^{T}\Upsilon\mathbf{S}(i)\mathbf{G}\Upsilon\mathbf{S}(i)P_3\right] \\ \Sigma' &= P_3\Sigma P^{T}{}_{3}-\mathbf{S}(i)\bar{\mathbf{D}}\Upsilon\mathbf{P}_3\Sigma P^{T}{}_{3}-P_3\Sigma P^{T}{}_{3}\Upsilon\mathbf{S}(i)\bar{\mathbf{D}} \\ &+E\left[\mathbf{D}(i)\Upsilon S(i)P_3\Sigma P^{T}{}_{3}\Upsilon S(i)\mathbf{D}(i)\right] \end{aligned} \tag{7.121}$$

By the notation widely used in corresponding literatures, we let

$$\sigma=\mathrm{vec}(\Sigma),\ \Sigma=\mathrm{vec}^{-1}(\sigma) \tag{7.122}$$

where the $\mathrm{vec}(\cdot)$ notation stacks the columns of its matrix argument on top of each other and $\mathrm{vec}^{-1}(\cdot)$ is the inverse operation. We further use the notation $\|\tilde{\mathbf{w}}(i)\|_{\sigma}^{2}$ to denote $\|\tilde{\mathbf{w}}(i)\|_{\Sigma}^{2}$. Using the property of the Kronecker product

$$\mathrm{vec}(U\Sigma V)=\left(V^{T}\otimes U\right)\mathrm{vec}(\Sigma) \tag{7.123}$$

And the fact that the expectation and vectorization operators commute, we vectorize expression (7.121) for Σ' as follows:

$$\sigma'=\mathrm{vec}(\Sigma')=F(i)\sigma \tag{7.124}$$

where the matrix is given by

$$F(i)=\left\{I-I\otimes\left(\mathbf{S}(i)\bar{\mathbf{D}}\Upsilon\right)-\left(\bar{\mathbf{D}}^{T}\mathbf{S}(i)\Upsilon\right)\otimes I+E\left[\left(\mathbf{D}^{T}(i)S(i)\Upsilon\right)\otimes\left(S(i)\mathbf{D}(i)\Upsilon\right)\right]\right\}(P_3\otimes P_3) \tag{7.125}$$

Using the property $Tr(\Sigma X)=\mathrm{vec}\left(X^{T}\right)^{T}\sigma$, the (7.121) can be reformulated as

$$E\|\tilde{\mathbf{w}}(i)\|_{\sigma}^{2}=E\|\tilde{\mathbf{w}}(i-1)\|_{F(i)\sigma}^{2}+\left[vec\left(P_3^{T}\Upsilon\mathbf{S}(i)\mathbf{G}^{T}\Upsilon\mathbf{S}(i)P_3\right)\right]^{T}\sigma \tag{7.126}$$

Starting from (7.126), we define $r(i) = \text{vec}\left(P_3^T \Upsilon \mathbf{S}(i) \bar{\mathbf{G}}^T \Upsilon \mathbf{S}(i) P_3\right)^T$. Then we obtain

$$E\|\tilde{\mathbf{w}}(i)\|_{\sigma}^2 = E\|\tilde{\mathbf{w}}(i-1)\|_{F(i)\sigma}^2 + r(i)\sigma \tag{7.127}$$

Employing the same method in [91] to expanding (7.127), one can obtain the MSD at node k as

$$MSD_K(i) = E\|\tilde{w}_k(i)\|^2 = E\|w_k(i) - w^o\|^2 = E\|\tilde{\mathbf{w}}(i)\|_{m_k}^2 \tag{7.128}$$

where $m_k = \text{vec}\left(\text{diag}(b_{k,N}) \otimes I_L\right)$, and $b_{k,N}$ is the k-th column vector of diagonal matrix I_N.

Then, we define the network MSD as the average MSD across all nodes in the network as

$$\text{MSD}^{\text{network}} = \frac{1}{N}\sum_{k=1}^{N} \text{MSD}_k = \frac{1}{N} E\|\tilde{w}_k(i)\|^2 \tag{7.129}$$

To obtain the real value for (7.128) and (7.129), F_i and $\mathbf{S}(i) = E[S(i)]$ should be calculated explicitly. We have

$$\begin{aligned} E[S_k(i)] &= E\left\{\omega|e_k(i)|^{p-2} + (1-\omega)|e_k(i)|^{q-2}\right\} \\ &= \omega E\left\{|e_k(i)|^{p-2}\right\} + (1-\omega) E\left\{|e_k(i)|^{q-2}\right\} \end{aligned} \tag{7.130}$$

Since the small fluctuation of $|e_k(i)|$, we get an approximation $E\left\{|e_k(i)|^{p-2}\right\} \approx \left\{E|e_k(i)|^2\right\}^{\frac{p-2}{2}}$. Although this approximation may not be true in general, this idea of approximation is customary in the context of adaptive filters, as it simplifies the analysis [383,384]. Then, (7.130) is changed to

$$\begin{aligned} &E[S_k(i)] \\ &\approx \omega\left\{E|e_k(i)|^2\right\}^{\frac{p-2}{2}} + (1-\omega)\left\{E|e_k(i)|^2\right\}^{\frac{q-2}{2}} \\ &\approx \omega((\sigma_{ek}^2(i))^{\frac{p-2}{2}} + (1-\omega)(\sigma_{ek}^2(i))^{\frac{q-2}{2}} \end{aligned} \tag{7.131}$$

where $\sigma_{ek}^2(i) = E|e_k(i)|^2 = \sigma_{\varepsilon k}^2(i) + \sigma_{vk}^2$, and $\sigma_{\varepsilon k}^2(i) = E|\varepsilon_k(i)|^2 = E|\mathbf{u}_k(i)\tilde{\mathbf{w}}_k(i-1)|^2$.

7.4.2.4 Steady-State Analysis

The step sizes approach the steady-state and will be small for all nodes when $i \to \infty$, thus from (7.126), we have

$$E\left\|\tilde{\mathbf{w}}(\infty)\right\|^2_{(I-F(\infty))\sigma}=\left[\mathrm{vec}\left(P_3^T\Upsilon\mathbf{S}^T(\infty)\bar{\mathbf{G}}^T\Upsilon P_3\right)\right]^T\sigma \tag{7.132}$$

where $F(\infty)=\left\{I-I\otimes\left(\mathbf{S}(\infty)\bar{\mathbf{D}}^T\Upsilon\right)-\left(\mathbf{S}^T(\infty)\bar{\mathbf{D}}^T\Upsilon\right)\otimes I\right\}(P_3\otimes P_3)$ and $\mathbf{S}(\infty)=E\left[S(\infty)\right]$.

Using (7.132), the steady-state MSD at node k and the network MSD can be given by

$$\mathrm{MSD}_k(\infty)=E\left\|\tilde{w}_k(\infty)\right\|^2_{m_k}=\left[\mathrm{vec}\left(P_3^T\Upsilon\mathbf{S}(\infty)\bar{\mathbf{G}}^T\mathbf{S}(\infty)\Upsilon P_3\right)\right]^T(I-F_\infty)^{-1}m_k \tag{7.133}$$

$$\mathrm{MSD}^{\mathrm{network}}(\infty)=\frac{1}{N}\sum_{k=1}^{N}\mathrm{MSD}_k(\infty)=\frac{1}{N}E\left\|\tilde{w}_k(\infty)\right\|^2 \tag{7.134}$$

To achieve $\mathbf{S}(\infty)$, $E\left[S_k(\infty)\right]$ for $k=1,2,\ldots,N$ should be analyzed, and it cannot be easily represented in a closed form. Instead of using a numerical method to derive an exact solution, we adopt Assumption 7.8 to obtain a closed-form solution which is more intuitive than that of the numerical method. According to (7.131) and Assumption 7.8, we obtain

$$\begin{aligned}E\left[S_k(\infty)\right]&\approx\omega\left(\left(\sigma_{e_k}^2(\infty)\right)^{\frac{p-2}{2}}+(1-\omega)\left(\sigma_{e_k}^2(\infty)\right)^{\frac{q-2}{2}}\right)\\&\approx\omega\left(\left(\sigma_{v_k}^2\right)^{\frac{p-2}{2}}+(1-\omega)\left(\sigma_{v_k}^2\right)^{\frac{q-2}{2}}\right)\end{aligned} \tag{7.135}$$

Following Assumption 7.8 and the approximation result above, a negligible discrepancy will be existed in the final values of the MSD of the mean square analysis and its steady-state analysis, but it is tolerated to hold the consistency between these two analyses.

According to (7.132), by introducing the small step size approximation, we have

$$E\left\|\tilde{\mathbf{w}}(i)\right\|^2_\Sigma=E\left\|\tilde{\mathbf{w}}(i-1)\right\|^2_{[I-\Upsilon\mathbf{S}(i)\mathbf{D}(i)]\Sigma(I-\Upsilon\mathbf{S}(i)\mathbf{D}(i))}+Tr\left[\Sigma P_3^T\Upsilon\mathbf{S}(i)\bar{\mathbf{G}}\mathbf{S}(i)\Upsilon P_3\right] \tag{7.136}$$

Furthermore, we assume that the step size is chosen such that $I-\Upsilon\mathbf{S}(i)\bar{\mathbf{D}}$ is stable. Then, the MSD of network is presented as

$$\mathrm{MSD}^{\mathrm{network}}(\infty)=\frac{1}{N}E\left\|\tilde{w}_k(\infty)\right\|^2\approx\frac{1}{N}\sum_{j=0}^{\infty}Tr\left[X^jY\left(X^T\right)^j\right] \tag{7.137}$$

where $X=P_3^T\left[I-\Upsilon\mathbf{S}(i)\bar{\mathbf{D}}\right]$, and $Y=P_3^T\Upsilon\mathbf{S}(i)\bar{\mathbf{G}}\mathbf{S}(i)\Upsilon P_3$

7.4.3 Simulation Results

This part presents some simulation results to demonstrate the performance of the DGMN algorithm for DEoN under Gaussian noise and non-Gaussian noise environments, respectively. The topology of the network with 20 nodes is generated as a realization of the random geometric graph model as shown in Figure 7.12, and the content associated with the topology construction can be seen in the contributions of Sayed in [90]. The input is zero-mean Gaussian, independent in time and space with size $L = 10$. The Metropolis rule is used for combination weights.

7.4.3.1 Gaussian Noise

In this case, the performance of the DGMN algorithm is studied under the Gaussian noise environments. The variances of the input signals and white Gaussian noises are illustrated in Figure 7.13.

First, Figure 7.14 shows the convergence curves in terms of MSD. One can see that the DGMN with different p and q work well in this case. Especially, the DGMN with $p = 2$ and $q = 4$ (i.e. DLMMN) shows excellent performance in convergence rate and accuracy compared with other cases. The results confirm that the DGMN algorithm exhibits a significant performance improvement for distributed estimation over network under the Gaussian noise environments. We also give the results of the steady-state MSDs at each node k, which can be shown in Figure 7.15. From this result, one can easily

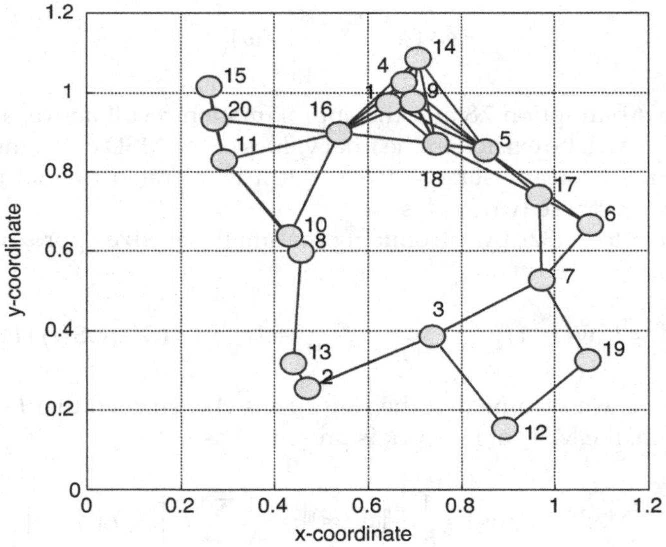

FIGURE 7.12
Network topology of the simulated diffusion network consisting of 20 nodes [268].

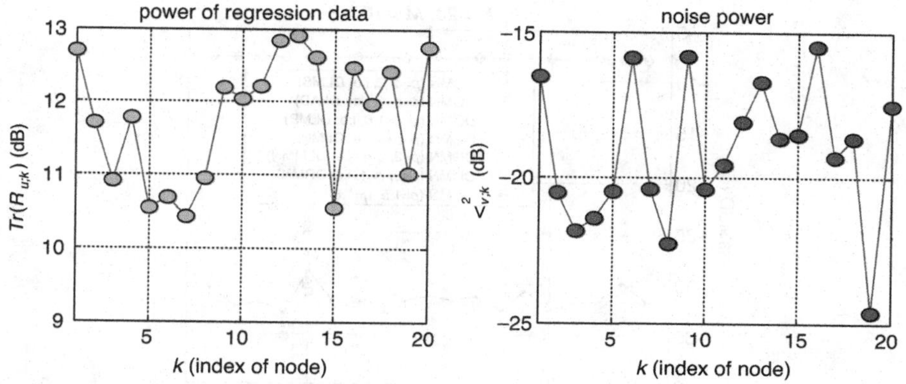

FIGURE 7.13
Variances of the input signals and white Gaussian noises at all nodes in the network [268].

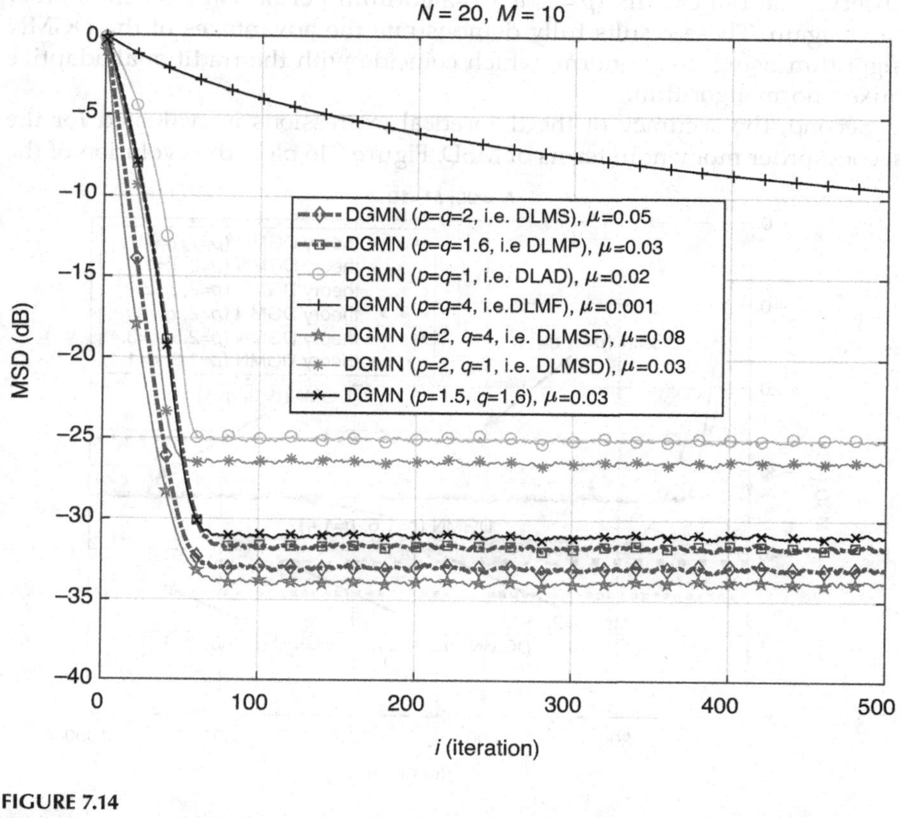

FIGURE 7.14
Transient network MSD for DGMN under Gaussian noise [268].

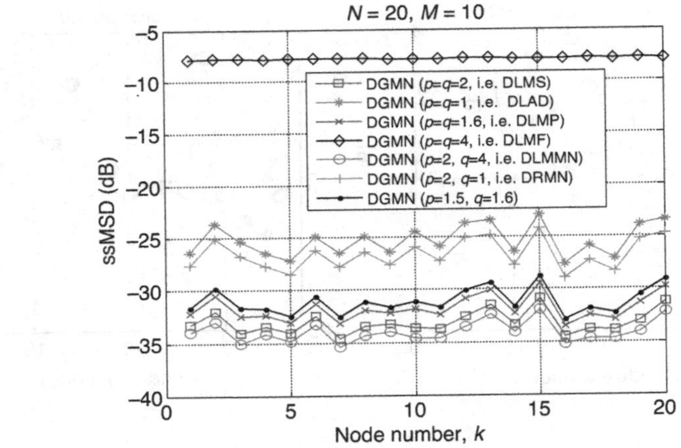

FIGURE 7.15
Steady-state MSDs of DGMN with different p and q at each node [268].

observe that the DGMN ($p = 2$, $q = 4$) algorithm performs better than other cases again. These results fully demonstrate the advantages of the DGMN algorithm using mixed norm, which coincide with the traditional adaptive mixed-norm algorithm.

Second, the accuracy of the theoretical expressions is evaluated for the second-order moment in terms of MSD. Figure 7.16 plots the evolution of the

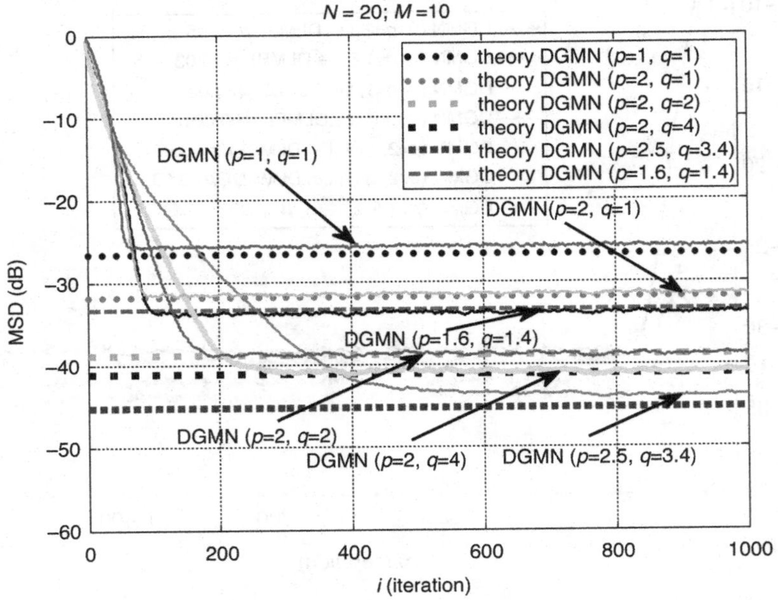

FIGURE 7.16
Evolution of the learning curves [268].

ensemble-average learning curves of DGMN with different p and q with all nodes and the steady-state MSD curves of the theory represented by the dotted lines for the DGMN algorithms with different p and q. The measure corresponds to the average MSD across all agents at time i. The learning curves are obtained by averaging the trajectories over 100 repeated experiments. It is observed in Figure 7.16 that the simulated learning curves tend to the corresponding MSD value predicted by the theoretical expression, which provides a desirable approximation for the performance of distributed strategies for small step-sizes.

Third, the accuracy of the theoretical expressions for steady-state MSDs with different step-sizes is tested. Here the results of the DGMN ($p = 2, q = 4$) and DGMN ($p = 1.6, q = 1.4$) are only given, and the mixing parameter is 0.65. Figure 7.17 shows the steady-state MSDs with different step-sizes. One can observe that the steady-state MSDs are increasing with step-size, and the steady-state MSDs computed by simulations match well with the theoretical values. This result demonstrates the consistency of the theoretical values and the simulation results, which means that the theoretical analysis for the DGMN method is reasonable.

7.4.3.2 Mixture Noise

In this part, the measurement noise with a mixture of long-tailed and short-tailed distributions is considered. Three kinds of additive noises with zero-mean are generated similar to that reported in [309] to show the performance of the DGMN: i) Mixture of a Laplace distributed noise of power 0.08 and a

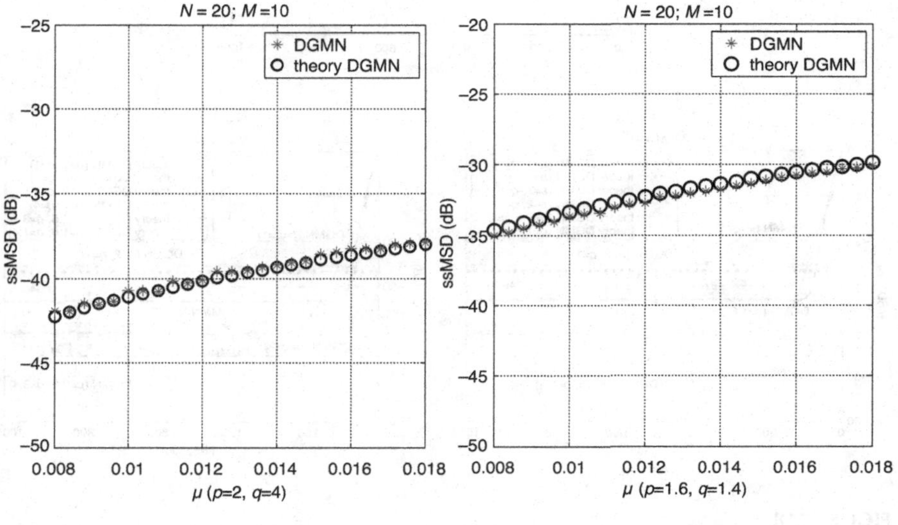

FIGURE 7.17
Theoretical and simulated steady-state MSDs with different step-sizes [268].

uniformly distributed noise of power 1.0; ii) Mixture of a Gaussian distributed noise of power 1.0 and a uniformly distributed noise of power 0.15; iii) Mixture of a Bernoulli distributed noise of power 1.0 and a Laplace distributed noise of power 0.2.

First, the steady-state MSDs of the theory and simulation results are evaluated for DGMN with different p and q. Figure 7.18a–c are the evolution of the ensemble-average learning curves of DGMN with all nodes and the steady-state MSD curves of the theory under the measurement noise i–iii above mentioned, respectively. As one can see, the theoretical results are in good agreement with simulation results within a certain tolerance range. Therefore, it is concluded that (7.137) is only an approximate solution.

Second, the misadjustment of the DGMN method is investigated in terms of the MSD. In this case, the input signal is used as a zero-mean uniformly distributed with unity power series similar to the reference in [309]. The mixing parameters are 0.55, 0.85, and 0.85, for case i–iii, respectively. The corresponding convergence behavior in terms of MSDs and the steady-state MSDs

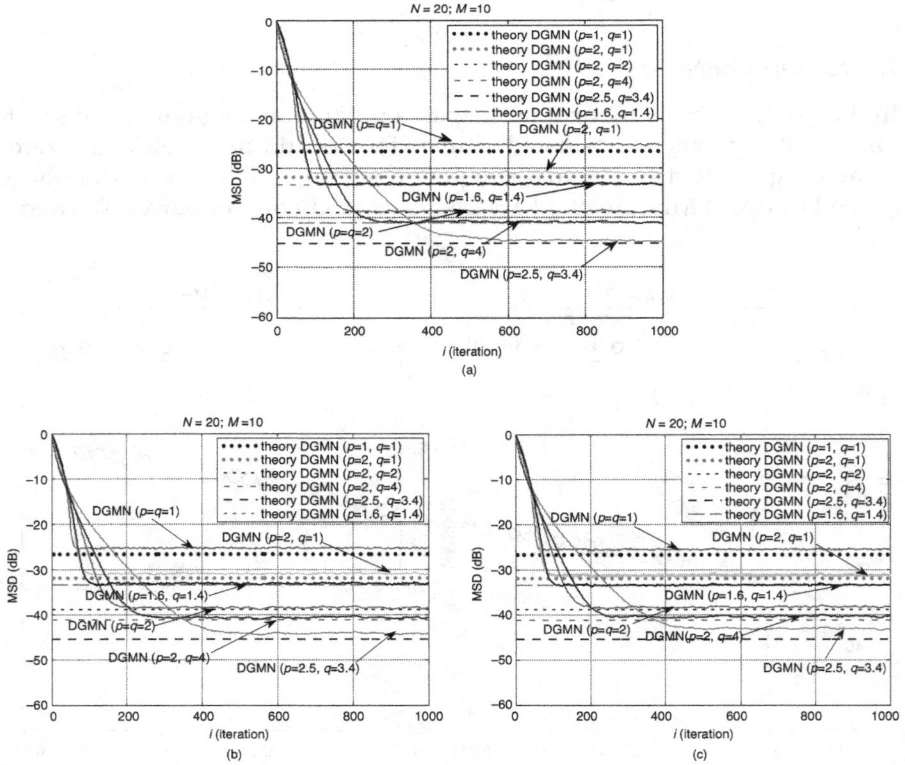

FIGURE 7.18
Evolution of the learning curves under different measurement noises [268].

at each node k are shown in Figure 7.19a–c and a1–c1, respectively. From these results, one can observe that the DGMN, especially for $p=2$, $q=4$, exhibits superior steady-state behavior than other algorithms, which demonstrates clearly the potential of the DGMN for practical problems when the measurement noise has mixed statistical features.

7.5 Kernel Adaptive Filters under Mixture MMPE

A convex combination of different norms, also known as the mixture MMPE criterion, can be used to address the issue of KAFs with single criterion. Two examples of this approach are the kernel least mean mixed-norm (KLMMN) [364,385,386] and kernel robust mixed-norm (KRMN) [363]. The KLMMN

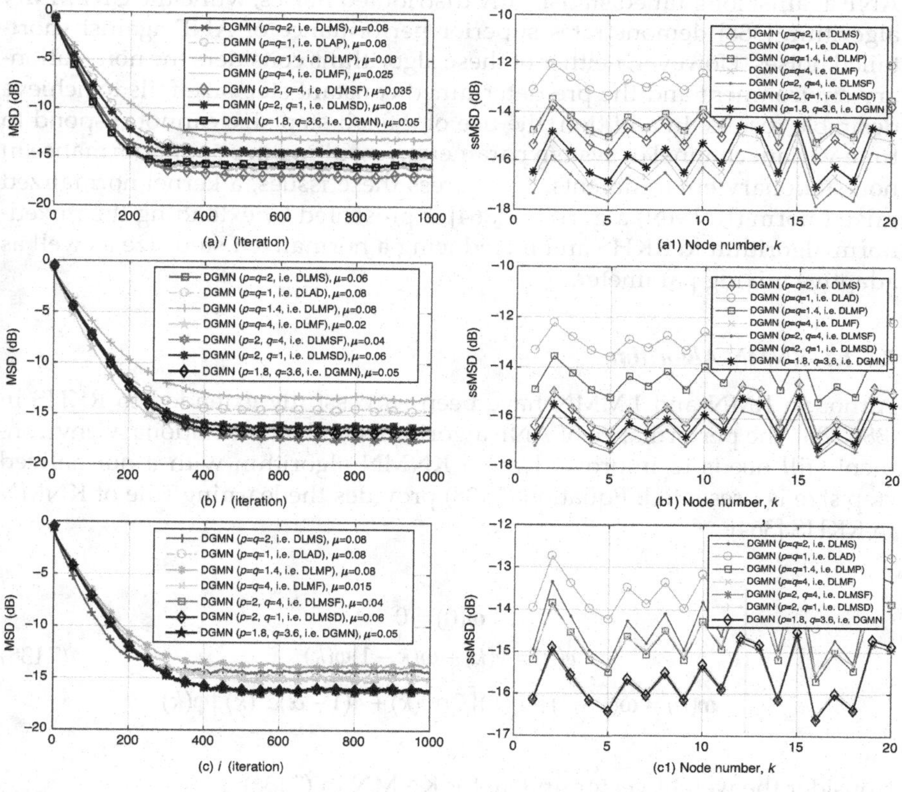

FIGURE 7.19
Transient network MSD and steady-state MSD at node k of the DGMN under different measurement noises mentioned above [268].

algorithm combines the strengths of the well-established LMS and LMF algorithms in kernel space, inspired by the LMMN algorithm. To enhance the robustness of the original KLMS algorithm and handle impulsive noises, the KRMN algorithm was developed, leveraging the advantages of the RMN algorithm. However, these algorithms still suffer from slow convergence and suboptimal steady-state performance due to their reliance on stochastic gradient-based methods for updating the solution. To address this issue, recursive update rules with superior tracking ability can be employed to improve the convergence performance of AFAs. In this subchapter, we introduce the kernel recursive generalized mixed-norm (KRGMN) algorithm [387], which achieves excellent convergence performance by appropriately combining l_p and l_q-norms of the error.

7.5.1 Kernel Normalized Mixed-Norm Algorithm

The KRMN [363] algorithm was designed to improve the performance of ANF against long-tailed statistically distributed noises, while the QKLMMN algorithm [388] demonstrates superior performance of ANF against short-tailed noises. However, neither of these algorithms considers the non-stationary environment and the pre-determined constant step size fails to achieve desirable results. In addition, the use of a fixed step size cannot respond to time-variant channel or system parameters, resulting in poor performance in non-stationary environments. To address these issues, a kernel normalized mixed-norm (KNMN) algorithm [364] is presented by extending the mixed-norm algorithm to RKHS and introducing a normalized step size as well as adaptive mixing parameter.

7.5.1.1 KNMN Algorithm

Although RMN and LMMN have been successfully extended to RKHS in [386,388], the performance of ANF algorithms in the non-stationary environment still needs to improve. Here a KNMN algorithm with a normalized step size is presented. Equation (7.138) provides the learning rule of KNMN in RKHS [364]:

$$\begin{cases} \omega(0) = 0 \\ e(k) = e(k) + \omega(k-1)\varphi(k) \\ \omega(k) = \omega(k-1) + \eta(k)\left[2\alpha e(k) + 4(1-\alpha)e^3(k)\right]\varphi(k) \end{cases} \tag{7.138}$$

Consider the weight vector update for KNMN in (7.138):

$$\omega(k) = \omega(k-1) + \eta(k)\left[2\alpha e(k) + 4(1-\alpha)e^3(k)\right]\varphi(k) \tag{7.139}$$

Multiplying $\varphi(k)$ on the transpose of both sides of (7.139), we have

$$\omega^T(k)\varphi(k) = \omega^T(k-1)\varphi(k) + \eta(k)\left[2\alpha e(k) + 4(1-\alpha)e(k)^3\right]\kappa\langle \mathbf{u}(k),\mathbf{u}(k)\rangle \quad (7.140)$$

Similar to [389], the posterior error is defined as

$$e_{\text{post}}(k) = d(k) - \omega(k)\varphi(k) \quad (7.141)$$

After subtracting $d(k)$ from both sides of (7.140), we have

$$e_{\text{post}}(k) = e(k)\left\{1 - \eta(k)\kappa\langle \mathbf{u}(k),\mathbf{u}(k)\rangle\left[2\alpha + 4(1-\alpha)e(k)^2\right]\right\} \quad (7.142)$$

To minimize the square of the posterior error $e_{\text{post}}(k)$, the term inside the large bracket in (7.142) is set to zero. The resulting normalization formula for $\eta(k)$ is given by

$$\eta(k) = \frac{e(k)}{\left[2\alpha e(k) + 4(1-\alpha)e(k)^3\right]\kappa\langle \mathbf{u}(k),\mathbf{u}(k)\rangle} \quad (7.143)$$

The coefficient update equation can be expressed as

$$a_k = \eta(k)\left[2\lambda e(k) + 4(1-\lambda)e(k)^3\right] \quad (7.144)$$

Therefore, the detailed algorithm is summarized in Table 7.3 [364].

TABLE 7.3

KNMN Algorithm

Initialization

Choose initial step size $\eta(1)$, mixing parameter λ

$a_1 = \eta(1)[2\lambda d(1) + 4(1-\lambda)d(1)^3]$

Computation

while $\{\mathbf{u}(k), d(k)\}$ available do

1. evaluate the output

$$y(k) = \sum_{j=1}^{k} a_j \kappa(\mathbf{u}(j), \mathbf{u}(k))$$

2. compute the error term

$e(k) = d(k) - y(k)$

3. compute the normalized step size

$$\eta(k) = \frac{e(k)}{[2\lambda e(k) + 4(1-\lambda)e(k)^3]\kappa\langle \mathbf{u}(k),\mathbf{u}(k)\rangle}$$

4. update the coefficient

$a_k = \eta(k)[2\lambda e(k) + 4(1-\lambda)e(k)^3]$

End while

For mixed-norm algorithms, there are two approaches to set the mixing parameter $A(k)$: a constant value which is fixed during learning and an adaptive value that has better tracking behavior to the nonstationarity of the operation environment. The optimal adaptive mixing parameter has been extensively studied before [390], which is given by:

$$\lambda(k)=\left\{\left(\mathbf{d}>|d(k)|\right)\cup\left(\mathbf{d}<-|d(k)|\right)\right\} \tag{7.145}$$

where $\mathbf{d}$ models the desired signal $\mathbf{d}(k)$ in the absence of additive noises. In the case $\mathbf{d}\sim N\left(0,\sigma_d^2\right)$, equation (7.145) becomes

$$\lambda(k)=2erfc\left\{\frac{|d(k)|}{\tilde{\sigma}_d}\right\} \tag{7.146}$$

where $\tilde{\sigma}_d$ is the estimator of the standard deviation of $d(k)$. The underlying concept of (7.146) is that the probability of the desired signal containing outliers can be approximated as $1-\lambda(k)$, i.e., the LMS is progressively replaced by LMF in RKHS as the probability of an outlier increases. Estimation of (7.146) is computationally consuming for real-time implementation [390]. Thus, an outlier trimmed sliding window [391] is applied to provide an effective simplification strategy. In this method, a threshold T is set as follows:

$$\tau=\frac{\theta\sum_{i=1}^{k}|d(i)|}{k} \tag{7.147}$$

where the coefficient θ is fixed to 1.5 and k is the window length (set to 10 in this case). If the incoming desired output d is larger than the threshold, small $\lambda(k)$ is selected, whereas if d is smaller than the threshold, then $\lambda(k)$ close to 1 is selected. The value of θ is determined heuristically to keep a balance between algorithm stability and convergence rate. Usually, a small value of θ tends to excite unstability during convergence while a large value may reduce the convergence rate. We denote KNMN with adaptive mixing parameter (AMP) as KNMN-AMP.

7.5.1.2 Mean Square Convergence Analysis

This part gives mean square convergence analysis of the KNMN algorithm based on fundamental energy conservation relation (ECR) [364].

Theorem 7.3

The mean weight vector in RKHS $E[\omega(k)]$ converges as $k\to\infty$ if $\eta(k)$ satisfies (7.143) and the noise variance is relatively small.

Proof: Consider the weight vector update for KNMN in (7.139):

$$\boldsymbol{\omega}(k)=\boldsymbol{\omega}(k-1)+\eta(k)\left[2\lambda e(k)+4(1-\lambda)e(k)^3\right]\boldsymbol{\varphi}(k) \tag{7.148}$$

We define weight error vector as $\tilde{\boldsymbol{\omega}}(k)=\boldsymbol{\omega}_o-\boldsymbol{\omega}(k)$, a priori error as $e_a(k)=\tilde{\boldsymbol{\Omega}}(k-1)\boldsymbol{\varphi}(k)$, and a posteriori error as $e_p(k)=\tilde{\boldsymbol{\omega}}(k)\boldsymbol{\varphi}(k)$. Here $\boldsymbol{\Omega}_o$ is the optimal weight vector, after subtracting it from and then multiplying $\varphi(k)$ to both sides of (7.148), we have

$$\tilde{\boldsymbol{\omega}}(k)=\tilde{\boldsymbol{\omega}}(k-1)-\eta(k)\left[2\lambda e(k)+4(1-\lambda)e(k)^3\right]\boldsymbol{\varphi}(k) \tag{7.149}$$

And

$$e_p(k)=e_a(k)-\eta(k)\kappa\langle\mathbf{u}(k),\mathbf{u}(k)\rangle\left[2\lambda(k)e(k)+4(1-\lambda(k))e(n)^3\right] \tag{7.150}$$

Combining (7.149) and (7.150), we get

$$\tilde{\boldsymbol{\omega}}(k)=\tilde{\boldsymbol{\omega}}(k-1)+\frac{\left(e_p(k)-e_a(k)\boldsymbol{\varphi}(k)\right)}{\kappa\langle\mathbf{u}(k),\mathbf{u}(k)\rangle} \tag{7.151}$$

Squaring both sides of (7.151) and after straightforward derivation, we have:

$$\|\tilde{\boldsymbol{\omega}}(k)\|_{\mathbf{F}}^2+\frac{e_a^2(k)}{\kappa\langle\mathbf{u}(k),\mathbf{u}(k)\rangle}=\|\tilde{\boldsymbol{\omega}}(k-1)\|_F^2+\frac{e_p^2(k)}{\kappa\langle\mathbf{u}(k),\mathbf{u}(k)\rangle} \tag{7.152}$$

where $\|\tilde{\boldsymbol{\omega}}(k)\|_{\mathbf{F}}^2=\tilde{\boldsymbol{\omega}}^T(k)\tilde{\boldsymbol{\omega}}(k)$ and subscript F refers to the high-dimensional feature space.

Equation (7.152) is the ECR for KNMN, which has the same expression for KLMS [122]. Substituting (7.150) into (7.152) and then taking expectations, we have

$$E\left[\|\tilde{\boldsymbol{\omega}}(k)\|_{\mathbf{F}}^2\right]=E\left[\|\tilde{\boldsymbol{\omega}}(k-1)\|_{\mathbf{F}}^2\right]-\frac{1}{\kappa\langle\mathbf{u}(k),\mathbf{u}(k)\rangle}\left\{2E\left[e_a(k)e(k)\right]-E\left[e^2(k)\right]\right\} \tag{7.153}$$

where $E\left[\|\tilde{\boldsymbol{\omega}}(k)\|_{\mathbf{F}}^2\right]$ is called the WEP in F.

We also have

$$\begin{aligned} e(k)&=d(k)-\boldsymbol{\omega}^T(k-1)\boldsymbol{\varphi}(k)\\ &=d(k)-\left[\boldsymbol{\omega}_o-\boldsymbol{\omega}^T(k-1)\right]^T\boldsymbol{\varphi}(k)\\ &=e_a(k)+v(k)\end{aligned} \tag{7.154}$$

where $v(k)$ is the additive noise, $e_a(k)$ is the *a priori* error. By combining (7.153) and (7.154) and assuming that the additive noise $v(k)$ is zero-mean, independent, identically distributed, and independent of the *a priori* error $e_a(k)$, we get

$$E\left[\|\tilde{\omega}(k)\|_{\mathbf{F}}^2\right]=E\left[\|\tilde{\omega}(k-1)\|_{\mathbf{F}}^2\right]-\frac{1}{\kappa\langle\mathbf{u}(k),\mathbf{u}(k)\rangle}\left\{E\left[e_a^2(k)\right]-\varsigma_v^2\right\} \quad (7.155)$$

where ς_v^2 is the noise variance.

To make KNMN converge after finite iterations, we have

$$E\left[\|\tilde{\omega}(k)\|_{\mathbf{F}}^2\right]\le E\left[\|\tilde{\omega}(k-1)\|_{\mathbf{F}}^2\right]\Leftrightarrow E\left[e_a^2(k)\right]\ge\varsigma_v^2 \quad (7.156)$$

According to (7.158), the convergence performance of KNMN depends on $v(k)$ through ς_v^2 only. As previously mentioned, we mainly focus on noises with short-tailed distributions, which always have small variances, thus (7.158) normally holds.

7.5.1.3 Simulation Results

Here some simulation results are given to compare the KAPA [20], KNLMS [364], KRMN [363], KNMN, and KNMN-AMP algorithms in terms of the performance in nonlinear system identification. A discrete-time nonlinear dynamic system is considered, which has already been investigated in [392,393]. The input signal is generated from Gaussian distributed input sequence with zero-mean and standard deviation 0.25, and the noise is a linear combination of Gaussian noise and Bernoulli noise with different power ratios. The learning curves are shown in Figure 7.20. One can see that KNMN and KNMN-AMP can always achieve faster convergence rate and lower steady-state MSE than their KNLMS, KAPA-2, and KRMN counterparts. Meanwhile, the performance of KNMN-AMP is slightly better than that of KNMN, although it has higher computational complexity.

7.5.2 Kernel Recursive Generalized Mixed-Norm Algorithm

7.5.2.1 KRGMN Algorithm

This part presents the KRGMN algorithm [387] employing the GMN and kernel method. A weighted cost function (7.157) is defined by an exponentially-weighted mechanism to put more emphasis on recent data and to deemphasize data from the remote past [387], i.e.,

$$J(\omega(k))=\lim_{\omega}\sum_{j=1}^{k}\gamma^{k-j}\{\frac{\lambda}{p}\left|d(j)-\omega^T(k)\varphi(j)\right|^p+\frac{1-\lambda}{q}\left|d(j)-\omega^T(k)\varphi(j)\right|^q\}+\frac{1}{2}\gamma^k\beta$$

$$\|\omega(k)\|^2 \quad (7.157)$$

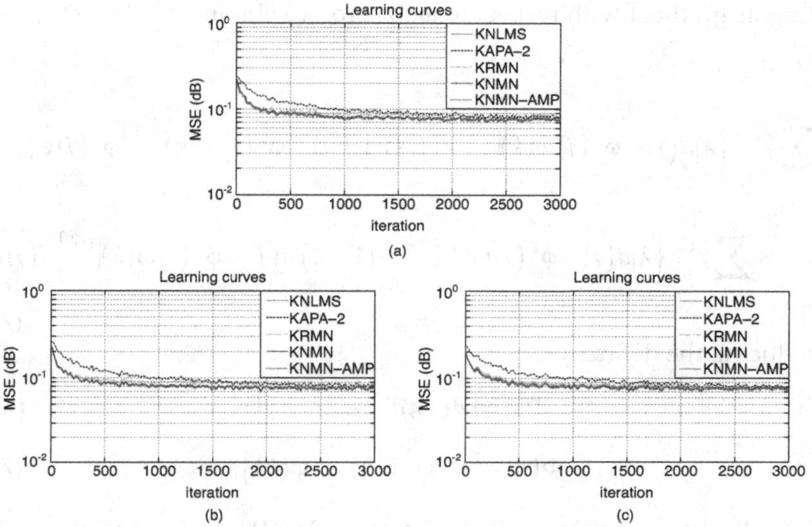

FIGURE 7.20
Simulation results: the learning curves of different algorithms in nonlinear system identification, where the power ratios for Gaussian noise and Bernoulli noise are set to be (a) 1: 1, (b) 1: 10, and (c) 1: 100, respectively [364].

where γ denotes the forgetting factor, usually $0 \le \gamma \le 1$, and β is the regularization factor. The second term is a norm penalizing term, which is designed to guarantee the existence of the inverse of the data auto-correlation matrix, especially during the initial update stage. Furthermore, the regularization term is weighted by γ, which deemphasizes regularization as time progresses. The gradient descent method is used to get the optimal solution as

$$\begin{aligned}
&\frac{\partial J(\omega(k))}{\partial \omega(k)} \\
&= -\sum_{j=1}^{k} \gamma^{k-j} \left\{ \lambda \,|\, d(j) - \varphi^T(j)\omega(k) \,|^{p-2} + (1-\lambda) \,|\, d(j) - \varphi^T(j)\omega(k) \,|^{q-2} \right\} \left(d(j) - \varphi^T(j)\omega(k) \right) \varphi^T \\
&\quad (j) + \gamma^k \beta \omega(k) \\
&= -\sum_{j=1}^{k} \gamma^{k-j} \left\{ \lambda \,|\, d(j) - \varphi^T(j)\omega(k) \,|^{p-2} + (1-\lambda) \,|\, d(j) - \varphi^T(j)\omega(k) \,|^{q-2} \right\} d(j)\varphi^T(j) + \gamma^k \beta \omega(k) \\
&\quad + \sum_{j=1}^{k} \gamma^{k-j} \left\{ \lambda \,|\, d(j) - \varphi^T(j)\omega(k) \,|^{p-2} + (1-\lambda) \,|\, d(j) - \varphi^T(j)\omega(k) \,|^{q-2} \right\} \varphi^T(j)\varphi(j) + \gamma^k \beta \omega(k)
\end{aligned} \tag{7.158}$$

Setting its gradient with respect to $\boldsymbol{\omega}$ to zero, we obtain

$$\boldsymbol{\omega}(k) = \left(\sum_{j=1}^{k} \gamma^{k-1}\left\{\lambda\left|d(j)-\boldsymbol{\varphi}^T(j)\boldsymbol{\omega}(k)\right|^{p-2}+(1-\lambda)\left|d(j)-\boldsymbol{\varphi}^T(j)\boldsymbol{\omega}(k)\right|^{q-2}\right\}\boldsymbol{\varphi}^T(j)\boldsymbol{\varphi}(j)+\gamma^i\beta\right)^{-1}$$
$$\times\sum_{j=1}^{k}\gamma^{k-j}\left\{\lambda\left|d(j)-\boldsymbol{\varphi}^T(j)\boldsymbol{\omega}(k)\right|^{p-2}+(1-\lambda)\left|d(j)-\boldsymbol{\varphi}^T(j)\boldsymbol{\omega}(k)\right|^{q-2}\right\}d(j)\boldsymbol{\varphi}^T(j) \quad (7.159)$$

Introducing the defines of

$$\mathbf{d}(i)=\left[d(1),d(2),\ldots,d(i)\right] \quad (7.160)$$

$$\boldsymbol{\Phi}(k)=\left[\boldsymbol{\varphi}(1),\boldsymbol{\varphi}(2),\ldots,\boldsymbol{\varphi}(k)\right] \quad (7.161)$$

$$\begin{aligned}\boldsymbol{\Psi}(k) = \mathrm{diag}[&\gamma^{k-1}\{\lambda\,|\,d(1)-\boldsymbol{\varphi}^T(1)\boldsymbol{\omega}(k)\,|^{p-2}+(1-\lambda)\,|\,d(1)-\boldsymbol{\varphi}^T(1)\omega(k)\,|^{q-2}\},\\ &\gamma^{k-2}\{\lambda\,|\,d(2)-\boldsymbol{\varphi}^T(2)\boldsymbol{\omega}(k)\,|^{p-2}+(1-\lambda)\,|\,d(2)-\boldsymbol{\varphi}^T(2)\boldsymbol{\omega}(k)\,|^{q-2}\},\ldots,\\ &\{\lambda\,|\,d(j)-\boldsymbol{\varphi}^T(j)\boldsymbol{\omega}(k)\,|^{p-2}+(1-\lambda)\,|\,d(j)-\boldsymbol{\varphi}^T(j)\boldsymbol{\omega}(k)\,|^{q-2}\}]\end{aligned} \quad (7.162)$$

Then, we obtain the matrix form of (7.159) as

$$\boldsymbol{\omega}(k)=\left(\boldsymbol{\Phi}(k)\boldsymbol{\Psi}(k)\boldsymbol{\Phi}^T(k)+\gamma^k\beta\mathbf{I}\right)^{-1}\boldsymbol{\Phi}(k)\boldsymbol{\Psi}(k)\mathbf{d}(k) \quad (7.163)$$

Now, applying the following matrix inversion lemma with the identifications $\beta\gamma^k\mathbf{I}\to A, \boldsymbol{\Phi}(k)\to B, \boldsymbol{\Psi}(k)\to C, \boldsymbol{\Phi}^T(k)\to D$, we have

$$\left(\boldsymbol{\Phi}(k)\boldsymbol{\Psi}(k)\boldsymbol{\Phi}^T(k)+\gamma^i\beta I\right)^{-1}\boldsymbol{\Phi}(k)\boldsymbol{\Psi}(k)=\boldsymbol{\Phi}(k)(\boldsymbol{\Phi}(k)\boldsymbol{\Phi}^T(k)+\gamma^k\beta\boldsymbol{\Psi}(k)^{-1})^{-1} \quad (7.164)$$

Substituting the above result into (7.163) yields

$$\boldsymbol{\omega}(k)=\boldsymbol{\Phi}(k)(\boldsymbol{\Phi}(k)\boldsymbol{\Phi}^T(k)+\gamma^k\beta\boldsymbol{\Psi}(k)^{-1})^{-1}\mathbf{d}(k) \quad (7.165)$$

Then, the weight vector can be expressed explicitly as a linear combination of the input data in kernel space as

$$\boldsymbol{\omega}(k)=\boldsymbol{\Phi}(k)\mathbf{a}(k) \quad (7.166)$$

where $\mathbf{a}(k)$ denotes the computable expansion coefficients vector of the weight by kernel trick and is defined by

$$\mathbf{a}(k)=(\boldsymbol{\Phi}(k)\boldsymbol{\Phi}^T(k)+\gamma^k\beta\boldsymbol{\Psi}(k)^{-1})\mathbf{d}(k) \quad (7.167)$$

We further denote

$$\mathbf{Q}(k)=\left(\mathbf{\Phi}(k)\mathbf{\Phi}^T(k)+\gamma^k\beta\mathbf{\Psi}(k)^{-1}\right)^{-1} \tag{7.168}$$

where $\mathbf{\Phi}(k)=\{\mathbf{\Phi}(k-1),\boldsymbol{\varphi}(k)\}$, we further have

$$\mathbf{Q}(k)=\begin{bmatrix}\mathbf{\Phi}(k-1)\mathbf{\Phi}^T(k-1)+\gamma^i\beta\mathbf{\Psi}(k-1)^{-1} & \mathbf{\Phi}^T(k-1)\boldsymbol{\varphi}(k)\\ \boldsymbol{\varphi}^T(k)\mathbf{\Phi}(k-1) & \boldsymbol{\varphi}^T(k)\boldsymbol{\varphi}(k)+\gamma^i\beta\left(\lambda\left|d(k)-\boldsymbol{\varphi}^T(k)\boldsymbol{\omega}(k)\right|^{p-2}+(1-\lambda)\left|d(k)-\boldsymbol{\varphi}(k)\boldsymbol{\omega}(k)\right|^{q-2}\right)^{-1}\end{bmatrix}^{-1}$$

$$=\begin{bmatrix}\mathbf{\Phi}(k-1)\mathbf{\Phi}^T(k-1)+\gamma^i\beta\mathbf{\Psi}(k-1)^{-1} & \mathbf{\Phi}^T(k-1)\boldsymbol{\varphi}(k)\\ \boldsymbol{\varphi}^T(k)\mathbf{\Phi}(k-1) & \boldsymbol{\varphi}^T(k)\boldsymbol{\varphi}(k)+\gamma^i\beta\theta(k)\end{bmatrix}^{-1} \tag{7.169}$$

where $\theta(k)=\left(\lambda\left|d(k)-\boldsymbol{\varphi}^T(k)\boldsymbol{\omega}(k)\right|^{p-2}+(1-\lambda)\left|d(k)-\boldsymbol{\varphi}^T(k)\boldsymbol{\omega}(k)\right|^{q-2}\right)^{-1}=H^{-1}(k)$.
Now we obtain

$$\mathbf{Q}(k)^{-1}=\begin{bmatrix}\mathbf{Q}(k-1)^{-1} & \mathbf{h}(k)\\ \mathbf{h}(k)^T & \gamma^k\beta\theta(k)+\boldsymbol{\varphi}^T(k)\boldsymbol{\varphi}(k)\end{bmatrix} \tag{7.170}$$

where $\mathbf{h}(k)=\mathbf{\Phi}^T(k-1)\boldsymbol{\varphi}(k)$. Using the following block matrix inversion identity, then, one can get the updating of the inverse of this growing matrix (7.170) as

$$\mathbf{Q}(k)=g^{-1}(k)\begin{bmatrix}\mathbf{Q}(k-1)g(k)+z(k)z^T(k) & -z(k)\\ -z^T(k) & 1\end{bmatrix} \tag{7.171}$$

where $z(k)=\mathbf{Q}(k-1)\mathbf{h}(k)$, and $g(k)=\gamma^k\beta\theta(k)+\boldsymbol{\varphi}^T(k)\boldsymbol{\varphi}(k)-z^T(k)\mathbf{h}(k)$.

Combining (7.168) and (7.171), we get

$$\mathbf{a}(k)=\mathbf{Q}(k)\mathbf{d}(k)=\begin{bmatrix}\mathbf{Q}(k-1)+z(k)z^T(k)g^{-1}(k) & -z(k)g^{-1}(k)\\ -z^T(k)g^{-1}(k) & g^{-1}(k)\end{bmatrix}\begin{bmatrix}\mathbf{d}(k-1)\\ d(k)\end{bmatrix}$$

$$=\begin{bmatrix}\mathbf{a}(k-1)-z(k)r^{-1}(k)e(k)\\ g^{-1}(k)e(k)\end{bmatrix} \tag{7.172}$$

Then the KRGMN algorithm is described in Table 7.4 [387].

TABLE 7.4

Kernel Recursive Generalized Mixed-Norm algorithm

Initialization:
$\mathbf{Q}(1) = (\beta\gamma + \kappa(\mathbf{u}(1), \mathbf{u}(1)))^{-1}, \ \mathbf{a}(1) = \mathbf{Q}(1)d(1), \gamma, \beta, \lambda, p, q$
Computation:
Iterate for $k > 1$:
$\mathbf{h}(k) = [\kappa(\mathbf{u}(k), \mathbf{u}(1)), \ldots, \kappa(\mathbf{u}(k), \mathbf{u}(k-1))]^T$
$e(k) = d(k) - \mathbf{h}(k)^T \mathbf{a}(k-1)$
$\mathbf{z}(k) = \mathbf{Q}(k-1)\mathbf{h}(k)$
$\theta(k) = (\omega \mid d(k) - \boldsymbol{\varphi}^T(k)\boldsymbol{\omega}(k) \mid^{p-2} + (1-\omega) \mid d(k) - \boldsymbol{\varphi}^T(k)\boldsymbol{\omega}(k) \mid^{q}$
$g(k) = \beta\gamma^k \theta(k) + \kappa(\mathbf{u}(k), \mathbf{u}(k)) - \mathbf{z}(k)^T \mathbf{h}(k)$
$\mathbf{Q}(k) = g(k)^{-1} \begin{bmatrix} \mathbf{Q}(k-1)g(k) + \mathbf{z}(k)\mathbf{z}(k)^T & -\mathbf{z}(k) \\ -\mathbf{z}(k)^T & 1 \end{bmatrix}$
$\mathbf{a}(k) = \begin{bmatrix} \mathbf{a}(k-1) - \mathbf{z}(k)g(k)^{-1}e(k) \\ g(k)^{-1}e(k) \end{bmatrix}$

Remark 7.5

In KRGMN algorithm, the term of $\Psi(k)$ that contains $p-2$ and $q-2$ order moment of absolute of the error sequence is included in the matrix form of the weight in $q-2$. This is the major difference between the KRGMN and KRLS.

Remark 7.6

The **g**(k) in KRGMN plays a key role, which guarantees the robustness of the algorithm against outliers. The error can be very large when outliers appear, and the value of $\theta(k)$ with $p<2$ and $q<2$ can also be very large. This will result in a very small value of $g^{-1}(k)$, which reduces the negative effects of the outliers on the update of the coefficients.

Remark 7.7

The KRGMN will reduce to the KRLS algorithm when $p = q = 2$. In addition, the computational complexity of the KRGMN is almost similar to that of KRLS. The extra computational burden is the $H(k)$ in $g^{-1}(k)$ at each iteration.

Remark 7.8

The KRGMN algorithm can be viewed as the generalized version of the KRLS. Furthermore, one can obtain the kernel recursive version of the LAD, LMF, LMMN, and RMN by setting the corresponding parameters denoted as KRLAD, KRLMF, KRLMMN, and KRRMN, respectively. Table 7.5 summarizes the choices of the p and q required to obtain different KAF algorithms mentioned above, where "–" denotes an arbitrary real value in Table 7.5.

Remark 7.9

Five parameters (i.e. $\gamma, \beta, \lambda, p, q$) will jointly affect the performance of the KRGMN algorithm. Below we give a brief discussion about the choice of free parameters.

Choice of p **and** q: The parameter p and q control the order of the error, and they influence the performance of the KRGMN algorithm. Under the Gaussian noise, they should be set at 2. They can be selected as different values to accelerate the convergence or result in a small steady-state misalignment under the noise mixed by two distribution models.

Choice of γ: The forgetting factor γ is also an important parameter for the KRGMN algorithm. Just like other recursive algorithms, the value of γ should be close to one.

Choice of β: The regularization factor β prevents the algorithm over fitting. In general, it should be less than one.

Choice of λ: The mixing parameter λ controls p and q norm which are the key components. Commonly, the value of λ can be an arbitrary value when $p = q$. Under the non-Gaussian and mixture of two non-Gaussian noise environments, the value of ω should be set bigger for $p = 1$ or $q = 1$. The distribution information of the noise should be usually known for specific applications.

TABLE 7.5

Suitable Choices for p and q Result in Different KAF Algorithms

	KRLS	KRLAD	KRLMF	KRLMMN	KRRMN	KRGMN
p	2	1	4	2	2	$p>1$
q	2	1	4	4	1	$Q>1$
λ	–	–	–	$\lambda\neq0$	$\lambda\neq0$	$\lambda\neq0$

7.5.2.2 Simulation Results

Here some simulation results are given to verify the performance of the KRGMN algorithm in nonlinear channel equalization (NCE) and nonlinear system identification (NSI) under different noise environments. The performance of the KRGMN is compared with that of the existing KAF (including KLMS, KRLS, KRMN, and KLMMN).

7.5.2.2.1 Nonlinear Channel Equalization

The basic structure of NCE is shown in Figure 7.21. For the setting of the NCE problem, a simple description of it is as follows: a binary signal{$s(1)$, $s(2)$, …, $s(N)$} is fed into the nonlinear channel. At the receiver end of the channel, the signal is corrupted by additive noise and is then observed as{$r(1)$, $r(2)$,…, $r(N)$}. In this case, the nonlinear channel model is defined by $x(i) = s(i) + 0.5(i-1), r(i) = x(i) - 0.9x^2(i-1) + v(i)$, where $v(i)$ denotes the measurement noise, which is generated by adding two noises $v_1(i)$ and $v_2(i)$ with different distributions. Three kinds of noises with zero-mean are selected to show the performance of the KRGMN algorithm: Adding a Gaussian distributed noise $v_1(i)$ of power 1 and a Bernoulli distributed noise $v_2(i)$ of power 0.45; Adding a Laplace distributed noise $v_1(i)$ of power 0.08 and a uniformly distributed noise $v_2(i)$ of power 1; Adding a Gaussian distributed noise $v_1(i)$ of power 1 and a uniformly distributed noise $v_2(i)$ of power 0.15.

First, the ensemble convergence curves of different algorithms are plotted. The specific parameters of all algorithms are adjusted such that the algorithms exhibit similar steady-state performance that can be seen in the legend. The common parameter for kernel bandwidth is 1, while the mixed noise total power is fixed here at 0.1. The KAF is a filtering method with a sequential online learning model. In the simulation, the training and testing data should be set first. The convergence curves are shown in Figure 7.22–7.24, which are obtained by the testing stage. From them, one can observe that: (1) the KAF based on mixed norm outperforms the traditional KAF in terms of convergence speed; (2) the results of KRGMN are better than that of the KLMMN and the KRMN algorithm in terms of the steady-state testing MSE; (3) the steady-state testing MSE of the KRGMN is smaller that of the KRLS. According to these results, we conclude that the recursive kernel

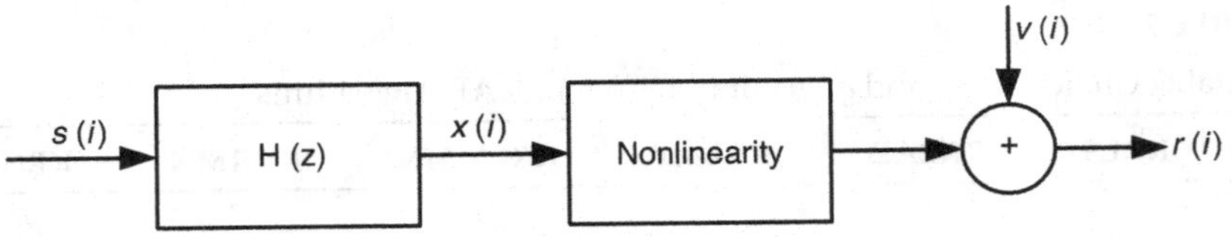

FIGURE 7.21
Basic structure of a nonlinear channel [387].

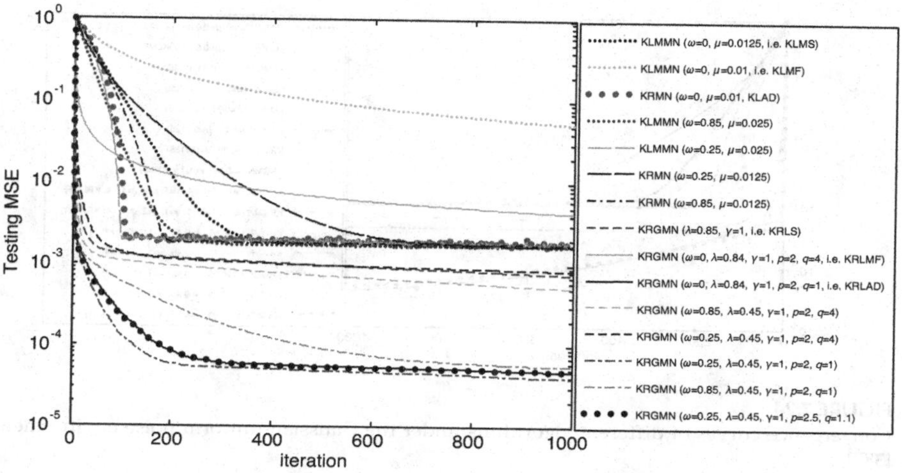

FIGURE 7.22
Convergence curves of different algorithms under the Gaussian-Bernoulli noise environment [387].

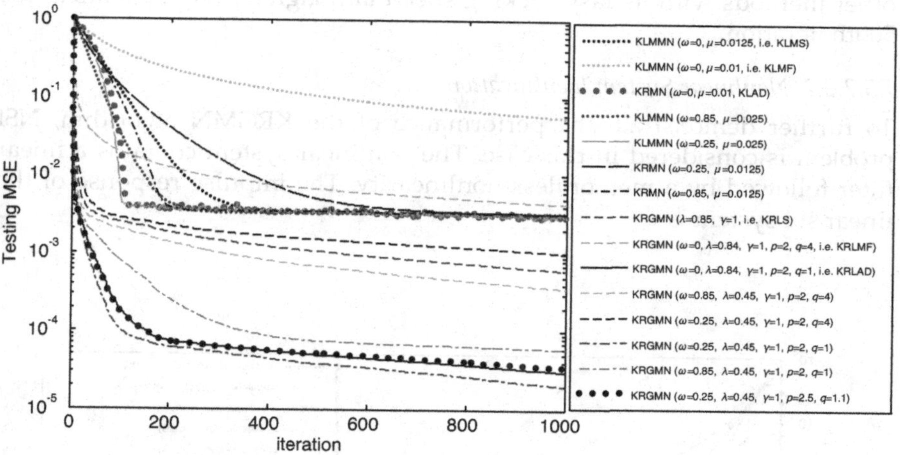

FIGURE 7.23
Convergence curves of different algorithms under the Laplace-uniform noise environment [387].

mixed-norm algorithms outperform the stochastic gradient-based mixed-norm algorithms under this situation although their complexity is relatively high.

Second, the tracking ability of the RGMN method is tested by introducing an abrupt channel change during training. For the first 500 iterations, the channel model is kept the same as before, but for the last 1000 iterations, the

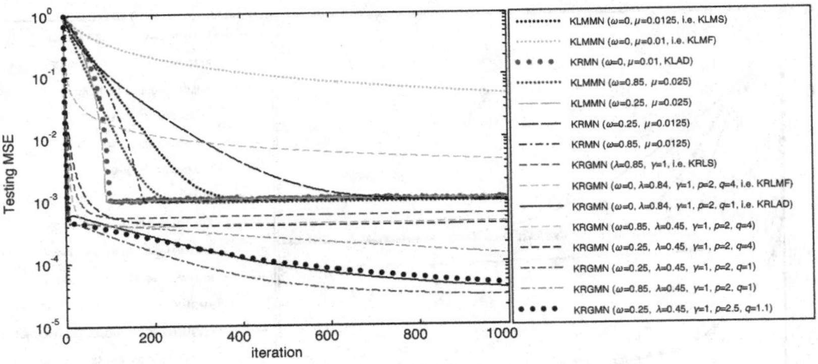

FIGURE 7.24
Convergence curves of different algorithms under the Gaussian-uniform noise environment [387].

nonlinearity of the channel is switched to $r(i) = -x(i) + 0.9x^2(i-1) + v(i)$. In this case, the noise model is selected as ii). The ensemble learning curves are given in Figure 7.25. It is observed that the KRGMN algorithm outperforms other methods with its fast tracking speed although a jump appears at the 500th iteration.

7.5.2.2.2 Nonlinear System Identification

To further demonstrate the performance of the KRGMN algorithm, NSI problem is considered in this case. The nonlinear system contains a linear filter followed by a memoryless nonlinearity. The impulse response of the linear subsystem is

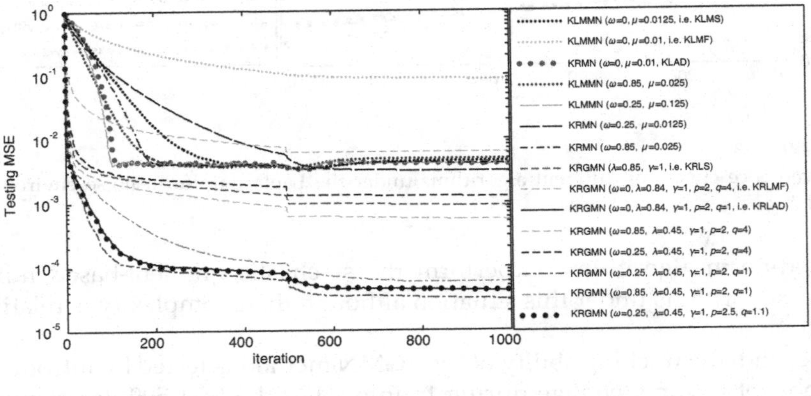

FIGURE 7.25
Ensemble learning curves of different algorithms with an abrupt change [387].

$$H(z) = 0.1 + 0.2z^{-1} + 0.3z^{-2} + 0.4z^{-3} + 0.5z^{-4} + 0.4z^{-5} + 0.3z^{-6} + 0.2z^{-7} + 0.1z^{-8}$$

The nonlinearity is considered as

$$d(i) = 0.2 \times \left[\cos\left((u(i)+1)^* \pi\right)^2 + \cos\left((u(i)-1)^* \pi\right)^2 \right] + v(i)$$

where $\{u(i)\}$ is the input sequence that is uniformly distributed with zero-mean and unity power, and $\{v(i)\}$ is the noise process that is independent of $\{u(i)\}$. In the following simulation, two kinds of noises with zero-mean are mainly considered to evaluate the performance of the KRGMN algorithm: (i) Adding a Gaussian distributed noise $v_1(i)$ of power 1 and a uniformly distributed noise $v_2(i)$ of power 0.15; (ii) Adding a Bernoulli distributed noise $v_1(i)$ of power 1 and a Laplace distributed noise $v_2(i)$ of power 0.2.

The convergence curves are shown in Figures 7.26 and 7.27 for the noise model (i) and (ii), respectively. One can see that the KRGMN-aware algorithms can achieve relatively notable steady-state testing MSE compared with the KLMMN and KRMN. All the testing MSE values at the final stage that are computed by the average of testing MSE at the last 100 iterations under the noise model (i) and (ii) are summarized in Table 7.6, which are in the form of "average standard deviation". One can clearly see that the performance of the KRGMN is better than that of the KLMMN and KRMN.

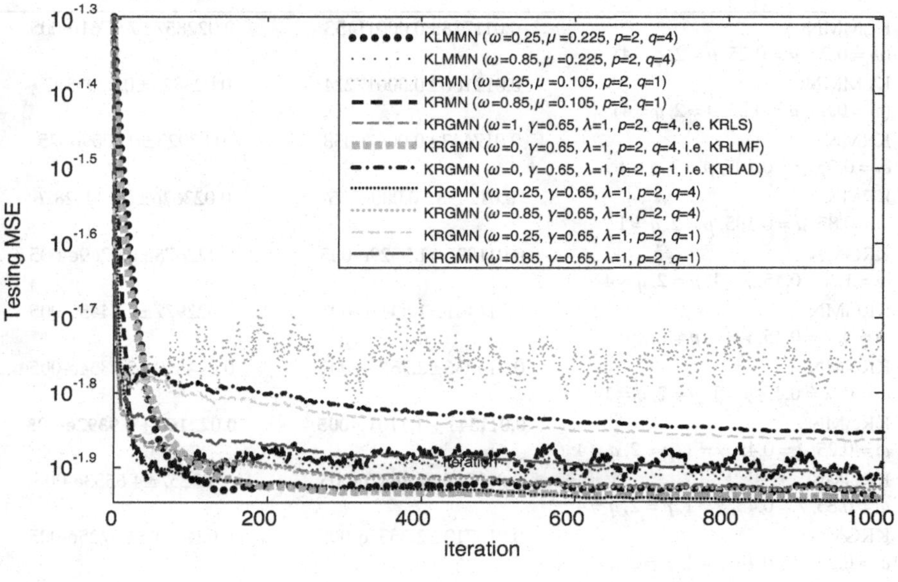

FIGURE 7.26
Convergence curves for Gaussian-Uniform noise [387].

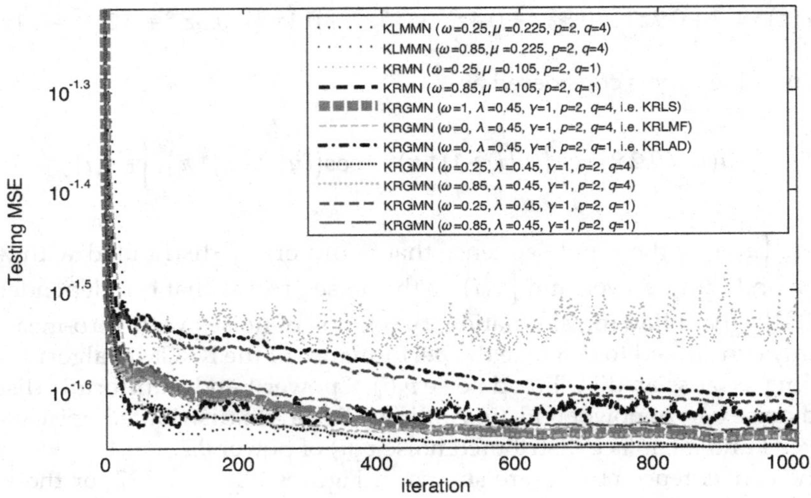

FIGURE 7.27
Convergence curves for Bernoulli-Laplace noise [387].

TABLE 7.6
Testing MSE at the Final Stage [387]

	Testing MSE	
Algorithm	**Noise Model (i)**	**Noise Model (ii)**
KLMMN ($\omega = 0.25, \mu = 0.25, p = 2, q = 4$)	0.011718±0.00011453	0.022857±7.9761e-005
KLMMN ($\omega = 0.85, \mu = 0.25, p = 2, q = 4$)	0.012166±0.00017224	0.023823±0.00038979
KRMN $\omega = 0.25, \mu = 0.105, p = 2, q = 1$	0.017443±0.00079058	0.02925±0.00096375
KRMN $\omega = 0.85, \mu = 0.105, p = 2, q = 1$	0.012498±0.0002555	0.023626±0.00032876
KRGMN $\omega = 1, \lambda = 0.45, \gamma = 1, p = 2, q = 4$	0.011835±1.5529e-005	0.022678±0.9269e-005
KRGMN $\omega = 0, \lambda = 0.45, \gamma = 1, p = 2, q = 4$	0.011461±1.8343e-005	0.022977±2.7445e-005
KRGMN $\omega = 0, \lambda = 0.45, \gamma = 1, p = 2, q = 1$	0.014036±2.7573e-005	0.024899±5.0854e-005
KRGMN $\omega = 0.25, \lambda = 0.45, \gamma = 1, p = 2, q = 4$	**0.011341 ± 1.1101e-005**	**0.022169 ± 0.9392e-005**
KRGMN $\omega = 0.85, \lambda = 0.45, \gamma = 1, p = 2, q = 4$	0.011748±1.4687e-005	0.02257±4.6533e-005
KRGMN $\omega = 0.25, \lambda = 0.45, \gamma = 1, p = 2, q = 1$	0.013712±2.3553e-005	0.024474±4.6725e-005
KRGMN $\omega = 0.85, \lambda = 0.45, \gamma = 1, p = 2, q = 1$	0.012441±1.5997e-005	0.023014±3.5183e-005

Appendix I

Regularity conditions (*B*1)–(*B*4): From the statement of TST, we know that $\|\mathbf{w}\| < r$, $\|\mathbf{w}_1 < r|$ and $\|\mathbf{w}_2\| < r$ We also need the following well-known inequalities:

$$\|\mathbf{w}_1 + \mathbf{w}_2\| \le \|\mathbf{w}_1\| + \|\mathbf{w}_2\| \tag{I.1}$$

$$\left|\mathbf{u}^T(k)\mathbf{w}\right| \le \left\|\mathbf{u}^T(k)\right\|\|\mathbf{w}\| \tag{I.2}$$

Using the boundedness Assumption 7.4, and $\|\mathbf{w}\| < r$ in (I.2),

$$\left|\mathbf{u}^T(k)\mathbf{w}\right| \le r\chi \tag{I.3}$$

(*B*1) is clearly satisfied by substituting $\mathbf{w}(k) = \mathbf{0}_N$ into (7.33). It is also easy to verify (*B*4) since the left-hand side of (7.30) is zero for any θ_1 and θ_2. Let us start by verifying (*B*3). Using the boundedness assumption (7.12) in (7.34), we have

$$\|L(k,\mathbf{w})\| \le \mu\varpi\left\{\delta + (1-\delta)\varpi^2\right\}\chi \tag{I.4}$$

To verify (*B*2), we use (7.12) in (7.33) and write

$$\begin{aligned}&\|H(k,\mathbf{w}_1) - H(k,\mathbf{w}_2)\| \le \mu(1-\delta)\\ &\times\left[3\varpi^2\chi + 3\varpi\left|\left\{\mathbf{u}^T(k)\mathbf{w}_1\right\}^2 - \left\{\mathbf{u}^T(k)\mathbf{w}_2\right\}^2\right| + \left|\left\{\mathbf{u}^T(k)\mathbf{w}_1\right\}^3 - \left\{\mathbf{u}^T(k)\mathbf{w}_1\right\}^3\right|\right]\end{aligned} \tag{I.5}$$

By using (I.1), (I.3) and (7.12) in (I.5), we obtain

$$\|H(k,\mathbf{w}_1) - H(k,\mathbf{w}_2)\| \le 3\mu(1-\delta)\chi(\varpi + r\chi)^2\|\mathbf{w}_1 - w_2\| \tag{I.6}$$

and thus (*B*1)–(*B*4) are satisfied and the TST is applied.

8

Adaptive Filtering Algorithms under KMPE Family Criteria

8.1 Adaptive Filtering Algorithm under Original KMPE

This subsection introduces KMPE-based AFAs for identifying system parameters in non-Gaussian noise scenarios. Firstly, the linear AFA under KMPE is developed by the gradient descent method. Secondly, we define a KMPE cost function with a forgetting factor and derive a recursive AFA based on KMPE with a gain matrix for system identification.

8.1.1 KMPE-Based Adaptive Filtering Algorithm

Under the KMPE criterion, one can derive a robust adaptive filtering algorithm by the steepest descent method. Using the instantaneous value $J_{\text{KMPE}}(e(k)) = (1-\kappa_\sigma(e(k)))^{p/2}$ to replace the ensemble averaged value, we have

$$\mathbf{w}(k+1) = \mathbf{w}(k) - \eta \frac{\partial J_{\text{KMPE}}(e(k))}{\partial \mathbf{w}(k)} \tag{8.1}$$

The following expression can be obtained

$$\begin{aligned}
\frac{\partial J_{\text{KMPE}}(e(k))}{\partial \mathbf{w}(k)} &= \frac{\partial (1-\kappa_\sigma(e(k)))^{p/2}}{\partial \mathbf{w}(k)} \\
&= -\frac{p}{2}(1-\kappa_\sigma(e(k)))^{-p/2} \frac{\partial \kappa_\sigma(e(k))}{\partial \mathbf{w}(k)} \\
&= \frac{p}{2\sigma}(1-\kappa_\sigma(e(k)))^{-p/2} \kappa_\sigma(e(k)) \frac{\partial e(k)}{\partial \mathbf{w}(k)}
\end{aligned} \tag{8.2}$$

Thus (8.1) becomes

DOI: 10.1201/9781003176114-8

$$\begin{aligned}\mathbf{w}(k+1) &= \mathbf{w}(k) - \eta \frac{\partial J_{\text{KMPE}}(e(k))}{\partial \mathbf{w}(k)} \\ &= \mathbf{w}(k) - \eta \left[\frac{p}{2\sigma} (1 - \kappa_\sigma(e(k)))^{-p/2} \kappa_\sigma(e(k)) e(k) \frac{\partial e(k)}{\partial \mathbf{w}(k)} \right]\end{aligned} \tag{8.3}$$

Moreover, the gradient vector

$$\frac{\partial e(k)}{\partial \mathbf{w}(k)} = -\mathbf{u}(k) \tag{8.4}$$

Then we obtain the following KMPE algorithm

$$\begin{aligned}\mathbf{w}(k+1) &= \mathbf{w}(k) + \eta \frac{p}{2\sigma} (1 - \kappa_\sigma(e(k)))^{-p/2} \kappa_\sigma(e(k)) e(k) \mathbf{u}(k) \\ &= \mathbf{w}(k) + \mu (1 - \kappa_\sigma(e(k)))^{-p/2} \kappa_\sigma(e(k)) e(k) \mathbf{u}(k)\end{aligned} \tag{8.5}$$

where $\mu = \frac{\eta p}{2\sigma}$ is the step size.

Remark 8.1

The KMPE algorithm can be viewed as a variable step size LMS or MCC algorithm. When setting $\mu(k) = \mu(1 - \kappa_\sigma(e(k)))^{-p/2} \kappa_\sigma(e(k))$, the KMPE algorithm will reduce to a variable step size LMS algorithm with $\mu(k)$ as

$$\mathbf{w}(k+1) = \mathbf{w}(k) + \mu(k) e(k) \mathbf{u}(k) \tag{8.6}$$

Moreover, when setting $\mu(k) = \mu(1 - \kappa_\sigma(e(k)))^{-p/2}$, the KMPE algorithm will become a variable step size MCC algorithm with $\mu(k)$ as

$$\mathbf{w}(k+1) = \mathbf{w}(k) + \mu(k) \kappa_\sigma(e(k)) e(k) \mathbf{u}(k) \tag{8.7}$$

8.1.2 Recursive Adaptive Filtering Algorithm under KMPE Criterion

This section introduces the recursive KMPE (RKMPE) algorithm for system identification. An empirical KMPE with a forgetting factor is first defined by [260]

$$J_{\text{RKMPE}}(k) = \sum_{j=1}^{k} \lambda^{k-j} \left(1 - k_\sigma\left(e(j)\right)\right)^{p/2} \tag{8.8}$$

where $0<\lambda<1$ is the forgetting factor. Then an optimal model can be obtained as

$$w_o = \arg_w \min J_{\text{RKMPE}}(k) \tag{8.9}$$

Setting the gradient of (8.8) to zero, we have

$$-\frac{p}{2}\sum_{j=0}^{k}\lambda^{k-j}\left(1-\kappa_\sigma(e(j))\right)^{(p-2)/2}\kappa_\sigma(e(j))\frac{e(j)}{\sigma^2}\mathbf{u}(j) = 0 \tag{8.10}$$

Equation (8.10) is equivalent to the following equation

$$\begin{aligned}&\sum_{j=0}^{k}\lambda^{k-j}\left(1-\kappa_\sigma(e(j))\right)^{(p-2)/2}\kappa_\sigma(e(j))\mathbf{u}(j)\mathbf{u}^T(j)\,\mathbf{w}(\mathrm{k})\\&=\sum_{j=0}^{k}\lambda^{k-j}\left(1-\kappa_\sigma(e(j))\right)^{(p-2)/2}\kappa_\sigma(e(j))\mathbf{u}(j)d(j)\end{aligned} \tag{8.11}$$

Define the weighted auto-correlation matrix $\Psi(k)$ and weighted cross-correlation vector $\Phi(k)$ as

$$\begin{aligned}\Psi(k)&=\sum_{j=0}^{k}\lambda^{k-j}\left(1-\kappa_\sigma(e(j))\right)^{(p-2)/2}\kappa_\sigma(e(j))\mathbf{u}(j)\mathbf{u}^T(j)\\\Phi(k)&=\sum_{j=0}^{k}\lambda^{k-j}\left(1-\kappa_\sigma(e(j))\right)^{(p-2)/2}\kappa_\sigma(e(j))\mathbf{u}(j)d(j)\end{aligned} \tag{8.12}$$

Note that there is an extra term $\left(1-\kappa_\sigma(e(j))\right)^{(p-2)/2}\kappa_\sigma(e(j))$ in $\Psi(k)$ and $\Psi(k)$. Then one can obtain a matrix form of (8.11) as

$$\Psi(k)\mathbf{w}(k)=\Phi(k) \tag{8.13}$$

Thus we have

$$\mathbf{w}(k)=\Psi^{-1}(k)\Phi(k) \tag{8.14}$$

In addition, $\Psi(k)$ can be expressed as

$$\begin{aligned}\Psi(k)=\lambda&\sum_{j=0}^{k}\lambda^{k-j-1}\left(1-\kappa_\sigma(e(j))\right)^{(p-2)/2}\kappa_\sigma(e(j))\mathbf{u}(j)\mathbf{u}^T(j)\\&+\left(1-\kappa_\sigma(e(k))\right)^{(p-2)/2}\kappa_\sigma(e(k))\mathbf{u}(k)\mathbf{u}^T(k)\end{aligned} \tag{8.15}$$

According to (8.15), we get

$$\Psi(k) = \lambda\Psi(k-1) + f\left(e(k)\right)\mathbf{u}(k)\mathbf{u}^T(k) \tag{8.16}$$

where

$$f\left(e(k)\right) = \left(1-\kappa_\sigma\left(e(k)\right)\right)^{(p-2)/2}\kappa_\sigma\left(e(k)\right) \tag{8.17}$$

Similarly, we obtain

$$\Phi(k) = \lambda\Phi(k-1) + f\left(e(k)\right)\mathbf{u}(k)d(k) \tag{8.18}$$

To solve (8.14), the following notations are defined

$$\mathbf{A} = \Psi(k), \mathbf{B}^{-1} = \lambda\Psi(k-1), \mathbf{C} = \sqrt{f\left(e(k)\right)}\ \mathbf{u}(k), \mathbf{D} = \mathbf{I} \tag{8.19}$$

By the matrix inversion lemma, the inverse of $\Psi(k)$ can be expressed as

$$\Psi^{-1}(k) = \lambda^{-1}\Psi^{-1}(k-1) - \frac{f\left(e(k)\right)\lambda^{-1}\Omega^{-1}(k-1)\mathbf{u}(k)\mathbf{u}^T(k)\Psi^{-1}(k-1)}{\lambda + f\left(e(k)\right)\mathbf{u}^T(k)\Psi^{-1}(k-1)\mathbf{u}(k)} \tag{8.20}$$

To describe simply, we define the following extended gain vectors:

$$\Omega(k) = \Psi^{-1}(k), \mathbf{K}(k) = \frac{\Omega(k-1)\mathbf{u}(k)}{\lambda + f\left(e(k)\right)\mathbf{u}^T(k)\Omega(k-1)\mathbf{u}(k)} \tag{8.21}$$

Then we have

$$\Omega(k) = \lambda^{-1}[\mathbf{I} - f(e(k))\kappa(k)\mathbf{u}^T(k)]\Omega(k-1) \tag{8.22}$$

After some simplified calculations, one can obtain the update equation

$$\mathbf{w}(k) = \mathbf{w}(k-1) + f\left(e(k)\right)\mathbf{K}(k)\left(d(k) - \mathbf{u}^T(k)\mathbf{w}(k-1)\right) \tag{8.23}$$

The derived recursive KMPE algorithm is described in Table 8.1 [260].

8.1.3 Simulation Results

This section presents some simulation results to demonstrate the performance of the developed algorithms in the presence of non-Gaussian noise. The input signal is a zero-mean Gaussian process with unit variance, and the weight vector of the system is set at $\mathbf{w}_o = \left[0.1, 0.2, 0.3, 0.4, 0.5, 0.6, 0.7, 0.8, 0.9\right]^T$.

TABLE 8.1

RKMPE Algorithm

RKMPE Algorithm
Initialization: $\mathbf{w}(0) = 0, \Omega(0) = \varepsilon^{-1} I_{,\lambda,p,\sigma}$
For $k = 1,2,\ldots$ Do
$y(k) = \mathbf{w}^T(k-1)\mathbf{u}(k)$
$e(k) = d(k) - y(k)$
$\mathbf{K}(k) = \dfrac{\Omega(k-1)\mathbf{u}(k)}{\lambda + f(e(k))\mathbf{u}^T(k)\Omega(k-1)\mathbf{u}(k)}$
$\mathbf{w}(k) = \mathbf{w}(k-1) + f(e(k))\mathbf{K}(k)\left(d(k) - \mathbf{u}^T(k)\mathbf{w}(k-1)\right)$
$\Omega(k) = \lambda^{-1}\left[I - f(e(k))\mathbf{K}(k)\mathbf{u}^T(k)\right]\Omega(k-1)$
End

8.1.3.1 Performance Comparison

To test the performance of the algorithms in different noise environments, the four noises below are considered. (1) Uniform noise distributed over $\left[-\sqrt{3}, \sqrt{3}\right]$; (2) Sine noise $2\sin(v)$, in which v is uniformly distributed over $[0, 2\pi]$; (2) Binary noise distributed over {–1,1} with probability mass $p[x=-1] = p[x=1] = 0.5$; (4) Poisson noise with parameter $\lambda = 0.1$. The convergence curves are shown

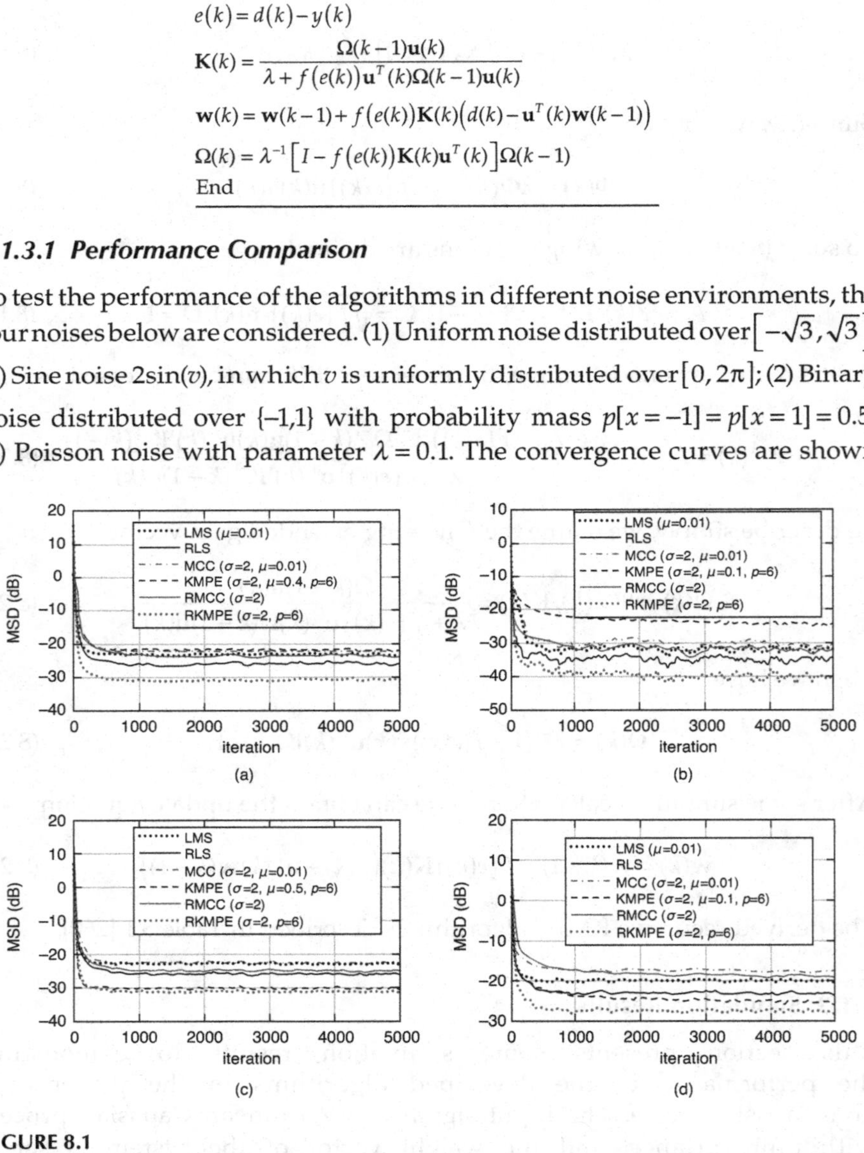

FIGURE 8.1
Convergence curves under different noises: (a) Uniform; (b) Poisson; (c) Binary; (d) Sine wave [260].

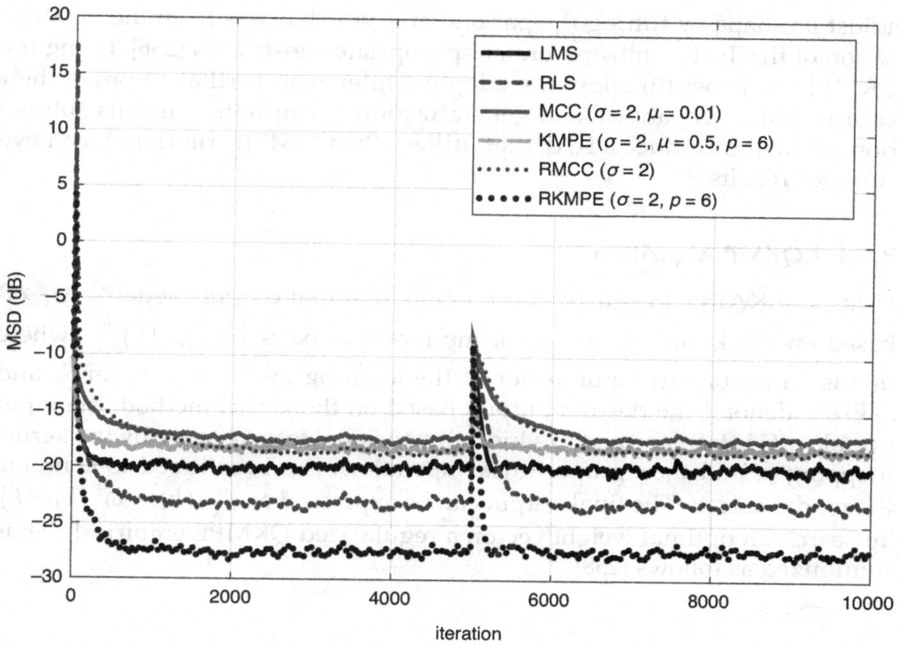

FIGURE 8.2
Convergence curves under the sudden changes [260].

in Figure 8.1. One can see that the performance of the RKMPE algorithm is better than that of other algorithms in four cases, which confirms the advantage of the KMPE loss and the recursive approach.

8.1.3.2 Performance Evaluation under Sudden Changes

To further verify the availability of the RKMPE algorithm in a non-stationary scenario, where the system weight vector is changed at the 5000th iteration. The convergence curves are shown in Figure 8.2. One can observe that the RKMPE algorithm achieves the best tracking performance in this case.

8.2 Kernel Adaptive Filtering Algorithm under q-Gaussian KMPE

The Gaussian kernel is the default choice in the original KMPE, but it may not always yield optimal results. To address this issue, the q-Gaussian distribution has been proposed as an alternative for modeling the impact of external data randomness, as the q-Gaussian density function can adaptively adjust its shape by tuning the parameter q, which arises from the maximization of the Tsallis entropy under appropriate constraints [286]. Using the QKMPE as a cost function, the adaptive filters can further improve their approximation capability and generalization performance. In this subsection, we will introduce a KAF that utilizes the QKMPE criterion to achieve superior results.

8.2.1 RQKMP Algorithm

Using the KAFs can construct a continuous input-output model $f:\mathbf{U}\to\mathbf{R}$ based on the known sequence of input-output pairs $\{\mathbf{u}(k),d(k)\}_{k=1}^{n}$, where $\mathbf{u}(k)$ is an m-column input vector of the learning model at iteration k, and $d(k)\in\mathbb{R}$ denotes the desired output. Based on the kernel method, the input sample $\mathbf{u}(k)$ is transformed to a high-dimensional feature space by the kernel mapping φ, and then proper linear operations can be utilized to compute the model output. The final output can be expressed as $f(\mathbf{u}(k))=\boldsymbol{\omega}^T(k)\varphi(k)$. To search an optimal weight vector, a regularized QKMPE empirical risk is minimized as follows [258]

$$\min_{\omega(k)} C_{\text{QKMPE}}(\omega(k))=\sum_{j=1}^{k}\left[1-\kappa_{q,\sigma}(e_{kj})\right]^{\frac{p}{2}}+\lambda\|\omega(k)\|^2 \tag{8.24}$$

where $\kappa_{q,\sigma}(e_{kj})$ is the q-Gaussian kernel function, $e_{kj}=C_{\text{QKMPE}}(\boldsymbol{\omega}(k))=\sum_{j=1}^{k}\left[1-\kappa_{q,\sigma}(e_{kj})\right]^{\frac{p}{2}}+\lambda\|\omega(k)\|^2$ $e_{kj}=d_j-\boldsymbol{\omega}^T(k)\varphi(j)$ denotes the error between the desired value d_j and the corresponding model output $\boldsymbol{\omega}^T(k)\varphi(j)$, and $\lambda>0$ is the regularization constant. Let the derivative of (8.24) with respect to $\boldsymbol{\omega}(k)$ to zero, one can derive

$$\begin{aligned}
&\frac{\partial C_{\text{QKMPE}}(\boldsymbol{\omega}(k))}{\partial\boldsymbol{\omega}(k)}=0\\
&\Rightarrow\sum_{j=1}^{i}\left[\left(1-\kappa_{q,\sigma}(e_{kj})\right)^{\frac{p-2}{2}}\left(\kappa_{q,\sigma}(e_{kj})\right)^{q}\varphi(j)e_{kj}\right]-\eta\boldsymbol{\omega}(k)=0\\
&\Rightarrow\sum_{j=1}^{i}\left[\varphi(j)g(j)\left(d(j)-\varphi^T(j)\boldsymbol{\omega}(k)\right)\right]-\eta\boldsymbol{\omega}(k)=0\\
&\Rightarrow\phi
\end{aligned}$$

$$\frac{\partial C_{\text{QKMPE}}(\boldsymbol{\omega}(k))}{\partial \boldsymbol{\omega}(k)}=0$$

$$\Rightarrow \sum_{j=1}^{i}\left[\left(1-\kappa_{q,\sigma}(e_{kj})\right)^{\frac{p-2}{2}}\left(\kappa_{q,\sigma}(e_{kj})\right)^{q}\varphi(j)e_{kj}\right]-\eta\boldsymbol{\omega}(k)=0$$

$$\Rightarrow \sum_{j=1}^{i}\left[\varphi(j)g(j)\left(d(j)-\varphi^{T}(j)\boldsymbol{\omega}(k)\right)\right]-\eta\boldsymbol{\omega}(k)=0 \tag{8.25}$$

$$\Rightarrow \boldsymbol{\Phi}(k)\mathbf{G}(k)\mathbf{D}(k)-\boldsymbol{\Phi}(k)\mathbf{G}(k)\boldsymbol{\Phi}^{T}(k)\boldsymbol{\omega}(k)-\eta\boldsymbol{\omega}(k)=0$$

$$\Rightarrow \boldsymbol{\omega}(k)=\left[\boldsymbol{\Phi}(k)\mathbf{G}(k)\boldsymbol{\Phi}^{T}(k)+\eta\mathbf{I}(k)\right]^{-1}\boldsymbol{\Phi}(k)\mathbf{G}(k)\mathbf{D}(k)$$

where $\eta=\frac{4\lambda\sigma^{2}}{p}$, $\boldsymbol{\Phi}(k)=\left[\varphi(1),\varphi(2),\ldots\varphi(k)\right]$ is a mapping vector matrix, $\mathbf{G}(k)=\text{diag}\left[g(1),g(2),\ldots g(k)\right]$ is a diagonal matrix with $g(j)=\left(1-\kappa_{q,\sigma}(e_{kj})\right)^{q}$, $\mathbf{D}(k)=\left[d(1),d(2),\ldots d(k)\right]^{T}$, is a vector for the desired output, and I_k denotes the $k \times k$ identity matrix. We can observe that the derived solution (8.25) is actually an iterative point function because the right side of (8.25) still depends on $\boldsymbol{\omega}_{(k)}$. We can rewrite (8.25) as

$$\boldsymbol{\omega}(k)=\eta^{-1}\boldsymbol{\Phi}(k)\left(\mathbf{G}(k)\mathbf{D}(k)-\mathbf{G}(k)\boldsymbol{\Phi}^{T}(k)\boldsymbol{\omega}(k)\right)=\boldsymbol{\Phi}(k)\boldsymbol{\Omega}(k) \tag{8.26}$$

where $\Omega(k)$ is the weight coefficient. It is obvious that (8.26) is a linear function with the recoded input data in the feature space. In addition, the $\Omega(k)$ can be expressed in a more computable form as

$$\begin{aligned}\Omega(k)&=\eta^{-1}(\mathbf{G}(\mathrm{k})\mathbf{D}(\mathrm{k})-\mathbf{G}(\mathrm{k})\boldsymbol{\Phi}_{i}^{T}\omega(k))\\&=\eta^{-1}(\mathbf{G}(k)\mathbf{D}(k)-\mathbf{G}(k)\boldsymbol{\Phi}^{T}(k)\boldsymbol{\Phi}(k)\Omega(k))\\&=[\eta I_{k}+\mathbf{G}(k)\boldsymbol{\Phi}^{T}(k)\boldsymbol{\Phi}(k)]^{-1}\mathbf{G}(k)\mathbf{D}(k)\end{aligned} \tag{8.27}$$

One can observe that $\boldsymbol{\Phi}^{T}(k)\boldsymbol{\Phi}(k)$ is a $k \times k$ kernel matrix with elements $K_{js}=K_{q,\sigma}(\mathbf{u}(\mathrm{j}),\mathbf{u}(s))$, $j,s\in\{1,2,\ldots k\}$. Let

$$\mathbf{H}(k)=\boldsymbol{\Phi}(k)=\begin{pmatrix}\mathbf{H}(k-I) & \mathbf{h}(k)\\ \mathbf{h}^{T}(k) & K_{kk}\end{pmatrix} \tag{8.28}$$

where $\mathbf{h}(k)=\boldsymbol{\Phi}^{T}(k-1)\varphi(k)$. Let $\mathbf{Q}(k)=[\eta\mathbf{I}_{k}+\mathbf{G}(k)\mathbf{H}(k)]^{-1}$. The following recursive formulations can be got

$$\mathbf{G}(k)\mathbf{H}(k) = \begin{bmatrix} \mathbf{G}(k-1)\mathbf{H}(k-1) & \mathbf{G}(k-1)\mathbf{h}(k) \\ g(\mathrm{k})\mathbf{h}^T(k) & k_{kk}g(k) \end{bmatrix} \tag{8.29}$$

$$\mathbf{Q}(k) \begin{bmatrix} \mathbf{Q}^{-1}(k-1) & \mathbf{G}(k-1)\mathbf{h}(k) \\ g(k)\mathbf{h}^T(k) & k_{kk}g(k)+\eta \end{bmatrix}^{-1} \tag{8.30}$$

Employing the block matrix inversion lemma to (8.30), the growing matrix can be expressed as

$$\mathbf{Q}(k) \begin{bmatrix} \mathbf{Q}(k-1)+r(k)g(k)\mathbf{z}(k)\mathbf{k}_k^T & -r(k)\mathbf{z}(k) \\ -r(k)g(k)\mathbf{z}_k^T & r(k) \end{bmatrix}^{-1} \tag{8.31}$$

where $\mathbf{z}(k)=\mathbf{Q}(k-1)\mathbf{G}(k-1)\mathbf{h}(k)$, $\mathbf{k}_k=\mathbf{Q}^T(k-1)\mathbf{h}(k)$, and $r(k)=(\eta+k_{kk}g(k)-g(k)\mathbf{z}^T(k)\mathbf{h}(k))^{-1}$. Substituting (8.31) into (8.30), the weight coefficient can be obtained as

$$\Omega(k)=\mathbf{Q}(k)\mathbf{G}(k)\mathbf{D}(k)=\begin{bmatrix} \Omega(k-1)-r(\mathrm{k})\,g(\mathrm{k})\mathbf{z}(\mathrm{k})\,e_k \\ r(k)g(k)e_k \end{bmatrix} \tag{8.32}$$

where $e_k=d(k)-\mathbf{h}^T(k)\Omega(k-1)$ is the error of the model prediction at the time step k. The above algorithm (also called RQKMP algorithm) [258] is an iterative fixed-point algorithm, whose convergence can be guaranteed by the Banach fixed-point theorem [349].

8.2.2 Simulation Results

This section presents simulation results for MG time series prediction, demonstrating the effectiveness of the RQKMP algorithm. The filtering performance is evaluated using the MSE metric. The training data are contaminated with mixed noise.

8.2.2.1 Testing the Effects under Different Parameters

The regularization factor λ is set as 0.1 in all the simulations. The kernel size σ is fixed at 1. The power p is set within [0.1,6], while the range of q is selected as (1,3). The filtering performance in terms of steady-state MSE is illustrated in Figure 8.3, from which one can observe the following results: (1) The filter can achieve desirable performance when p is selected within the range (0.1,2) or else q is large enough; (2) The filtering accuracy will drop gradually with the increase of p; (3) The parameter q has slight influence on the filtering accuracy when p is small; while p increases, the larger q can reduce the steady-state MSE obviously.

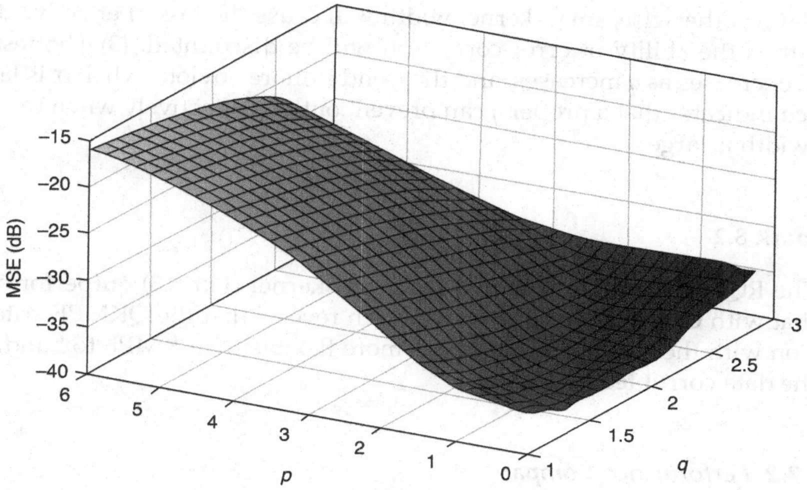

FIGURE 8.3
Steady-state MSE of RQKMP with different p and q [258].

In addition, how the shape parameter q affects the performance of RQKMP with different kernel widths is tested. Figure 8.4 shows the surface of the steady-state MSE, from which one can observe that: (1) The RQKMP can achieve high filtering accuracy when $\sigma \in (0.5, 1.5)$ or else q is large enough. (2) The filtering performance degrades when σ is either too small or too large. It is reasonable that the filters are sensitive to outliers when the kernel width is

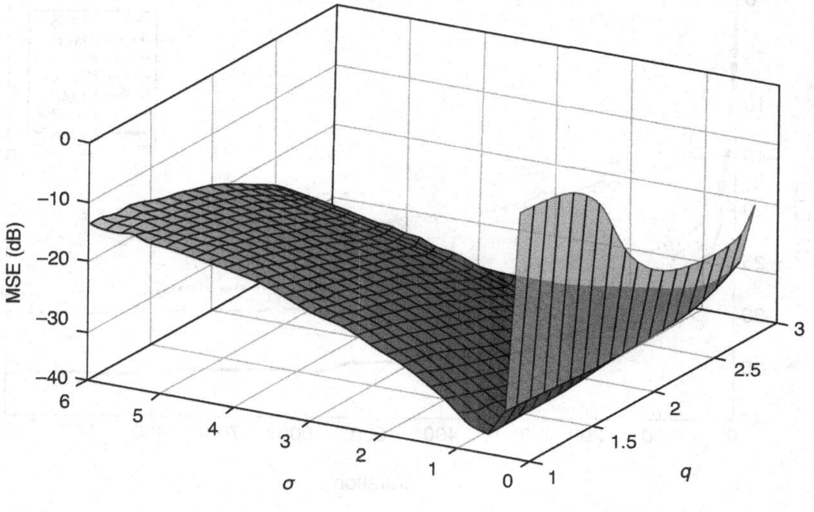

FIGURE 8.4
Steady-state MSE of RQKMP with different σ and q [258].

too large; otherwise, small kernel width will cause the loss of effective data, and thus the ability of error-correction will be discounted. (3) The testing MSE decreases as q increases, and the trend is more obvious when σ is large, which indicates that a proper q can prevent outliers effectively when the kernel width is large.

Remark 8.2

The RQKMP algorithm with q-Gaussian kernel ($1<q<3$) outperforms that with Gaussian kernel($q \rightarrow 1$), which reveals that the QKMPE criterion with the shape parameter q is more flexible than KMPE to handle the data corrupted by noise.

8.2.2.2 Performance Comparison

The performance of the RQKMP is compared with that of some other KAF algorithms (i.e., KLMS [122], KRLS [116], KMC [394], KRMC [395], KLMP, and KRLP [241]). Based on the simulation results above, the parameters of RQKMP are set at $q = 2$, $p = 1.2$, $\sigma = 1$, and the parameters of other algorithms are experimentally selected to achieve desirable performance. The learning curves in terms of the steady-state MSE are shown in Figure 8.5. From the results, one can observe the following results: (1) In comparison with other KAFs in this case, the RQKMP algorithm can achieve higher filtering

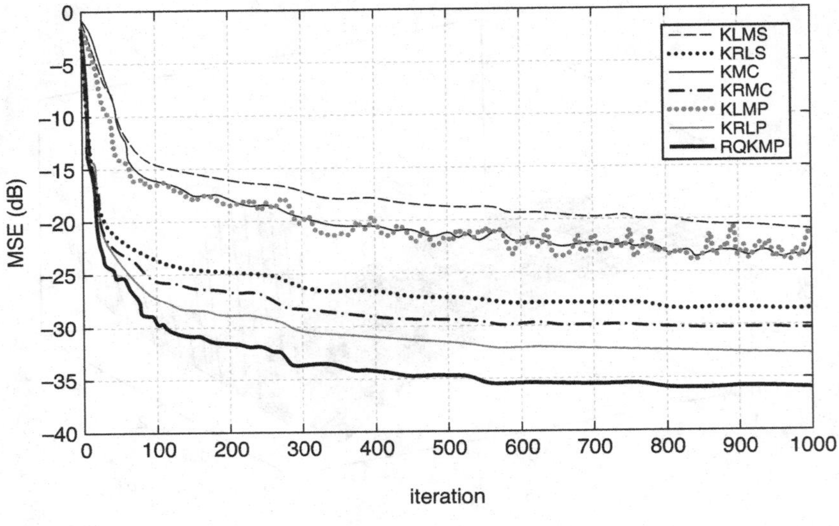

FIGURE 8.5
Learning curves of different algorithms [258].

accuracy. (2) The recursive algorithms (i.e., KRLS, KRMC, KRLP, and RQKMP) outperform the stochastic gradient-based algorithms (i.e., KLMS, KMC, and KLMP) both in the convergence rate and the steady-state MSE. (3) The MPE-criterion-based algorithms (i.e., KLMP and KRLP) and the MCC-based algorithms (i.e., KMC and KRMC) outperform those algorithms under MSE criterion (i.e., KLMS and KRLS), and the RQKMP algorithm combining the advantages of q-Gaussian kernel and QKMPE criterion shows superior learning performance.

8.2.2.3 Testing the Performance of the RQKMP Algorithm under Larger Outliers

Here the scale parameter of the noise process $B(i)$ is increased from 0.1 to 1, which will enlarge the size of outliers. The average learning curves are shown in Figure 8.6. In this case, the RQKMP algorithm still keeps satisfactory forecasting ability, while others exhibit poorer performance, especially the MSE-criterion-based algorithms (i.e., KLMS and KRLS) deteriorate significantly. The results confirm again that the RQKMP algorithm has strong robustness against large outliers.

Experimental results show that the QKMPE criterion with the shape parameter q is more flexible than KMPE to handle the signal corrupted by Gaussian and non-Gaussian noises, especially when data contain large outliers, and the RQKMP algorithm can achieve better filtering performance than some reported algorithms.

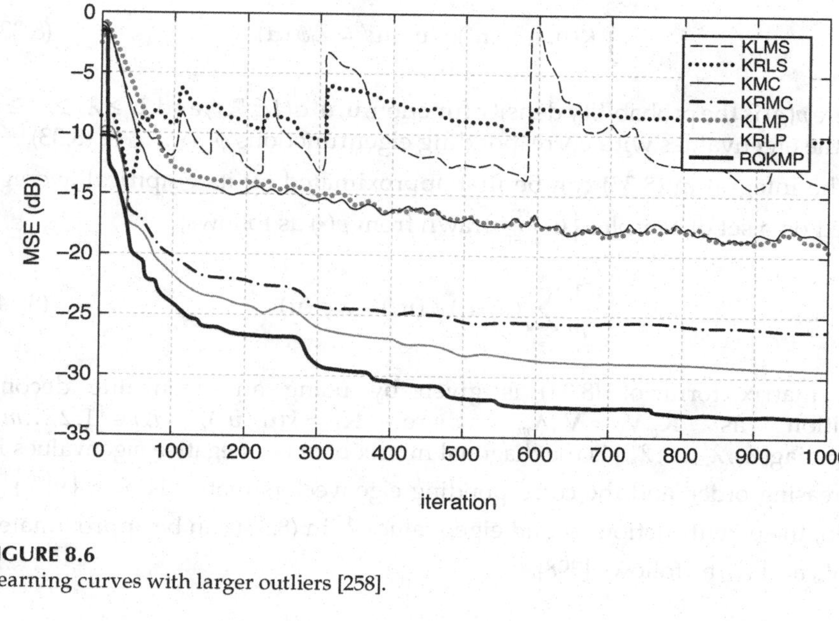

FIGURE 8.6
Learning curves with larger outliers [258].

8.3 Kernel Adaptive Filtering Algorithm under KMMPE

The KMMPE criterion, when paired with an appropriate p-value, can yield more accurate results than MC for robust learning. To prevent the network size of KAFs from growing too large, the Nyström method is a highly efficient solution [396]. Additionally, the recursive update form can enhance the tracking ability of KAFs. In light of those, the Nyström method and recursive update form have been applied to the KMMPE criterion, resulting in the development of a recursive minimum KMMPE algorithm called NysRMKMP [259]. This subsection will focus on presenting the NysRMKMP algorithm and its performance.

8.3.1 NysRMKMP Algorithm

This section introduces the NysRMKMP algorithm in which the Nyström method is applied to improve computational and storage efficiency.

8.3.1.1 Nyström Method

In general, the kernel method is known to be time-consuming and space-intensive, which makes it unsuitable for online applications. However, the Nyström method offers an effective solution by approximating eigenfunctions of RKHS in a fixed-dimensional space [397]. Specifically, the Nyström method is designed to address the kernel eigenfunction problem as follows [259]:

$$\int K(\mathbf{u},\mathbf{u}')p(\mathbf{u}')\phi_i(\mathbf{u}')d\mathbf{u}' = \lambda_i\phi_i(\mathbf{u}) \tag{8.33}$$

where $p(g)$ is the probability density function, $\mathbf{u},\mathbf{u}' \in U \subseteq \mathbb{R}^D$, and $\lambda_1 \geq \lambda_2 \geq \ldots \geq 0$ are the eigenvalues with corresponding eigenfunctions $\phi_1, \phi_2, \ldots,$ of (8.33).

The integral in (8.33) can be first approximated by its empirical average based on a set of samples $\{\mathbf{u}_j\}_{j=1}^m$ drawn from $p(\bullet)$ as follows:

$$\frac{1}{m}\sum_{j=1}^{m}(\mathbf{u},\mathbf{u}_j)\,\phi_i(\mathbf{u}_j) \approx \lambda_i\phi_i(\mathbf{u}) \tag{8.34}$$

The matrix form of (8.34) is given by using an eigenvalue decomposition as $\mathbf{K}_m\mathbf{V}_m = \mathbf{V}_m\mathbf{\Lambda}_m$ where $\mathbf{K}_m = k(\mathbf{u}_i,\mathbf{u}_j)$, $i,j \in \{1,2,\ldots m\}$, $\Lambda_m = \text{diag}[\lambda_1,\lambda_2,\ldots,\lambda_m]$ is the diagonal matrix of m nonnegative eigenvalues in decreasing order, and the corresponding eigenvectors matrix is $\mathbf{V}_m = \{\mathbf{v}_i^{(m)}\}_{i=1}^m$. Then, the eigenfunctions ϕ_i and eigenvalues λ_i in (8.33) can be approximated by $\mathbf{V}_m$ and Λ_m as follows [398]:

$$\phi_i(\mathbf{u}_j) \approx \sqrt{m}\mathbf{v}_{ij}^{(m)}, \tag{8.35}$$

$$\lambda_i \approx \frac{\lambda_i^{(m)}}{m}. \tag{8.36}$$

The transformed input $\varphi(\mathbf{u})$ in RKHS can be therefore constructed as:

$$\varphi(\mathbf{u}) = [\sqrt{\lambda_1}\phi_1(\mathbf{u}), \sqrt{\lambda_2}\phi_2(\mathbf{u}), \ldots\ldots,]^T \tag{8.37}$$

where the dimension of $\varphi(\mathbf{u})$ can be infinite when a Gaussian kernel is used. Hence, to avoid the calculation in an infinite dimension space, by using the approximations of (8.35) and (8.36) in a fixed-dimensional space, the eigenfunctions $\varphi_i(\mathbf{u})$ in (8.34) can be approximated as

$$\hat{\phi}_i(\mathbf{u}) = \frac{\sqrt{m}}{\lambda_i^{(m)}} \sum_{j=1}^{m} (\mathbf{u}, \hat{\mathbf{u}}_j)\mathbf{v}_{ij}^{(m)} \tag{8.38}$$

Finally, combining (8.35)–(8.38) generates the Nyström method in a finite dimension space as follows:

$$\begin{aligned}\mathbf{z}(\mathbf{u}) &= [\sqrt{\lambda_1}\hat{\phi}_1(\mathbf{u}), \sqrt{\lambda_2}\hat{\phi}_2(\mathbf{u}), \ldots \sqrt{\lambda_m}\hat{\phi}_m(\mathbf{u})]^T \\ &= \Lambda_m^{-\frac{1}{2}}\mathbf{V}_m^T[k(\mathbf{u}, \hat{\mathbf{u}}_1), k(\mathbf{u}, \hat{\mathbf{u}}_2), \ldots k(\mathbf{u}, \hat{\mathbf{u}}_m)]^T \\ &= \mathbf{P}_m\mathbf{\Phi}_m(\mathbf{u})\end{aligned} \tag{8.39}$$

where $\hat{\mathbf{u}}_i (i = 1, 2, \ldots m)$ are sampled points, $\mathbf{P}_m = \Lambda_m^{-\frac{1}{2}}\mathbf{V}_m^T$, and $\mathbf{\Phi}_m(\mathbf{u}) = [k(\mathbf{u}, \hat{\mathbf{u}}_1), k(\mathbf{u}, \hat{\mathbf{u}}_2), \ldots, k(\mathbf{u}, \hat{\mathbf{u}}_m)]^T$ is the kernel vector. The Nyström method generates a fixed m-dimensional feature space, which can be used to approximate a Gaussian kernel function $K_\sigma(\bullet)$.

8.3.1.2 NysRMKMP Algorithm

The recursive update form and Nyström method are combined into the KMMPE cost function to develop Nyström recursive minimum kernel mixture mean p-power error (NysRMKMP) algorithm. Given a sequence of training samples $\{\mathbf{u}(k), d(k)\}_{k=1}^N$ with $\mathbf{u}(k) \in \mathbb{R}^D$ being the D-dimensional input vector and $d(k) \in \mathbb{R}$ being the desired output at discrete time k in the fixed m-dimensional Nyström feature space, the cost function of NysRMKMP is denoted as [259]

$$J(k) = \sum_{j=1}^{k} \beta^{k-j}(1 - (S(j) + F(j)))^{p/2} + \frac{1}{2}\beta^k\gamma\|\mathbf{\Omega}_z(k)\|^2 \tag{8.40}$$

where $\gamma \in \mathbb{R}$ is a regularization factor for avoiding overfitting, $\beta \in [0,1]$ is a forgetting factor, $S(j) = \alpha \exp\left(-\frac{e^2(j)}{2\sigma_1^2}\right)$, $F(j) = (1-\alpha)\exp\left(-\frac{e^2(j)}{2\sigma_2^2}\right)$, and $e(j) = d(j) - \Omega_z^T(k)\mathbf{z}(\mathbf{u}(j))$ with coefficient $\Omega_z(k) \in \mathbb{R}^m$. According to Property 3.16 in 3.5.3, the empirical KMMPE is convex regarding e at any point if $\max|e(j)| \le \sigma_1, j = 1,2,\ldots,k$, and $p \ge 2$ are satisfied.

To obtain the minimum of (8.40), its gradient is calculated by

$$\frac{\partial J(k)}{\partial \Omega_z(k)} = -\frac{p}{2}\sum_{j=1}^{k} \beta^{k-j}\rho(j)\mathbf{z}(\mathbf{u}(j))e(j) + \beta^k \gamma \Omega_z(k)$$

$$= -\frac{p}{2}\sum_{j=1}^{k} \beta^{k-j}\rho(j)\mathbf{z}(\mathbf{u}(j))d(j) \quad (8.41)$$

$$+\frac{p}{2}\sum_{j=1}^{k} \beta^{k-j}\rho(j)\mathbf{G}(j)\Omega_z(k) + \beta^k \gamma \Omega_z(k)$$

where $\rho(j) = (1-(S(j)+F(j)))^{(p-2)/2}(S(j)/\sigma_1^2 + F(j)/\sigma_2^2$ and $\mathbf{G}(j) = \mathbf{z}(\mathbf{u}(j))\mathbf{z}^T(\mathbf{u}(j))$.

Because of the convexity of (8.40), $\Omega_z(k)$ is solved by setting (8.41) to zero, i.e.,

$$\Omega_z(k) = \mathbf{P}(k)\Theta(k) \quad (8.42)$$

where $\mathbf{P}(k) = \mathbf{R}^{-1}(k)$ with $\mathbf{R}(k) = \sum_{j=1}^{k} \beta^{k-j}\rho(j)\mathbf{G}(j) + \beta^k \gamma 2/p\mathbf{I}$, $\Theta(k) = \mathbf{Z}(k)\mathbf{Q}(k)\mathbf{d}(k) = \sum_{j=1}^{k} \beta^{k-j}\rho(j)\mathbf{z}(\mathbf{u}(j))d(j)$ with $\mathbf{Q}(k) = \text{diag}[\beta^{k-1}\rho(1),$ $\beta^{k-2}\rho(3),\ldots,\rho(k)]$, $\mathbf{Z}(k) = [\mathbf{z}(\mathbf{u}(1)), \mathbf{z}(\mathbf{u}(2)),\ldots,\mathbf{z}(\mathbf{u}(k))]$, and $\mathbf{d}(k) =$ $[d(1), d(2), k\ldots, d(k)]^T$.

Then to further derive an online form, $\mathbf{R}(k)$ and $\Theta(k)$ can be updated recursively by

$$\begin{cases} \mathbf{R}(k) = \beta\mathbf{R}(k-1) + \rho(k)\mathbf{G}(k) \\ \Theta(k) = \beta\Theta(k-1) + \rho(k)\mathbf{z}(\mathbf{u}(k))d(k) \end{cases} \quad (8.43)$$

To avoid the inverse matrix operation, $\mathbf{P}(k)$ can be updated by using the matrix inversion lemma. By letting $\mathbf{A} = \beta\mathbf{R}(k-1), \mathbf{B} = \mathbf{z}(\mathbf{u}(k)), \mathbf{C} = \rho(k)\mathbf{I}$, and $\mathbf{D} = \mathbf{z}^T(\mathbf{u}(k))$ in the matrix inversion lemma for (8.42), we have

$$
\begin{aligned}
P(k) &= \beta^{-1}\left(\frac{\rho(k)\mathbf{P}(k-1)}{\beta+\rho(k)\mathbf{z}^T(\mathbf{u}(k))\mathbf{P}(k-1)\mathbf{z}(\mathbf{u}(k))}\right) \\
&\beta^{-1}\left(\mathbf{P}(k-1)-\frac{\rho(k)\mathbf{P}(k-1)\mathbf{G}(k)\mathbf{P}(k-1)}{\beta+\rho(k)\mathbf{z}^T(\mathbf{u}(k))\mathbf{P}(k-1)\mathbf{z}(\mathbf{u}(k))}\right)
\end{aligned}
\tag{8.44}
$$

Substituting (8.44) into (8.42) yields

$$
\begin{aligned}
\mathbf{\Omega}_z(k) &= \mathbf{\Omega}_z(k-1)+Y(k)e(k) \\
&= \mathbf{\Omega}_z(k-1)+\frac{\rho(k)\beta^{-1}\mathbf{z}^T(\mathbf{u}(k))\mathbf{P}(k-1)\mathbf{z}(\mathbf{u}(k))}{1+\rho(k)\beta^{-1}\mathbf{z}^T(\mathbf{u}(k))\mathbf{P}(k-1)\mathbf{z}(\mathbf{u}(k))} \\
&= \mathbf{\Omega}_z(k-1)+\mathbf{P}(k)\mathbf{z}(\mathbf{u}(k))e(k)
\end{aligned}
\tag{8.45}
$$

with

$$
\begin{cases}
\mathbf{Y}(k)=\rho(k)\beta^{-1}\mathbf{P}(k-1)\mathbf{z}(\mathbf{u}(\mathrm{k}))\,/\,q(k) \\
q(k)=1+\rho(k)\beta^{-1}\mathbf{z}^T(\mathbf{u}(k))\mathbf{P}(k-1)\mathbf{z}(\mathbf{u}(k)) \\
e(k)=d(k)-\mathbf{\Omega}_z^T(k-1)\mathbf{z}(\mathbf{u}(k))
\end{cases}
\tag{8.46}
$$

Finally, combining (8.39)–(8.41) generates the NysRMKMP algorithm.

8.3.2 Simulation Results

In this section, the short-term prediction of MG-chaotic time series is used to validate the superior performance of the NysRMKMP algorithm. The impulsive noise used in this example is modeled by the mixture of two independent noise processes.

8.3.2.1 Parameter Selection

For parameter selection, the steady-state MSEs and the averaged consumed time of NysRMKMP with different sampled points m are discussed by simulations in mixed noise model, where the steady-state MSEs are obtained as averages over the last 100 iterations. The simulation results are shown in Figure 8.7. As one can see from Figure 8.7, one can obtain that: (1) the consumed time of NysRMKMP increases with m; (2) in the initial stage, the filtering accuracy increases obviously while as m increases; (3) when m exceeds a fixed value, i.e., ($m \approx 50$), the steady-state MSEs almost keep unchanged. In addition, the steady-state MSEs with different power parameters $p \in [2,16]$ are shown in Figure 8.8, where the steady-state MSEs are obtained as averages over the last 100 iterations. As one can see from Figure 8.8, NysRMKMP

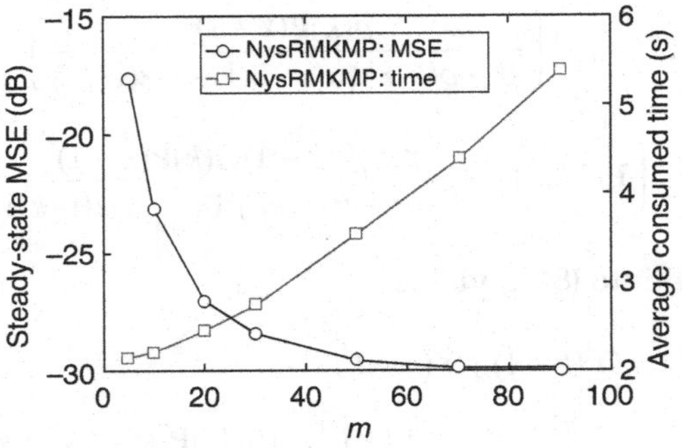

FIGURE 8.7
Steady-state MSEs and averaged consumed time of NysRMKMP with different sampled points m in MG time series prediction [259].

has the highest filtering accuracy when $p = 10$. Similarly, the suitable values of σ_1, σ_2, and α are obtained for NysRMKMP by trial and error to achieve the desirable filtering performance. These optimal parameters are not presented here to conserve space.

8.3.2.2 Performance Comparison

The performance comparison of RFFRLS [337], KRMC [395], NysKRMC [399], KRMC-NC [395], and NysRMKMP is conducted. Figure 8.9 shows the testing MSEs of NysRMKMP and the compared algorithms. For detailed comparison, steady-state MSEs are given in Figure 8.9. As one can see from Figure 8.9, NysRMKMP achieves the highest filtering accuracy for combating impulsive noises among all the compared algorithms.

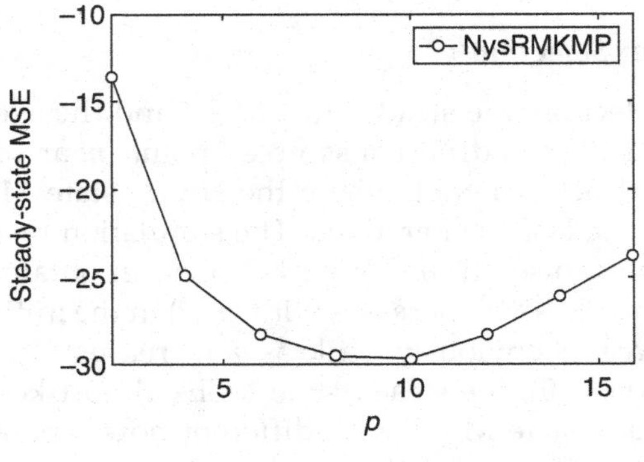

FIGURE 8.8
Influence of p on the performance of NysRMKMP in MG time series prediction [259].

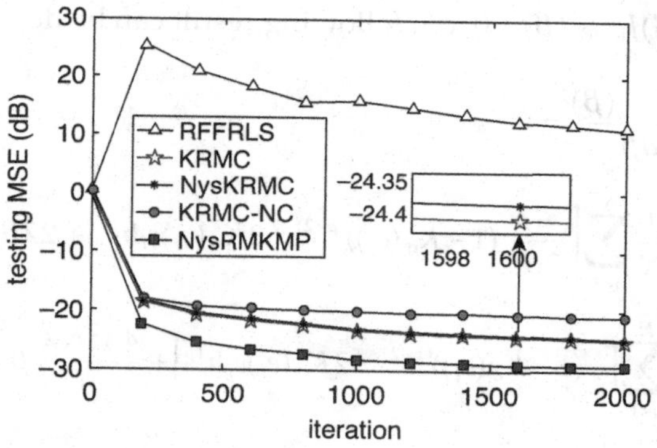

FIGURE 8.9
Testing MSEs of RFFRLS, KRMC, NysKRMC, KRMC-NC, and NysRMKMP in MG time series prediction [259].

8.4 Extreme Learning Machine under KMPE

The fundamental framework of learning theory typically involves learning from examples through the optimization of a specific cost function, which enables the learned model to identify the underlying structures or dependencies in the data-generating system, despite the presence of uncertainty caused by noise or lack of knowledge about the system. To address non-Gaussian data or noise, this subsection introduces an ELM model with KMPE [257].

8.4.1 ELM-KMPE

To obtain an ELM that is robust with respect to large outliers, we consider the following KMPE-based cost function using some definitions of the variables in 6.2.1 [257]:

$$\begin{aligned} J_{\text{KMPE}}(\boldsymbol{\beta}) &= \hat{C}_p(\mathbf{T},\mathbf{H}\boldsymbol{\beta}) + \lambda\|\boldsymbol{\beta}\|_2^2 \\ &= \frac{1}{N}\sum_{i=1}^{N}(1-K_\sigma(e_i))^{p/2} + \lambda\|\boldsymbol{\beta}\|_2^2 \\ &= \frac{1}{N}\sum_{i=1}^{N}\left(1-\exp\left(-\frac{e_i^2}{2\sigma^2}\right)\right)^{p/2} + \|\boldsymbol{\beta}\|_2^2 \end{aligned} \tag{8.47}$$

Different from the cost function in (6.38), the above cost function will be little influenced by large errors since the term $(1-K_\sigma(e_i))^{p/2}$ is upper bounded by 1.0.

Setting $(\partial/\partial\boldsymbol{\beta})J_{\text{KMPE}}(\boldsymbol{\beta})=0$, the following result can be derived

$$\begin{aligned}
&\frac{\partial J_{\text{KMPE}}(\boldsymbol{\beta})}{\partial\boldsymbol{\beta}}=0\\
&\Rightarrow\frac{1}{N}\sum_{i=1}^{N}\left[\frac{-p}{2\sigma^2}(1-K_\sigma(e_i))^{(p-2)}/2K_\sigma(e_i)e_i\boldsymbol{h}_i^T\right]+2\lambda\boldsymbol{\beta}=0\\
&\Rightarrow\sum_{i=1}^{N}\left[-(1-K_\sigma(e_i))^{(p-2)}/2K_\sigma(e_i)e_i\boldsymbol{h}_i^T\right]+\frac{4\sigma^2N\lambda}{p}\boldsymbol{\beta}=0\\
&\Rightarrow\sum_{i=1}^{N}(\varphi(e_i)\boldsymbol{h}_i^T\boldsymbol{h}_i\boldsymbol{\beta}-\varphi(e_i)\boldsymbol{h}_i^T)+\lambda'\boldsymbol{\beta}=0\\
&\Rightarrow\sum_{i=1}^{N}(\varphi(\mathrm{e}_i)\boldsymbol{h}_i^T\boldsymbol{h}_i\boldsymbol{\beta})+\lambda'\boldsymbol{\beta}=\sum_{i=1}^{N}\varphi(e_i)t\boldsymbol{h}_i^T\\
&\Rightarrow\boldsymbol{\beta}=[\mathbf{H}^T\Lambda\mathbf{H}+\lambda'\mathbf{I}]^{-1}\mathbf{H}^T\Lambda\mathbf{T}
\end{aligned}\tag{8.48}$$

where $\lambda'=(4\sigma^2N/p)\lambda$, $\boldsymbol{h}_i$ is the ith row of $\mathbf{H}$, and Λ is a diagonal matrix with diagonal elements $\Lambda_{ii}=\varphi(e_i)$, with $\varphi(e_i)=(1-K_\sigma(e_i))^{(p-2)/2}K_\sigma(e_i)$.

The derived optimal solution $\boldsymbol{\beta}=[\mathbf{H}^T\Lambda\mathbf{H}+\lambda'\mathbf{I}]^{-1}\mathbf{H}^T\Lambda\mathbf{T}$ is not a closed-form solution since the matrix Λ on the right-hand side depends on the weight vector $\boldsymbol{\beta}$ through $e_i=t_i-h_i\boldsymbol{\beta}$. So it is actually a fixed-point equation. The true optimal solution can thus be solved by a fixed-point iterative algorithm, as summarized in Table 8.2. This algorithm is referred to as the ELM-KMPE algorithm [257].

TABLE 8.2

ELM-KMPE Algorithm

Parameters setting: number of hidden nodes L, regularization parameter λ', maximum iteration number M, kernel width σ, power parameter p and termination tolerance ε

Input: samples $\{\mathbf{X}_i,t_i\}_{i=1}^N$
Output: weight vector β
Initialization: Set $\beta_0=\mathbf{0}$ and randomly initialize the parameters $\mathbf{w}_j$ and b_j $(j=1,2,\ldots,L)$

1: **for** $k=1,2,\ldots,M$ **do**
2: Compute the error based on $\boldsymbol{\beta}_{k-1}:e_i=t_i-\boldsymbol{h}_i\boldsymbol{\beta}_{k-1}$
3: Compute the diagonal matrix $\Lambda:\Lambda_{ii}=\varphi(e_i)$
4: Update the weight vector $\boldsymbol{\beta}:\boldsymbol{\beta}_k=[\mathbf{H}^T\Lambda\mathbf{H}+\lambda'\mathbf{I}]^{-1}\mathbf{H}^T\Lambda\mathbf{T}$
5: **Until** $|J_{\text{KMPE}}(\beta_k)-J_{\text{KMPE}}(\beta_{k-1})|<\varepsilon$
6: end for

8.4.2 Simulation Results

Some simulation results are presented to verify the advantages of the ELM-KMPE algorithm.

8.4.2.1 Function Estimation with Synthetic Data

In this section, the SinC function estimation is used to evaluate the performance of the ELM-KMPE and other ELM algorithms, such as ELM [169], regularized extreme learning machine (RELM) [173], and ELM-regularized correntropy criterion ELM-RCC [400]. The synthetic data are generated by $y(i) = k.\,\mathrm{sinc}(x(i)) + v(i)$, where $k = 8$, the definition of sinc() can be seen in 6.2.3.2, and $v(i)$ is generated by a mixed noise model. Two background noises are considered: (1) uniform distribution over [–1.0,1.0] and (2) sine wave noise $\sin(x)$, with x uniformly distributed over $[0,2\pi]$. In addition, the input data $x(i)$ are drawn uniformly from [–10,10]. Here 200 samples are used for training and another 200 noise-free samples are used for testing. The root mean square error (RMSE) is employed to measure the performance. The estimation results and testing RMSEs are illustrated in Figure 8.10 and Table 8.3. It is evident that the ELM-KMPE achieves the best performance among the four algorithms.

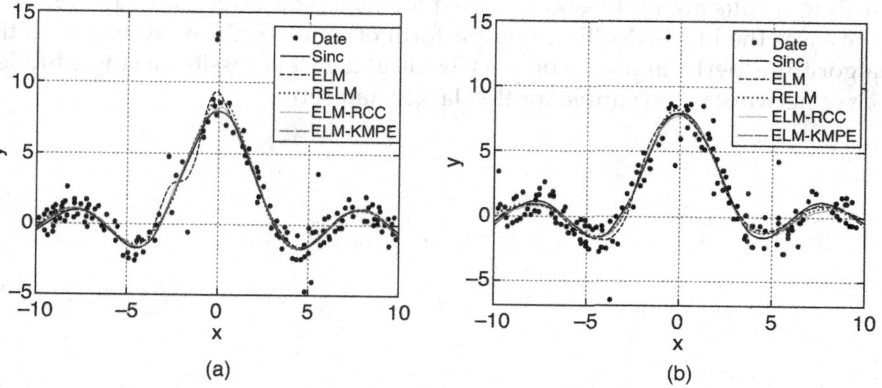

FIGURE 8.10
SinC function estimation results with different background noises. (a) Uniform. (b) Sine wave [257].

TABLE 8.3
Testing RMSEs of Four Algorithms [257]

	ELM	RELM	ELM-RCC	ELM-KMPE
Uniform	0.5117	0.2234	0.1671	**0.1079**
Sine wave	0.3340	0.2498	0.2335	**0.1156**

8.4.2.2 Regression and Classification on Benchmark Datasets

Here the performance of aforementioned four algorithms in regression and classification problems is compared by using benchmark datasets from the University of California, Irvine (UCI) machine learning repository [397] and the CIFAR-10 dataset [401]. For each UCI dataset, the training and testing samples are randomly selected from the dataset. In particular, the data for regression are normalized to the range [0,1]. For the CIFAR-10 dataset, 5000 training data are randomly selected from the training dataset and 10000 testing data are used for testing. For each algorithm, the parameters are experimentally chosen by fivefold cross-validation. The RMSE is used as the performance measure for regression. For classification, the performance is measured by the accuracy (ACC). Let p_i and t_i be the predicted and target labels of the ith sample. The ACC is defined by

$$\mathrm{ACC} = \frac{1}{n}\sum_{i=1}^{n}\delta(t_i, \mathrm{map}(p_i))$$

where $\delta(x,y)$ is an indicator function, $\delta(x,y)=1$ if $x = y$, otherwise $\delta(x,y)=0$, and map(·) maps each predicted label to the equivalent target label. The "mean ± standard deviation" results of the RMSE and ACC during training and testing over 100 runs are shown in Tables 8.4 and 8.5, where the best testing results are represented in bold for each dataset. As one can see, in all the cases the ELM-KMPE can outperform other algorithms, although all the algorithms perform poorly on the CIFAR-10 dataset (possibly a single hidden layer network is too simple for this larger dataset).

TABLE 8.4

Performance Comparison of Four Algorithms with Benchmark Regression Datasets [257]

Datasets	ELM		RELM		ELM-RCC		ELM-KMPE	
	Training RMSE	Testing RMSE	Training RMSE	Testing RMSE	Training RMSE	Testing RMSE	Training RMSE	Testing RMSE
Servo	0.0741 ± 0.0126	0.1183 ± 0.0204	0.0720 ± 0.0106	0.1036 ± 0.0152	0.0739 ± 0.0106	0.1032 ± 0.0148	0.0570 ± 0.0108	**0.10221 ± 0.0184**
Concrete	0.0612 ± 0.0026	0.0994 ± 0.0013	0.0742 ± 0.0025	0.0914 ± 0.0042	0.0559 ± 0.0018	0.0879 ± 0.0077	0.0577 ± 0.0021	**0.0864 ± 0.0058**
Wine red	0.1280 ± 0.0031	0.1312 ± 0.0032	0.1282 ± 0.0031	0.1309 ± 0.0031	0.1264 ± 0.0031	0.1306 ± 0.0032	0.1198 ± 0.0028	**0.1302 ± 0.0035**
Housing	0.0728 ± 0.0070	0.0994 ± 0.0120	0.0502 ± 0.0044	0.0835 ± 0.0100	0.0493 ± 0.0046	0.0832 ± 0.0099	0.0554 ± 0.0045	**0.0821 ± 0.0101**
Airfoil	0.0664 ± 0.0027	0.0942 ± 0.0099	0.0967 ± 0.0061	0.1025 ± 0.0058	0.0742 ± 0.0026	0.0896 ± 0.0050	0.0695 ± 0.0029	**0.0880 ± 0.0058**
Slump	0 ± 0	0.0429 ± 0.0091	0.0066 ± 0.0041	0.0424 ± 0.0097	0.0001 ± 0	0.0423 ± 0.0130	0.0028 ± 0.0004	**0.0410 ± 0.0107**
Yacht	0.0040 ± 0.0004	0.0740 ± 0.1267	0.0370 ± 0.0079	0.0530 ± 0.0086	0.0126 ± 0.0008	0.0333 ± 0.0086	0.0051 ± 0.0009	**0.0250 ± 0.0147**

TABLE 8.5

Performance Comparison of Four Algorithms with Benchmark Classification Datasets [257]

Datasets	ELM		RELM		ELM-RCC		ELM-KMPE	
	Training ACC	Testing ACC	Training ACC	Testing ACC	Training ACC	Testing ACC	Training ACC	Testing ACC
Glass	95.32 ± 3.39	77.62 ± 9.55	96.04 ± 1.72	92.50 ± 4.31	96.80 ± 2.71	93.24 ± 3.38	97.85 ± 2.03	**94.54 ± 3.12**
Wine	99.55 ± 0.70	96.91 ± 2.07	99.65 ± 0.67	96.92 ± 2.07	99.85 ± 0.44	97.43 ± 1.95	99.91 ± 0.35	**97.58 ± 1.73**
Ecoli	90.45 ± 3.46	80.65 ± 3.51	92.61 ± 3.44	82.19 ± 3.11	92.48 ± 3.24	82.27 ± 2.93	93.50 ± 3.39	**82.35 ± 2.77**
User-Modeling	92.80 ± 1.99	84.17 ± 3.63	93.47 ± 1.68	85.47 ± 3.26	94.02 ± 1.63	85.57 ± 3.42	93.17 ± 1.82	**86.29 ± 3.08**
Wdbc	93.07 ± 2.11	84.81 ± 3.43	93.52 ± 2.07	85.86 ± 3.31	92.34 ± 2.16	86.63 ± 3.27	91.01 ± 2.02	**87.09 ± 3.20**
Leaf	93.63 ± 1.28	68.86 ± 3.92	95.91 ± 1.15	71.51 ± 3.71	95.01 ± 1.53	71.56 ± 3.88	95.21 ± 1.48	**73.87 ± 4.20**
Vehicle	92.87 ± 1.03	81.14 ± 1.91	94.01 ± 0.97	81.51 ± 1.99	94.88 ± 0.80	81.61 ± 2.03	95.80 ± 0.79	**82.23 ± 2.25**
Seed	98.48 ± 1.07	92.26 ± 2.65	98.75 ± 0.93	94.40 ± 2.13	98.30 ± 1.03	94.65 ± 1.95	98.65 ± 1.05	**95.01 ± 1.96**
CIFAR-10	57.67 ± 0.66	32.09 ± 0.53	59.30 ± 0.66	32.91 ± 0.46	56.33 ± 0.62	33.25 ± 0.49	54.60 ± 0.75	**34.65 ± 0.43**

References

[1] J. S. Goldstein, I. S. Reed, L. L. Scharf. A multistage representation of the Wiener filter based on orthogonal projections. *IEEE Trans. Inform. Theory*, 1998, 44(7):2943–2959.

[2] V. Tim, S. Doclo, J. Wouters, et al. Speech enhancement with multichannel Wiener filter techniques in multimicrophone binaural hearing aids. *J. Acoust. Soc. Am.*, 2009, 125(1):360.

[3] G. Welch. Kalman Filter. Siggraph Tutorial, 2001.

[4] G. Welch, G. Bishop. An introduction to the Kalman filter (Technical Report No. TR 95–041). 1995.

[5] S. Haykin. *Adaptive Filter Theory*, 3rd ed. Upper Saddle River, NJ: Prentice-Hall, 1996.

[6] A. H. Sayed. *Adaptive Filters*, 1st ed. Hoboken, NJ: Wiley, 2009.

[7] B. Ma, D. Qu, Y. Zhu. A novel adaptive filtering approach for genomic signal processing. *IEEE International Conference on Signal Processing*, IEEE, 2010. Beijing, China

[8] B. Mulgrew. Orthonormal functions for nonlinear signal processing and adaptive filtering. *Proceedings - ICASSP, IEEE International Conference on Acoustics, Speech and Signal Processing*, 1994, 6, pp. 509–512. Adelaide, SA, Australia

[9] C. C. Min, E. Zahedi. *Limb Cardiovasculature System Identification Using Adaptive Filtering*. Berlin, Heidelberg: Springer, 2007.

[10] J. Yang, H. Sakai. Performance analysis of the ICA-based adaptive filtering algorithm for system identification. *Proceedings of the ISCIE International Symposium on Stochastic Systems Theory & Its Applications*, 2009, 2009, pp. 85–90.

[11] R. W. Christiansen, D. M. Chabries, D. Lynn. Noise reduction in speech using adaptive filtering I: Signal processing algorithms. *J. Acoust. Soc. Am.*, 1982, 71(S1):S7–S8.

[12] J. Gao, H. Sultan, H. Jing, et al. Denoising nonlinear time series by adaptive filtering and wavelet shrinkage: A comparison. *IEEE Signal Process. Lett.*, 2009, 17(3):237–240.

[13] P. Regalia. Adaptive IIR filtering in signal processing and control. *J. Geophys. Res.*, 1994, 70(2):451–458.

[14] G. C. Goodwin, K. S. Sin. Adaptive filtering prediction and control. *IEEE Trans. Acoust. Speech Signal Process.*, 2003, 33(1):337–338.

[15] H. Lorenz, G. M. Richter, M. Capaccioli, et al. Adaptive filtering in astronomical image processing - Part One - Basic considerations and examples. *Astron. Astrophys.*, 1993, 277:321.

[16] K. Egiazarian, A. Foi, V. Katkovnik. Compressed sensing image reconstruction via recursive spatially adaptive filtering. *2007 IEEE International Conference on Image Processing*, 2007.

[17] W. A. Frank. An efficient approximation to the quadratic Volterra filter and its application in real-time loudspeaker linearization. *Signal Process.*, 1995, 45(1):97–113.

[18] C. E. Davila, A. J. Welch, H. R. Iii. A second-order adaptive Volterra filter with rapid convergence. *IEEE Trans. Acoust. Speech Signal Process.*, 1987, 35(9):1259–1263.

[19] W. Liu, J. C. Principe, S. Haykin. Kernel adaptive filtering: A comprehensive introduction [Book Review]. *IEEE Comput. Intell. Mag.*, 2010, 5(3):52–55.

[20] W. Liu, J. C. Principe, S. Haykin. *Kernel Adaptive Filtering: A Comprehensive Introduction*. Hoboken, NJ: John Wiley & Sons, 2011.

[21] H. Kong, L. Guan. A neural network adaptive filter for the removal of impulse noise in digital images. *Neural Netw.*, 1996, 9(3):373–378.

[22] H. Zhao, X. Zeng, Z. He. Low-complexity nonlinear adaptive filter based on a pipelined bilinear recurrent neural network. *IEEE Trans. Neural Netw.*, 2011, 22(9):1494.

[23] R. Chassaing. *Adaptive Filters*. Hoboken, NJ: John Wiley & Sons, Inc. 2005.

[24] H. H. Dam, A. Cantoni, K. L. Teo, et al. Variable digital filter with least-square criterion and peak gain constraints. *IEEE Trans. Circuits Syst. II Exp. Briefs.*, 2007, 54(1):24–28.

[25] L. Jia, W. A. Krzymien. A minimum mean-square error criterion based non-linear joint transmitter-receiver processing algorithm for the downlink of multi-user MIMO systems. *2006 IEEE 63rd Vehicular Technology Conference (VTC-Spring),* Melbourne, 2006.

[26] C. R. Rao. Some comments on the minimum mean square error as a criterion of estimation. In *Statistics & Related Topics*, M. Csiirgo, D. Dawson, J. Rao, A. Saleh, Eds. New York: Elsevier, pp. 123–143, 1980.

[27] P. Bloomfield, W. L. Steiger. *Least Absolute Deviations. Theory, Applications, and Algorithms*. Basel: Birkhauser, 1983.

[28] A. Sovic, D. Sersic. Efficient least absolute deviation adaptive wavelet filter bank. *IEEE Trans. Signal Process.*, 2014, 62(14):3631–3642.

[29] S. Guan, L. Zhi. Noise constrained least mean absolute third algorithm. *Syst. Control*, 2018. doi:10.48550/arXiv.1805.01305.

[30] W. Shang. Adaptive filter based on third-order cumulants in fault diagnosis of rolling bearing. *Chin. J. Constr. Mach.*, 2004, 2(4):464–468

[31] E. Eweda, A. Zerguine. New insights into the normalization of the least mean fourth algorithm. *Signal Image Video Process.*, 2011, 7(2):255–262.

[32] P. I. Hubscher, J. C. M. Bermudez. An improved statistical analysis of the least mean fourth (LMF) adaptive algorithm. *IEEE Trans. Signal Process.*, 2003, 51(3):664–671.

[33] J. R. Wolberg, J. Wolberg, *Data Analysis Using the Method of Least Squares: Extracting the Most Information from Experiments, 1*, Berlin, Germany: Springer, 2006.

[34] L. R. Vega, H. Rey, J. Benesty, S. Tressens. A fast robust recursive least-squares algorithm. *IEEE Trans. Signal Process.*, 2009, 57(3):1209–1216.

[35] M. Z. A. Bhotto, A. Antoniou. Robust recursive least-squares adaptive-filtering algorithm for impulsive-noise environments. *IEEE Signal Process. Lett.*, 2011, 18(3):185–188.

[36] Y. Zakharov, G. P. White, J. Liu. Low-complexity RLS algorithms using dichotomous coordinate descent iterations. *IEEE Trans. Signal Process.*, 2008, 56(7):3150–3161.

[37] L. Chi, H. Young, J. Sum, et al. On the regularization of forgetting recursive least square. *IEEE Trans. Neural Networks.*, 1999, 10(6):1482–1486.

[38] S. Haykin, B. Widrow, *Least-Mean-Square Adaptive Filters*. Hoboken, NJ: Wiley, 2003.

[39] S. Zhao, Z. Man, S. Khoo, H. Wu. Stability and convergence analysis of transform-domain LMS adaptive filters with second-order autoregressive process. *IEEE Trans. Signal Process.*, 2009, 57(1):119–130.
[40] J. B. Foleyakd, F. M. Boland. A note on the convergence analysis of LMS adaptive filters with Gaussian data. *IEEE Trans. Acoust. Speech Signal Process.*, 1988, 36(7):1087–1089.
[41] R. R. Bitmead. Convergence properties of LMS adaptive estimators with unbounded dependent inputs. *IEEE Trans. Autom. Control*, 1984, 29(5):477–479.
[42] O. Dabeer, E. Masry. Analysis of mean-square error and transient speed of the LMS adaptive algorithm. *IEEE Trans. Inform. Theory*, 2002, 48(7):1873–1894.
[43] Y. Li, G. R. Arce. A maximum likelihood approach to least absolute deviation regression. *EURASIP J. Adv. Signal Process.*, 2004, 2004(12):1–8.
[44] C. Yang, Y. Dong. Cluster-based least absolute deviation regression for dimension reduction. *J. Statistic. Theory Pract.*, 2016, 10(1):121–132.
[45] J. M. Ramirez, J. L. Paredes. A fast normalization method of cDNA microarray data based on LAD. *IV Latin American Congress on Biomedical Engineering 2007*, Bioengineering Solutions for Latin America Health, 2008, pp. 583–587.
[46] H. Zheng, X. Dan, Y. Yang, et al. The design of adaptive filter based on minimum dispersion. *Signal Process.*, 1998, 14(1):43–46.
[47] M. S. Prakash, R. A. Shaik. A distributed arithmetic based approach for the implementation of the sign-LMS adaptive filter. *2015 International Conference on Signal Processing And Communication Engineering Systems (SPACES)*, IEEE, 2015. Guntur, India
[48] S. E. Kim, Y. S. Choi, J. W. Lee, et al. Partial-update normalized sign LMS algorithm employing sparse updates. *IEICE Trans. Fundam. Electron. Commun. Comput. Sci.*, 2013, 96(6):1482–1487.
[49] P. I. Hubscher, J. Bermudez, V. H. Nascimento. A mean-square stability analysis of the least mean fourth adaptive algorithm. *IEEE Trans. Signal Process.*, 2007, 55(8):4018–4028.
[50] E. Eweda. Dependence of the stability of the least mean fourth algorithm on target weights non-stationarity. *IEEE Trans. Signal Process.*, 2014, 62(7):1634–1643.
[51] E. Walach, B. Widrow. The least mean fourth (LMF) adaptive algorithm and its family. *IEEE Trans. Inform. Theory*, 1984, IT-30:275–283.
[52] S. L. Gay. *The Fast Affine Projection Algorithm*. Boston, MA: Kluwer Academic Publishers, 2000.
[53] L. Li, J. A. Chambers. A new incremental affine projection-based adaptive algorithm for distributed networks. *Signal Process.*, 2008, 88(10):2599–2603.
[54] T. Tanaka, A. Cichocki. Subband decomposition independent component analysis and new performance criteria. *2004 IEEE International Conference on Acoustics, Speech, and Signal Processing*, IEEE, 2004, Montreal, Canada
[55] Z. Chen, G. Yue. LMS algorithm based on subband decomposition. *International Conference on Communication Technology Proceedings, ICCT'98*, 1998, Beijing, China
[56] S. C. Manapragada. Subband-based adaptive algorithms for acoustic echo cancellation. 1995. 1 January 1995, Northern Illinois University.
[57] S. Zhang, J. Zhang, H. C. So. Mean square deviation analysis of LMS and NLMS algorithms with white reference inputs. *Signal Process.*, 2017, 131(Feb.):20–26.

[58] V. A. Gholkar. Mean square convergence analysis of LMS algorithm (adaptive filters). *Electron. Lett.*, 1990, 26(20):1705–1706.
[59] J. Chen, C. Richard, J. Bermudez, et al. Variants of non-negative least-mean-square algorithm and convergence analysis. *IEEE Trans. Signal Process.*, 2014, 62(15): 3990–4005.
[60] H. Sakai, J. M. Yang, T. Oka. Exact convergence analysis of adaptive filter algorithms without the persistently exciting condition. *IEEE Trans. Signal Process.*, 2007, 55(5):2077–2083.
[61] S. Bittanti, M. Campi. Adaptive RLS algorithms under stochastic excitation-L2 convergence analysis. *IEEE Trans. Autom. Control*, 2002, 36(8):963–967.
[62] T. Chen, Y. Zakharov. Convergence analysis of RLS-DCD algorithm. *IEEE/SP Workshop on Statistical Signal Processing*, IEEE, 2009.
[63] B. G. Y. Wei, J. V. Krogmeier. The stability of variable step-size LMS algorithms. *IEEE Trans. Signal Process.*, 1999, 47(12):3277–3288.
[64] T. Aboulnasr, K. Mayyas. A robust variable step-size LMS-type algorithm: Analysis. *IEEE Trans. Signal Process.*, 1997, 45(3):631–639.
[65] S. H. Leung, C. F. So. Gradient-based variable forgetting factor RLS algorithm in time-varying environments. *IEEE Trans. Signal Process.*, 2005, 53(8):3141–3150.
[66] B. Qin, Y. Cai, B. Champagne, M. Zhao, S. Yousefi. Low-complexity variable forgetting factor constrained constant modulus RLS algorithm for adaptive beamforming. *Signal Process.*, 2014, 105(Dec.):277–282.
[67] Y. J. Chu, C. M. Mak. A variable forgetting factor diffusion recursive least squares algorithm for distributed estimation. *Signal Process.*, 2017, 140(11):219–225.
[68] Y. Chen, Y. Gu, A. O. Hero. Sparse LMS for system identification. *Proceedings of IEEE ICASSP*, Apr. 2009, pp. 3125–3128, Taipei, Taiwan.
[69] Y. Gu, J. Jin, S. Mei. l0 norm constraint LMS algorithm for sparse system identification. *IEEE Signal Process. Lett.*, 2009, 16(9):774–777.
[70] K. Shi, P. Shi. Convergence analysis of sparse LMS algorithms with *l1*-norm penalty based on white input signal. *Signal Process.*, 2010, 90(12):3289–3293.
[71] G. Su, J. Jin, Y. Gu, J. Wang. Performance analysis of l0 norm constraint least mean square algorithm. *IEEE Trans. Signal Process.*, 2012, 60(5):2223–2235.
[72] F. Y. Wu, F. Tong. Non-uniform norm constraint LMS algorithm for sparse system identification. *IEEE Commun. Lett.*, 2013, 17(2):385–388.
[73] R. C. de Lamare, R. Sampaio-Neto. Sparsity-aware adaptive algorithms based on alternating optimization and shrinkage. *IEEE Signal Process. Lett.*, 2014, 21(2):225–229.
[74] C. Wang, Y. Zhang, Y. Wei, N. Li. A new l0-LMS algorithm with adaptive zero attractor. *IEEE Commun. Lett.*, 2015, 19(12):259–269.
[75] B. Babadi, N. Kalouptsidis, V. Tarokh. SPARLS: The sparse RLS algorithm. *IEEE Trans. Signal Process.*, 2010, 58(8):4013–4025.
[76] D. Angelosante, J. A. Bazerque, G. B. Giannakis. Online adaptive estimation of sparse signals: Where RLS meets the norm. *IEEE Trans. Signal Process.*, 2010, 58(7):3436–3447.
[77] E. M. Eksioglu. Sparsity regularized recursive least squares adaptive filtering. *IET Signal Process.*, 2011, 5(5):480–487.
[78] E. M. Eksioglu, A. Korhan Tanc. RLS algorithm with convex regularization. *IEEE Signal Process. Lett.*, 2011, 18(8):470–473.
[79] D. L. Duttweiler. Proportionate normalized least-mean-squares adaptation in echo cancelers. *IEEE Trans. Speech Audio Process.*, 2000, 8(5):508–518.

[80] S. L. Gay. An efficient, fast converging adaptive filter for network echo cancellation. *Proceedings of the 32nd Asilomar Conference on Signals, Systems, and Computers*, vol. 1., Nov. 1998, pp. 394–398.

[81] J. Benesty, S. L. Gay. An improved PNLMS algorithm. *Proceedings of the IEEE International Conference on Acoustics, Speech and Signal Processing (ICASSP)*, vol. 2., May 2002, pp. 1881–1884.

[82] H. Deng, M. Doroslovacki. Proportionate algorithms for echo cancellation. *IEEE Trans. Signal Process.*, 2006, 54(5):1794–1803.

[83] H. Deng, M. Doroslovacki. Improving convergence of the PNLMS algorithm for sparse impulse response identification. *IEEE Signal Process. Lett.*, 2005, 12(3):181–184.

[84] K. Wagner, M. Doroslovacki, H. Deng. Convergence of proportionate-type LMS adaptive filters and choice of gain matrix. *Proceedings of the 40th Asilomar Conference on Signals, Systems, and Computers*, Nov. 2006, pp. 243–247, Pacific Grove, CA.

[85] H. Deng. Adaptive algorithms for sparse impulse response identification. Ph.D. dissertation, School Eng. Appl. Sci., George Washington Univ., Washington, DC, 2005.

[86] K. Wagner, M. Doroslovacki. Probability density of weight deviations given preceding weight deviations for proportionate-type LMS adaptive algorithms. *IEEE Signal Process. Lett.*, 2011, 18(11):667–670.

[87] K. Pelekanakis, M. Chitre. Natural gradient-based adaptive algorithms for sparse underwater acoustic channel identification. *Proceedings of the 4th Underwater Acoustic Measurement Conference (UAM)*, Jun. 2011, pp. 1403–1410, Montreal, Canada.

[88] Y. Huang, J. Benesty, J. Chen. *Acoustic MIMO Signal Processing*, 1st ed. New York: Springer-Verlag, 2006.

[89] Z. Qin, J. Tao, Y. Xia. A proportionate recursive least squares algorithm and its performance analysis. *IEEE Trans. Circuits Syst. II Exp. Briefs.*, 2021, 68(1):506–510.

[90] A. H. Sayed, *Adaptation, Learning and Optimization over Networks*, Foundations and Trends® in Machine Learning Series. Cambridge England: Cambridge University Press, 2014.

[91] C. G. Lopes, A. H. Sayed. Diffusion least-mean squares over adaptive networks: Formulation and performance analysis. *IEEE Trans. Signal Process.*, 2008, 56(7):3122–3136.

[92] F. S. Cattivelli, A. H. Sayed. Diffusion LMS strategies for distributed estimation. *IEEE Trans. Signal Process.*, 2010, 58(3):1035–1048.

[93] S. Zhang, H. C. So, W. Mi, H. Han. A family of adaptive decorrelation NLMS algorithms and its diffusion version over adaptive networks. *IEEE Trans. Circuits Syst. I*, 2017, 65(2):638–649.

[94] W. Huang, L. Li, Q. Li, X. Yao. Diffusion robust variable stepsize LMS algorithm over distributed networks. *IEEE Access*, 2018, 6:47511–47520.

[95] N. Takahashi, I. Yamada, A. H. Sayed. Diffusion least-mean squares with adaptive combiners: Formulation and performance analysis. *IEEE Trans. Signal Process.*, 2010, 58(9):4795–4810.

[96] H. S. Lee, S. E. Kim, J. W. Lee, W. J. Song. A variable step-size diffusion LMS algorithm for distributed estimation. *IEEE Trans. Signal Process.*, 2015, 63(7):1808–1820.

[97] J. Chen, C. Richard, A. H. Sayed. Diffusion LMS over multitask networks. *IEEE Trans. Signal Process.*, 2015, 63(11):2733–2748.

[98] Y. Yu, H. Zhao, W. Wang, L. Lu. Robust diffusion Huber-based normalized least mean square algorithm with adjustable thresholds. *Circuits Syst. Signal Process.*, 2020, 39:2065–2093.

[99] Z. Hadi. Robust minimum disturbance diffusion LMS for distributed estimation. *IEEE Trans. Circuits Syst. II Exp. Briefs.*, 2021, 68(1):521–525.

[100] S. Cattivelli, C. G. Lopes, A. H. Sayed. Diffusion recursive least squares for distributed estimation over adaptive networks. *IEEE Trans. Signal Process.*, 2008, 56(5):1865–1877.

[101] V. Vahidpour, A. Rastegarnia, A. Khalili, S. Sanei. Analysis of partial diffusion recursive least squares adaptation over noisy links. *IET Signal Process.*, 2017, 11(6):749–758.

[102] G. Mateos, I. D. Schizas, G. B. Giannakis. Distributed recursive least-squares for consensus-based in-network adaptive estimation. *IEEE Trans. Signal Process.*, 2009, 57(11):4583–4588.

[103] Z. Wang, Z. Yu, Q. Ling, D. Berberidis, G. B. Giannakis. Decentralized RLS with data-adaptive censoring for regressions over large-scale networks. *IEEE Trans. Signal Process.*, 2018, 66(6):1634–1648.

[104] A. Reza, D. Kutluyıl, W. Stefan, Y. Huang. Adaptive distributed estimation based on recursive least-squares and partial diffusion. *IEEE Trans. Signal Process.*, 2014, 62(14): 3510–3522.

[105] Y. Yu, H. Zhao, R. C. de Lamare, Y. Zakharov. Robust diffusion recursive least squares estimation with side information for networked agents. *Proceedings of the IEEE International Conference on Acoustics, Speech and Signal Processing (ICASSP)*, Apr. 2018, pp. 4099–4103, Calgary, Canada.

[106] A. Bertrand, M. Moonen, A. H. Sayed. Diffusion bias-compensated RLS estimation over adaptive networks. *IEEE Trans. Signal Process.*, 2011, 59(11):5212–5224.

[107] S. Zhang, H. C. So. Diffusion average-estimate bias-compensated LMS algorithms over adaptive networks using noisy measurements. *IEEE Trans. Signal Process.*, 2020, 68:4643–4655.

[108] M. L. R. de Campos, S. Werner, J. A. Apolinário. Constrained adaptive filters. In *Adaptive Antenna Arrays: Trends and Applications (Signals and Communication Technology)*, S. Chandran, Ed. Berlin, Germany: Springer, 2004.

[109] O. L. Frost. An algorithm for linearly constrained adaptive array processing. *Proc. IEEE*, 1972, 60(8):926–935.

[110] R. Arablouei, K. Doğançay. On the mean-square performance of the constrained LMS algorithm. *Signal Process.*, 2015, 117:192–197.

[111] L. S. Resende, J. M. T. Romano, M. G. Bellanger. A fast least-squares algorithm for linearly constrained adaptive filtering. *IEEE Tran. Signal Process.*, 1996, 44(5):1168–1174.

[112] R. Arablouei, K. Dogancay. Linearly-constrained recursive total least-squares algorithm. *IEEE Signal Process. Lett.*, 2012, 19(12):821–824.

[113] R. Arablouei, K. Dogancay, Reduced-complexity constrained recursive least–squares adaptive filtering algorithm,. *IEEE Trans. Signal Process*, 2012, 60(12):6687–6692.

[114] S. Werner, J. A. Apolinario, M. L. R. de Campos, P. S. R. Diniz. Low-complexity constrained affine-projection algorithms. *IEEE Trans. Signal Process.*, 2005, 53(12):4545–4555.

[115] K. Lee, Y. Baek, Y. Park. Nonlinear acoustic echo cancellation using a nonlinear postprocessor with a linearly constrained affine projection algorithm. *IEEE Trans. Circuits Syst. II Exp. Briefs.*, 2015, 62(9):881–885.
[116] Y. Engel, S. Mannor, R. Meir. The kernel recursive least-squares algorithm. *IEEE Trans. Signal Process.*, 2004, 52(8):2275–2285.
[117] S. Van Vaerenbergh, J. Via, I. Santamaria. A sliding-window kernel RLS algorithm and its application to nonlinear channel identification. *2006 IEEE International Conference on Acoustics Speech and Signal Processing Proceedings*, 2006, Toulouse, France.
[118] W. Liu, I. Park, Y. Wang, et al. Extended kernel recursive least squares algorithm. *IEEE Trans. Signal Process.*, 2009, 57(10):3801–3814.
[119] W. Liu, I. Park, J. C. Principe. An information theoretic approach of designing sparse kernel adaptive filters. *IEEE Trans. Neural Netw.*, 2009, 20(12):1950–1961.
[120] B. Chen, S. Zhao, P. Zhu, et al. Quantized kernel recursive least squares algorithm. *IEEE Trans. Neural Netw. Learn. Syst.*, 2013, 24(9):1484–1491.
[121] S. Wang, W. Wang, S. Duan, L. Wang. Kernel recursive least squares with multiple feedback and its convergence analysis. *IEEE Trans. Circuits Syst. II Exp. Briefs.*, 2017, 64(10):1237–1241.
[122] W. Liu, P. Pokharel, J. C. Principe. The kernel least mean square algorithm. *IEEE Trans. Signal Process.*, 2008, 56(2):543–554.
[123] W. D. Parreira, J. C. M. Bermudez, C. Richard, J. Y. Tourneret. Stochastic behavior analysis of the Gaussian Kernel least-mean-square algorithm. *IEEE Trans. Signal Process.*, 2012, 60(5):2208–2222.
[124] W. Gao, J. Huang, J. Han, Q. Zhang. Theoretical convergence analysis of complex Gaussian kernel LMS algorithm. *J. Syst. Eng. Electron.*, 2016, 27(1):39–50.
[125] B. Chen, S. Zhao, P. Zhu, et al. Quantized kernel least mean square algorithm, *IEEE Trans. Neural Netw. Learn. Syst.*, 2012, 23(1):22–32.
[126] P. P. Pokharel, W. Liu, J. C. Principe. Kernel least mean square algorithm with constrained growth. *Signal Process.*, 2012, 89(3):257–265.
[127] S. Zhao, B. Chen, P. Zhu, et al. Fixed budget quantized kernel least-mean-square algorithm. *Signal Process.*, 2013, 93(9):2759–2770.
[128] S. Wang, Y. Zheng, C. Ling, Regularized kernel least mean square algorithm with multiple-delay feedback. *IEEE Signal Process. Lett.*, 2016, 23(1):98–101.
[129] J. Zhao, X. Liao, S. Wang, et al. Kernel least mean square with single feedback. *IEEE Signal Process. Lett.*, 2015, 22(7):953–957.
[130] X. Xu, H. Qu, J. Zhao, X. Yang, B. Chen. Quantized kernel least mean square with desired signal smoothing. *Electron. Lett.*, 2015, 51(18):1457–1459.
[131] B. Chen, J. Liang, N. Zheng, et al. Kernel least mean square with adaptive kernel size. *Neurocomputing*, 2016, 191:95–106.
[132] A. Gersho. Some aspects of linear estimation with non mean-square error criteria. *Proceedings of the Asilomar Circuits Systems Conference*, 1969.
[133] S. H. Cho, S. D. Kim, K. Y. Jeon. Statistical convergence of the adaptive least mean fourth algorithm. *Proceedings of the ICSP*, pp. 610–613, 1996.
[134] T. Y. Al-Naffouri, A. Sayed. Transient analysis of adaptive filters. *Proceedings of the International Conference on Acoustics, Speech, and Signal Processing*, vol. 6, May 2001, pp. 3869–3872, Salt Lake City, UT.
[135] E. Eweda. Global stabilization of the least mean fourth algorithm. *IEEE Trans. Signal Process.*, 2012, 60(3):1473–1477.

[136] E. Eweda. A stable normalized least mean fourth algorithm with improved transient and tracking behaviors. *IEEE Trans. Signal Process.*, 2016, 64(18):4805–4816.

[137] E. Eweda. Mean-square stability analysis of a normalized least mean fourth algorithm for a Markov plant. *IEEE Trans. Signal Process.*, 2014, 62(24):6454–6553.

[138] E. Eweda. Stochastic analysis of a stable normalized least mean fourth algorithm for adaptive noise canceling with a white Gaussian reference. *IEEE Trans. Signal Process.*, 2012, 60(12):6235–6244.

[139] G. Gui, F. Adachi. Adaptive sparse system identification using normalized least mean fourth algorithm. *Int. J. Commun. Syst.*, 2010, 28(1):38–48.

[140] G. Gui, L. Xu, F. Adachi. Extra gain: Improved sparse channel estimation using reweighted *L1*-norm penalized LMS/F algorithm. *IEEE/CIC International Conference on Communications*, Oct 13–15, 2014, pp. 1–5, Shanghai, China.

[141] G. Gui, B. Zheng, L. Xu. Correntropy induced metric penalized NLMF algorithm to improve sparse system identification. *IEEE/CIC ICCC 2015 Symposium on Signal Processing for Communications*, 2015.

[142] Y. Chen, G. Guan, L. Xu, Nobuhiro Shimoi. Improved adaptive sparse channel estimation using re-weighted *L1*-norm normalized least mean fourth algorithm. *SICE Annual Conference*, July 28–30, 2015, Hangzhou, China.

[143] W. Ma, J. Qiu, D. Zheng, Z. Zhang, X. Hu. Bias compensated normalized least mean fourth algorithm with correntropy induced metric constraint. *Proceedings of the 37th Chinese Control Conference*, July 25–27, 2018, Wuhan, China.

[144] I. Naveed, B. Murwan, Z. Azzedine, A. E. B. Abdeldjalil. Convex combination of LMF and ZA-LMF for variable sparse system identification. *2019 53rd Asilomar Conference on Signals, Systems, and Computers*, 2019, Pacific Grove, CA.

[145] Y. Yasin, S. K. Suleyman. An extended version of the NLMF algorithm based on proportionate Krylov subspace projections. *2009 International Conference on Machine Learning and Applications*, 2009, Miami, FL.

[146] O. S. Muhammed, Y. Yasin, D. Alper, S. K. Suleyman. The Krylov-proportionate normalized least mean fourth approach: Formulation and performance analysis. *Signal Process.*, 2015, 109:1–13.

[147] Z. Jiang, W. Shi, X. Huang, Y. Li. An enhanced proportionate NLMF algorithm for group-sparse system identification. *AEU-Int. J. Electron. Commun.*, 2020, 119:153178.

[148] J. Ni, J. Yang. Variable step-size diffusion least mean fourth algorithm for distributed estimation. *Signal Process.*, 2016, 122:93–97.

[149] H. Mojtaba, Z. J. Hossein. Distributed adaptive LMF algorithm for sparse parameter estimation in Gaussian mixture noise. *2014 7th International Symposium on Telecommunications (IST'2014)*, 2014.

[150] W. Wang, H. Zhao. Performance analysis of diffusion least mean fourth algorithm over network. *Signal Process.*, 2017, 141:32–47.

[151] Y. Xi, S. K. Mohamed. A generalized least absolute deviation method for parameter estimation of autoregressive signals. *IEEE Trans. Neural Netw.*, 2008, 19(1):107–118.

[152] Z. Wang, S. P. Bradley. Constrained least absolute deviation neural networks. *IEEE Trans. Neural Netw.*, 2008, 19(2):273–283.

[153] S. Dasgupta, J. S. Garnett, C. Richard Johnson, Jr. Convergence of an adaptive filter with signed filtered error. *IEEE Trans. Signal Process.*, 1994, 42(4):946–950.

[154] E. Eweda. Transient performance degradation of the LMS, RLS, sign, signed regressor, and sign-sign algorithms with data correlation. *IEEE Trans. Circuits Syst. II Analog Digital Signal Process.*, 1999, 46(8):1055–1062.
[155] Y. Zong, J. Ni, J. Chen. A family of normalized dual sign algorithms. *Digital Signal Process.*, 2021, 110:102954.
[156] R. Y. Nabil, A. H. Sayed. Steady-state and tracking analyses of the sign algorithm without the explicit use of the independence assumption. *IEEE Signal Process. Lett.*, 2000, 7(11):307–309.
[157] H. Chen, G. Yin. Asymptotic properties of sign algorithms for adaptive filtering. *IEEE Trans. Autom. Control*, 2003, 48(9):154–156.
[158] W. Gao, J. Chen. Transient analysis of signed LMS algorithms with cyclostationary colored Gaussian inputs. *IEEE Trans. Circuits Syst. II Exp. Briefs.*, 2020, 67(12):3562–3566.
[159] Y. Yu, T. Yang, H. Chen, Rodrigo C. de Lamarec, Yingsong Li. Sparsity-aware SSAF algorithm with individual weighting factors: Performance analysis and improvements in acoustic echo cancellation. *Signal Process.*, 2021, 178:107806
[160] Z. Yang, Y. R. Zheng, S. L. Grant. Proportionate affine projection sign algorithms for network echo cancellation. *IEEE Trans. Audio Speech Lang. Process.*, 2011, 19(8):2273–2284.
[161] J. Ni. Diffusion sign subband adaptive filtering algorithm for distributed estimation. *IEEE Signal Process. Lett.*, 2015, 22(11):2029–2033.
[162] J. Ni, J. Chen, X. Chen. Diffusion sign-error LMS algorithm: Formulation and stochastic behavior analysis. *Signal Process.*, 2016, 129:142–149.
[163] J. Ni, Y. Zhu, J. Chen. Multitask diffusion affine projection sign algorithm and its sparse variant for distributed estimation. *Signal Process.*, 2020, 172:107561.
[164] Y. Gao, J. Ni, J. Chen, X. Chen. Steady-state and stability analyses of diffusion sign-error LMS algorithm. *Signal Process.*, 149, 2018:62–67.
[165] R. Meir, V. E. Maiorov. On the optimality of neural-network approximation using incremental algorithms. *IEEE Trans. Neural Netw.*, 2000, 11(2):323–337.
[166] S. Ferrari, R. F. Stengel. Smooth function approximation using neural networks. *IEEE Trans. Neural Netw.*, 2005, 16(1):24–38.
[167] D. J. C. MacKay. A practical Bayesian framework for backpropagation networks. *Neural Comput.*, 1992, 4(3):448–472.
[168] G. Kechriotis, E. Zervas. Using recurrent neural networks for adaptive communication channel equalization. *IEEE Trans. Neural Netw.*, 1994, 5(2):267–278.
[169] G. B. Huang, Q. Y. Zhu, C. K. Siew. Extreme learning machine: Theory and applications. *Neurocomputing*, 2006, 70(1):489–501.
[170] O. Castillo, P. Melin. *Adaptive Noise Cancellation Using Type-2 Fuzzy Logic and Neural Networks*. Berlin, Heidelberg: Springer, 2007.
[171] K. R. Mulller. Improving network models and algorithmic tricks. In *Neural Networks: Tricks of the Trade*, C. Yann, B. Leon, B. O. Genevieve, K. R. Muller, Eds. Berlin, Germany: Springer, 1998, pp. 139–141.
[172] G. B. Huang, L. Chen, C. K. Siew. Universal approximation using incremental constructive feedforward networks with random hidden nodes. *IEEE Trans. Neural Netw.*, 2006, 17(4):879–892.
[173] G. B. Huang, H. Zhou, X. Ding, R. Zhang. Extreme learning machine for regression and multiclass classification. *IEEE Trans. Syst. Man Cybern. B Cybern.*, 2012, 42(2):513–529.

[174] N. Liang, G. Huang, P. Saratchandran, N. Sundararajan. A fast and accurate online sequential learning algorithm for feedforward networks. *IEEE Trans. Neural Netw.*, 2006, 17(6):1411–1423.

[175] H. J. Rong, G. B. Huang, N. Sundararajan, P. Saratchandran. Online sequential fuzzy extreme learning machine for function approximation and classification problems. *IEEE Trans. Syst., Man, Cybern. B, Cybern.*, 2009, 39(4):1067–1072.

[176] Y. Ye, S. Squartini, F. Piazza. Online sequential extreme learning machine in nonstationary environments. *Neurocomputing*, 2013, 116:94–101.

[177] W. Deng, Q. Zheng, Z. Wang. Cross-person activity recognition using reduced kernel extreme learning machine. *Neural Netw.*, 2014, 53:1–7.

[178] J. S. Lim, S. Lee, H.-S. Pang. Low complexity adaptive forgetting factor for online sequential extreme learning machine (OS-ELM) for application to nonstationary system estimations. *Neural Comput. Appl.*, 2013, 22:569–576.

[179] S. G. Soares, R. Araújo. An adaptive ensemble of on-line extreme learning machines with variable forgetting factor for dynamic system prediction. *Neurocomputing*, 2016, 171:693–707.

[180] J. Luo, C.-M. Vong, P.-K. Wong. Sparse Bayesian extreme learning machine for multi-classification. *IEEE Trans. Neural Netw. Learn. Syst.*, 2014, 25(4):836–843.

[181] J. Ma, L. Yang. Robust supervised and semi-supervised twin extreme learning machines for pattern classification. *Signal Process.*, 2021, 180:107861.

[182] S. Liao, A. C. S. Chung. Feature based nonrigid brain MR image registration with symmetric alpha stable filters. *IEEE Trans. Med. Imag.*, 2010, 29(1):106–119.

[183] C. L. P. Chen, Z. Liu. Broad learning system: An effective and efficient incremental learning system without the need for deep architecture. *IEEE Trans. Neural Netw. Learn. Syst.*, 2018, 29(1):10–24.

[184] C. L. P. Chen, Z. Liu, S. Feng. Universal approximation capability of broad learning system and its structural variations. *IEEE Trans. Neural Netw. Learn. Syst.*, 2019, 30(4):1191–1204.

[185] C. M. Bishop. *Pattern Recognition and Machine Learning*. New York: Springer-Verlag, 2006.

[186] G. E. Hinton, S. Osindero, Y.-W. The. A fast learning algorithm for deep belief nets. *Neural Comput.*, 2006, 18(7):1527–1554.

[187] P. Vincent, H. Larochelle, I. Lajoie, Y. Bengio, P.-A. Manzagol. Stacked denoising autoencoders: Learning useful representations in a deep network with a local denoising criterion. *J. Mach. Learn. Res.*, 2010, 11(12):3371–3408.

[188] J. W. Jin, C. L. Philip Chen. Regularized robust broad learning system for uncertain data modelling. *Neurocomputing*, 2018, 322:58–69.

[189] Y. Kong, X. Wang, Y. Cheng, C. L. P. Chen. Hyperspectral imagery classification based on semi-supervised broad learning system. *Remote Sens.*, 2018, 10(5):585.

[190] T. L. Zhang, R. Chen, X. Yang, S. Guo. Rich feature combination for cost-based broad learning system. *IEEE Access*, 2019, 7:160–172.

[191] Q. Zhou, X. He. Broad learning model based on enhanced features learning. *IEEE Access*, 2019, 7:42536–42550.

[192] J. Zou, Q. She, F. Gao, M. Meng. Multi-task motor imagery EEG classification using broad learning and common spatial pattern. *Proceedings of the International Conference on Intelligence Science*, 2018, pp. 3–10., vol 539, Springer, Cham.

[193] H. Guo, B. Sheng, P. Li, C. L. P. Chen. Multiview high dynamic range image synthesis using fuzzy broad learning system. *IEEE Trans. Cybern.*, 2019, doi: 10.1109/TCYB.2019.2934823.

[194] J. Han, L. Xie, J. Liu, X. Li. Personalized broad learning system for facial expression. *Multimedia Tools Appl.*, 2019, 79(23–24):16627–16644.
[195] W. Yu, C. Zhao. Broad convolutional neural network based industrial process fault diagnosis with incremental learning capability. *IEEE Trans. Ind. Electron.*, 2020, 67(6):5081–5091.
[196] S. Feng, C. L. P. Chen. Fuzzy broad learning system: A novel neurofuzzy model for regression and classification. *IEEE Trans. Cybern.*, 2020, 50(2):414–424.
[197] J. Jin, Z. Liu, C. L. P. Chen. Discriminative graph regularized broad learning system for image recognition. *Sci. China Inf. Sci.*, 2018, 61(11):112209.
[198] M. Xu, M. Han, C. L. P. Chen, T. Qiu. Recurrent broad learning systems for time series prediction. *IEEE Trans. Cybern.*, 2020, 50(4):1405–1417.
[199] M. Han, S. Feng, C. L. P. Chen, M. Xu, T. Qiu. Structured manifold broad learning system: A manifold perspective for large-scale chaotic time series analysis and prediction. *IEEE Trans. Knowl. Data Eng.*, 2019, 31(9):1809–1821.
[200] M. Shao, C. L. Nikiw. Signal processing with fractional lower order moments: Stable processes and their applications. *Proc. IEEE*, 1993, 81:986–1009.
[201] S.-C. Pei, C.-C. Tseng. Least mean p-power error criterion for adaptive FIR filter. *IEEE J. Sel. Areas Commun.*, 1994, 12(9):1540–1547.
[202] W. Liu, P. P. Pokharel, J. C. Principe. Correntropy: Properties and applications in non-Gaussian signal processing. *IEEE Trans. Signal Process.*, 2007, 55(11):5286–5298.
[203] A. Singh, J. C. Principe. Using correntropy as a cost function in linear adaptive filters. *IEEE 2009 International Joint Conference on Neural Networks*, IEEE, 2009, Atlanta, GA.
[204] B. Chen, X. Wang, N. Lu , S. Wang, J. Cao, J. Qin. Mixture correntropy for robust learning. *Pattern Recognit.*, 2018, 79:318-327.
[205] Y. Tian, Y. Lu, X. Fu. Principle and application of minimum error entropy estimation in discrete case. *J. China Instit. Commun.*, 1994, 15(2):38–45
[206] B. Chen, L. Dang, Y. Gu, et al. Minimum error entropy Kalman filter. *IEEE Trans. Syst. Man Cybern. Syst.*, 2019, 51(9):5819–5829.
[207] F. Solms, P. V. Rooyen, J. S. Kunicki. Maximum entropy and minimum relative entropy in performance evaluation of digital communication systems. *IEE Proc. Commun.*, 1995, 142(4):250–254.
[208] L. S. Lambert, L. Zwald. Robust regression through the Huber's criterion and adaptive lasso penalty. *Electron. J. Stat.*, 2011, 5:1015–1053.
[209] E. Ollila, A. Mian. Block-wise minimization-majorization algorithm for huber's criterion: Sparse learning and Applications. *IEEE International Workshop on Machine Learning For Signal Processing*, 2020.
[210] L. Chang, B. Hu, G. Chang, et al. Huber-based novel robust unscented Kalman filter. *IET Sci. Meas. Technol.*, 2012, 6(6):502–509.
[211] B. Weng, E. B. Kenneth. Nonlinear system identification in impulsive environments. *IEEE Trans. Signal Process.*, 2005, 53(7):2588–2594.
[212] B. Lin, R. He, X. Wang, B. Wang. The steady-state mean-square error analysis for least mean p-order algorithm. *IEEE Signal Process. Lett.*, 2009, 16(3):176–179.
[213] O. Arikan, M. Belge, A. E. Cetin, E. Erzin. Adaptive filtering approaches for non-Gaussian stable process. *International Conference on Acoustics, Speech, and Signal Processing, ICASSP-95*, vol. 2, 2000, pp. 1400–1403, Detroit, MI.
[214] S. C. Pei, C. C. Tseng. Adaptive IIR Notch filter based on least mean p-power error criterion. *IEEE Trans. Circuits Syst. II Analog Digital Signal Process.*, 1993, 40(8):525–528.

[215] S. Maha. Steady state analysis of the p-power algorithm for constrained adaptive IIR Notch Filters. *Inf. Technol. J.*, 2007, 6(3):353–358.

[216] A. G. Deczky. Synthesis of recursive digital filters using the minimum p-error criterion. *IEEE Trans. Audio Electroacoust.*, 1972, 20(4):257–263.

[217] W. S. Lu, Y. Cui, R. L. Kirlin. Least pth optimization for the design of 1-D filters with arbitrary amplitude and phase responses. In *Proceedings of the IEEE International Conference on Acoustics, Speech and Signal Processing (ICASSP)*, 1993, 3, pp. 61–64.

[218] C. C. Tseng. Design of stable IIR digital filter based on least p-power error criterion. *IEEE Trans. Circuits Syst. I Regular Papers*, 2004, 51(9):1879–1888.

[219] Y. Xiao, Y. Tadokoro, K. Shida. Adaptive algorithm based on least mean - power error criterion for Fourier analysis in additive noise. *IEEE Trans. Signal Process.*, 1999, 47(4):1172–1181.

[220] R. Leahy, Z. Zhou, Y. C. Hsu. Adaptive filtering of stable processes for active attenuation of impulsive Noise. *Proceedings of IEEE International Conference on Acoustics Speech and Signal Processing*, 1995, 5, pp. 2983–2986, Detroit, MI.

[221] T. A. Muhammad, M. Wataru. Improving robustness of filtered-x least mean p-power algorithm for active attenuation of standard symmetric-a-stable impulsive noise. *Appl. Acoust.*, 2011, 72:688–694.

[222] L. Godara, C. Antonio. Analysis of constrained LMS algorithm with application to adaptive beamforming using perturbation sequences. *IEEE Trans. Antennas. Propagat.*, 1986, 34(3):368–379.

[223] S. Peng, B. Chen, L. Sun, Wee Ser, Z. Lin. Constrained maximum correntropy adaptive filtering. *Signal Process.*, 2017, 140:116–126.

[224] G. Qian, F. He, S. Wang, et al. Robust constrained maximum total correntropy algorithm. *Signal Process.*, 2021, 181(8):107903.

[225] S. Peng, Z. Wu, B. Chen. Constrained least mean P-power error algorithm. *Proceedings of the 35th Chinese Control Conference*, July 27–29, 2016, Chengdu, China.

[226] S. H. Yim, H. S. Lee, W. J. Song. A proportionate diffusion LMS algorithm for sparse distributed estimation. *IEEE Trans. Circuits Syst. II Exp. Briefs.*, 2015, 62(10):992–996.

[227] F. Wen. Diffusion least-mean P-power algorithms for distributed estimation in alpha-stable noise environments. *Electron. Lett.*, 2013, 49(21):1355–1356.

[228] L. Lu, H. Zhao, W. Wang, Y. Yu. Performance analysis of the robust diffusion normalized least mean p-power algorithm. *IEEE Trans. Circuits Syst. II Exp. Briefs.*, 2018, 65(12): 2047– 2051

[229] P. Konstantinos, C. Mandar. Adaptive sparse channel estimation under symmetric alpha-stable noise. *IEEE Trans. Wireless Commun.*, 2014, 13(6):3183–3195.

[230] W. Ma, H. Qu, G. Gui, et al. Maximum correntropy criterion based sparse adaptive filtering algorithms for robust channel estimation under non-Gaussian environments. *J. Frank. Inst.*, 2015, 352(7):2708–2727.

[231] W. Ma, J. Duan, B. Chen, et al. Robust proportionate adaptive filter based on maximum correntropy criterion for sparse system identification in impulsive noise environments. *Signal, Image Video Process.*, 2017(3):1–8.

[232] Z. Wu, S. Peng, B. Chen, et al. Proportionate minimum error entropy algorithm for sparse system identification. *Entropy*, 2015, 17(9):5995–6006.

[233] S. Seth, J. C. Príncipe. Compressed signal reconstruction using the correntropy induced metric. *IEEE International Conference on Acoustics, Speech and Signal Processing (ICASSP)*, pp. 3845–3848, 2008, Las Vegas, NV.

[234] W. Ma, B. Chen, H. Qu, J. Zhao, Sparse least mean p-power algorithms for channel estimation in the presence of impulsive noise., *Signal, Image Video Process.*, 2016, 10(3):503–510.

[235] X. Zhang, S. Peng, Z. Wu, Y. Zhou, Y. Fu. An improved proportionate least mean p -power algorithm for adaptive filtering. *Signal, Image Video Process.*, 2018, 12:59–66.

[236] B. Murat, L. M. Eric. A sliding window RLS-Like adaptive algorithm for filtering alpha-stable noise. *IEEE Signal Process. Lett.*, 2000, 7(4):86–89.

[237] N. V. Ángel, A. G. Jerónimo. Combination of recursive least-norm algorithms for robust adaptive filtering in alpha-stable noise. *IEEE Trans. Signal Process.*, 2012, 60(3):1478–1482.

[238] S. Zhang, J. Zhang. Enhancing the tracking capability of recursive least p-norm algorithm via adaptive gain factor. *Digital Signal Process.*, 2014, 30:67–73

[239] J. Sun, S. Peng, Q. Liu, R. Zhao, Z. Lin. Robust constrained recursive least P-power algorithm for adaptive filtering. *2018 IEEE 23rd International Conference on Digital Signal Processing (DSP)*, 2018, Shanghai, China.

[240] W. Gao, J. Chen. Kernel least mean p-power algorithm. *IEEE Signal Process. Lett.*, 2017, 24(7):996–1000.

[241] W. Ma, J. Duan, W. Man, H. Zhao, B. Chen. Robust kernel adaptive filters based on mean p-power error for noisy chaotic time series prediction. *Engin. Appl. Artif. Intel.*, 2017, 58:101–110.

[242] W. Gao P. Ruan J. Li, T. Xu. Exponentially weighted kernel recursive least P-power algorithm. *2019 IEEE International Conference on Signal Processing, Communications and Computing (ICSPCC)*, 2019, Dalian, China.

[243] W. Gao, Y. Xu, L. Huang. Random Fourier features extended kernel recursive least P-power algorithm. *2019 11th International Conference on Wireless Communications and Signal Processing (WCSP)*, 2019, Xi'an, China.

[244] J. Zhao, H. Zhang, G. Wang, J. A. Zhang. Projected kernel least mean p-power algorithm: Convergence analyses and modifications. *IEEE Trans. Circuits Syst. I Regul. Pap.*, 2020, 67(10):3498–3511.

[245] X. Huang, K. Xiong, L. Wang, S. Wang. The robust kernel conjugate gradient least mean p-power algorithm. *2019 2nd China Symposium on Cognitive Computing and Hybrid Intelligence (CCHI)*, 2019, Xi'an, China.

[246] W. Gao, J. Chen, L. Zhang. Diffusion approximated kernel least mean P-power algorithm. *2019 IEEE International Conference on Signal Processing, Communications and Computing (ICSPCC)*, 2019, Dalian, China.

[247] S. Zhou, X. Liu, Q. Liu, S. Wang, C. Zhu, J. Yin. Random Fourier extreme learning machine with ℓ2;1-norm regularization. *Neurocomputing*, 2016, 174:143–153.

[248] M. Jiang, Z. Pan, N. Li. Multi-label text categorization using L21-norm minimization extreme learning machine. *Neurocomputing*, 2017, 261:4–10.

[249] J. Yang, P. Chen, H. Rong, B. Chen. Least mean P-power extreme learning machine for obstacle avoidance of a mobile robot. *2016 International Joint Conference on Neural Networks (IJCNN)*, 2016, Vancouver, Canada.

[250] J. Yang, F. Ye, H. Rong, B. Chen. Recursive least mean p-power extreme learning machine. *Neural Netw.*, 2017, 91:22–33

[251] J. Yang, Y. Xu, H. Rong, S. Du, B. Chen. Sparse recursive least mean p-power extreme learning machine for regression. *IEEE Access*, 2018, 6:16022–16034.

[252] F. Chu, T. Liang, C. L. P. Chen, X. Wang, X. Ma. Weighted broad learning system and its application in nonlinear industrial process modelling. *IEEE Trans. Neural Netw. Learn. Syst.*, 2020, 31(8):3017–3031.

[253] S. Feng, W. Ren, M. Han, Y. W. Chen. Robust manifold broad learning system for large-scale noisy chaotic time series prediction: A perturbation perspective. *Neural Netw.*, 2019, 117:179–190.

[254] Y. Zheng, B. Chen. Least p-norm based broad learning system. *Pattern Recognit. Artif. Intel.*, 2019, 32(1):51–57.

[255] Y. Zheng, X. Qin, Z. Xi, B. Chen. Mixed-norm based broad learning system for EEG classification. *2019 41st Annual International Conference of the IEEE Engineering in Medicine and Biology Society (EMBC)*, July 23–27, 2019, Berlin, Germany.

[256] Y. Zheng, B. Chen, S. Wang, W. Wang. Broad learning system based on maximum correntropy criterion. *IEEE Trans. Neural Netw. Learn. Syst.*, 2020, 32(7):3083–3097.

[257] B. Chen, L. Xing, X. Wang, J. Qin, N. Zheng. Robust learning with kernel mean p-power error loss. *IEEE Trans. Cyber.*, 2018, 48(7):2101–2113.

[258] L. Peng, X. Li, D. Bi, Y. Xie. Robust adaptive filtering with q-Gaussian kernel mean p-power error. *IEEE Signal Process. Lett.*, 2018, 25(9):1335–1339.

[259] T. Zhang, F. He, Z. Zheng, S. Wang. Recursive minimum kernel mixture mean p-power error algorithm based on the Nyström method. *IEEE Trans. Circuits Syst. II Exp. Briefs.*, 2020, 67(11):2772–2776.

[260] W. Ma, J. Qiu. Recursive kernel MPE loss algorithm. *2019 IEEE 2nd International Conference on Electronic Information and Communication Technology (ICEICT)*, IEEE, 2019, Harbin, China.

[261] A. H. Sayed, *Fundamentals of Adaptive Filtering*. New York: Wiley-Interscience, 2003.

[262] Y. Gu, K. Tang, H. Cui, et al. Modifier formula on mean square convergence of LMS algorithm. *Electron. Lett.*, 2002, 38(19):1147–1148.

[263] D. Feng, D. Xiao, T. Ding. Convergence of least mean squares algorithm under attenuating excitation conditions. *Control Theory App.*, 2003, 20(1):109–112.

[264] S. Marcos, O. Macchi. Tracking capability of the least mean square algorithm: Application to an asynchronous echo canceller. *IEEE Trans. Acoust. Speech Signal Process.*, 1987, 35(11):1570–1578.

[265] S. Bittanti, M. Campi. Mean square convergence of an adaptive RLS algorithm with stochastic excitation. *Proceedings of the 28th IEEE Conference on Decision and Control*, 1989, Tampa, FL.

[266] Y. Zhu, B. Chen, J. Hu. Adaptive filtering with adaptive p-power error criterion. *Int. J. Innovat. Comput. Inform. Control.*, 2011, 7(4):1725–1737.

[267] B. Chen, L. Xing, Z. Wu, J. Liang, J. C. Principe, N. Zheng. Smoothed least mean p-power error criterion for adaptive filtering. *Digital Signal Process.*, 2015, 40:154–163.

[268] W. Ma, J. Duan, W. Man, J. Liang, B. Chen. General mixed-norm-based diffusion adaptive filtering algorithm for distributed estimation over network. *IEEE Access*, 2017, 5:1090–1102.

[269] B. Chen, J. Hu, Y. Zhu, Z. Sun, Information theoretic interpretation of error criteria. *Acta Automat. Sinica*, 2009, 35(10):1302–1309.

[270] M. K. Varanasi, B. Aazhang, Parametric generalized Gaussian density estimation. *J. Acoust. Soc. Am.*, 1989, 86(4):1404–1415.
[271] T. M. Cover, J. A. Thomas, *Element of Information Theory*. Chichester: Wiley-Interscience, 1991.
[272] T. Y. Al-Naffouri, A. H. Sayed, Adaptive filters with error nonlinearities: Mean-square analysis and optimum design. *EURASIP J. Appl. Signal Process.*, 2001, 4:192–205.
[273] K. Kokkinakis, A. K. Nandi, Exponent parameter estimation for generalized Gaussian probability density functions with application to speech modeling. *Signal Process.*, 2005, 85:1852–1858.
[274] J. A. Chambers, O. Tanrikulu, A. G. Constantinides. Least mean mixed-norm adaptive filtering. *Electron. Lett.*, 1994, 30(19): 1574–1575.
[275] J. Chambers, A. Avlonitis. A robust mixed-norm adaptive filter algorithm. *IEEE Signal Process. Lett.*, 1997, 4(2):46–48.
[276] Y. Wang, Y. Li, R. Yang. Sparse adaptive channel estimation based on mixed controlled *l2* and *lp*-norm error criterion. *J. Frank. Inst.*, 2017, 354(15):7215–7239.
[277] B. Chen, J. C. Principe. On the smoothed minimum error entropy criterion. *Entropy*, 2012, 14(11):2311–2323
[278] J. C. Principe. *Information Theoretical Learning*. Berlin, Germany: Springer.
[279] A. Singh, R. Pokharel, J. Principe. The C-loss function for pattern classification. *Pattern Recognit.*, 2014, 47(1):441–453.
[280] L. Chen, H. Qu, J. Zhao, B. Chen, J. C. Principe. Efficient and robust deep learning with correntropy-induced loss function. *Neural Comput. Appl.*, 2016, 27(4):1019–1031.
[281] R. He, B. Hu, W. Zheng, X. Kong. Robust principal component analysis based on maximum correntropy criterion. *IEEE Trans. Image Process.*, 2011, 20(6):1485–1494.
[282] B. Chen, X. Liu, H. Zhao, J. C. Principe. Maximum correntropy Kalman filter. *Automatica*, 2017, 76:70–77.
[283] B. Chen, L. Xing, H. Zhao, N. Zheng, J. C. Príncipe. Generalized correntropy for robust adaptive filtering. *IEEE Trans. Signal Process.*, 2016, 64(13):3376–3387.
[284] B. Chen, J. Wang, H. Zhao, N. Zheng, J. C. Príncipe. Convergence of a fixed-point algorithm under maximum correntropy criterion. *IEEE Signal Process. Lett.*, 2015, 22(10):1723–1727.
[285] B. Chen, L. Xing, J. Liang, N. Zheng, J. C. Princípe. Steady-state mean-square error analysis for adaptive filtering under the maximum correntropy criterion. *IEEE Signal Process. Lett.*, 2014, 21(7):880–884.
[286] C. Tsallis. Possible generalization of Boltzmann-Gibbs statistics. *J. Stat. Phys.*, 1988, 52(1):479–487.
[287] F. Colombo, G. Gentili, I. Sabadini. A Cauchy kernel for slice regular functions. *Ann. Glob. Anal. Geom.*, 2010, 37(4):361–378.
[288] T. M. Nguyen, Q. M. Wu. Robust student's-t mixture model with spatial constraints and its application in medical image segmentation. *IEEE Trans. Med. Imaging*, 2012, 31(1):103–116.
[289] A. D'Onofrio, *Bounded Noises in Physics, Biology, and Engineering*. Basel, Switzerland: Birkhäuser, 2013.
[290] R. Tinós, S. Yang. Use of the q-Gaussian mutation in evolutionary algorithms. *Soft Comput.*, 2011, 15(8):1523–1549.
[291] C. Rogers, T. Ruggeri. q-Gaussian integrable Hamiltonian reductions in anisentropic gas dynamics. *Discrete Cont. Dyn.-B*, 2017, 19(7):2297–2312.

[292] K. Xiong, Y. Zhang, S. Wang. Robust variable normalization least mean p-power algorithm. *Chin. Sci. Inform. Sci.*, 2020, 63(9):3.
[293] B. Lin, R. He, X. Wang, B. Wang. The excess mean square error analyses for Bussgang algorithm. *IEEE Signal Process. Lett.*, 2008, 15:793–796.
[294] D. H. Brandwood. A complex gradient operator and its application in adaptive array theory. *Proc. Inst. Electr. Eng. H*, 1983, 130:11–16.
[295] N. R. Yousef, A. H. Sayed. A unified approach to the steady-statec and tracking analyses of adaptive filters. *IEEE Trans. Signal Process.*, 2001, 49(2):314–324.
[296] E. T. Jaynes, Information theory and statistical mechanics. *Phys. Rev.*, 1957, 106:620–630.
[297] J. N. Kapur, H. K. Kesavan, *Entropy Optimization Principles with Applications*. San Diego, CA: Academic Press, 1992.
[298] R. L. Joshi, T. R. Fisher, Comparison of generalized Gaussian and Laplacian modeling in DCT image coding. *IEEE Signal Proc. Lett.*, 1995, 2(5):81–82.
[299] K. Sharifi, A. Leon-Garcia, Estimation of shape parameter for generalized Gaussian distributions in subband decompositions of video. *IEEE Trans. Circuits Syst. Video Technol.*, 1995, 5(1):52–56.
[300] F. Wang, H. Li, R. Li. Unified parametric and nonparametric ICA algorithms for hybrid source signals and stability analysis. *Int. J. Innov. Comput. Inf. Control*, 2008, 4(4):933–942.
[301] T. Y. Al-Naffouri, A. H. Sayed. Transient analysis of data-normalized adaptive filters. *IEEE Trans. Signal Process.*, 2003, 51:639–652.
[302] T. Y. Al-Naffouri, A. H. Sayed, Transient analysis of adaptive filters with error nonlinearities. *IEEE Trans. Signal Process.*, 2003, 51:653–663.
[303] E. Eweda. Stabilization of high-order stochastic gradient adaptive filtering algorithms. *IEEE Trans Signal Process*, 2017, 65:3948–3959
[304] S. Al-Sayed, A. M. Zoubir, A. H. Sayed. Robust adaptation in impulsive noise. *IEEE Trans. Signal Process.*, 2016, 64:2851–2865
[305] S. Zhang, W. X. Zheng, J. S. Zhang, et al. A family of robust M-shaped error weighted least mean square algorithms: Performance analysis and echo cancellation application. *IEEE Access*, 2017, 5:14716–14727.
[306] F. Huang, J. Zhang, S. Zhang. Maximum versoria criterion-based robust adaptive filtering algorithm. *IEEE Trans. Circuits Syst. CAS II.*, 2017, 64:1252–1256.
[307] E. Santana, A. K. Barros, R. C. S. Freire. On the time constant under general error criterion. *IEEE Signal Process. Lett.*, 2007, 14(8):533–536.
[308] J. Weston, A. Elisseeff, B. Schölkopf, et al. Use of the zero norm with linear models and kernel methods. *J. Mach. Learn. Res.*, 2003, 3:1439–1461.
[309] M. S. Salman. Sparse leaky-LMS algorithm for system identification and its convergence analysis. *Int. J. Adapt. Cont. Signal Process.*, 2014, 28(10):1065–1072.
[310] M. N. S. Jahromi, M. S. Salman, A. Hocanin. Convergence analysis of the zero-attracting variable step-size LMS algorithm for sparse system identification. *Signal, Image Video Process.*, 2015, 9:1353–1356.
[311] M. L. Aliyu, M. A. Alkassim, M. S. Salman. A p-norm variable step-size LMS algorithm for sparse system identification. *Signal, Image Video Process.*, 2014, 9(7):1559–1565.
[312] T. Omid, S. A. Vorobyov. Reweighted *l1*-norm penalized LMS for sparse channel estimation and its analysis. *Signal Process.*, 2014, 104:70–79.
[313] http://www.ee.ic.ac.uk/hp/staff/dmb/voicebox/voicebox.html
[314] A. M. NRSC. Preemphasis/Deemphasis and Broadcast Audio Transmission Bandwidth Specifications (ANSI/EIA-549-88).

[315] F. C. Souza, O. J. Tobias, R. Seara, D. R. Morgan. A PNLMS algorithm with individual activation factors. *IEEE Trans. Signal Process.*, 2010, 58(4):2036–2047.

[316] S. Zhang, J. Zhang. Transient analysis of zero attracting NLMS algorithms without Gaussian input signal. *Signal Process.*, 2013, 97:100–109.

[317] E. V. Kuhn, F. d. C. de Souza, R. Seara, D. R. Morgan. On the steady-state analysis of PNLMS-type algorithms for correlated Gaussian input data. *IEEE Signal Process. Lett.*, 2014, 21(11):1433–1437.

[318] R. L. Das, M. Chakraborty. On convergence of proportionate-type normalized least mean square algorithms. *IEEE Trans. Circuits Syst. II Exp. Briefs.*, 2015, 62(5):491–495.

[319] P. Tsakalides, C. L. Nikias. Maximum likelihood localization of sources in noise modeled as α stable process. *IEEE Trans. Signal Process.*, 1995, 43(11):2700–2713

[320] C. Li, P. Shen, Y. Liu, Z. Zhang. Diffusion information theoretic learning for distributed estimation over network. *IEEE Trans. Signal Process.*, 2013, 61(16):4011–4024.

[321] G. Sun, M. Li, T. C. Lim. A family of threshold based robust adaptive algorithms for active impulsive noise control. *Appl. Acoust.*, 2015, 97:30–36.

[322] S. C. Chan, Y. Zou. A recursive least M-estimate algorithm for robust adaptive filtering in impulsive noise: Fast algorithm and convergence analysis. *IEEE Trans. Signal Process.*, 2004, 52(4):975–991.

[323] L. Weruaga, S. Jimaa. Exact NLMS algorithm with *p*-norm constraint. *IEEE Signal Process. Lett.*, 2015, 22(3):366–370.

[324] D. Zha. Robust multiuser detection method based on least p-norm state space criterion. *Wirel. Pers. Commun.*, 2007, 40:191–204.

[325] P. J. Rousseeuw, A. M. Leroy. *Robust Regression and Outlier Detection*. New York: Wiley, 1987.

[326] Y. Zou, S. C. Chan, T. S. Ng. Least mean M-estimate algorithms for robust adaptive filtering in impulse noise. *IEEE Trans. Circuits Syst. II*, 2000, 47(12):1564–1569.

[327] S. Koike. Recursive least absolute error algorithm: Analysis and simulations. *IEICE Trans. Fundam. Electron. Commun. Comput. Sci.*, 2002, E85-A(12):2886–2893.

[328] S. Zhang, J. Zhang. New steady-state analysis results of variable step-size LMS algorithm with different noise distributions. *IEEE Signal Process. Lett.*, 2014, 21(6):653–657.

[329] E. E. Kuruoglu, P. J. W. Rayner, W. J. Fitzgerald. Least lp-norm impulsive noise cancellation with polynomial filters. *Signal Process.*, 1998, 69(1):1–14.

[330] W. Ma, J. Duan, G. Gui, et al. Robust diffusion recursive adaptive filtering algorithm based on l_p-norm. *The 35th China Control Conference*, 2016, Chengdu, China.

[331] S. J. Lim, J. G. Haris, Combined LMS/F algorithm. *Electron. Lett.*, 1997, 33(6):467–468

[332] J. Zhao, H. Zhang, G. Wang. Projected kernel recursive maximum correntropy. *IEEE Trans. Circuits Syst. II Exp. Briefs.*, 2018, 65(7):963–967.

[333] J. Zhao, H. Zhang. Projected kernel recursive least squares algorithm. *Proceedings of International Conference on Neural Information Processing (ICONIP)*, 2017, pp. 356–365.

[334] Y. Zheng, S. Wang, J. Feng, C. K. Tse. A modified quantized kernel least mean square algorithm for prediction of chaotic time series. *Digit. Signal Process.*, 2016, 48:130–136.

[335] C. J. C. Burges. A tutorial on support vector machines for pattern recognition. *Data Mining Knowl. Discov.*, 1998, 2(2):121–167.

[336] Paulo Cortez, 2009, http://www3.dsi.uminho.pt/pcortez/series/

[337] Z. Qin, B. Chen, N. Zheng. Random Fourier feature kernel recursive least squares. *Proceedings of the IEEE IJCNN*, May 2017, pp. 2881–2886.

[338] A. Rahimi, B. Recht. Random features for large-scale kernel machines. *NIPS*, 2007, 3:1–8.

[339] W. Rudin, *Fourier Analysis on Groups*. New York: Courier Dover Publications, 2017.

[340] P. Bouboulis, S. Chouvardas, S. Theodoridis. Online distributed learning over networks in RKH spaces using random Fourier features. *IEEE Trans. Signal Process.*, 2018, 66(7):1920–1932.

[341] P. Bouboulis, S. Pougkakiotis, S. Theodoridis. Efficient KLMS and KRLS algorithms: A random Fourier feature perspective. *Proceedings of the IEEE SSP*, 2016, pp. 1–5.

[342] Y. K. B. Widrow, A. Greenblatt, D. Park. The no-prop algorithm: A new learning algorithm for multilayer neural networks. *Neural Netw.*, 2013, 37:182–188.

[343] G. B. Huang, L. Chen. Convex incremental extreme learning machine. *Neurocomputing*, 2007, 70(16–18):3056–3062.

[344] G. B. Huang, L. Chen. Enhanced random search based incremental extreme learning machine. *Neurocomputing*, 2008, 71(16–18):3460–3468.

[345] G. Feng, G. B. Huang, Q. Lin, R. Gay. Error minimized extreme learning machine with growth of hidden nodes and incremental learning. *IEEE Trans. Neural Netw.*, 2009, 20(8):1352–1357.

[346] Y. Yang, Y. Wang, X. Yuan. Bidirectional extreme learning machine for regression problem and its learning effectiveness. *IEEE Trans. Neural Netw. Learn. Syst.*, 2012, 23(9):1498–1505.

[347] Z. Bai, G. B. Huang, D. Wang, H. Wang, M. B. Westover. Sparse extreme learning machine for classification. *IEEE Trans. Cybern.*, 2014, 44(10):1858–1870.

[348] G.-B. Huang, N.-Y. Liang, H.-J. Rong, P. Saratchandran, N. Sundararajan. On-line sequential extreme learning machine. *Proceedings of the IASTED International Conference on Computational Intelligence (CI)*, 2005, pp. 232–237, Calgary, Canada.

[349] R. P. Agarwal, M. Meehan, D. O'Regan. *Fixed Point Theory and Applications*. Cambridge, UK: Cambridge University Press, 2001.

[350] C. W. Hsu, C. J. Lin. A comparison of methods for multiclass support vector machines. *IEEE Trans. Neural Netw.*, 2002, 13(2):415–425.

[351] N. Y. Liang, P. S. Aratchand, G. B. Huang, et al. Classification of mental tasks from EEG signals using extreme learning machine. *Int. J. Neural Syst.*, 2006, 16(1):29–38.

[352] R. Blanke, R. Tomioka, R. Lemma, et al. Optimizing spatial filters for robust EEG single-trial analysis. *IEEE Signal Process. Mag.*, 2008, 25(1):41–56.

[353] H. Han, Z. Wang, J. Liu, et al. Gait recognition based on linear discriminant analysis and support vector machine. *Pattern Recogn. Artif. Intell.*, 2005, 18(2):160–164.

[354] J. M. Keller, Graymr, J. A. Givens. A fuzzy K-nearest neighbor algorithm. *IEEE Trans. Syst. Man Cybern.*, 1985, 15(4):580–585.

[355] C. Cortes, V. Vapnik. Support-vector networks. *Machine Learning*, 1995, 20(3):273–297.

[356] M. L. Zhang, J. M. Pea, V. Robles. Feature selection for multi-label naive Bayes classification. *Inf. Sci.*, 2009, 179(19):3218–3229.
[357] M. Naeem, C. Brunner, R. Leeb, B. Graimann, G. Pfurtscheller. Seperability of four-class motor imagery data using independent components analysis. *J. Neural Eng.*, 2006, 3(3):208–216.
[358] G. Liu, G. Huang, X. Zhu. Application of CSP method in multiclass classification. *Chin. J. Biomed. Eng.*, 2006, 28(6):935–938.
[359] O. Tanrikulu, J. A. Chambers. Convergence and steady-state properties of the least-mean mixed-norm (LMMN) adaptive algorithm. *IEE Proc.-Vision, Image Signal Process., IET*, 1996, 143(3):137–142.
[360] A. Zerguine, C. F. N. Cowan, M. Bettayeb. Adaptive echo cancellation using least mean mixed-norm algorithm. *IEEE Trans. Signal Process.*, 1997, 45(5):1340–1343.
[361] M. Srinivas, I. Hussain, B. Singh. Combined LMS-LMF based control algorithm of DSTATCOM for power quality enhancement in distribution system. *IEEE Trans. Ind. Electron.*, 2016, 63(7):4160–4168.
[362] E. V. Papoulis, T. Stathaki. A normalized robust mixed-norm adaptive algorithm for system identification. *IEEE Signal Process. Lett.*, 2004, 11(1):56–59.
[363] J. Liu, H. Qu, B. Chen, W. Ma. Kernel robust mixed-norm adaptive filtering. *2014 International Joint Conference on Neural Networks (IJCNN)*, IEEE, 2014, pp. 3021–3024.
[364] S. Yu, X. You, K. Zhao, et al. Kernel normalized mixed-norm algorithm for system identification. *2015 International Joint Conference on Neural Networks (IJCNN)*, IEEE, 2015, pp. 1–6.
[365] L. Shi, H. Zhao. Generalized variable step-size diffusion continuous mixed p-norm algorithm. *Circuits Syst. Signal Process.*, 2021, 40:3690–3620.
[366] B. D. O. Anderson, R. R. Bitmead, C. R. Jr. Johnson, et al. *Stability of Adaptive Systems: Passivity and Averaging Analysis*. Cambridge, MA: MIT Press, 1986.
[367] W. A. Sethares, B. D. O. Anderson, C. R Johnson. Adaptive algorithms with filtered regressor and filtered error. *Math. Control Signals Syst.*, 1989, 2:381–403.
[368] W. A. Sethares. Adaptive algorithms with nonlinear data and error functions. *IEEE Trans. Signal Process.*, 1992, 40(9):2199–2206.
[369] L. Ljung. Analysis of recursive stochastic algorithms. *IEEE Trans. Autom. Control*, 1977, 22(4):551–575.
[370] V. Solo, X. Kong. *Adaptive Signal Processing Algorithms: Stability and Performance*. Englewood,NJ: Prentice Hall, 1995.
[371] O. Macchi. *Adaptive Processing: The Least Mean Squares Approach with Applications in Transmission*. New York: Wiley, 1995.
[372] V. J. Mathews, S. H. Cho. Improved convergence analysis of stochastic gradient adaptive filters using the sign algorithm. *IEEE Trans. Acoust., Speech, Signal Process.*, 1987, 35(4):450–454.
[373] S. Alliney, S. A. Ruzinsky. An algorithm for the minimization of mixed *l1* and *l2* norms with application to Bayesian estimation. *IEEE Trans. Signal Process.*, 1994, 42(3):618–627.
[374] H. Zhang, Y. Peng. Lp-norm based minimisation algorithm for signal parameter estimation. *Electron. Lett.*, 1999, 35(20):1704–1705.
[375] A. Zidouri. Convergence analysis of a mixed controlled L2-Lp adaptive algorithm. *EURASIP J. Adv. Signal Process.*, 2010, 893809:1–10.

[376] Y. Li, Y. Wang, T. Jiang. Sparse least mean mixed-norm adaptive filtering algorithms for sparse channel estimation applications. *Int. J. Commun. Syst.*, 2016, 30(8):1–12.

[377] E. J. Candes, M. B. Wakin, S. P. Boyd. Enhancing sparsity by reweighted *l1* minimization, *J. Fourier Anal. Appl.*, 2007, 14(5):877–905.

[378] D. L. Donoho. Compressed sensing. *IEEE Trans. Inf. Theory.*, 2006, 52(4):1289–1306.

[379] K. Mayyas, F. Momani. An LMS adaptive algorithm with a new step size control equation. *J. Frankl. Inst.*, 2011, 348(4):589–605.

[380] R. Price. A useful theorem for non-linear devices having Gaussian inputs. *IEEE Trans. Inf. Theory*, 1958, 4:69–72.

[381] Y. P. Li, T. S. Lee, B. F. Wu. A variable step-size sign algorithm for channel estimation. *Signal Process.*, 2014, 102:304–312.

[382] S. Koike. A class of adaptive step-size control algorithms for adaptive filters. *IEEE Trans. Signal Process.*, 2002, 50(6):1315–1326.

[383] Y. Yu, H. Zhao. Novel sign subband adaptive filter algorithms with individual weighting factors. *Signal Process.*, 2016, 122:14–23.

[384] Y. Yu, H. Zhao, B. Chen. Steady-state mean square deviation analysis of the sign subband adaptive filter algorithm. *Signal Process.*, 2016, 120:36–42.

[385] S. Yu, X. You, X. Jiang, et al. Generalized kernel normalized mixed-norm algorithm: Analysis and simulations. *International Conference on Neural Information Processing. Springer International Publishing*, 2015, pp. 61–70, Istanbul, Turkey.

[386] X. Luo, J. Deng, J. Liu, et al. A novel entropy optimized kernel least-mean mixed-norm algorithm. *2016 International Joint Conference on Neural Networks (IJCNN)*, IEEE, 2016, pp. 1716–1722, Vancouver, Canada.

[387] W. Ma, X. Qiu, J. Duan, Y. Li, B. Chen. Kernel recursive generalized mixed norm algorithm. *J. Frank. Inst.*, 2018, 355(4):1596–1613.

[388] S. Yu, Z. Fan, Y. Zhao, J. Zhu, K. Zhao, D. Wu. Quantized kemel least mean mixed-norm algorithm. *Proceedings of the IEEE ICSP*, October 2014, pp. 199–204, Hangzhou, China.

[389] B. Chen, S. Zhao, P. Zhu, J. C. Principe. Mean square convergence analysis for kernel least mean square algorithm. *Signal Process.*, 2012, 92(11):2624–2632.

[390] D. P. Mandic, E. Y. Papoulis, C. G. Boukis. A normalized mixed-norm adaptive filtering algorithm robust under impulsive noise interference. *Proceedings of the IEEE ICASSP*, April 2003, Hong Kong.

[391] C. H. Lee, C. R. Lin, M. S. Chen. Sliding-window filtering: An efficient algorithm for incremental mining. *Proceedings of the ACM CIKM*, 2001, pp. 263–270, Atlanta, GA.

[392] C. Richard, M. Bermudez, J. Carlos, P. Honeine. Online prediction of time series data with kernels. *IEEE Trans. Signal Process.*, 2009, 57(3):1058–1067.

[393] C. Richard, J. Carlos, M. Bermudez. Affine projection algorithm applied to nonlinear adaptive filtering. *Statis. Signal Process*, 2007.

[394] S. Zhao, B. Chen, J. C. Principe. Kernel adaptive filtering with maximum correntropy criterion. *Proceedings of the IEEE International Joint Conference on Neural Networks*, 2011, pp. 2012–2017.

[395] Z. Wu, J. Shi, X. Zhang, W. Ma, B. Chen. Kernel recursive maximum correntropy. *Signal Process.*, 2015, 117:11–16.

[396] S. Wang, W. Wang, L. Dang, Y. Jiang. Kernel least mean square based on the Nyström method. *Circuits Syst. Signal Process.*, 2019, 38(7):3133–3151.

[397] A. Frank, A. Asuncion. UCI Machine Learning Repository, 2010 [Online]. Available: https://archive.ics.uci.edu/ml.
[398] C. K. I. Williams, M. Seeger. Using the Nyström method to speed up kernel machines. *Proceedings of the 14th Annual Neural Information Processing Systems Conference*, 2001, pp. 682–688.
[399] S. Wang, L. Dang, G. Qian, Y. Jiang. Kernel recursive maximum correntropy with Nyström approximation. *Neurocomputing*, 2019, 329:424–432.
[400] H. Xing, X. Wang. Training extreme learning machine via regularized correntropy criterion. *Neural Comput. Appl.*, 2013, 23(7–8):1977–1986.
[401] A. Krizhevsky, G. Hinton. Learning multiple layers of features from tiny images. University of Toronto, Toronto, Canada, Tech. Rep., 2009.

Index

Printed in the United States
by Baker & Taylor Publisher Services